MICROWAVE
RECEIVERS
and
RELATED
COMPONENTS

by

James Bao-yen Tsai

Electronic Engineering Series

Wexford Press
2008

CONTENTS

CONTENTS

CONTENTS

CONTENTS

CONTENTS

CONTENTS

CHAPTER 1
INTRODUCTION

1.1 HISTORICAL REVIEWS(1-3)

In 1864 James Clerk Maxwell collected and unified a few previously known relationships to generate four famous equations often referred to as the Maxwell's equations. From these equations electromagnetic energy propagating through free space was postulated. Between 1884 to 1886, Heinrich Rudolph Hertz verified this phenomena through a laboratory experimental setup. In his experiment he used a rapidly oscillating electric spark to produce electromagnetic waves which caused similar electrical oscillations in a distant wire loop. His discovery opened the field of wireless communications. Since Marconi produced a practical wireless telegraph system in 1895, many new devices (i.e., vacuum tubes and transistors) and technology have been developed. Today, radio communications and television are expanding continuously at such tremendous speed that they almost become an integral part of everyday life.

During World War II radar was invented and used in military operations. Today radar is used in airports to guide airplanes to safe landings in fog or storms. Airplanes and ships also use radar as a guidance system through poor visibility conditions. Weathermen use radar in weather forecasting. In military applications radar has become a vital equipment. It is not only used in the search and detection of hostile aircrafts, ships, and vehicles, but also to guide weaponry. The extension of the applications of electrical communication and radar systems in the future is almost beyond imagination.

All of the communication and radar systems discussed above contain two major parts: the transmitter and receiver. The transmitter and the receiver form an inseparable unit. They must be designed together as a pair. In a communication system, the transmitter and the receiver are installed at different locations. The message is sent from the transmitter to the receiver. In a radar system the transmitter and the receiver are located very close together. The transmitted signal is reflected by a target (or targets) to the receiver, thus the information about the target (i.e., range and velocity) is obtained by the radar. In both cases, the frequency, bandwidth, and special codes, if used, must be properly matched between the transmitter and the receiver in order to optimize the performance. That is why the transmitter and receiver are discussed as a unit. If the transmitter signal is specially coded such as frequency modulation (FM) chirp, biphase shift keying, the receiver must have the same coded waveform to correlate with the input signal to produce the desired processing gain. There are few articles and books dealing solely with the receiver subject. ''Microwave Receivers'' by Van Voorhis contains a great deal of information; however, it was written before the invention of transistors (Reference 1). There are

1

other books that have chapters dedicated to receivers (References 2 and 3). Most of the receivers discussed in these books are radar receivers.

Along with the development of communication and radar systems for use in military application, a highly interesting area called electronic warfare (EW) was also developed. The purpose of EW is to defeat or disturb the hostile military operation. In EW applications, receivers are used to intercept signals of limited information from a hostile transmitter, while a jamming transmitter is used to generate false information or noise to mask the true signal received by the possible hostile radar receiver. The received information could be used to establish jamming criterion (i.e., priority and jamming modes). In this kind of operation, although some information on the radar transmitter is available, the intercept receiver cannot be designed according to the transmitter parameters because the radiating source is a noncooperative one. The parameters of these systems may be alternated intentionally in order to avoid the reception of the intercept receiver. Under such circumstances, the receiver must be designed somewhat independently of the transmitter. In other words, the receiver must be able to receive signals from a number of transmitters. Therefore, the receiver itself has become an independent subject.

1.2 RECENT DEVELOPMENTS IN RECEIVERS[4]

Intercept receivers are often operated in dense electronic environments. There are various signals in a battlefield such as hostile and friendly radar signals, as well as communication signals in the environment. The receiver must be able to receive these signals in a short time interval, encode, and sort them. Decisions must be made and action taken against the lethal hostile radar. These desired capabilities constitute some stringent requirements on the receiver system. Receivers designed with these capabilities have received considerable attention.

Recent advances in component development have made many different types of receiver systems possible. For example, solid-state amplifiers up to the 20 GHz frequency range with relative low noise figures made microwave receivers with high sensitivity possible. Microwave integrated circuits (MIC) made complicated, bulky receivers portable. Research and accomplishments in surface acoustic wave (SAW) filters have rekindled the practicality of building a channelized receiver which contains a large number of filters (Reference 4). The SAW filters have the potential to make a channelized receiver in a manageable size and weight. The accomplishment in SAW dispersive delay lines have made the compressive receiver very attractive, and the development in SAW nondispersive delay lines created a number of new approaches in receiver designs. The development in high speed and versatile logic circuits also have direct impact on receivers. Some of the wide bandwidth receivers can collect an enormous amount of information in a short time. A powerful digital processor must be used at the output of the receiver to analyze the information.

Without a digital processor, the information collected by the receiver will not be utilized effectively. The logic circuits and digital control also add versatility to the receiver.

Microwave receivers can be divided into the following groups according to their structures: crystal video, superheterodyne (superhet), instantaneous frequency measurement (IFM), channelized, compressive (also called microscan), and Bragg cell receivers. Most research and development is concentrated on the last three types of receivers, because their performances are not fully reaching the predicted levels. For electronic warfare applications, the performance of the latter three types of receivers will be far better than that of the former three types in terms of bandwidth, probability of intercept and simultaneous signal handling capability.

Recent developments in superhet receivers are in the area of digital control. The tuning speed of the receiver has been improved because of advances in oscillator capability. A superhet receiver is shown in Figure 1.1. Most of the modern receivers have digital display to indicate the frequency and other parameters of the input signal,

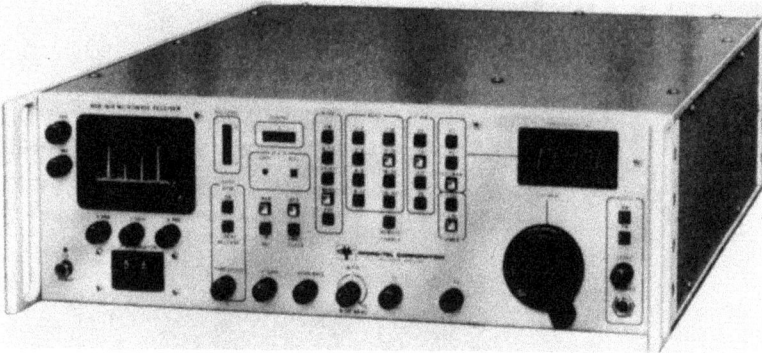

Figure 1.1. Manually Tuned Superheterodyne Receiver (Courtesy of Micro-Tel Corp.)

Figure 1.2 shows a channelized receiver. The idea of building channelized receivers is not new. In the past, it was not practical to build such a receiver because the large number of filters used in the receiver made it bulky. The MIC and SAW filter developments made this kind of receiver feasible. The channelized receivers provide good performance in terms of intercepting and encoding radar signals. Thus, they have the high potential to be used as electronic warfare receivers. Figure 1.3 shows a wide-band intercept system. The system basically contains a channelized receiver, preprocessor, some display units, and a recording unit.

Compressive receivers are becoming realistic because of high speed logic developments. The outputs from the receiver are closely packed short pulses (nanoseconds in width). The density of the output pulses depends on the input signal conditions. Handling these short pulses requires fast digitizing circuits. Developments in microwave components and SAW technology also have positive impacts on the compressive receivers. New approaches to build the delay lines and oscillators will reduce the complexity of the receiver and improve its performance.

Figure 1.2. *Channelized Receiver (Courtesy of U.S. Air Force Wright Aeronautical Laboratories)*

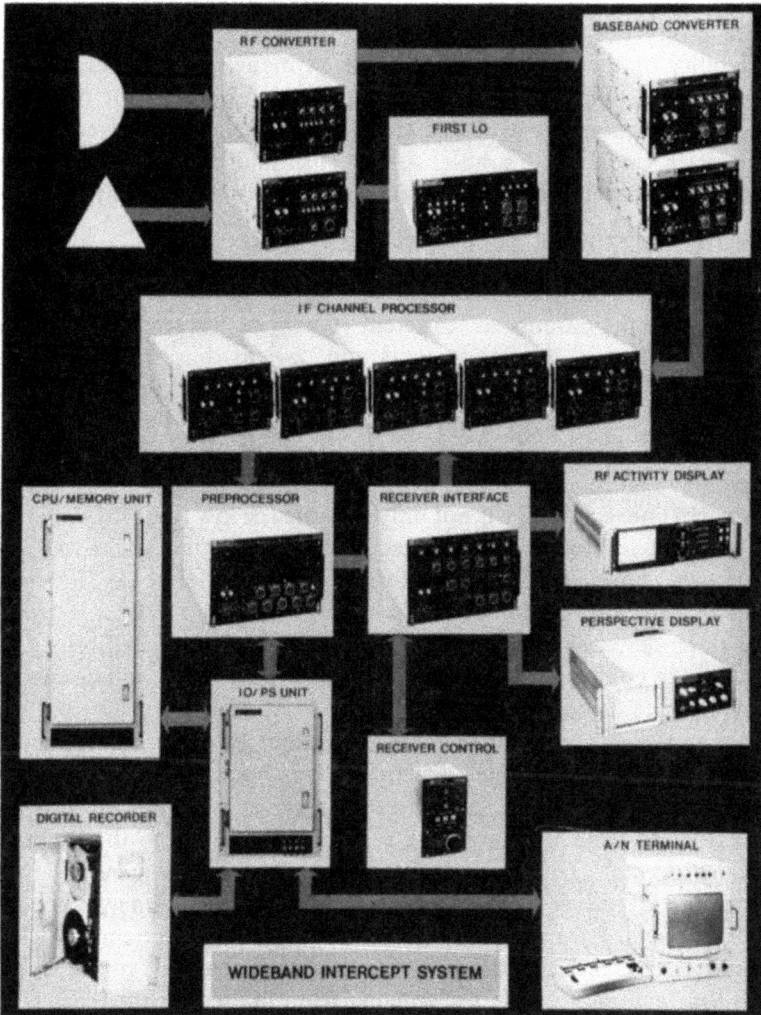

Figure 1.3. Wide-band Intercept System (Courtesy of Watkins-
Johnson Co.)

The Bragg cell receiver uses a new approach to solve the receiver problem. The achievement in effectively changing electrical signals to acoustic signals in bulk and surface modes in the last decade made this kind of receiver possible. The development in solid-state lasers and photo-detector arrays made this receiver more than a laboratory curiosity. Although much research and development is needed to make the Bragg cell receiver useful, it has the potential of being a simple channelized receiver with less hardware. The Bragg cell receiver has this potential application in an airborne system because of its extremely small size and light weight.

Most research and development in channelized, compressive, and Bragg cell receivers is in the frequency range below 4 GHz, because of the availability of the components. Signals at higher frequencies are down converted to the lower frequency and processed by a proper receiver. If these components were available at higher frequencies, receivers at those frequencies would be desirable. As for IFM and crystal video receivers, research and development at higher frequencies are underway.

1.3 ORGANIZATION OF THE BOOK[4, 5]

This book is basically divided into two parts. Chapters 2 through 7 contain the definitions of different types of receivers. Chapters 8 through 15 discuss the components and devices that are used in these receivers. Since these components and devices are used in more than one kind of receiver, it is appropriate to devote separate chapters to the discussion of them. This book will concentrate on strip line or microstrip line technology, since this is the trend in receiver design. Strip line and microstrip line technology is not only suitable for low frequency operation but is advancing toward higher frequencies. Waveguides and waveguide components have already been discussed in many other books (References 4 and 5). Since the trend in device technology used for receivers is toward a solid-state device because of the compact size, better efficiency, and reliability in comparison with gaseous devices, this book will discuss only the solid-state devices.

Microwave receivers can be classified either by their input frequency (VHF and UHF) or by their structures. This book will classify them according to their structure. A microwave receiver is usually considered as the hardware from the output of a receiving antenna to the input of a digital processor. The input to the receiver is radio frequency (RF); the output of the receiver is digital information. This definition may not apply to all receivers. For example, in some crystal video or extremely high frequency (EHF) receivers, the antenna is an integrated part of the receiver. Thus, one cannot separate the antenna from the rest of the receiver.

In this book the discussion of receivers will range from the RF input circuits to the video detector. The digitizing circuit beyond the detector will not be discussed because its design depends on the input requirements of the processor following the receiver. Therefore, the digitizing circuits following the detectors might be quite different. If the processor following the receiver is changed, the digitizing circuits must be redesigned; however, the rest of the receiver from the RF input to the video detector can remain the same. Of course, from the RF input to the detectors of each receiver, the designs can also be different even for the receivers of the same structure. However, there are enough similarities among these designs that a general discussion on a certain type of receiver will provide enough information to understand how they work.

Chapter 2 will discuss the general characteristics related to receiver operation. Chapters 3 through 7 will discuss different types of receivers. Chapters 8 through 15 will present all the devices and components used in receivers. Chapter 16 will discuss some preliminary work on laboratory evaluations of EW receivers.

1.4 FREQUENCY RANGE OF RECEIVERS[6]

The receiver frequency range must match the transmitter frequency range no matter whether the transmitter is a cooperative or noncooperative one. The frequency range of microwave receivers can be extremely wide. While an AM radio utilizes frequency below 1 MHz, the space-to-earth communication can operate above 100 GHz. Radar frequency ranges from 25 MHz up to 100 GHz. These frequency ranges are by no means the limits of communication or radar applications. Generating frequency up to 300 GHz is presently under development. The components used in different frequency ranges have different characteristics. At low frequency ranges, lumped circuit elements are used. At high frequency ranges, distributed circuit elements and waveguides are used. In the hundred gigahertz range, optical techniques can be used. However, it is not clear cut as to what techniques should be applied to which frequency ranges.

The microwave frequency range has been divided into bands, with each band designated by a letter (or letters). The original purpose was to guard military secrecy, then it was used in peacetime. Since these band designations, such as L, S, and C, are difficult to remember by new engineers starting in this field, new designations have been assigned to frequencies below 40 GHz. Both the new and old designations are presently used. Table 1.1 shows the band designations below 40 GHz. The frequency bands above 40 GHz are listed in Table 1.2. They are classified by the size of the waveguides. It should be noted that under the current band designations the same letter is used for bands below 40 GHz as well as above 40 GHz. However, frequency ranges can always be used to resolve the ambiguity in frequency bands.

TABLE 1.1 FREQUENCY BANK DESIGNATION BELOW 40 GHz (IN GHz)

UHF	L	S		C		X		Ku	K	Ka
(Previous Frequency Designations)										
C	D	E	F	G	H	I		J		K
			(Current Frequency Designations)							

0.5   1.0   2.0   3.0   4.0   6.0   8.0   10.0   12.4   18.0   20.0   26.5   40.0

TABLE 1.2 FREQUENCY BAND DESIGNATIONS ABOVE 40 GHz (IN GHz)

Q	U	V	E	W	F	D	G
33-50	40-60	50-75	60-90	75-110	90-140	110-170	140-220

The current trend in microwave receivers is that strip lines and microstrip lines are replacing the waveguides, because the dielectric substrate material is improved and has lower loss. Today, receivers with frequencies below 20 GHz are almost all strip line/microstrip line techniques. Between 20 to 40 GHz, waveguide and strip lines/microstrip line are both used. Above 40 GHz, waveguides are the main transmission lines. However, some other transmission lines (i.e., fin line and suspended lines) are also used.

REFERENCES

1. Van Voorhis, S.N., *Microwave receivers,* Radiation Laboratory Series, Vol. 23, McGraw-Hill, 1948.
2. Boyd, J.A., Harris, D.B., King, D.D., Welch, H.W., Jr., Eds., "Electronic countermeasures," Prepared by University of Michigan for U.S. Army Signal Corps under Contract DA-36-039, SC-71204, 1961.
3. Skolnik, M.E., *Radar handbook,* McGraw-Hill, 1970.
4. Collins, J.H., Grant, P.M., "A review of current and future components for electronic warefare receivers," IEEE Trans., Microwave Theory and Techniques, Vol. MTT-29, pp. 395-403, May 1981.
5. Reich, H.J., Ordung, P.F., Krauss, H.L., Skolnik, J.G., *Microwave theory and techniques,* D. Van Nostrand Co., 1953.
6. Ishii, T.K., *Microwave engineering,* The Ronald Press Co., 1966.

CHAPTER 2
CHARACTERISTICS OF MICROWAVE RECEIVERS

In this chapter, the commonly used terms by receiver engineers will be discussed. Although these terms are used on a daily basis, some of their meanings are not clearly defined and others have too many definitions. They are: frequency accuracy, sensitivity and dynamic range, pulse width (PW), pulse amplitude (PA), time of arrival (TOA), angle of arrival (AOA) measurements, simultaneous signals handling capability, probability of intercept, and throughput rate.

2.1 FREQUENCY MEASUREMENTS[1]

For a radar receiver, the frequency of the input signal is known. The receiver is designed to receive the anticipated frequency. The only frequency shift is caused by the Doppler effect. The Doppler frequency shift for a radar receiver is

$$f_d = \frac{2v_r f_o}{C} \cos \theta \qquad (2.1)$$

where v_r is the relative velocity of the target with respect to the radar, f_o is the transmitted frequency, and C is the velocity of light (3×10^8 m/sec). θ is the angle between direction of the target velocity and the radar to target line of sight.

In deriving Equation 2.1, the signal is considered traveling from the transmitter (colocated with the receiver) to the target and reflected back. If the distance between the target and the transmitter is D, the total distance traveled by the signal will be 2D. The derivation of the Doppler frequency will be discussed in Section 2.8. Since v_r is much less than C, the percentage of the frequency shift caused by the Doppler effect is small. Therefore, for all practical purposes, the input frequency to the radar receiver can be considered known. In many radars, the Doppler frequency is measured accurately to determine the velocity of the target.

In the electronic warfare (EW) field, a broad input frequency range can be expected. The frequency information of the input signal is one of the most important parameters for signal sorting and jamming. Therefore, measuring the frequency is one of the most important tasks. With the exception of crystal video receivers, all receivers discussed in this book have the capability to measure the frequency of the input signal. The basic differences among generic receivers are in the frequency measurement schemes. All other parameters (i.e., AOA and TOA) are measured in somewhat similar manners in different receivers. If the frequency of the input signal is f_s and the receiver intercepts and reports this frequency as f_o, then the error frequency is

$$\Delta f = f_o - f_s \qquad (2.2)$$

If the input frequency is unknown, a number of measurements must be made on one input frequency, and the average value can be used to represent f_s. When the input frequency varies, the output frequency may not change accordingly. Thus, a different error frequency may be generated. The error frequency depends not only on the input frequency but also on the signal strength, also called pulse amplitude (PA) for pulsed signal and the duration of the signal, referred to as pulse width (PW) for pulsed signal. Therefore, many different values of Δf can be obtained. In some receivers, the worst error is specified, which means the error frequency will never exceed the specified value. In other receivers, a more meaningful definition is used and that is the root mean square (rms) value of the error frequency which is defined as

$$\Delta f_{rms} = \sqrt{\frac{1}{N} \sum_{i=1}^{N} (\Delta f_i)^2} \qquad (2.3)$$

where i is an integer $= 1, 2, \ldots N$, and Δf_i is the error frequency of the ith measurement among a total of N measurements.

The minimum Δf_{rms} of a perfectly rectangular pulse modulated wave can be expressed as (Reference 1)

$$\Delta f_{rms} = \frac{\sqrt{3}}{\pi \tau (2E/N_0)^{1/2}} \qquad (2.4)$$

where τ is the pulse width, E is the signal energy which is related to the signal power S through $S = E/\tau$, and N_0 is the noise power per unit bandwidth which is related to the total noise N and bandwidth B through $N = N_0 B$.

Equation 2.4 shows that the longer the PW, the wider the bandwidth; and the higher the signal-to-noise ratio, the better the frequency measurement accuracy. As an example, a receiver with a bandwidth of 10 MHz, PW $= 100$ ns and S/N $= 40$ dB (i.e., B $= 10^7$ Hz, $\tau \doteq 10^{-7}$ sec, and S/N $= 10^4$).

$$\Delta f_{rms} = \frac{\sqrt{3}}{\pi \sqrt{2} \times 10^{-5}} = 39 \text{ KHz}$$

which is better than most practical receivers can realize on one single 100 ns pulse. However, Equation 2.4 predicts the limiting case of the root mean square frequency measurement which contains a large number of measurements.

If the receiver output is in digital format, the least significant bit represents the resolution of the parameters measured. Let f_{LSB} be the frequency resolution, the best frequency accuracy measurement will be f_{LSB}. Often the accuracy is many times worse than the resolution. For example, if a receiver has a frequency resolution of 1 MHz, the frequency accuracy of the receiver is greater than or equal to 1 MHz; it could even be 10 MHz.

When two signals are present at the input of the receiver simultaneously, an EW receiver should read both frequencies correctly. If the two signals are closing in frequency, at a certain frequency separation, the receiver is no longer able to read both frequencies. The minimum frequency separation at which the receiver still can identify both signals is called the frequency resolution of the receiver on simultaneous signals which may be different from the frequency resolution limited by the f_{LSB}. The two signals used to measure to the simultaneous signal resolution are usually continuous waves (cw) of the same amplitudes; thus, the spectrum spreading effect produced by limited pw can be neglected.

Another common term used in connection with the frequency measurement capability of the receiver is instantaneous bandwidth of the receiver. In the instantaneous bandwidth of the receiver, if multiple signals reach the receiver simultaneously, the receiver will receive all of them provided they are in the dynamic range of the receiver and are separated far enough that the receiver can distinguish their frequencies. In EW applications this bandwidth is often time shared in the entire frequency of interest. It is impractical to build a receiver to cover the entire frequency band instantaneously because the receiver will be very complicated, bulky, and expensive.

2.2 THERMAL NOISE AND NOISE FIGURE

The sensitivity of a receiver is limited by the internally generated noise level. There is always the thermal noise, or Johnson noise, which is generated by the thermal motion of electrons in all components used in the receiver. The available thermal noise power at the input of a receiver can be expressed as

$$P_{thermal} = kTB \qquad (2.5)$$

where k is Boltzman's constant $= 1.38 \times 10^{-23}$ joule/deg, T is temperature in Kelvin (T = 290 °k for room temperature), and B is the bandwidth of the receiver.

In receiver field the power is often measured in dBm which is defined as

$$P(dBm) = 10 \log (P) \qquad (2.6)$$

where P is the power in mw. For P = 1 mw, P(dBm) = 0. When P > 1 mw, the corresponding P(dBm) is positive. When P < 1 mw, the corresponding P(dBm) is negative. The thermal noise can be represented in dBm as

$$P_{thermal} = -174 \text{ dBm/Hz}$$

$$P_{thermal} = -114 \text{ dBm/MHz}$$

These two numbers are often used in determining sensitivity of the receiver. The noise power of a practical receiver is always higher than the thermal noise, because noise is introduced from every component. The noise figure is defined as

$$F = \frac{N_o}{GN_i} = \frac{\text{noise out of practical receiver}}{\text{noise out of ideal receiver at temperature T}} \qquad (2.7)$$

where N_o is the noise output from the receiver, G is the gain of the receiver, and N_i is the input thermal noise = kTB.
   However, the gain of the receiver is

$$G = \frac{S_o}{S_i} \qquad (2.8)$$

where S_o is the output signal level and S_i is the input signal level.
   Combining Equations 2.7 and 2.8,

$$F = \frac{S_i/N_i}{S_o/N_o} \qquad (2.9)$$

It should be noted that the input signal-to-noise ratio (S_i/N_i) is always greater than (or equal to) the output signal-to-noise ratio (S_o/N_o), thus F is always greater than unity. F is often expressed in dB as

$$F(dB) = 10 \log F \qquad (2.10)$$

which is always positive.
   If there are n components connected in cascade, the overall noise figure F_T can be obtained from the following equation (also referred to as the Friis formula). The derivation of this equation is in Chapter 13.

$$F_T = F_1 + \frac{F_2 - 1}{G_1} + \frac{F_3 - 1}{G_1 G_2} + \ldots + \frac{F_n - 1}{\prod\limits_{i=1}^{n-1} G_i} \qquad (2.11)$$

where $F_1, F_2, \ldots F_n$ are the noise figure of the first, second, and nth elements in the RF signal path of the receiver. $G_1, G_2, \ldots G_n$ are the gain of the first, second, and nth element. If the first element is an attenuator (say 3 dB), then $F_1 = 2$ (corresponds to 3 dB) and $G_1 = \frac{1}{2}$. The overall gain G_T of the system is

$$G_T = G_1 G_2 \ldots G_n \qquad (2.12)$$

A minimum discernible signal (MDS) is a signal level where the signal power equals the noise power. In other words, when the input signal power is at the MDS level, the receiver output is 3 dB over its noise level.

2.3 TANGENTIAL SENSITIVITY (TSS)[2-9]

The tangential sensitivity (TSS) is determined through visual display on an oscilloscope at the output of a diode detector or the output of the video amplifier following the detector, when the input signal is modulated by rectangular pulses. On the scope display, when the bottom of the noise edge of a pulse is leveled with the top of the noise edge without the input pulse, as shown in Figure 2.1, the receiver is at its TSS. It is generally agreed that at TSS, the signal is 8 dB above the noise level at the output of the detector (References 2 and 3). However, a ±1 dB variation in the mean has been observed among different observers. Although TSS is a relatively easy measurement quantity, it does not provide the information on where the threshold should be set in a receiver. If the threshold is set at TSS level, the false alarm rate is rather high. Here the discussion on TSS is basically based on Lucas' article.

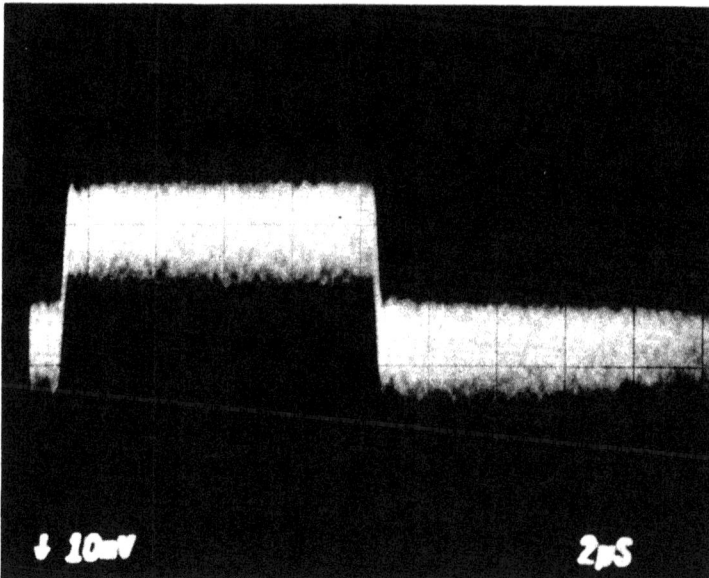

Figure 2.1. Tangential Sensitivity Display

The TSS depends on the bandwidth, noise figure, and the characteristics of the detector. In a practical receiver, the RF bandwidth is almost always greater than the video bandwidth. The TSS from Lucas' paper (Reference 3) can be rewritten here in a different format as

$$\text{TSS} = -114 + 10 \log F_T + 10 \log \left[3.15\, B_R \right.$$

$$\left. + 2.5 \sqrt{2\, B_R B_V - B_V{}^2 + \frac{A B_V}{(G_T F_T)^2}} \right] \text{dBm} \qquad (2.13)$$

$$\text{for } B_V \leqslant B_R \leqslant 2\, B_V$$

$$\text{TSS} = -114 + 10 \log F_T + 10 \log \left[6.31\, B_V \right.$$

$$\left. + 2.5 \sqrt{2\, B_R B_V - B_V{}^2 + \frac{A B_V}{(G_T F_T)^2}} \right] \text{dBm} \qquad (2.14)$$

$$\text{for } B_R \geqslant 2\, B_V$$

where F_T is the overall noise figure from the input of the receiver to the detector, B_V is the video bandwidth in MHz, B_R is the RF bandwidth in MHz, G_T is the total gain from the input of the receiver to the diode, -114 dBm is the thermal noise floor of 1 MHz bandwidth as discussed in the above section, and A is a constant related to the diode parameter and the noise figure of the video amplifier following it. It can be expressed as (Reference 3)

$$A = \frac{4 F_V}{k T M^2} \times 10^{-6} \qquad (2.15)$$

where F_V is the noise figure of the video amplifier expressed in power ratio rather than in dB, and M is the figure of merit of the diode which can be expressed as (Reference 8)

$$M = \frac{\gamma}{\sqrt{R}} \qquad (2.16)$$

where γ is the detector sensitivity in mv/mw, and R is the dynamic impedance of the diode in ohms. A detailed discussion on detectors can be found in Chapter 11.

The value of A can be determined in one of the following ways:

1. The figure of merit M and noise figure F_V are given, then A can be

obtained by Equation 2.15.
2. The sensitivity γ, dynamic impedance R, and noise figure F_v are given, then A can be obtained by Equations 2.15 and 2.16.
3. It can be measured experimentally. If a video detector is considered as the only element in a microwave receiver, there is no gain or loss in front of the detector, then $G_T = F_T = 1$ in Equations 2.13 and 2.14. Under this condition, the only dominant term in these equations is that containing A. These two equations can be approximated by

$$TSS = -110 + 10 \log \sqrt{AB_V} \text{ dBm} \qquad (2.17)$$

By measuring the TSS experimentally, A can be determined by Equation 2.17. The error of this approximation will be demonstrated in an example later in this section.

The sensitivity of a receiver can be considered under two different situations: (1) RF gain limited case, and (2) noise limited case. In the former case there is not enough signal gain (or no gain at all) in front of the detector, the sensitivity of the receiver is determined primarily by the detector. This condition occurs when, in Equations 2.13 and 2.14,

$$\frac{AB_V}{(G_T F_T)^2} \geqslant 2 B_R B_V - B_V^2 \qquad (2.18)$$

Filter   Amplifier   Video amplifier

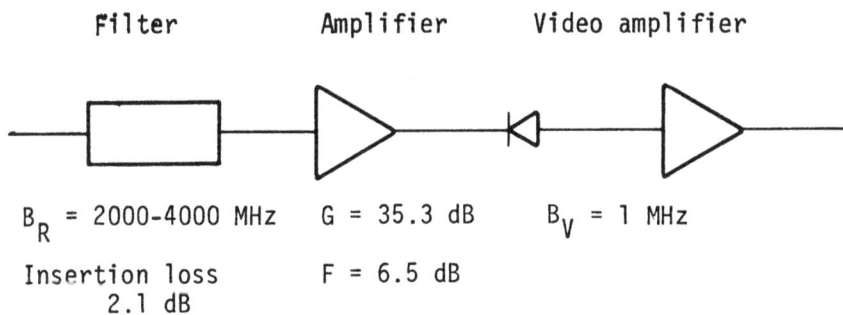

B_R = 2000-4000 MHz   G = 35.3 dB B_V = 1 MHz

Insertion loss   F = 6.5 dB
     2.1 dB

Figure 2.2. A Simple Crystal Video Receiver

The following example will demonstrate the application of these equations. Figure 2.2 shows a receiver with a RF filter followed by a RF amplifier. The bandwidth of the filter is 2000 MHz, with an insertion loss of 2.1 dB (1.62). The RF amplifier gain is 35.3 dB (3388) with a noise figure F = 6.5 dB (4.47). The video bandwidth B_V = 1 MHz. The constant A of the detector is measured experimentally. The TSS is −38 dBm without the RF amplifier. From Equation 2.14,

$$TSS = -38 = -114 + 2.1 + 10 \log (6.31 + 2.5 \sqrt{4000 - 1 + A})$$

one can obtain $A = 9.64 \times 10^{13}$.

If Equation 2.17 (the simplified one) is used,

$$TSS = -38 = -110 + 2.1 + 10 \log \sqrt{A}$$

which corresponds to an A value of 9.55×10^{13}. The difference in A reflects only 0.04 dB in TSS (which is smaller than the possible measurement error) by neglecting B_R, $B_R B_V$, and B_V^2 in Equation 2.14. After the value of A is determined, the next step is to calculate the TSS with the RF amplifier and filter. The overall noise figure will be calculated first from Equation 2.11

$$F_T = F_1 + \frac{F_2 - 1}{G_1} = 1.62 + \frac{4.47 - 1}{0.62} = 7.22 \ (8.6 \text{ dB})$$

$$G_T = G_1 G_2 = 2089 \ (33.2 \text{ dB})$$

Substituting A, F_T and G_T into Equation 2.14

$$TSS = -114 + 8.6$$

$$+ 10 \log \left(6.31 + 2.5 \sqrt{4000 - 1 + \frac{9.64 \times 10^{13}}{(7.22 \times 2089)^2}} \right)$$

$$= -73.3 \text{ dBm}$$

When there is adequate RF in front of the detector, the sensitivity of the receiver is not limited by the input signal but rather by the noise of the receiver. If

$$\frac{AB_V}{(G_T F_T)^2} < 0.2 \ (2 B_R B_V - B_V^2)$$

or   (2.19)

$$G_T > \frac{2.24}{F_T} \sqrt{\frac{A}{2 B_R - B_V}}$$

then the term $AB_V/(G_T F_T)^2$ can be neglected in Equations 2.13 and 2.14 and the error induced will be less than 0.4 dB. The above situation means the TSS is independent of the characteristics of the detector. The overall noise figure of the receiver becomes a dominant factor.

In a receiver, it is highly desirable to use one bandwidth to calculate the sensitivity. This bandwidth is generally referred to as the "effective bandwidth." However, from Equations 2.13 and 2.14, it is obvious that an effective bandwidth cannot be defined easily. It is only when the receiver is noise limited and the RF bandwidth B_R is much greater than the video bandwidth B_V, then Equation 2.14 can be written as

$$TSS = -114 + 10 \log F_T + 10 \log (2.5 \sqrt{2 B_R B_V})$$
$$= -114 + 10 \log F_T + 4 + 10 \log \sqrt{2 B_R B_V} \qquad (2.20)$$

The 4 dB in Equation 2.20 can be regarded as the input signal-to-noise ratio required to produce the TSS of the receiver (References 8 and 9). The $\sqrt{2 B_R B_V}$ can then be regarded as the "effective bandwidth" of the receiver which agrees with Klipper's result (Reference 2).

2.4 OPERATIONAL SENSITIVITY[1, 4, 5, 10-12]

In order to operate the receiver in a satisfactory manner, a certain threshold must be set up to keep the false alarm rate at a desirable level. The threshold must be set above the noise floor. Although the TSS is easily measured and is related to the threshold, it does not provide the necessary information such as false alarm rate and probability of detection. The operational sensitivity is used by most microwave receiver engineers oriented toward system applications. This subject is discussed in References 1, 5, and 10. Detailed analysis of random noise and detection capability can be found in References 11 and 12. Only the results will be presented here. These results are based on Reference 1 (Chapter 2). The noise voltage (V) entering the RF filter is assumed Gaussian with a variance ψ_0. The mean value of V is equal to 0. Let threshold voltage be V_T. Then the average time interval between crossings of the threshold by noise alone is defined as the false alarm time T_{fa}.

$$T_{fa} = \frac{1}{B_R} \exp \frac{V_T^2}{2\psi_0} \quad \text{sec} \qquad (2.21)$$

The narrower the RF bandwidth B_R and the higher the threshold, the longer the false alarm time. The probability of false alarm is defined as the ratio of the time the noise crossing the threshold to the overall observation time. It can be expressed as

$$P_{fa} = \frac{1}{T_{fa} B_R} \qquad (2.22)$$

If a sine wave signal of amplitude A is present along the noise in the receiver, the probability of detection P_d is

$$P_d = \int_{V_T}^{\infty} \frac{E_0}{\psi_0} \exp\left(- \frac{E_0^2 + A^2}{2\psi_0}\right) I_0\left(\frac{E_0 A}{\psi_0}\right) dE_0 \qquad (2.23)$$

where E_0 is the amplitude of the envelope of the filter output.

$I_0(z)$ is the modified Bessel function of zero order and augmented z defined by

$$I_0(z) = \sum_{n=0}^{\infty} \frac{z^{2n}}{2^{2n} n! n!} \qquad (2.24)$$

The integration of Equation 2.23 can be approximated by numerical approach. When $E_0/\psi_0 \gg 1$ and $A \gg |E_0 - A|$ then Equation 2.23 can be written as

$$P_d = \frac{1}{2}\left(1 - \text{erf}\, \frac{V_T - A}{\sqrt{2\psi_0}}\right) + \frac{\exp - \dfrac{(V_T - A)^2}{2\psi_0}}{2\sqrt{2\pi}\,(A/\sqrt{\psi_0})}$$

$$\bullet \left[1 - \frac{V_T - A}{4A} + \frac{1 + (V_T - A)^2/\psi_0}{8A^2/\psi_0} \cdots\right] \qquad (2.25)$$

where the error function is

$$\text{erf}\, y = \frac{2}{\sqrt{\pi}} \int_0^y e^{-u^2} du \qquad (2.26)$$

The signal amplitude A and noise variance ψ_0 can be related to signal noise ratio as

$$\frac{A}{\sqrt{\psi_0}} = \sqrt{\frac{2S}{N}} \qquad (2.27)$$

The probability of detection P_d as a function of signal-to-noise ratio and probability of false alarm can be presented in Figure 2.3. In this figure $V_T^2/2\psi_0$ is replaced by $(1/P_{fa})$. The value of P_d is calculated from Equation 2.23 through numerical integration by letting $A = 1$, and changing variable E_0 to $1/x$ to bound the limits of integration. The value of $I_0(R/\psi_0)$ is evaluated from Equation 2.24. The scale of P_d is generated through the normal distribution

$$\phi(y) = \int_{-\infty}^{y} \frac{1}{\sqrt{2\pi}} \exp\left(-\frac{x^2}{2}\right) dx \qquad (2.28)$$

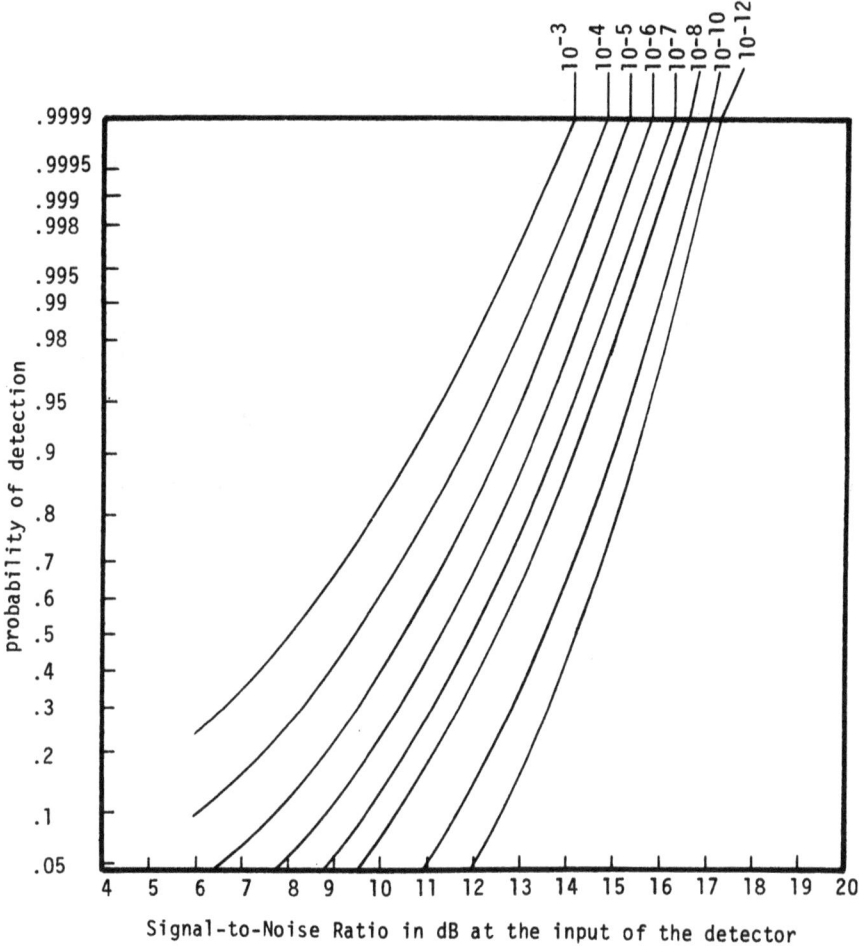

Figure 2.3. *Probability of Detection for a Sine Wave in Noise as a Function of the Signal-to-Noise (Power) Ratio and the Probability of False Alarm*

The value of $\phi(y)$ can be found in mathematical tables. For a different value of V_T in Equation 2.23, the corresponding probability of false alarm can be determined from Equation 2.21 and 2.22.

The above discussions can be utilized in the following manner in a receiver design or analysis. The RF bandwidth B_R and the false alarm time T_{fa} in the equation must be determined first. Then from Equation 2.22 the false alarm rate P_{fa} is calculated. The probability of detection P_d versus the required signal-to-noise rate can be read directly from Figure 2.3. For example, a receiver with $B_R = 1$ MHz can have only one false alarm every 100 seconds. ($T_{fa} = 100$ sec), the corresponding $P_{fa} = 10^{-8}$ (from Equation 2.22). If a 90 percent probability of detection is desirable, a signal-to-noise ratio of approximately 14 dB can be obtained from Figure 2.3. It should be noted that the S/N is at the input of the detector not at the input of the receiver. If the noise figure of the receiver is know, the operational sensitivity under the specified conditions can be determined. If the receiver noise figure is 10 dB, then the operational sensitivity of the receiver is $-114 + 10 + 14 = -90$ dBm.

In the above discussion, the video bandwidth B_V is assumed to be greater than $B_R/2$ but less than B_R. However, the effect of B_V is not included in this discussion. If the RF bandwidth B_R is much wider than B_V, the probability of false alarm and detection will be quite different. The operational sensitivity problem with the video bandwidth B_V taken into consideration was discussed by Emerson (Reference 4). Although useful curves are not generated in the article, foundations have been laid to obtain these curves. A summary of the results will be presented here. The probability density functions are approximated by Emerson as

$$p(E_0) \cong \frac{1}{\sqrt{K_2}} \sum_{j=0}^{\infty} a_j \phi^{(j)} \left(\frac{E_0 - E_1}{\sqrt{K_2}} \right) \text{ for } \gamma > 2 \qquad (2.29)$$

where

$$\phi^{(j)}(\chi) = \frac{d^j}{d\chi^j} \left\{ \frac{1}{(2\pi)^{1/2}} \exp\left(-\frac{\chi^2}{2} \right) \right\} \qquad (2.30)$$

and

$$\left.\begin{array}{l} \alpha_0 = 1 \\[2mm] \alpha_1 = \alpha_2 = 0 \\[4mm] \alpha_3 = -\dfrac{K_3}{6K_2^{3/2}} \end{array}\right\} \qquad (2.31)$$

$$p(E_0) \cong \frac{K_1}{K_2} \sum_{j=0}^{\infty} \beta_j \psi^{(j)} \left(\frac{K_1 E_0}{K_2} \right) \text{ for } \gamma < 2 \qquad (2.32)$$

where

$$\psi^{(j)}(\chi) = \frac{d^j}{d\chi^j} \left\{ \chi^{(K_1^2/K_2) - 1} \exp(-\chi) \right\} \qquad (2.33)$$

and

$$
\left.\begin{aligned}
\beta_0 &= \frac{1}{\Gamma(K_1^2/K_2)} \\
\beta_1 &= \beta_2 = 0 \\[2ex]
\beta_3 &= -\frac{K_1^2}{6K_2}\,\frac{2 - (K_1 K_3/K_2^2)}{[\Gamma(K_1^2/K_2) + 3]}
\end{aligned}\right\}
\tag{2.34}
$$

where $\Gamma(n)$ is the Gamma function defined as

$$
\Gamma(n) = \int_0^\infty \chi^{n-1}\exp(-\chi)\,d\chi
\tag{2.35}
$$

E_0 is the output voltage from video amplifier following the detector

$$
K_1 = N(1 + X)
\tag{2.36}
$$

$$
K_2 = \frac{N^2}{(1+2\gamma^2)^{1/2}}\left[1 + 2X\left(\frac{1+2\gamma^2}{1+\gamma^2}\right)^{1/2}\right]
\tag{2.37}
$$

$$
K_3 = \frac{4N^3}{2+3\gamma^2}\left[1 + 3X\left(\frac{2+3\gamma^2}{2+\gamma^2}\right)^{1/2}\right]
\tag{2.38}
$$

$$
N = \phi_0 \Delta f
\tag{2.39}
$$

$$
X = \frac{S}{N}
\tag{2.40}
$$

$$
\gamma = \frac{B_R}{2B_V}
\tag{2.41}
$$

ϕ_0 is the IF input noise power per unit frequency, and Δf is the equivalent IF noise bandwidth.

In Equations 2.29 and 2.32, the convergence is fairly rapid and three terms are the most required for graphical accuracy. The probability density normalized with respect to Y where $Y = E_0/N$ are shown in Figures 2.4 and 2.5 for several γ values. Figure 2.4 shows the case when there is no

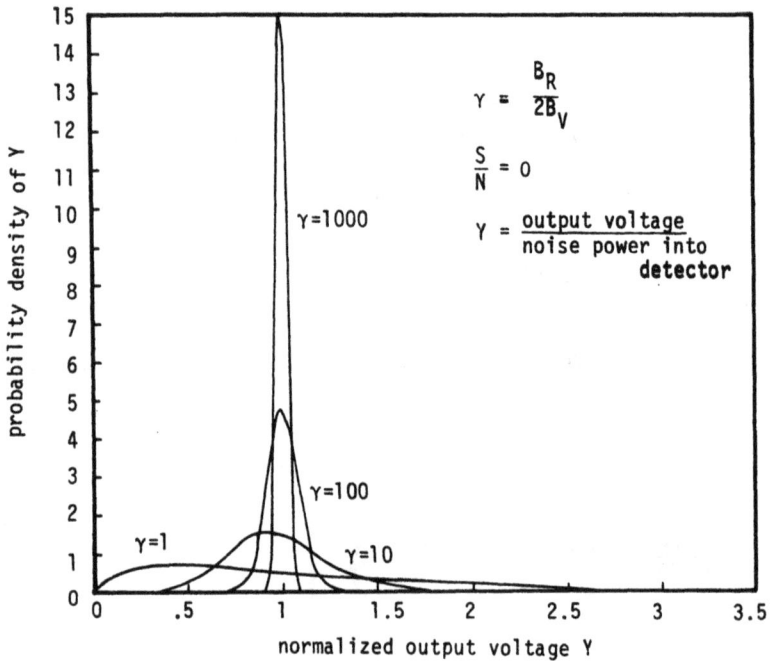

Figure 2.4. Output Probability Densities for the Gaussian System (Signal Absent)

input signal or substituting $X = 0$ in Equations 2.36 through 2.38. Figure 2.5 shows the results of $X = 2$ (3 dB). It should be noted that the X-axis and Y-axis scales are different for the plots in Figures 2.4 and 2.5. Using these two curves, the false alarm rate and probability of detection can be determined. The false alarm rate can be calculated from $p(E_0)$ in Equations 2.29 and 2.32 as

$$P_{fa} = \frac{\int_{V_T}^{\infty} p(E_0) \, dE_0}{\int_{0}^{\infty} p(E_0) \, dE_0} \qquad (2.42)$$

where $N = 1$ and $X = 0$ are substituted into $p(E_0)$ for this calculation.

The probability of detection can be calculated as

$$P_d = \frac{\int_{V_T}^{\infty} p(E_0)\, dE_0}{\int_{0}^{\infty} p(E_0)\, dE_0} \qquad (2.43)$$

where $N = 1$ and fixed value of X will be substituted in $p(E_0)$ for the calculation in Equation 2.43. The value of X is the signal-to-noise ratio. The value of y in both equations is the same.

For $y > 2$ these integrals can be calculated as

$$\int_{0}^{\infty} p(E_0)\, dE_0 = -\frac{1}{\sqrt{2\pi}} \frac{K_3}{6K_2^{3/2}}\left(1 - \frac{K_1^2}{K_2}\right)\exp\left(-\frac{K_1^2}{2K_2}\right)$$

$$+ \frac{1}{\sqrt{2\pi}} \int_{-\frac{K_1}{\sqrt{K_2}}}^{\infty} \exp\left(-\frac{X^2}{2}\right) dX \qquad (2.44)$$

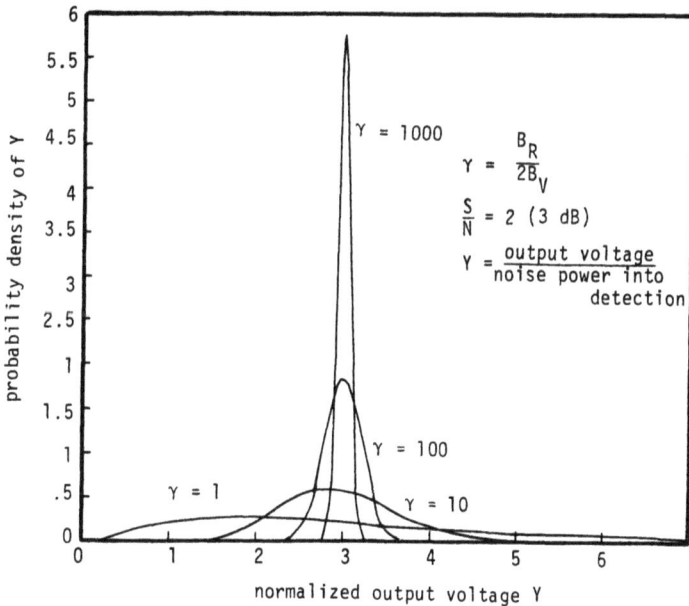

Figure 2.5. Output Probability Densities for the Gaussian System (Signal-to-Noise Ratio = 2)

$$\int_{V_T}^{\infty} p(E_0)\, dE_0 =$$

$$-\frac{1}{\sqrt{2\pi}}\frac{K_3}{6K_2^{3/2}}\left[1 - \left(\frac{V_T - K_1}{\sqrt{K_2}}\right)^2\right] \exp\left[-\frac{1}{2}\left(\frac{V_T - K_1}{\sqrt{K_2}}\right)^2\right]$$

$$+\frac{1}{\sqrt{2\pi}} \int_{\frac{V_T - K_1}{\sqrt{K_2}}}^{\infty} \exp\left(-\frac{X^2}{2}\right) dX \qquad (2.45)$$

The integrals in Equations 2.44 and 2.45 are closely related to the normal probability integral

$$\mathrm{erf}\frac{y}{\sqrt{2}} = \frac{1}{\sqrt{2\pi}} \int_{-y}^{y} \exp\left(-\frac{X^2}{2}\right) dX \qquad (2.46)$$

The erf is the error function. The two integrals in Equations 2.44 and 2.45 can be evaluated from the following relations.

$$\frac{1}{\sqrt{2\pi}} \int_{y}^{\infty} \exp\left(-\frac{X^2}{2}\right) dX =$$

$$\frac{1}{\sqrt{2\pi}}\left[\int_{0}^{\infty}\exp\left(-\frac{X^2}{2}\right) dX - \int_{0}^{y}\exp\left(-\frac{X^2}{2}\right) dX\right]$$

$$= \frac{1}{2} - \frac{1}{2\sqrt{2\pi}} \int_{-y}^{y} \exp\left(-\frac{X^2}{2}\right) dX \qquad (2.47)$$

where the integral in Equation 2.47 can be expressed through a series

$$\frac{1}{\sqrt{2\pi}} \int_{-y}^{y} \exp\left(-\frac{X^2}{2}\right) dX$$

$$= y\left(\frac{2}{\pi}\right)^{1/2}\left[1 - \frac{y^2}{2\cdot 1!\cdot 3} + \frac{y^4}{2^2\cdot 2!\cdot 5} - \frac{y^6}{2^3\cdot 3!\cdot 7} + \ldots\right] \qquad (2.48)$$

or for a large value of y the following asumptotic series may be used.

$$\frac{1}{\sqrt{2\pi}} \int_{-y}^{y} \exp\left(-\frac{X^2}{2}\right) dX$$

$$\cong 1 - \left(\frac{2}{\pi}\right)^{\frac{1}{2}} \frac{\exp\left(-\frac{y^2}{2}\right)}{y} \left[1 - \frac{1}{y^2} + \frac{1\cdot 3}{y^4} - \frac{1\cdot 3\cdot 5}{y^6} + \dots \right] \quad (2.49)$$

It should be noted that Equation 2.49 does not converge.

Using Equation 2.49, the result with the minimum absolute value should be used to determine the total number of terms required.

For the case of $y < 2$

$$\int_{0}^{\infty} p\,(E_0)\,dE_0 = 1 \quad (2.50)$$

and for

$$\int_{V_T}^{\infty} p\,(E_0)\,dE_0$$

$$= \frac{-K_4}{6} \frac{2 - \frac{K_1 K_3}{K_2^{\,2}}}{\Gamma(K_4 + 3)} \left[(K_4 - 1)(K_4 - 2)X^{(K_4 - 3)} \right.$$

$$\left. - 2(K_4 - 1)X^{(K_4 - 2)} + X^{(K_4 - 1)} \right] \exp\,(-X) \Bigg|_{X = \frac{K_1}{K_2} V_T}^{\infty}$$

$$+ \frac{1}{\Gamma(K_4)} \int_{\frac{K_1}{K_2} V_T}^{\infty} X^{(K_4 - 1)} \exp(-X)\,dX \quad (2.51)$$

where

$$K_4 = \frac{K_1^2}{K_2} \quad (2.52)$$

The integral in Equation 2.51 can be evaluated from results in integral tables or through numerical integral.

From the above calculations it is shown that, given a P_{fa} value instead of one curve as shown in Figure 2.3, a family of curves with a different value of γ can be generated. Figure 2.6 shows an example of $P_{fa} = 10^{-8}$ and $y = 4$ to 128. It shows that for the same P_{fa} and P_d values, the higher the γ value, the less the signal to noise is required. The explanation of this phenomena is that the video bandwidth B_V following the detector has a smoothing effect. The video filter reduces the noise resulting from the wide RF bandwidth B_R of the receiver. Similar curves could be generated for different values of P_{fa} and different γ values.

The application of the curves is quite similar to that discussed above. Equation 2.22 will be used to calculate the false alarm rate, then the signal-to-noise ratio can be read directly from curves similar to Figure 2.6 if γ is known. The measurement on the sensitivity of a microwave receiver will be discussed in Chapter 16.

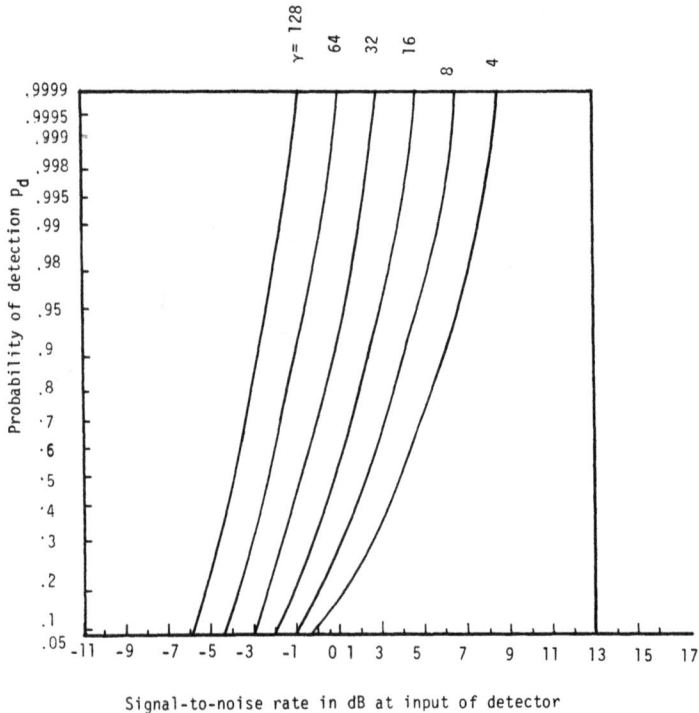

Figure 2.6. *Probability of Detection for a Sine Wave in Noise as a Function of Signal-to-Noise (Power) Ratio and* $y = (B_R/2B_v)$ *at* $P_{fa} = 10^{-8}$

2.5 DYNAMIC RANGE (13-15)

Dynamic range is a term commonly used to indicate the input signal range that the receiver can process properly. Unfortunately there is no universal definition given to this quantity. There is no agreeable sensitivity level. Among the different standards used are minimum discernible signal, tangential sensitivity, operational sensitivity level, etc. The upper limit of the input signal that a receiver can handle is not well defined either. In this section some of the commonly used dynamic ranges will be discussed. Some of these discussions can also be found in the amplifier section (Chapter 13) since the dynamic range is also used to specify amplifier performance.

A. One dB Compression Point

If the receiver provides amplitude information on the input signal, theoretically the amplitude can be measured for any single valued input-output relation. The output amplitude information is often linearly proportional to the input signal level for easy calibration. However, if the input signal keeps increasing, some of the linear components in the receiver will start to saturate; and the output no longer increases linearly. When the output of the receiver is deviated 1 dB from its linear region, the input level (or the output level) is called the 1 dB compression point. Beyond this point, the output signal level will deviate more from linear. As shown in Figure 2.7, the dynamic range of the receiver is from a specified input sensitivity level to the 1 dB compression point with respect to the output level.

B. Dynamic Range From Frequency Measurement Capacity of the Receiver

When the receiver provides only the frequency information and does not measure the amplitude information on the input signal, the dynamic range can be determined by its frequency measurement capability. Since the frequency measurement is considered to be the primary performance of a receiver, the dynamic range of the receiver is often determined by this capability, no matter whether the receiver measures the amplitude of the input signal or not. Under this condition the dynamic range is usually defined from a weak signal where the frequency error is within a certain predetermined range to a strong signal level where the receiver can still measure it properly. For example, an instantaneous frequency measurement (IFM) receiver measures both the frequency and amplitude of the input signal. If the receiver can measure an input signal from -65 dBm to $+15$ dBm with ± 3 MHz accuracy, the dynamic range of this receiver can be called 80 dB, although the amplitude measurement circuit may have only a 30 dB linear range.

If there is only one signal at the input of a receiver, there should be only one signal at the output of the receiver. However, when some linear components in the receiver are driven into the nonlinear region, additional signals

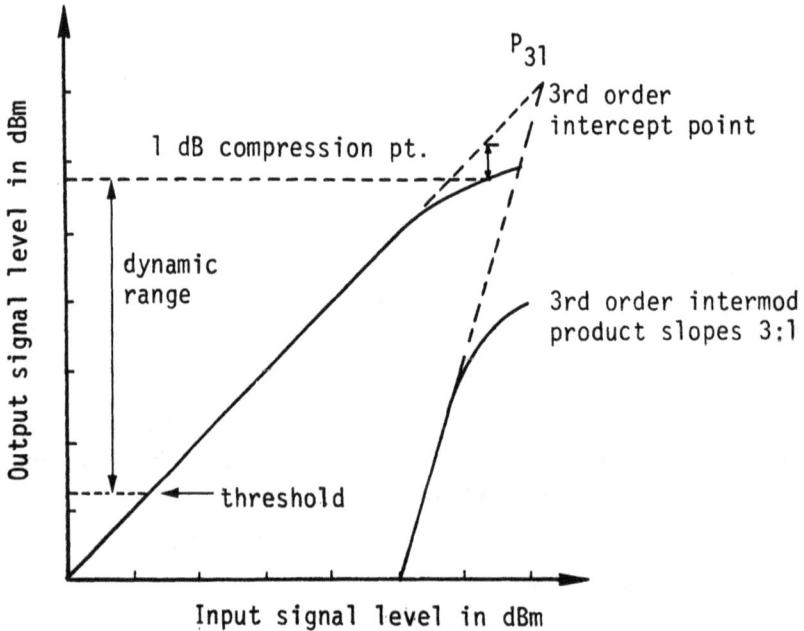

Figure 2.7. One dB Compression Point of a Receiver

(spurs) may appear at the output of the receiver. The range from a certain sensitivity level to the upper limit of the input signal and just before the generation of the spurs is often referred to as the single signal spur free dynamic range.

2.6 SIMULTANEOUS SIGNAL DYNAMIC RANGE (INSTANTANEOUS DYNAMIC RANGE) AND TWO-TONE SPUR FREE DYNAMIC RANGE (15)

An instantaneous dynamic range indicates the capability of the receiver to handle two simultaneous signals: one strong and one weak. In general, the strong signal can suppress the weak signal and cause the receiver to miss it. The instantaneous dynamic range is defined as the maximum amplitude separation between the two signals, such that the receiver can measure both of them correctly. This dynamic range depends on the frequency separation between the two signals. The signals used to evaluate the dynamic range are often cw. When pulsed signals are used, the weak one must not be buried in the side-lobes of the strong one. A typical result of instantaneous dynamic range is shown in Figure 2.8. The two signals must be separated more than the frequency resolution of the receiver, so that the receiver can read both signals correctly. The farther the two signals are separated in frequency, the wider the dynamic range because filters are usually used to separate signals close in frequency, and the dynamic range versus the frequency separation plot is generally related to the composite filter shape. When the two signals are separated far apart, the filter effect will no longer show and the dynamic range is more or less a constant value determined by the linear active components used in the receiver.

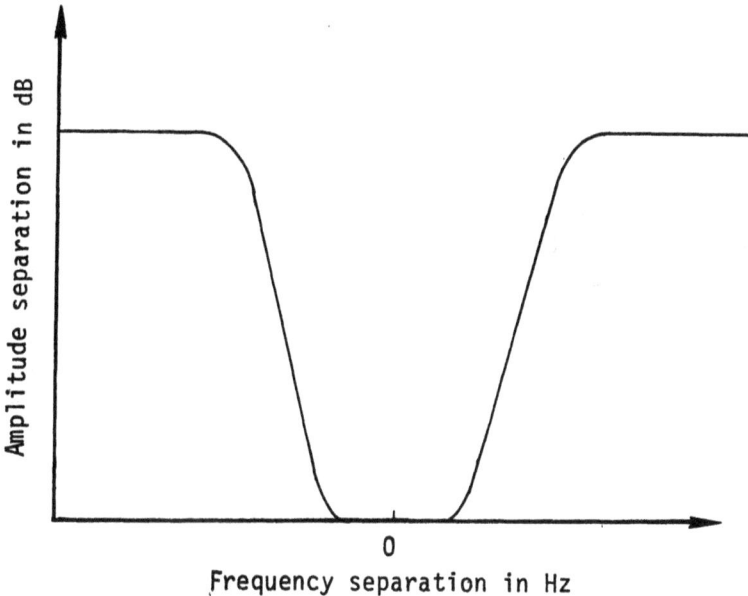

Figure 2.8. *Instantaneous Dyanmic Range to Receive Strong and Weak Signals*

Another requirement on simultaneous signals handling capability is that the receiver does not generate spurious responses. When two strong signals are applied to a receiver, they may drive the receiver into a nonlinear region and generate some undesirable output which are referred to as spurs. In Figure 2.9 A and B are two strong signals, while C and D are the spurious products generated by A and B. C and D are referred to as the third order intermodulation products when A and B are equal in amplitude. The frequency separations between C-A, A-B, and B-D are all equal to Δf. The spurs must be below or equal to the noise level, otherwise false alarm rate will increase. This is sometimes called the two-tone spur free dyanmic range. This

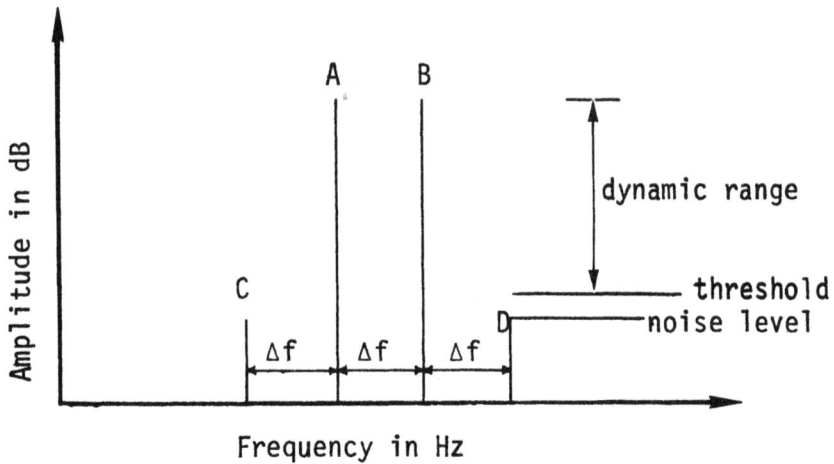

Figure 2.9. *Spurs Generated by Two Strong Signals*

dynamic range is closely related to the nonlinear characteristic of the receiver. If the characteristics of all the components used in the receiver are known, the two-tone dynamic range of the receiver can be calculated theoretically. The two-tone dynamic range will be further discussed in Chapter 13 under linear amplifiers, but the results will be discussed here. The spurs generated in Figure 2.9 can be incorporated in the input-output relation of a receiver as shown in Figure 2.7. The third order intercept point (P_{31}) is the intercept point of the gain curve of slope 1 and the third order product curve of slope 3. Given the third order intercept point P_{31}, the total gain of the receiver G_T and the input power P_i, the amplitude C and D (third order intermod products in Figure 2.9) can be calculated as (see the derivations of Equation 13.17)

$$IM_3 = 3(P_i + G_T) - 2P_{31} \qquad (2.53)$$

When IM_3 is approximately equal to the noise floor, the input level P_i to a certain predetermined threshold is the two-tone spur-free dynamic range (see Figure 2.9).

If the third order intercept point of each component is known, the overall third order intercept point $P_{31,T}$ can be written as

$$\frac{1}{P_{31,T}} = \frac{1}{P_{31,1}} + \frac{G_1}{P_{31,2}} + \frac{G_1 G_2}{P_{31,3}} + \dots \qquad (2.54)$$

where $P_{31,i}$ is the third order intercept point of element i. If the element is a passive device (i.e., an attenuator), a large value of P_{31} can be assigned to it (i.e., $P_{31,i} = 99$ dBm) for the calculation of $P_{31,T}$.

Equations 2.11 and 2.54 are two of the most important calculations in receiver design. The former one is used to calculate the sensitivity of the receiver, while the latter one is used to calculate the dynamic range. Computer programs are available to perform these calculations (Reference 15).

In all the dynamic ranges mentioned above, the absolute level of the dynamic range can be translated to different input levels by adding an attenuator in the proper stage of the receiver. For example, the instantaneous dynamic range of a certain receiver is 30 dB (from -70 to -40 dBm). This 30 dB can be placed anywhere above the -70 dBm level by adding an attenuator. A 20 dB attenuator will place the dynamic range in -50 to -20 dBm range. This attenuator can be switched in and out on a pulse-by-pulse basis. Thus, for a single signal the receiver can handle a very wide dynamic range. However, this kind of dynamic range should be considered as a fictitious one for receivers that can handle simultaneous signals. Thus the instantaneous signal dynamic range, the two tone spur free dynamic range and the single signal dynamic range should all be specified in order to reduce confusion.

2.7 PULSE AMPLITUDE (PA), PULSE WIDTH (PW), AND TIME OF ARRIVAL (TOA) MEASUREMENTS[1, 16]

If the input signal is pulse modulated RF, the PW and PA are often desirable information. The PA information may be required on a non-cooperative emitter to obtain a gross range information on the emitter. PA can also be used as a sorting parameter to generate the scan pattern of the hostile radar and more importantly to provide information for the jammer. The PA is measured at the peak of the pulse. Since there is a multipath problem which means the signal reaches the receiver through a different path length (i.e., reflected from some objects), the pulse shape will be distorted and is not always a rectangular one. After the leading edge, the pulse may start to droop with time because of the multipath. The peak of the pulse is close to the leading edge. A peak detecting circuit is often used to hold the peak value temporarily for the measurement. Since the PA may spread over a wide range, logarithmic amplifiers are often used to compress the PA to a manageable range. The relative error in PA measurement is (Reference 1, Chapter 2)

$$\frac{\delta A}{A} = \frac{1}{(2 \, S/N)^{\frac{1}{2}}} \qquad (2.55)$$

where δA is the rms value of amplitude error which is defined similar to that of Equation 2.3, A is the amplitude of the signal, and S/N is the signal-to-noise ratio in power. Equation 2.55 can be obtained from a sinusoidal wave of amplitude A while the signal power $S = A^2/2$ the noise power $N = (\delta A)^2$.

In a practical receiver, the amplitude is usually measured with approximately

1 dB resolution and accuracy. The range of the amplitude that can be measured is up to 60 dB.

The PW and TOA are required in an EW system as sorting parameters and provide the necessary information for a sophisticated jamming system. In a radar receiver, the TOA difference among various antenna locations can provide the angle information on the target.

To measure the PW, there are two major concerns: first is how to define the PW, and secondly the bandwidth of the receiver must be wide enough to preserve the pulse shape. If the pulse is measured on a linear scale, the 50 percent and 90 percent points are often chosen. The 50 percent points will generally provide the steepest slope which is easy to measure on a scope. If the logarithm of the PA is measured, the points that are 3 dB below the peak value are generally used to define the PW. The leading edge of the pulse is used to generate the TOA information. The TOA information, in turn, can be used to generate the pulse repetition frequency (PRF) of the radar which is one of the important sorting parameters. The rms error of the leading edge of the pulse can be expressed as (Reference 1)

$$\delta T = \frac{t_r}{(2 \, S/N)^{1/2}} \qquad (2.56)$$

where t_r is the rising time of the pulse, often defined approximately from 10 percent to 90 percent of the pulse amplitude on the leading edge. Since the RF bandwidth B_R is usually larger than the video bandwidth B_V, the value of B_V usually limits the accuracy of the PW measurement. The B_V must be wide enough to accomodate the leading edge. The minimum required valued of B_V can be approximated by (Reference 16)

$$B_V \sim \frac{0.35}{t_r} \qquad (2.57)$$

The leading edge measurement is more critical in a radar receiver than an EW receiver, because the range information is obtained from it.

2.8 ANGLE OF ARRIVAL (AOA) MEASUREMENTS [17, 18]

The importance of AOA information does not need to be stressed in a radar system. Even in an EW system, the AOA information is extremely important, because this is the only parameter the hostile emitter cannot change easily. Thus, the AOA becomes the most dependable sorting parameter in an EW receiving system, especially when the hostile radar intentionally varies its RF and pulse repetition interval (PRI).

The easiest way to measure AOA is to use a narrow beam antenna point- ing at the direction of the emitter. The beamwidth of the antenna will pro- vide AOA information (Reference 17). This approach is not suitable fo EW applications in general, because in an EW system, it is desirable t cover 360 degrees azimuth instantaneously. The two common ways t measure AOA in EW are through amplitude and phase comparisons. In both approaches, a multiple number of antennas and receivers are needed. Thus, the AOA information becomes the most costly parameter to be measured. The AOA information actually contains two angles: the azimuth and elevation. For simplicity, only the azimuth will be considered here.

A. Amplitude Comparison System[18]

A four quadrant amplitude comparison direction finding (DF) system is shown in Figure 2.10. Four antennas point in four directions. This system can be frequency independent which means it may cover a wide bandwidth.

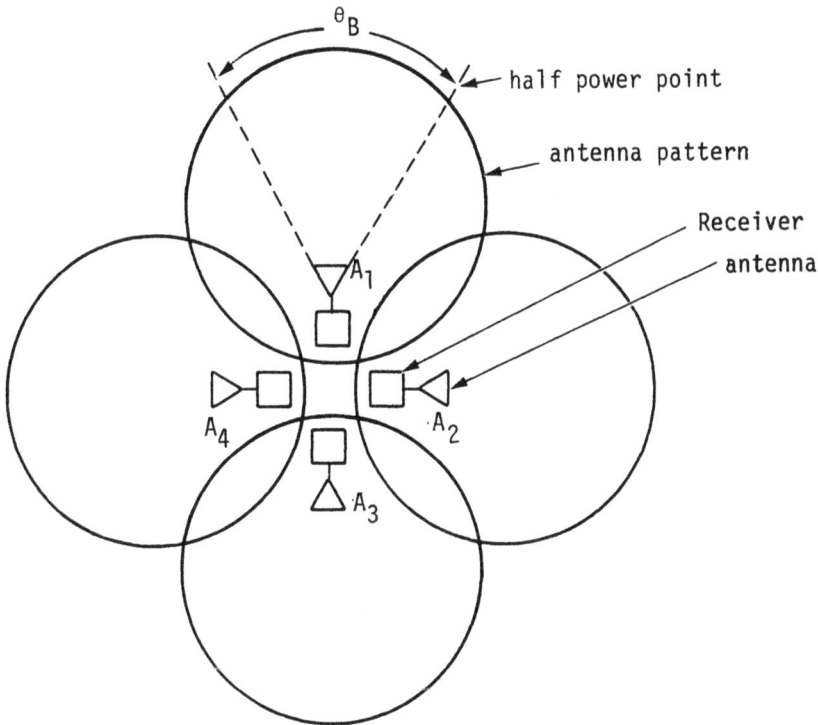

Figure 2.10. A Four Quandrant Amplitude Direction Finding System

However, the actual frequency bandwidth of this system depends on the four receivers used in the system and the amplitude tracking among them. Wider bandwidth will improve the probability of intercept (POI), but it also increases probability of contamination by signals arriving at the receivers simultaneously. Let us assume that the amplitudes from the four antennas can be written as

$$A \exp\left[-k^2 (\theta - \alpha)^2\right] \qquad (2.58)$$

where A is the relative amplitude of antenna peak power, $k^2 = 2.776/(\theta_B^2)$; where θ_B is the antenna beamwidth measured between half-power points of the same antenna pattern; θ is the bearing angle measured from the antenna axis; and α is the beam offset angle. The constant 2.776 comes from the relation that when $\alpha = 0$ and $\theta = \theta_B/2$, the amplitude from the antenna gain is 0.5A. For the system in Figure 2.10,

$$\alpha_1 = 0, \ \alpha_2 = 90 \ \text{deg}, \ \alpha_3 = 180 \ \text{deg}, \ \text{and} \ \alpha_4 = 270 \ \text{deg}$$

The AOA information is obtained from the ratio of the amplitudes of two adjacent antennas (i.e., through antennas 1 and 2)

$$R = \frac{A_1 \exp\left(-k^2 \theta^2\right)}{A_2 \exp\left[-k^2 (\theta - \alpha_2)^2\right]}$$

$$= \frac{A_1}{A_2} \exp\left[-2k^2 \alpha_2 \theta + k^2 \alpha_2^2\right] \qquad (2.59)$$

where $\alpha_1 = 0$.

R can be expressed in dB as

$$R(\text{dB}) = 10 \log \frac{A_1}{A_2} - 8.68 \, k^2 \alpha_2 \theta + 4.34 \, \alpha_2^2 \qquad (2.60)$$

The error slope is, by definition, the rate of R with respect to θ; therefore, differentiation of Equation 2.60 give the error slope as

$$\frac{d}{d\theta}[R(\text{dB})] = -8.68 \, k^2 \alpha_2 \qquad (2.61)$$

The error slope is a function of antenna beamwidth and squint angle between the two antennas. It is independent of the incident angle θ. In other words, the error slope is constant for all θ provided the antenna pattern may be approximated by the exponential form. If the antenna amplitude imbalance is measured in dB, then the angular error is

$$\Delta\theta = \frac{A_2/A_1}{8.68 \, k^2 \alpha_2} \qquad (2.62)$$

Therefore, the system accuracy is directly proportional to the amplitude imbalance between the two antennas and inversely proportional to the error slope. In a practical system, not only is the antenna not perfectly balanced, but the amplitude tracking among the four receivers is also not perfect. Thus in Equation 2.62 the amplitude imbalance should include the receiver amplitude mismatch. In a practical four quadarant amplitude comparison AOA system, the accuracy is in the range of 10 degrees to 15 degrees. If more antennas with narrow beamwidth are used, the angular accuracy can be improved at the cost of system complexity.

B. Phase Comparison (Interferometry) AOA System [19-21]

A simple two-element phase comparison system is shown in Figure 2.11.

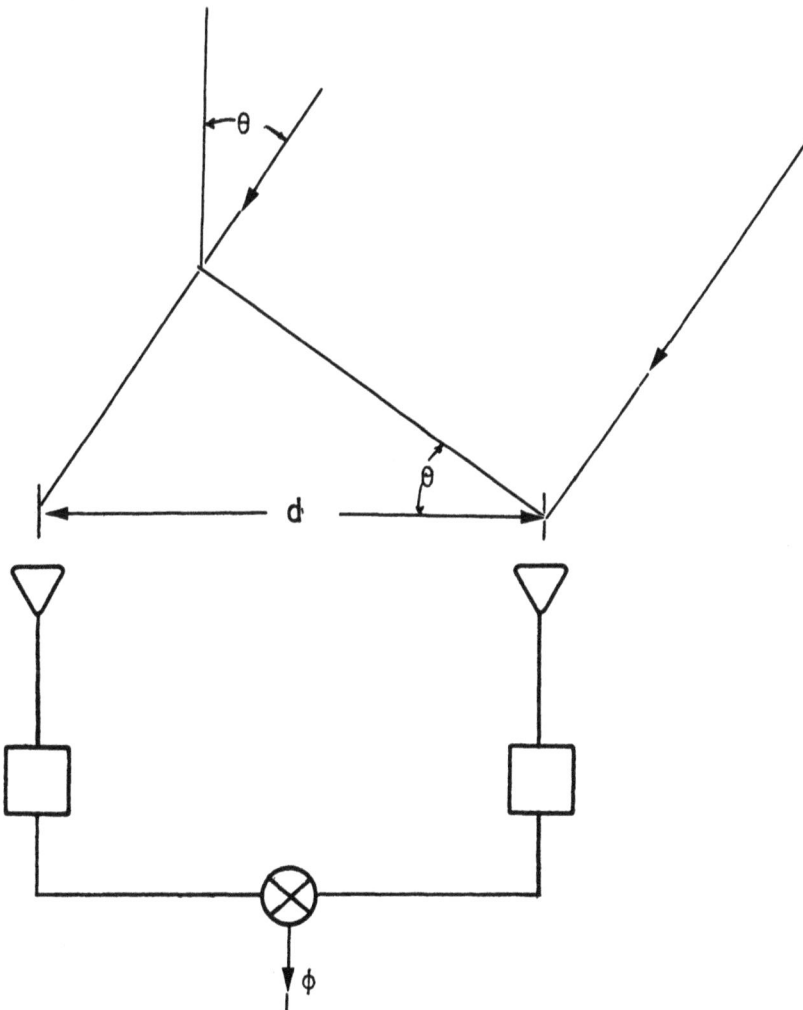

Figure 2.11. Two-Element Phase Comparison for AOA Measurement.

The two antennas are separated by distance d. If a wave is coming in at angle θ from the boreside, the phase difference ϕ can be expressed as

$$\phi = \frac{2\pi d \sin \theta}{\lambda} \tag{2.63}$$

where d is the distance between the two antennas, θ is the incident angle as shown in Figure 2.11, and λ is the wavelength of the incident wave.

From Equation 2.63, it is obvious that the frequency (or wavelength) must be known. For a fixed ϕ value, the longer the distance d, the more accurate the angle θ can be measured. However, there is a maximum distance between the two antennas. Beyond this value, it will cause ambiguity in θ. The maximum phase shift ϕ must be within 2π in order to avoid the ambiguity. The maximum distance can be obtained by changing θ from +90 to −90 degrees and keeping the change of ϕ within 2π.

$$\Delta\phi = 2\pi = \frac{2\pi d_{max} [\sin 90° - \sin (-90°)]}{\lambda} \tag{2.64}$$

or

$$d_{max} = \frac{\lambda}{2} \tag{2.65}$$

If more accuracy is desirable without ambiguity, a multiple number of antennas can be used with inequal spacings between them as shown in Figure 2.12. The long distances provide the AOA accuracy while the short one can resolve the ambiguity. The AOA error is contributed from three possible sources: the frequency, phase measurements error and noise. The DF error for a two channel phase comparison system is discussed in References 18 and 19. The results will be presented here.

The contribution of phase errors to rms DF error is

$$\varepsilon_\phi = \frac{\lambda \sigma_\phi}{\sqrt{2}\pi d} \tag{2.66}$$

where σ_ϕ is the rms channel phase tracking error.

The contribution of DF error by frequency inaccuracy is

$$\varepsilon_f = \sin \theta \frac{\Delta f}{f} \tag{2.67}$$

where Δf is the rms frequency error.

The contribution of the thermal noise to rms DF error is

$$\varepsilon_n = \frac{\lambda}{\sqrt{2}\pi d \sqrt{S/N}}$$

(2.68)

where S/N is the signal to noise ratio.

The overall rms DF error can be written as

$$\varepsilon = \sqrt{\varepsilon_\phi^2 + \varepsilon_f^2 + \varepsilon_n^2}$$

(2.69)

The smaller the phase error (σ_ϕ) and frequency error (Δf) and the higher the S/N, the less the overall error.

If a long baseline antenna is used, the accuracy of the system can be very good. AOA accuracy of approximately 1 degree is an achievable value.

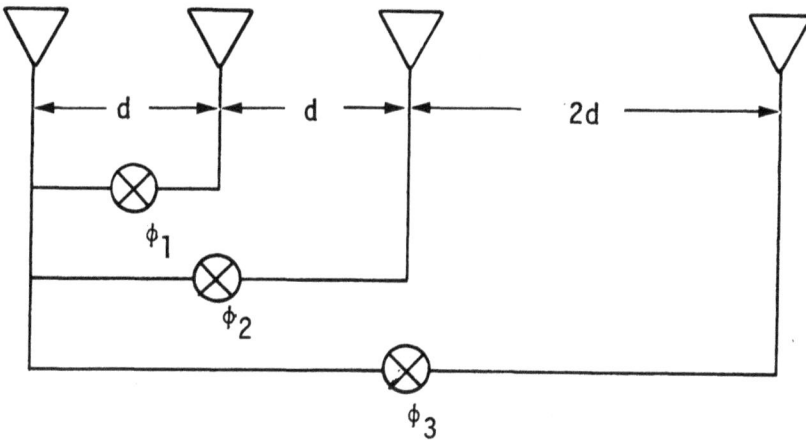

Figure 2.12. Multiple Element Phase Comparison AOA System

C. AOA Measurement Through Doppler Frequency Shift (22-24)

If the antenna of a receiver is moving in a certain direction (say x) with velocity v, the frequency of the input signal will be changed by the Doppler frequency shift. The Doppler frequency shift depends on the direction of the emitter with respect to the line of motion as shown in Figure 2.13. If the distance between the emitter and receiver is D, this distance can be expressed in terms of phase by

$$\phi = \frac{2\pi D}{\lambda}$$

(2.70)

where λ is the wavelength. The Doppler frequency f_d is generated by the time change of phase angle ϕ along the direction D

$$f_d = \frac{1}{2\pi}\frac{d\phi}{dt} = \frac{1}{\lambda}v\cos\theta = \frac{f_0 v}{C}\cos\theta \tag{2.71}$$

where v is the velocity of the antenna moving, f_0 is the frequency of the emitter, C is the velocity of light, and θ is the angle between v and line of sight D. This equation is different from Equation 2.1 by a factor of 2, because in a radar system the signal travels distance 2D while in the case of an EW receiver the signal travels the distance D.

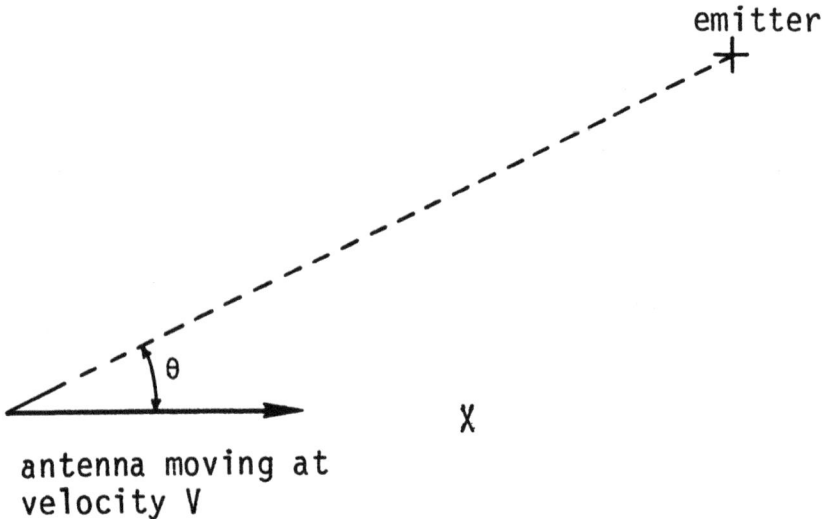

emitter

θ

X

antenna moving at
velocity V

Figure 2.13. Doppler Frequency Shift Induced by Moving the Receiving Antenna

The signal frequency f_0 can be measured by a stationary receiver. If the velocity v is known, then the AOA information can be obtained from Equation 2.71. Of course, it is impractical to move the antenna physically to generate any noticeable Doppler frequency shift. However, the antenna motion can be stimulated by switching the receiver rapidly among an array of antennas as shown in Figure 2.14. Receiver A is connected to a fixed antenna, while receiver B is switching from antenna 1 to N, through the N to 1 switch, to simulate a linear motion. The frequency difference between the two receivers can be used to determine the AOA information. In practical application, this system has been used in VHF to UHF frequency range and the antennas are arranged in a circle. This system can be used at microwave frequency with the advancements in receiver and switching technology.

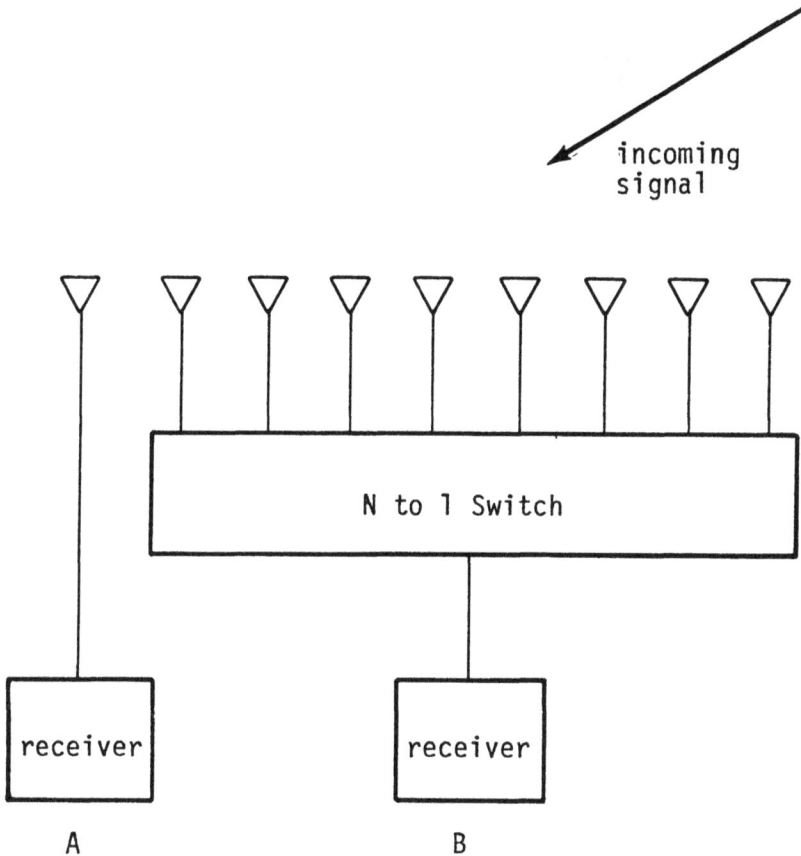

Figure 2.14. *High-Speed Switching Scheme to Simulate an Antenna Motion*

D. AOA Measurement Through Differential TOA

If the TOA of a pulsed signal can be measured very accurately, this information can be used to obtain AOA. As shown in Figure 2.15, the time difference Δt for the pulsed signal to reach the two antennas is related to the incident angle θ by

$$\Delta t = \frac{d}{C} \cos \theta \qquad (2.72)$$

where d is the distance between the two antennas and C is the speed of light.

If the angle θ is limited between 0 and 180 degrees, there is no ambiguity problem in this measurement scheme. From Equation 2.72 it is obvious that the longer the distance d, the more accurate the measurement. The main difficulty in this scheme is the accuracy of the Δt measurement. Since the leading edge of a radar pulse has finite rising time, it is not only difficult to measure an accurate TOA but also hard to define it. Therefore, this approach is not widely adapted for relatively short baseline system.

Another AOA measurement scheme is using a microwave lens. Its operational principle will be discussed in Chapter 12.

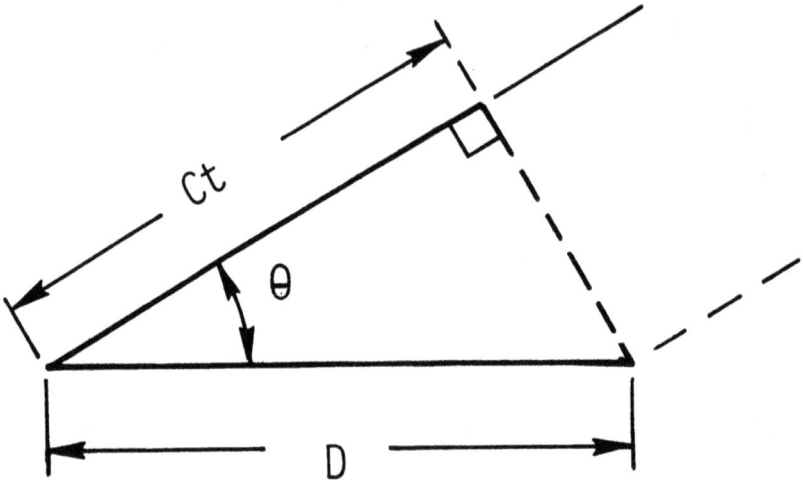

Figure 2.15.　AOA Measurement Through TOA Difference

E. Antennas (25-27)

A radar system usually uses a high-gain, high-directional antenna to direct the radar signal in a certain direction. At the same time, the high gain antenna will improve the radar detection range. These antennas are quite large in size; their linear dimensions are many times the operational wavelength. Most of these antennas have large mechanically driven reflectors. Other have many active elements to form an array of antennas that can be scanned electronically.

The antennas used with the EW receivers are relatively simple in structure and small in size. One of the primary requirements of EW applications is the high probability of intercepts and small size, especially for airborne applications. The antennas used in all the AOA measurement systems mentioned in this section usually fulfill these requirements. In order to have a high probability of intercept, the antenna should have broad frequency coverage and wide angular coverage. An antenna with a broad beam is usually small in size. The most commonly used antenna is a kind of spiral antenna as shown in Figure 2.16. The antenna is fabricated

Figure 2.16. Spiral Antenna

through printed circuit technology on a dielectric board. The feeding point is at the center of the antenna. The frequency coverage is often over octave bandwidth. It can cover approximately 90 degrees in spatial angle. The diameter of the antenna is approximately a half wavelength of the lowest operating frequency. This antenna usually works in the 2 to 18 MHz frequency range. The gain of the spiral antenna is slightly higher at high frequency. A spiral antenna with lower operating frequency is very desirable, and research is required in this area. The flush mounted spiral antenna is suitable for airborne applications.

For lower frequency, a stub antenna of $\lambda/4$ of $\lambda/2$ length is sometimes used. The analysis of this antenna can be found in many books (References 26 and 27). The antenna will have broad angular coverage but relatively narrow bandwidth in frequency. At higher frequencies, a horn antenna with waveguide feed is rather popular. The gain of the antenna is rather high. In other words, the angular coverage of the antenna is relatively narrow. Sometimes special designs will be adapted to broaden the angular coverage of the antenna.

2.9 PROBABILITY OF INTERCEPT (POI), THROUGHPUT RATE, AND SHADOW TIME

The POI is a term used in connection with EW receivers. It tells the percentage of pulses the receiver will collect in a dense environment. Of course, this is a complicated problem. It depends on the distribution of the pulses in the environment. The probability can be determined from three aspects: spatial, frequency range, and time. If the receiver antenna covers only a section of the whole area of interest, it will miss most of the signals outside the section. If the instantaneous bandwidth of the receiver does not cover the entire frequency range of interest, the receiver must be time shared among the different frequency ranges. The POI of the above situations can be easily determined. The POI discussed here is basically in the time domain. However, what happens to the receiver under dense signal environment conditions is a very difficult problem to answer, because it depends heavily on the signal condition. It is almost always possible to put up certain signal conditions so that a receiver will miss pulses.

Therefore, the POI is a very loosely defined term which will generate some qualitative data on the performance of a receiver. In general, if a receiver is referred to as having 100 percent POI in a certain frequency range, it does not mean the receiver will not miss signals in the bandwidth, it means it only reflects the instantaneous bandwidth of the receiver.

A more meaningful approach is to use throughput rate to evaluate the time domain performance of a receiver. The throughput rate is applicable to receivers dealing primarily with pulsed signals. It tells how fast the receiver can process a pulse and be ready for the next one. Of course, the throughput rate depends on the PW of the input signal, especially when the receiver has the capability to measure PW, since it must wait until the end of the pulse to generate the PW information. One easy way to find the throughput rate is to apply a signal with a short pulse so that the PW does not affect the measurement. The throughput rate is usually measured with PW signals slightly above the minimum value the receiver can process faithfully. Increase the PRF until the receiver starts to drop pulses, then the highest PRF before the receiver begins missing pulses is the throughput rate. The time between two pulses is the throughput time.

For example, if the throughput time of a receiver is 1 μs, PW of 200-300 ns may not affect throughput time. To test such a receiver, one can increase the input pulse rate to the receiver. When the PRF is below 1 MHz, every input pulse will generate a set of output data. When the PRF is above 1 MHz, the output data rate will drop to half of the input data rate because every other pulse will be missed by the receiver. The measurement of throughput rate will be further discussed in Chapter 16.

Another useful way to represent the signal processing rate of the receiver is called shadow time, which is quite similar to throughput time. It is usually specified as the minimum time from the trailing edge of first pulse to the leading edge of a second pulse that the receiver can process the second signal. The shadow time, in general, depends on the PW also, when the PW is below a certain value. When the PW is above this certain value, the shadow time usually becomes a constant. The shadow time is PW dependent, because most of the parameter measurements are made at the leading edge of the pulse and it requires a certain time to encode them.

If logarithmic (log) amplifiers are used in front of the video detectors, the throughput time and shadow time may also depend upon the PA of the signals. The recovery time of the log amplitude is longer for stronger input. This effect should be considered in determining the throughput rate.

The throughput rate and the shadow time are often used to replace the POI to represent the performance of the receiver.

REFERENCES

1. Skolnik, M.I., *Introduction to radar systems,* Chapter 10, McGraw-Hill Book Co., 1962.
2. Klipper, L., "Sensitivity of crystal video receivers with RF preamplification," Microwave Journal 8, pp. 85-92, 1965.
3. Lucas, W.J., "Tangential sensitivity of a detector video system with V.F. pre-amplification," Proc. Institution of Electrical Engineers, Vol. 113, No. 8, pp. 1321-1330, August 1966.
4. Emerson, R.C., "First probability densities for receivers with square law detectors," J. of Applied Physics, Vol. 24, No. 9, pp. 1168-1176, September 1953.
5. Harp, J.C., "What does receiver sensitivity mean?," Microwave Systems News, p. 54, July 1978.
6. Lipsky, S.E., "Calculate the effects of noise on ECM receivers,"

Microwaves, p. 65, October 1974.

7. Adamy, D.L., "Calculate receiver sensitivity," Electronic Design, p. 118, December 6, 1973.

8. Sareen, S., "Threshold detectors for nanosecond fault or level detection in RF systems," Application Note, Aertech Industries.

9. "The Criterion for the Tangential Sensitivity Measurement." Hewlett Packard Application Note 956-1.

10. DiFranco, J.V. and Rubin, W.L., *Radar detection,* Prentice Hall, 1968.

11. Rice, S.O., "Mathematical analysis of random noise," Bell System Technical Journal, Vol. 23, pp. 282-332, 1944; and Vol. 24, pp. 46-156, 1945.

12. Marcum, J.E., "A statistical theory of target detection by pulsed radar, mathematical appendix," IRE Trans. Information Theory, Vol. IT-6, pp. 145-267, April 1960.

13. Hayward, W. and Demaw, D., "Solid-state design for the radio amateur," Chapter 6, American Radio Relay League, 1977.

14. Norton, D.E., "The cascading of high dynamic range amplifiers," Application Note, Anzac Electronics, Waltham, MA, June 1973.

15. Tsui, J.B.Y., Johnson, S.J., Brumfield, W.T., "Receiver Analysis Program," Air Force Avionics Technical Report, AFAL-TR-76-199, Wright-Patterson Air Force Base, December 1976.

16. "Hot carrier diode video detector," Hewlett Packard Application Note 932.

17. Harper, T., "Airborne rotary DF antenna systems," Watkins Johnson Co., Tech-notes, Vol. 2., No. 2, March/April 1975.

18. Bullock, L.G., Oeh, G.R., and Sparagna, J.J., "An analysis of wideband microwave monopulse direct-finding techniques," IEEE Trans. Aerospace and Electronic System, Vol. AES-7, pp. 188-202, January 1971.

19. Howard, J.E, "Application of digital parallel processing, DIPPA, techniques to passive direction finding," Technical Report AFAL-TR-73-263, Air Force Wright Aeronautical Laboratories, June 1973.

20. Howard, J.E., "VHF digital parallel processing array (DIPPA) direction-finding techniques," Technical Report AFAL-TR-75-101, Air Force Wright Aeronautical Laboratories, September 1975.

21. Goodwin, R.L., "Ambiguity-resistant three- and four-channel interferometers," NRL Report 8005, Naval Research Laboratory, September 9, 1976.

22. Fantoni, J.A. and Benoit, R.C., Jr., "Applying the Doppler effect to direction finder design," Electronic Industries and Tele-Tech, p. 75, January 1957; p. 66, February 1957.

23. Davies, D.E.N., "A fast electronically scanned radar receiving system," Journal Brit. IRE, Vol. 21, pp. 305-318, April 1961.

24. Baghdady, E.J., "Induced directional FM (IPFM) multiple emitter

location techniques,'' Technical Report RADC-TR-80-157, Rome Air Development Center, May 1980.

25. Weeks, W.L., ''Antenna Engineering'' McGraw-Hill Book Company, 1968.

26. Jorden, E.C., *Electromagnetic waves and radiating systems,* Prentice-Hall, Inc., 1964.

27. Krause, J.D., *Antennas,* McGraw-Hill Book Co., Inc., 1950.

CHAPTER 3
CRYSTAL VIDEO AND SUPERHETERODYNE RECEIVERS

3.1 CRYSTAL VIDEO RECEIVERS(1-3)

A crystal video receiver is the simplest microwave receiver. The basic component in the receiver is a microwave detector which converts microwave energy into a video signal. Usually the radio frequency (RF) input bandwidth of this kind of receiver is very wide; (in gigahertz range) it is primarily determined by the bandwidth of the detector or by the antenna bandwidth. This receiver does not have the capability to determine the input signal frequency. It can only tell whether there is microwave energy in the input bandwidth. In order to keep the loss at a minimum, the detector is located very close to the antenna. The schematic of a simple crystal video receiver is shown in Figure 3.1. The sensitivity of the crystal video receiver is low. The sensitivity of the receiver is determined by the characteristic of the detector and the antenna gain if the antenna is considered part of the receiver. To improve the sensitivity of the receiver, the following approaches are used:

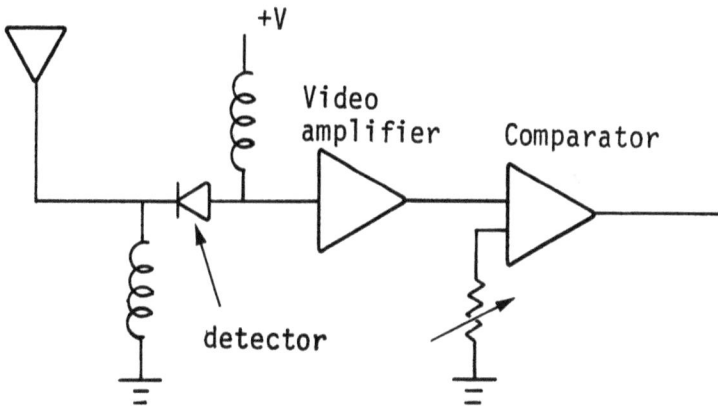

Figure 3.1. A Basic Crystal Video Receiver

1. Add a video amplifier at the output of the detector. For a weak input signal, the output of the detector is low. It is impractical to feed this signal directly into a threshold comparator, because the sensitivity will be low. A video amplifier will improve the sensitivity of the receiver. In Equations 2.13, 2.14, and 2.15, it is also mentioned that

the tangential sensitivity (TSS) depends on the bandwidth and the noise figure of the video amplifier. The video amplifier must have the proper bandwidth to accomodate the anticipated input signals. The shorter pulse width (PW) of the input signal, the wider the video bandwidth is required. However, the wider the video bandwidth, the more the noise contained in the bandwidth and the less the sensitivity of the receiver. Thus, video bandwidth is generally designed just wide enough to pass the shortest pulse anticipated.

2. Bias current can be applied to the detector diode to improve the receiver sensitivity. In general the optimum bias current is frequency dependent and proper biasing current must be determined according to the RF frequency chosen. More discussion of this subject is included in Chapter 11.

3. Impedance matching network between the antenna and the detector over a specific RF bandwidth ideally will make all the RF power deliverable to the diode. Thus, a properly designed matching network will improve the receiver sensitivity.

4. Sometimes a tunable RF filter is added in front of the detector to reduce the RF noise bandwidth. In general, this approach is not very effective since the filter will increase the RF loss. The noise figure (in dB) of the receiver is linearly related to the input loss but is not linearly related to the RF bandwidth. It is approximately related to the square root of the bandwidth. However, the filter in front of the detector will limit the RF bandwidth and provide coarse frequency information on the signal, since signals that can be detected by the detector must be in the passing band of the filter. The frequency resolution depends on the bandwidth of the filter. It is difficult to have a tunable filter with high-Q at RF frequency. Thus, this approach will not generate fine frequency resolution. Of course, the disadvantage of inserting a filter in front of the detector will reduce the probability of intercept on this signal. This kind of receiver is often called the tuned RF (TRF) receiver.

5. Adding RF amplifiers in front of the detector is the most effective way to improve the sensitivity. The relationship between the sensitivity and RF gain is discussed in Chapter 2. Adding RF amplifiers is also the most expensive way to do it, since RF amplifiers with wide bandwidth, high-gain, and low-noise figure are difficult to fabricate.

In addition to improving the sensitivity of the crystal video receiver, it is often required to extend the crystal video receiver capability to measure continuous wave (cw) signals and generate angle of arrival (AOA) information. Most of the video amplifier following the detector is ac coupled, since a dc coupled amplifier has drift problems. In addition, at the output of the detector, there is dc noise which will make the receiver sensitivity low if the video amplifier is dc coupled. A cw signal will appear at the output of the detector as a dc voltage which cannot pass through the ac coupled video amplifier. To avoid this problem, an amplitude modulator can be added in front of the detect. The purpose of the modulation is to make an amplitude modulated signal out of a cw input and to produce video pulses at the output of the video detector which can be amplified by the ac coupled video amplifier. The actual modulator could be a pin diode. The commonly used modulation waveform is simply on and off to produce a square wave train at a fixed frequency. If a band-pass filter with center frequency matching the modulator frequency is placed after the video amplifier, a cw signal will be "chopped" by the modulator and be detected at the output of the band-pass filter.

The crystal video can be used to measure PW, pulse amplitude (PA), and time of arrival (TOA) of the input signal. Multiple receivers can be used to measure AOA. The crystal receivers can only measure the AOA through amplitude comparison, because in a phase interferometric AOA system, the frequency of the input signal must be measured and a crystal video receiver does not have this capability. Logarithmic (log) amplifiers are added after the video amplifiers to measure AOA. In an amplitude comparison AOA system, the ratio of two adjacent antenna outputs is used to measure the angle information. However, a simple mathematic dividing circuit is difficult to design, but a subtraction circuit is rather common. The ratio of the antenna outputs can be obtained through sub-tracting the outputs of two log amplifiers. These log amplifiers will have an adverse effect on the throughput rate, because they have relatively slow recovery time. The general performance of a crystal video is listed below:

- Wide instantaneous RF bandwidth
- Poor frequency measurement capability
- Low to moderate sensitivity, the sensitivity is often RF gain limited
- Cannot differentiate simultaneous signals

Figure 3.2. Crystal Video Receiver in an Airborne System (Courtesy of Applied Technology, Div. of Itek)

Since the receivers can only provide the information of whether there are signals present at its input bandwidth but are not able to measure the signal frequency accurately, they have very limited usage in the communication area. Their primary application is in the radar warning system. The most commonly used ones are the warning receivers against police traffic radars. They are also commonly used in tactical or strategic aircrafts to warn against enemy radars, because they are simple in design, lightweight, and very compact. A crystal video receiver is shown in Figure 3.2. This is an airborne system and is used for radar warning. In this figure, there are four spiral antennas with four crystal video receivers to cover the four quadrants. A stub antenna, a processor, two different display units, and a power supply are also included.

3.2 BASIC SUPERHETERODYNE RECEIVERS[4-8]

A superheterodyne receiver (often referred to as a superhet receiver) is a relatively narrow band receiver (from a few kilohertz to over ten megahertz) which converts the input frequency through a mixer to an intermediate frequency (IF) and the IF signal is further processed. The primary reason for the superhet receiver is to improve the sensitivity of a crystal video receiver and at the same time provide fine frequency measurement capability. A basic superhet receiver is shown in Figure 3.3. The input signal with frequency f_s is mixed with the local oscillator (LO) frequency f_L in the mixer. The mixer output is the down converted signal frequency which can be written as $f_I = (\pm f_s \mp f_L)$ (where f_I is the desired output frequency). In some special applications, the up converted frequency $f_s + f_L = f_I$ is used. An IF band-pass filter is usually added after the mixer to reduce the spurious responses generated from the mixer. The

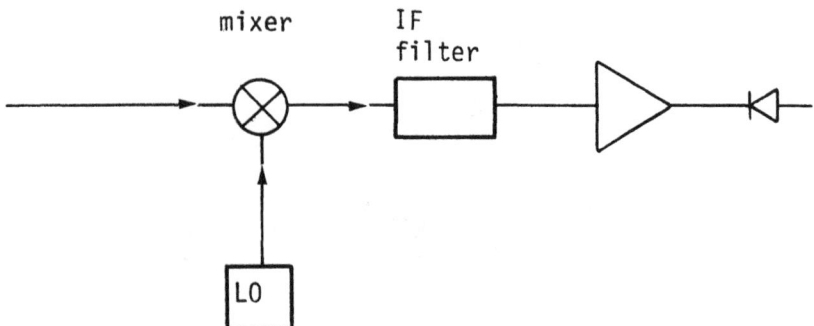

Figure 3.3.   Basic Superheterodyne Receiver

IF filter is also part of the frequency measurement circuit. Detailed analysis of spurs generated by mixers is discussed in Chapter 14. IF amplifiers are almost always used to amplify the mixed signal to improve the signal strength before the detector.

The input signal translating from one frequency to another through the mixer is a linear operation, although the mixer is basically a nonlinear device. This means that the information in the down converted signal is the same as the input signal. Thus, processing the down converted signal is equivalent to processing the input signal, but it is easier to process the down converted frequency, since the frequency is lower than the input signal. Through this superheterodyne arrangement, the following advantages can be realized:

1. The superhet receiver will provide high sensitivity. As discussed in the crystal video receiver, the most effective method to improve the sensitivity of the receiver is to add RF amplifiers in front of the detector. However, in many high frequency ranges, the RF amplifier is just not available. The only way to amplify the input signal is to down convert the input frequency to a lower range where amplifiers are readily available. Fortunately, high frequency mixers are much easier to obtain than amplifiers at the same frequency range. Although the mixer and filter will add extra insertion loss in the RF circuit, the amplification in the IF circuit will more than compensate for it. The sensitivity can be changed from an RF gain limited case to a noise limited case. The sensitivity of the receiver will improve drastically. If RF amplifiers can be added before the mixer, the sensitivity of the amplifier can be further improved. The detailed discussion on receiver sensitivity can be found in Section 2.3.

2. The superhet receiver can have very fine frequency resolution and accuracy. The filter following the mixer can be a very high-Q, fixed frequency, band-pass filter such as a crystal filter, which will have a very narrow bandwidth. This bandwidth determines the resolution of the receiver. The center frequency of the filter is usually below 1 GHz. The input signal frequency is equal to $f_L + f_I$ (or $f_L - f_I$) depending on whether f_L is less or greater than f_s. The LO frequency can be measured accurately, since it is generated at the receiver set. Thus the signal frequency can be determined accurately. In order to receive a signal with a certain frequency, the LO frequency f_L must be tunable so that the down converted signal can be placed at the center of the pass band of the IF filter.

3. The narrow IF bandwidth in the superhet receiver not only provides fine frequency resolution, but also reduces the noise in the RF bandwidth to much narrower IF bandwidth which further improves the sensitivity of the receiver. The narrow IF bandwidth can also reject the spurious responses generated outside the IF bandwidth from the front end of the receiver. The spurious responses are generated from

the mixers or from the RF amplifiers whenever the amplifiers are driven in the nonlinear region. The improvement of spur rejection will improve the dynamic range of the receiver.

4. If a receiver is operated in a dense signal environment with signal frequencies close to each other, it is difficult to separate a single signal from the environment and perform a detailed analysis on it. This is a common problem in the commercial broadcast band. The narrow IF bandwidth can provide good selectivity of signals. Of course, the IF bandwidth must be wide enough to cope with the signal spectrum it desires to measure.

From the above discussions, one can conclude that a superhet receiver can provide high sensitivity, wide dynamic, and excellent selectivity. These important factors make the superhet receiver a very important kind of receiver. These are some other important factors that should be considered in a practical superhet receiver and they will be discussed in the following section.

3.3 PRESELECTOR (TRACKING RF FILTERS)

In a practical superhet receiver, a filter is almost always required in front of the mixer. For example, a superhet receiver with an IF frequency centering at 100 MHz and LO frequency of 1000 MHz can receive a signal with a frequency of 1100 MHz as well as a signal with 900 MHz. These two signals are images of each other. At the output of the IF filter, one cannot differentiate between these two signals. Therefore, a decision must be made beforehand as to which signal is of interest. The filter in front of the mixer will eliminate the unwanted band. In some very special application, in order to improve the probability of intercept of the receiver, both the upper and lower bands can be folded in the same IF channel. In this special case additional circuitry is needed to identify whether the signal comes from the lower or higher RF band. This mode of operation is basically used to receive pulsed signals and is not for cw signals. In addition to this image problem, there are many other different combinations between the RF signal and the LO frequency produced in the mixer that the output signal is in the IF filter bandwidth. For the example used above, if the input signal is at 1050 MHz, then $2f_s - 2f_L$ (2100 − 2000 = 100 MHz) is also in the pass band of the IF filter. Not all of these products can be eliminated, but if the superhet receiver is properly designed, the amplitudes of these signals might be kept at a minimum (see mixer chart in Chapter 14).

The RF filter in front of the receiver must be tunable. For example, if the upper RF band is of interest, the center frequency (f_s) of the RF filter is related to the LO frequency by $f_s = f_L + f_I$. When the input signal is changed to a different value, the RF filter center frequency must be changed to receive the signal. Of course, the LO frequency must be changed also, such that the relation $f_s = f_L + f_I$ is fulfilled. In other words, the tuning of the LO and the tuning of the RF filter must be synchronized as shown in Figure 3.4. The RF filter bandwidth must be wide enough to accommodate the IF bandwidth and the possible tracking error between the RF filter and the LO. The tracking can be accomplished in one of the following ways:

1. For manually tuned superhet receivers, variable capacitors are commonly used. Two variable capacitors are controlled by the same tuning control: one capacitor changes the frequency of the LO, while the other controls the center frequency of the RF filter. The capacitors could be mechanically tunable moving vane type or electrically tuned varactors. This scheme is often used in AM and FM radios.

2. For receivers with a fixed number of channels, fixed filters and fixed LO frequencies are used. Each time the receiver is tuned to a new channel, a matched filter and LO frequency pair are chosen. This scheme is generally used in a television set.

3. The RF filter (or preselector) can be a yttrium iron garnet (YIG) filter while the oscillator is an YIG oscillator. The tracking between the filter and the oscillation must be carefully adjusted, such that the misalignment between the filter and the oscillator will be small compared to the RF filter bandwidth. The preselector is typically a multistage YIG filter with an instantaneous bandwidth of between 20 MHz and 70 MHz depending on the operation frequency. Wider bandwidth is available by staggering the frequency of each YIG

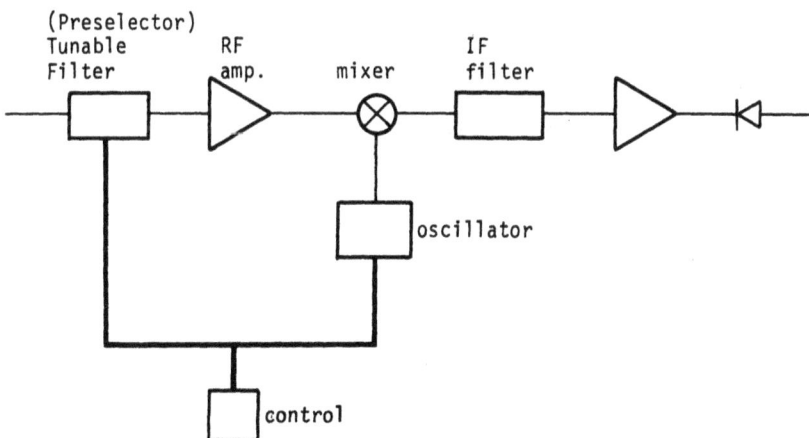

Figure 3.4. Superhet Receiver with Tracking RF Filter

sphere. If the preselector bandwidth is 40 MHz, a mistracking of 20 MHz between the LO and the preselector will result in an approximately 3 dB increase in the front end noise figure, because the RF signal will be at the 3 dB point on the preselector's pass band. The mistracking can be caused by a number of factors. Among them are tuning nonlinearity (RF frequency versus tuning current) and different ambient temperatures of the two devices. Temperature compensation schemes can be applied to the filter and the oscillator to keep them in close temperature range. In some cases, the filter can be adjusted externally through a dc bias current. Periodic alignment of the filter is required to ensure correct filter center frequency relative to the LO frequency. In some special designs, the YIG spheres of the filter and the YIG spheres of the YIG oscillator are controlled by one single tuning coil in a single housing unit as shown in Figure 3.5. The magnetic circuit consists of a magnetic shell with two pole pieces, but only a single main tuning coil. The preselector YIG sphere is mounted in one magnetic gap while the oscillator YIG sphere is located in the other. The main tuning coil tunes the two devices together across the frequency band. A small offset coil can be wound around one of the poles. Current supplied to the offset coil can be used to adjust the tracking between the two devices. Since the YIG spheres are in the same housing, the temperature difference between them is kept at a minimum. This arrangement can cover an operating frequency from 1 to 18 GHz.

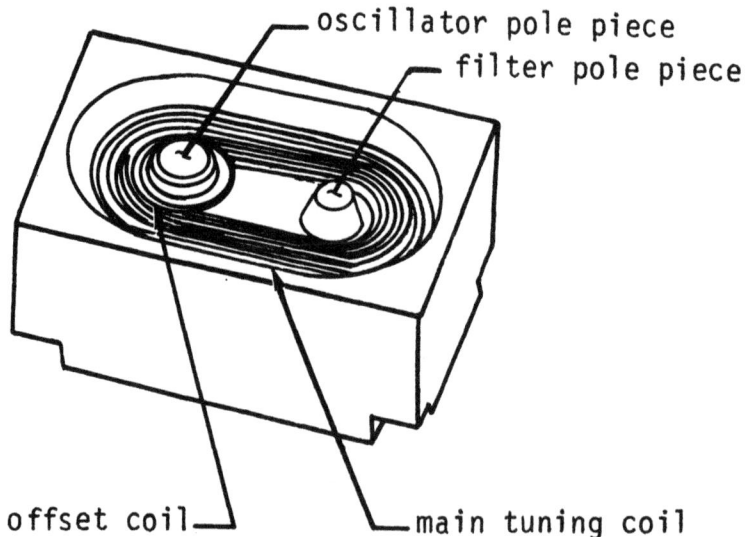

Figure 3.5. YIG Preselector and YIG Oscillator in One Magnetic Housing (Based on Papp and Jackson, Reference 6)

3.4 REQUIREMENTS ON THE LOCAL OSCILLATORS

The general requirements on a local oscillator are: accurate frequency settings, low-noise, low-spurs, wide tuning range, and rapid tuning speed. If one wants to measure a signal at 10 GHz with 100 KHz accuracy by a superhet receiver with an IF frequency at 100 Mhz, the LO frequency must be close to 10,100 MHz or 9,900 MHz with an accuracy better than 1 MHz. This means that the oscillator must have an accuracy better than $10^{-4}[(10^6/(10 \times 10^9)]$. If the LO does not provide such a frequency accuracy, its frequency inaccuracy will be translated to the IF bandwidth and contaminate the input signal. The tuning steps of the LO must be less than 100 KHz per step to provide the required frequency resolution. Crystal controlled oscillators have to be used to fulfill these requirements. At higher frequencies various phase locked schemes have been used to stabilize the LO frequency against a stable source (i.e., a crystal oscillator).

Since the receiver must cover the entire frequency of interest, the LO frequency must also cover that range. Using the example above, if the receiver is required to cover 8-12 GHz with 100 KHz frequency resolution, the LO must step a minimum of 4,000 ($4 \times 10^9/10^6$) 1 MHz step. This means the LO is very complicated and hard to design. To reduce the complexity of the LO, a multiple-stage down converter can be used. For example, a two-stage superhet receiver is shown in Figure 3.6. The first mixer down converts the input signals to an appropriate first IF (say 1 GHz) with a bandwidth of 50 MHz. It should be emphasized here that the IF frequency must be determined through the mixer chart to minimize the spurious responses. The IF bands used here are only to demonstrate the basic idea; they are not chosen for minimum spur products. The second converter shifts the frequency to 100 MHz. The first LO should have eighty 50 MHz steps to cover the 4 GHz bandwidth, while the second LO must have fifty 1 MHz steps to cover a 50 MHz bandwidth. The complexity of the LOs is reduced from one LO with 4,000 different frequencies to two LOs; one with 80 output frequencies, the other with 50 frequencies. The total number of down converter stages is seldom over three, because other hardware required will offset the design simplicity of the oscillators.

In case a superhet receiver is required to cover 1-18 GHz, the LO and the preselector must be able to cover the entire frequency range. The simplest design is to use one preselector and one LO to cover the entire frequency range. However, if the devices of such a wide range are not available, parallel channels can be used. For example, the 1 to 18 GHz range can be divided into any number of sections (i.e., 1-2 GHz, 2-4 GHz, and 4-18 GHz).

Usually the requirement on the tuning speed from one frequency to another is the faster, the better. This will cut down the time between measurements. In some special applications, extremely fast tuning speed is desirable. For example, a cueing receiver can be used to obtain the course

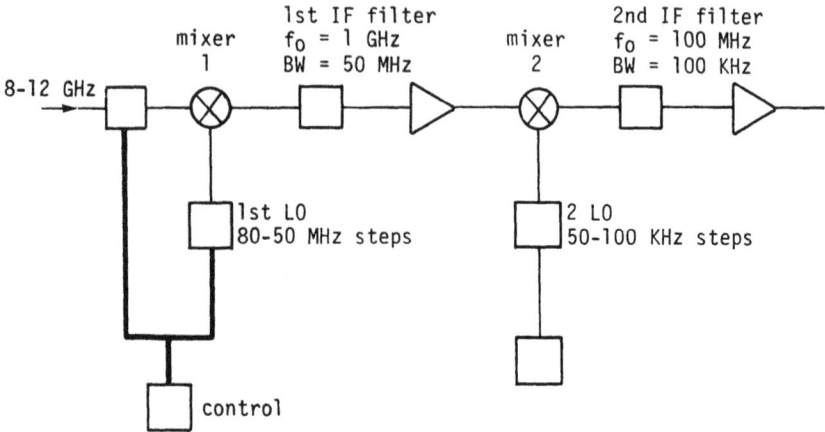

Figure 3.6. Two-Stage Superhet Receiver

frequency of the signal, then a superhet receiver is switched to the course frequency range to search and measure the detail information on the signal. In this application, delay time is often used in front of the superhet receiver to accommodate the switching speed of the receiver. The faster the switch time, the shorter the delay line can be which implies less insertion loss in the delay line. The sweeping speed of the LO is also limited by the bandwidth of the IF filter. It takes approximately $1/B_I$ seconds for the signal to build up in the IF filter due to the transient effect in the filter where B_I is the narrowest IF bandwidth. There are two kinds of LO tuning schemes. One is continuous [i.e., YIG oscillator, or voltage controlled oscillators (VCOs)], the other one is step scan (i.e., synthesizers). For the continuous sweeping LO, the maximum tuning speed is B_I^2, since it takes $1/B_I$ seconds to sweep the bandwidth B_I. If the LO is a fast sweeping/step scanning VCO, the frequency can be changed in a very short time (from ten to several hundred nanoseconds, 10^{-9} seconds). No preselector can track the LO frequency change with such a speed, because most of the preselectors are YIG tuned devices and have relatively slow response time. Under such a situation, fixed filters may be added in front of the mixer. The spur products are determined primarily by the mixer performance. Care should be emphasized to separate the spurs from the desired signals.

3.5 OTHER CONSIDERATIONS IN A SUPERHET RECEIVER

Impedence mismatch at the mixer input ports can cause the superhet receiver to function improperly. For example, the mismatch between the YIG filter and the mixer can cause gain ripple in the RF chain, and mismatch between the oscillator and mixer can cause frequency pulling of the oscillator so that the oscillator frequency is shifted due to the change of load impedance. Isolators are often at the two input ports of the mixers to improve the impedance matching. The isolator in the RF chain can also reduce the oscillator radiating through the input terminal. If wide bandwidth isolators are not available, sometimes fixed attenuators are used in place of the isolators. However, the attenuator will cause an undesirable noise figure increasing in RF channel, and wasting of LO power.

Output from the last IF filter of a superhet receiver is detected by a crystal detector. The dc output from the detector is compared with some predetermined threshold to decide whether the input RF signal is in the IF bandwidth. If the down converted signal is in the IF bandwidth, the input frequency is determined by the LO frequency and IF filter center frequency. This measurement scheme sounds very simple. However, the transient phenomena in the IF filter must be properly treated.

If the input is a cw signal, the receiver can be tuned continuously (or in fine steps) to observe the output signal level from the detector. If only one signal is present in the IF filter, the highest output from the detector implies that the down converted signal is at the center of the IF bands. The frequency of the input signal can then be read. If the signal is a pulse, the IF filter bandwidth must be greater than or equal to the inverse of the shortest pulse width or

$$B_I \geqslant \frac{1}{(PW)\,min}$$

where (PW) min represents the shortest pulse width anticipated.

When the pulsed signal is not down converted to the center of the IF band but at the edge of the filter, the output will have two "rabbit ears" as shown in Figure 3.7a. When the signal is tuned to the center of the filter, the two "ears" will disappear and the output is shown in Figure 3.7b. The "ears" are caused by the leading and trailing edges transient of the pulsed signal. Physically, it can be explained as follows: The leading and trailing edges of the signal contain a relative wide spectrum of frequency components compared to the middle of the pulse. Many of the frequency components are in the pass band of IF filter which will pass through the filter and appears as the "ears." The decision circuit must monitor the change of the detector output while the LO sweep across the signal. The maximum output corresponds to the down converted signal at the center of the IF filter. In many cases, the decision logic will sample the amplitude of the pulse after the leading edge transient and before the trailing edge transient. If the receiver cannot be tuned fast enough to bring the pulsed signal frequency to the center of the IF filter, the receiver cannot measure the frequency of the single pulses. Therefore, the probability of intercept on pulsed signals with a superhet receiver is rather low.

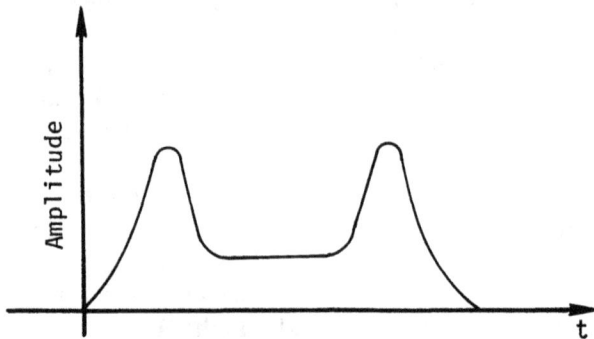

(a) Signal is at the Edge of the Filter

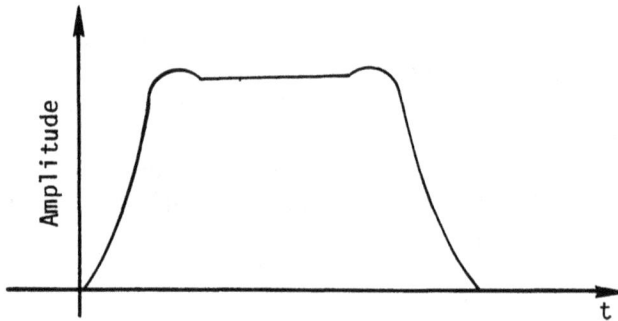

(b) Signal is at the Center of the Filter

Figure 3.7. Output from the Detector Output After the IF Filter

In a superhet receiver, besides the frequency information, other parameters of the signal are often desirable (i.e., amplitude modulation on the signal). In order to keep the fidelity of the signal, the receiver should not be saturated by the strongest signal anticipated. In other words, the receiver must be operated in the linear region. However, if frequency or phase are the only information of interest, the receiver can be driven into saturation, because it is easier to measure the receiver output with fixed amplitude to determine frequency. The dynamic range of a linear receiver is often limited by the detector. Logarithmic amplifiers can be inserted in front of the detector to improve the dynamic range of the receiver.

If the receivers are used to measure the AOA of the signal by phase comparison scheme, the phase relations among these receivers must be properly tracked. These requirements on receivers are difficult to fulfill, but superhet receivers are best suited for this application since it is easier to match phase between receivers of narrow bandwidth than those with wide bandwidth.

3.6 HOMODYNE RECEIVERS

A homodyne receiver can be considered as a special superhet receiver. The receiver also has a mixer and LO to down convert the input signal. The main difference between a superhet and a homodyne receiver is the LO. In the homodyne receiver, the LO frequency is derived from the input signal as shown in Figure 3.8. The input signal is divided into two paths. One of these signals is used to mix with an IF signal f_I to generate the LO frequency $f_R + f_I$. This LO frequency is then mixed with the input signal f_R to obtain the IF signal f_I. If there is no input signal, the mixer will isolate the fixed frequency f_I from reaching the IF filter. When there is an input signal of any arbitrary frequency f_R, there is an output at the output of the IF filter. It should be noted that this receiver could have a wide input frequency bandwidth; however, it does not provide frequency information on the input signal. This receiver is suitable to detect the existence of wide-band input signals (i.e., FM chirp signals, phase coded signals). No matter how the input signal frequency varies, the LO frequency always has the same variation. Therefore, the frequency information in the RF signal will be cancelled out at the output of the IF filter. This receiver will basically perform like a crystal video receiver which can only detect the existence of a signal. The application of this receiver is not very popular, because the design is relatively complicated and the information obtained is rather limited.

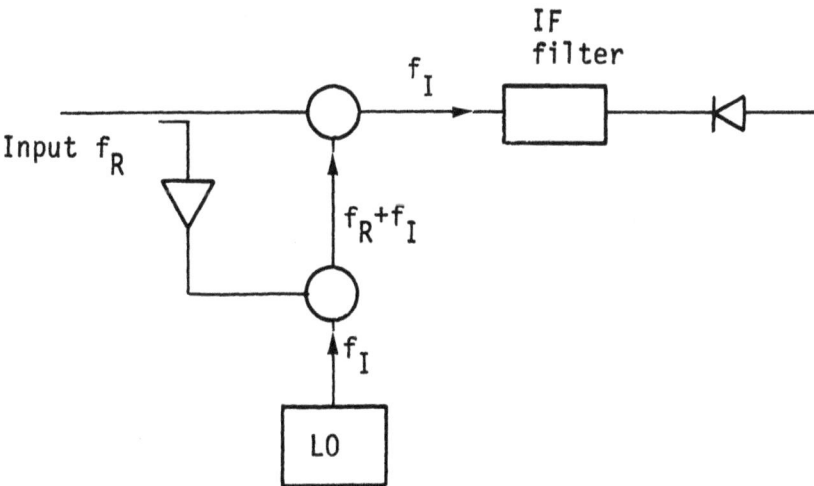

Figure 3.8. Basic Homodyne Receiver

3.7 APPLICATIONS OF SUPERHET RECEIVERS[9 and 10]

The performance of a superhet receiver is summarized as follows:
- Very high sensitivity
- Very fine frequency, accuracy, and resolution
- Very wide dynamic range
- High accuracy in measuring PA, PW, TOA, and AOA
- Poor probability of intercept: can receive only one signal at a time and miss all the other signals arriving simultaneously.

The superhet receiver is often used as a communication receiver. Since the frequency of the signal is already known, the receiver will have good selectivity to isolate and analyze the desired signal. In this application, the sensitivity of the receiver is very important because the input signal might be very weak. If the signal is coded, the same code should be injected through the LO to correlate with the input signal to produce a processing gain. For such application, the synchronization between the LO and the transmitter must be carefully maintained.

The superhet receiver is often used as a radar receiver. The requirements are quite similar to that of the communication receiver. The input bandwidth of the receiver matches the spectrum of the transmitted pulse. The sensitivity of the receiver relates to the radar detection range. Improving the receiver sensitivity can increase the radar detection range with the same transmitted power which is the more economic way to improve the performance of a radar system.

The superhet receiver can also be used as an analysis receiver, which measures the detailed information on an input signal. In an EW system a cueing receiver can be used to determine the coarse frequency of a certain signal, then a superhet receiver can be assigned to scan the course frequency range and lock on the signal. Detailed information on the signal can be obtained with less interference from other signals, because the relatively narrow IF bandwidth will reject unwanted signals. One of these applications is to measure the AOA of a signal through phase interferometry. Multiple receivers are required for this measurement. These receivers must be tuned to receive the same signal and only one signal can appear in the receiver, otherwise erroneous phase relation may be produced which corresponds to erroneous AOA information. All of the receivers used in the phase interferometry system must be phase tracked, and it is easier to match the phase of narrow bandwidth receivers than wide bandwidth receivers. The superhet receivers are also used to obtain AOA information through amplitude comparison, because it is easier to make them amplitude tracked among different receivers. Usually, a superhet receiver is the only receiver that is suitable to measure cw signals in a dense signal environment, especially when several cw signals are close in frequency.

If the superhet receiver is used as an electronic warfare receiver, it is commonly used in conjunction with another kind of receiver with wide instantaneous bandwidth. Since the superhet receiver has a relatively narrow IF bandwidth, the probability of intercept of the receiver is low. It will take a considerable amount of time to search over a wide frequency range. In electronic warfare applications, time is often a very critical factor. Although a superhet receiver can provide fine analysis on a signal, it usually does not have the desired probability of intercept to intercept the signal in time.

REFERENCES

1. Ayer, W.E., "Characteristics of crystal video receivers employing RF preamplification," Technical Report No. 150-3. Stanford Electronics Laboratories, Stanford University, September 20, 1956,
2. "Hot carrier diode video detectors," Application Note 923, Hewlett-Packard Co.
3. "Impedance matching techniques for mixers and detectors," Application Note 963, Hewlett-Packard Co.
4. Meier, P.J., O'Kean, H.E., Sard, E.W., "Integrated X-band sweeping superheterodyne receiver," IEEE trans Microwave Theory and Techniques, Vol. MTT-19, pp. 600-609, July 1971.
5. Crescenzi, E.J., Jr., Oglesbee, R.W., Chappell, R.A., "Integrating components for new front-end design," Microwaves, p. 35, August 1974.
6. Papp, J.C., Jackson, R.V., "YIG-tuned integrated devices," Watkins-Johnson Co., Tech Notes, Vol. 4, No. 5, September/October 1977.
7. Dexter, C.E., Glaz, R.D., "HF receiver design," Watkins-Johnson Co., Tech Notes, Vol. 5, No. 2, March/April 1978.
8. Dexter, C.E., "Digitally controlled VHF/UHF receiver design," Watkins-Johnson Co., Tech Notes, Vol. 7, No. 3, May/June 1980.
9. Dixon, R.C., "Spread Spectrum Systems," John Wiley & Sons, New York, 1976.
10. "Spread Spectrum Communications," Distributed by National Technical Information Service, U.S. Department of Commerce, AD 766914, July 1973.

CHAPTER 4
INSTANTANEOUS FREQUENCY MEASUREMENT (IFM) RECEIVERS

4.1 INTRODUCTION[1]

An Instantaneous Frequency Measurement (IFM) receiver uses delay lines to compare the phase of incoming signal to measure its frequency. An IFM receiver has a wide radio frequency (RF) bandwidth (possible over octave) and can measure the signal frequency accurately. It has the capability to measure frequency on very short pulses. The receiver can handle only one signal at a time. If more than one signal is arriving at the receiver simultaneously, the receiver will only report one frequency. Under a simultaneous signal condition, the reported frequency data can be one of the input frequencies or just an erroneous data.

An IFM receiver was discussed as early as 1948 by Earp (Reference 1). In the early stages, IFM was used to measure both frequency and amplitude of the signal. The output of the measured results were displayed on an oscilloscope in a polar form; the angle displacement represents the frequency information while the amplitude represents the signal strength. On such a display a single pulse is difficult to read, because the eyes are not very sensitive to a short display on the scope. A signal with the same repetitive frequency information can be detected from this arrangement.

In a modern IFM receiver, the frequency is measured on a pulse-by-pulse basis with the digital output. The frequency measurement circuit and the amplitude measurement circuit are, in general, separated. This design approach will produce more accurate frequency information. The bandwidth of an IFM receiver can be over an octave. The RF bandwidth of the receiver is limited only by the bandwidth of the components used to fabricate the receiver. An IFM receiver is considered as a wide-band receiver even though it measures frequency to a very fine resolution. This means that when calculating the sensitivity of an IFM receiver, the bandwidth used in the calculation is the total RF bandwidth and not the frequency resolution bandwidth and video bandwidth. The frequency resolution plays no role in the sensitivity of the receiver. In other words, the sensitivity of an IFM receiver is equivalent to a crystal video receiver with the same RF and video bandwidths.

One very important question to be answered in an IFM receiver is the output from the receiver when simultaneous incoming signals arrive at the receiver. The answer to this question is crucial to the design of the digital processor which follows the IFM receiver to sort and analyze the input signals. However, this question is still not fully answered and additional research is needed. This subject will be discussed in Section 4.7.

In most receiver designs, it is general practice to keep the bandwidth of the receiver below octave. In the amplifier chain at the front end of a receiver, amplifiers or mixers with over octave bandwidth may generate very high level second harmonics. The second harmonic can be rather high (approximately 10-15 dB below) in comparison with the real signal under normal condition. A wide bandwidth receiver with high dynamic range will receive both the signal and its harmonic which will be a spurious response of the receiver. In channelized, compressive, and Bragg cell receivers, the bandwidth is usually kept less than octave for this very reason. However, in IFM receivers, the over octave bandwidth does not cause any serious problems, since the receiver can report only one signal at a time, and will report the strong one if the two input signals are separated by more than 3-6 dB (depending on the receiver used for this test). Since the second harmonic is generally more than 10 dB below the real signal, the IFM receiver will report the real frequency without the disturbance from the weak one. This is an advantage of the IFM receiver which will work over octave bandwidth without adding more hardware to distinguish the spurious response generated by the second harmonic.

4.2 PRINCIPLE OF OPERATION[2-7]

Consider a sinusoidal wave that is split into two paths with one path delayed a constant time with respect to the other one. As shown in Figure 4.1a, there is a phase difference between the outputs caused by the time delay. The phase difference varies for various frequencies of the sinusoidal wave. Figures 4.1b and c show two sinusoidal waves of different frequency. Each of the sinusoidal waves is accompanied by another sinusoidal wave of the same frequency but delayed a constant time τ. The relative delayed phase angles are

$$\theta_1 = \omega_1 \tau \qquad (4.1)$$

$$\theta_2 = \omega_2 \tau \qquad (4.2)$$

Since $\omega_1 > \omega_2$, then $\theta_1 > \theta_2$.

Equations 4.1 and 4.2 show that the phase angle is directly proportional to the signal frequency. In an IFM receiver, a proper delay time τ is introduced. By measuring the phase delay between the undelayed and delayed signal, the frequency of the signal is obtained. Since the amplitude of the signal does not affect the relative phase angle, the signal strength can be measured at the same time.

In a practical IFM receiver, the frequency measurement circuit is arranged as in Figure 4.2. The incoming signal is divided into two paths; delayed and undelayed. Both of these signals are then fed into a phase correlator which has four outputs, and each of them is followed by a crystal detector. The outputs of the detectors are connected to the inputs of two differential amplifiers. The differential amplifer amplifies the difference of its two input signals. If one input is V_1 and the other one is V_2, then the

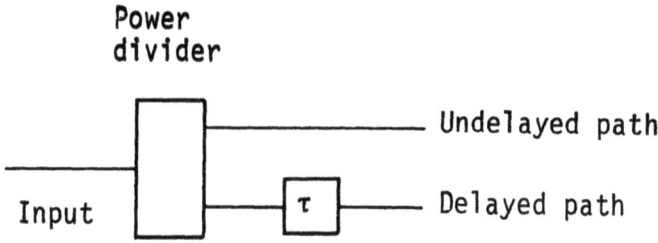

(a) Schematic diagram

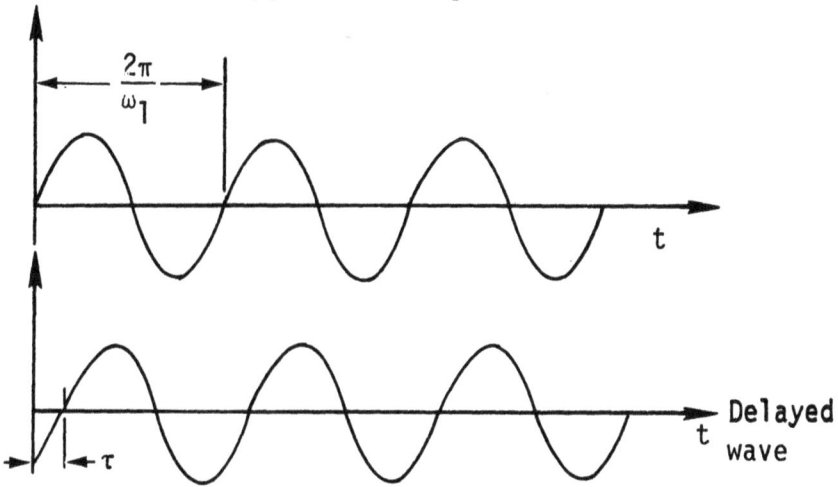

(b) For Signal with Angular Frequency ω_1

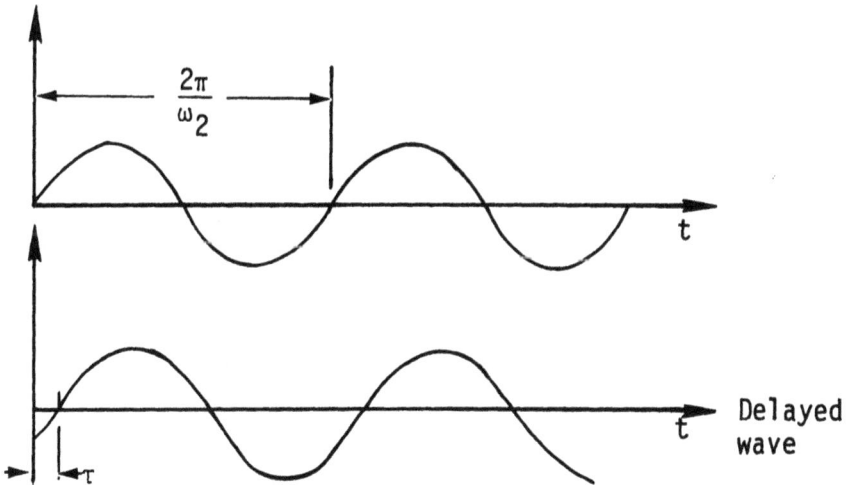

(c) For Signal with Angular Frequency $\omega_2 (\omega_1 > \omega_2)$

Figure 4.1. Phase Relation of Sinusoidal Waves with Constant Time Delay

output of the differential amplifier $V_0 = A(V_1 - V_2)$ where A is the amplification factor. The outputs of the differential amplifiers can be either digitized or directly displayed on a scope to obtain the frequency information.

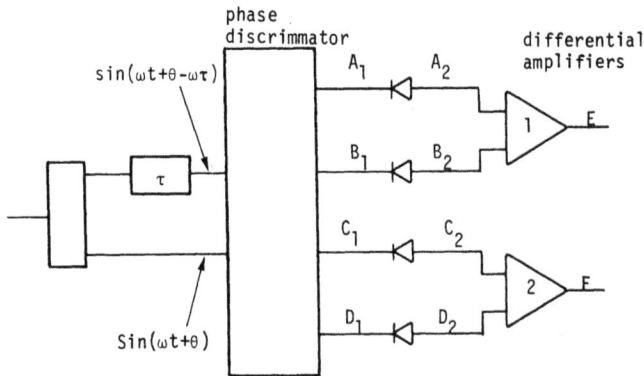

Figure 4.2. Basic Frequency Measurement Circuit of an IFM Receiver

A phase correlator is a passive component which contains four hybrid circuits. Its function is to introduce constant phase angles to each of the two input signals and to combine them in a certain way. Phase discriminators may have many different designs. A general discussion will be presented here by using Figure 4.2 to refer to the signal at each point. In order to provide a universal concept on the phase discriminator, the following steps are taken:

1. The input signals are known.
2. The desired outputs at E and F are specified.
3. The detectors are square law devices followed by low-pass filters.
4. With the input and output specified, signals at A_1, B_1, C_1, and D_1 are assigned so that after the detectors and differential amplifiers, the outputs will match the desired ones at E and F.
5. For simplicity, neglect the amplitude on each signal and assume that all amplitudes are units. Then assume that the input signal is

$$\sin(\omega t + \theta) \tag{4.3}$$

where ω is the angular frequency and θ is its phase angle. At the input of the phase discriminator, the two signals are $\sin(\omega t + \theta)$ and $\sin(\omega t - \omega \tau + \theta)$, where τ is the delay time introduced on one of the two outputs from the power divider. Assume that the desired E and F are

$$E = \sin\omega\tau \qquad (4.4)$$

and

$$F = \cos\omega\tau \qquad (4.5)$$

With these E and F, the relative amplitude of E and F can be measured to determine $\omega\tau$, which in turn determines the frequency of the input signal, because τ is a known value. In order to fulfill Equations 4.4 and 4.5, the terms $\sin\omega\tau$ and $\cos\omega\tau$ must be present at the inputs of the differential amplifiers E and F. The required outputs from the discriminator should be designated as follows. The choice of these inputs will become clear after the square law detectors and the differential amplifiers. The input that can produce $\sin\omega\tau$ after a square law detector is

$$A_1 = \pm [\sin(\omega t + \phi_A) + \cos(\omega t - \omega\tau + \phi_A)] \qquad (4.6a)$$

or

$$A_1 = \pm [\sin(\omega t - \omega\tau + \phi_A) - \cos(\omega t + \phi_A)] \qquad (4.6b)$$

The input that will produce $-\sin\omega\tau$ is

$$B_1 = \pm [\sin(\omega t - \omega\tau + \phi_B) + \cos(\omega t + \phi_B)] \qquad (4.7a)$$

or

$$B_1 = \pm [\sin(\omega t + \phi_B) - \cos(\omega t - \omega\tau + \phi_B)] \qquad (4.7b)$$

The input that will produce $\cos\omega\tau$ is

$$C_1 = \pm [\cos(\omega t - \phi_C) + \cos(\omega t - \omega\tau + \phi_C)] \qquad (4.8a)$$

or

$$C_1 = \pm [\sin(\omega t + \phi_C) + \sin(\omega t - \omega\tau + \phi_C)] \qquad (4.8b)$$

The input that will produce $-\cos\omega\tau$ is

$$D_1 = \pm [\cos(\omega t + \phi_D) - \cos(\omega t - \omega\tau + \phi_D)] \qquad (4.9a)$$

or

$$D_1 = \pm [\sin(\omega t + \phi_D) - \sin(\omega t - \omega\tau + \phi_D)] \qquad (4.9b)$$

where ϕ_A, ϕ_B, ϕ_C, and ϕ_D are phase angles at the output of the phase discriminator. Equations 4.6 through 4.9 can be treated actually as two sets of equations: 4.6 and 4.7 form one set, 4.8 and 4.9 form another one.

The outputs from detectors are the square of the input signals. Thus,

$$
\begin{aligned}
A_2 &= A_1^2 \\
&= 2 \sin (\omega t + \phi_A) \cos (\omega t - \omega \tau + \phi_A) + \sin^2 (\omega t + \phi_A) \\
&\quad + \cos^2 (\omega t - \omega \tau + \phi_A) \\
&= \sin \omega \tau + \sin (2\omega t - \omega \tau + 2\phi_A) + 1 - \tfrac{1}{2}\cos 2(\omega t + \phi_A) \\
&\quad + \tfrac{1}{2}\cos 2(\omega t - \omega \tau + \phi_A) \qquad\qquad\qquad (4.10a)
\end{aligned}
$$

or

$$
\begin{aligned}
A_2 &= A_1^2 \\
&= -2 \sin (\omega t - \omega \tau + \phi_A) \cos (\omega t + \phi_A) + \sin^2 (\omega t - \omega \tau + \phi_A) \\
&\quad + \cos^2 (\omega t + \phi_A) \\
&= \sin \omega \tau - \sin (2\omega t - \omega \tau + 2\phi_A) + 1 - \tfrac{1}{2}\cos 2(\omega t - \omega \tau + \phi_A) \\
&\quad + \tfrac{1}{2}\cos 2(\omega t + \phi_A) \qquad\qquad\qquad (4.10b)
\end{aligned}
$$

Similarly

$$
\begin{aligned}
B_2 &= B_1^2 \\
&= -\sin \omega \tau + \sin (2\omega t - \omega \tau + 2\phi_B) + 1 - \tfrac{1}{2}\cos 2(\omega t - \omega \tau + \phi_B) \\
&\quad + \tfrac{1}{2}\cos 2(\omega t + \phi_B) \qquad\qquad\qquad (4.11a)
\end{aligned}
$$

or

$$
\begin{aligned}
B_2 &= B_1^2 \\
&= -\sin \omega \tau - \sin (2\omega t - \omega \tau + 2\phi_B) + 1 - \tfrac{1}{2}\cos 2(\omega t + \phi_B) \\
&\quad + \tfrac{1}{2}\cos 2(\omega t - \omega \tau + \phi_B) \qquad\qquad\qquad (4.11b)
\end{aligned}
$$

$$
\begin{aligned}
C_2 &= C_1^2 \\
&= \cos \omega \tau + \cos (2\omega t - \omega \tau + 2\phi_C) + 1 + \tfrac{1}{2}\cos 2(\omega t + \phi_C) \\
&\quad + \tfrac{1}{2}\cos 2(\omega t - \omega \tau + \phi_C) \qquad\qquad\qquad (4.12a)
\end{aligned}
$$

or

$$
\begin{aligned}
C_2 &= C_1^2 \\
&= \cos \omega \tau - \cos (2\omega t - \omega \tau + 2\phi_C) + 1 - \tfrac{1}{2}\cos 2(\omega t + \phi_C) \\
&\quad - \tfrac{1}{2}\cos 2(\omega t - \omega \tau + \phi_C) \qquad\qquad\qquad (4.12b)
\end{aligned}
$$

$$
\begin{aligned}
D_2 &= D_1^2 \\
&= -\cos \omega \tau - \cos (2\omega t - \omega \tau + 2\phi_D) + 1 + \tfrac{1}{2}\cos 2(\omega t + \phi_D) \\
&\quad + \tfrac{1}{2}\cos 2(\omega t - \omega \tau + \phi_D) \qquad\qquad\qquad (4.13a)
\end{aligned}
$$

or

$$D_2 = D_1^2$$
$$= -\cos \omega\tau + \cos (2\omega t - \omega\tau + 2\phi_D) + 1 - \tfrac{1}{2}\cos 2(\omega t + \phi_D)$$
$$- \tfrac{1}{2}\cos 2(\omega t - \omega\tau + \phi_D) \qquad (4.13b)$$

In Equations 4.10 through 4.13, the terms $\sin\omega\tau$ and $\cos\omega\tau$ are constants (or dc voltages from the output of the detector), since ω is a constant (angular frequency of the incoming signal) and τ is a constant (the delay time in the phase discriminator). All of the other terms (e.g., $\sin (2\omega t - \omega\tau + 2\phi)$ and $\cos (2\omega t - \omega\tau + 2\phi)$, etc.) are the high frequency terms, and their frequency doubles that of the input signal. These high frequency terms can be filtered out by the low-pass filters formed at the outputs of the detectors. Therefore, the actual signals at the inputs of the differential amplifiers are

$$A_2 = 1 + \sin\omega\tau \qquad (4.14)$$

$$B_2 = 1 - \sin\omega\tau \qquad (4.15)$$

for the top amplifier (amplifier 1 in Figure 4.2) and

$$C_2 = 1 + \cos\omega\tau \qquad (4.16)$$

$$D_2 = 1 - \cos\omega\tau \qquad (4.17)$$

for the bottom amplifier (amplifier 2 in Figure 4.2). At the outputs of the amplifiers, these constant terms cancel out and

$$E = 2 \sin\omega\tau \qquad (4.18)$$

$$F = 2 \cos\omega\tau \qquad (4.19)$$

These are the desired results. They can be either displayed in a polar form or digitized and fed into a digital processor. For a polar display, $x = E = 2 \sin \omega\tau$ and $y = F = 2 \cos \omega\tau$. Then the angle of the polar display is

$$\theta = \tan^{-1}\frac{x}{y} = \omega\tau \qquad (4.20)$$

Thus, the frequency ω is linearly proportional to the angle θ. The important relation obtained from this section is that when a signal of $\sin (\omega\tau + \theta)$ is present at the input of the power divider (Figure 4.2), the outputs of the phase discriminator must fulfill Equations 4.6 through 4.9. The next section will present some actual network that will provide these relations. It should be emphasized here that the input/output relation of the phase discriminator is established from the requirements of the outputs of the two differential amplifiers.

4.3 PHASE CORRELATORS[8-10]

In this section an actual phase correlator and its signal paths are discussed. Figure 4.3 shows a basic IFM receiver which contains one power divider, one delay line, three 90 degree hybrid couplers, one 180 degree hybrid coupler, four detectors, and two differential amplifiers. In order to simplify the discussion, the input is assumed as $2\sqrt{2} A \cos \omega t$, where the phase angle θ is assumed zero. For the 90 degree coupler, a signal going through the direct path (the coupled path) does not change phase. While going through the diagonal path, the phase delay is 90 degrees. For the 180 degree coupler, a signal passing through the direct path retains the original phase; but while passing through the diagonal path, a 180 degree phase shift is induced.

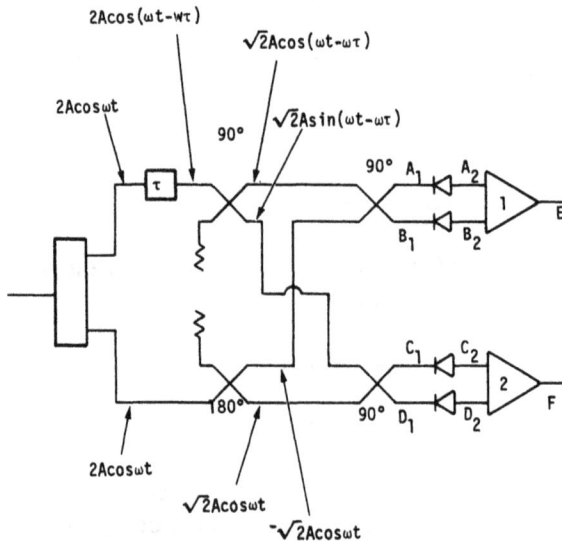

Figure 4.3. An Example of a Phase Discriminator

The amplitude of the signal decreases by factor of $\sqrt{2}$ whenever it passes either a power divider or a phase shifter. The signal flow can be traced as follows: after the power divider and the delay line, delayed and undelayed signals are present; after they pass through the 90 degree and 180 degree hybrid couplers: four terms are obtained. They are $\sqrt{2}A\cos (\omega t - \omega \tau)$, $\sqrt{2}A\sin (\omega t - \omega \tau)$, $\sqrt{2}A\cos \omega t$ and $-\sqrt{2}A\cos \omega t$ as shown in Figure 4.3. The next two 90 degree couplers combine the signals into

$$A_1 = A\cos(\omega t - \omega\tau) - A\sin\omega t \qquad (4.21)$$

$$B_1 = A\sin(\omega t - \omega\tau) - A\cos\omega t \qquad (4.22)$$

$$C_1 = A\sin(\omega t - \omega\tau) + A\sin\omega t \qquad (4.23)$$

$$D_1 = -A\cos(\omega t - \omega\tau) + A\cos\omega t \qquad (4.24)$$

where Equation 4.21 is equivalent to Equation 4.7b, Equation 4.22 is equivalent to Equation 4.6b, Equation 4.23 is equivalent to Equation 4.8b and Equation 4.24 is equivalent to Equation 4.9a. At outputs of the detectors, the only terms of interest are the square law outputs. They can be written as

$$A_2 = A^2\cos^2(\omega t - \omega\tau) - 2A^2\cos(\omega t - \omega\tau)\sin\omega t + A^2\sin^2\omega t \quad (4.25)$$

$$B_2 = A^2\sin^2(\omega t - \omega\tau) - 2A^2\sin(\omega t - \omega\tau)\cos\omega t + A^2\cos^2\omega t \quad (4.26)$$

$$C_2 = A^2\sin^2(\omega t - \omega\tau) - 2A^2\sin(\omega t - \omega\tau)\sin\omega t + A^2\sin^2\omega t \quad (4.27)$$

$$D_2 = A^2\cos^2(\omega t - \omega\tau) - 2A^2\cos(\omega t - \omega\tau)\cos\omega t + A^2\cos^2\omega t \quad (4.28)$$

Therefore, the outputs at the detectors considering the filter action can be written as

$$A_2 = A^2 - A^2\sin\omega\tau \qquad (4.29)$$

$$B_2 = A^2 + A^2\sin\omega\tau \qquad (4.30)$$

$$C_2 = A^2 + A^2\cos\omega\tau \qquad (4.31)$$

$$D_2 = A^2 - A^2\cos\omega\tau \qquad (4.32)$$

At the outputs of the differential amplifiers 1 and 2

$$E = B_2 - A_2 = 2A^2\sin\omega\tau \qquad (4.33)$$

$$F = C_2 - D_2 = 2A^2\cos\omega\tau \qquad (4.34)$$

If the outputs E and F are displayed on a x-y axis, then the angle θ

$$\theta = \tan^{-1}\frac{E}{F} = \tan^{-1}(\tan\omega\tau) = \omega\tau \qquad (4.35)$$

In order to keep the frequency in the unambiguous region, θ must be kept within 2π. For example, if the receiver covers frequency range from f_1 to f_2 then

$$\theta_1 = 2\pi f_1 \tau \qquad (4.36)$$

$$\theta_2 = 2\pi f_2 \tau \qquad (4.37)$$

where the angle θ_1 corresponds to f_1, and θ_2 corresponds to f_2.

When $\theta_2 = \theta_1 + 2k\pi$ where k is a constant. If $k \leqslant 1$, there is no frequency ambiguity among the frequency range f_1 and f_2. If $k = 1$, the receiver will provide the maximum frequency resolution. For example, if $f_1 = 2$ GHz, $f_2 = 4$ GHz then from Equations 4.36 and 4.37 one obtains

$$\theta_2 - \theta_1 = 2\pi = 2\pi\ (f_2 - f_1)\tau$$

$$\tau = \frac{1}{2 \times 10^9} = 0.5 \text{ ns} \qquad (4.38)$$

The maximum delay time without ambiguity is 0.5 ns. The amplitude of the display is

$$\varsigma = (E^2 + F^2)^{\frac{1}{2}} \qquad (4.39)$$

Substituting Equations 4.33 and 4.34 into Equation 4.39

$$\varsigma = 2\sqrt{2}A^2 \qquad (4.40)$$

which means that the amplitude is proportional to the input power. The results from Equations 4.33 and 4.34 can also be digitized and the frequency is displayed in digital form.

Various kinds of phase discriminators (correlator) can be fabricated using different combinations of power dividers and phase shifters. One more example is shown in Figure 4.4. In this arrangement, two power dividers, a 90 degree and a 180 degree hybrid couplers are used. However, it is easier to fabricate a hybrid coupler than a power divider in the microwave frequency range.

From the above discussion of IFM receivers, note that in the receiver design there is no special filter to limit the RF bandwidth. The frequency resolution generated by the receiver is not from any narrow bandwidth system such as in a superhet receiver. The frequency information is generated by comparison of the phase relation of the delayed and undelayed signals. Both paths are wide bandwidth; therefore, the receiver is basically referred to as a "wide bandwidth" system. For such a system, the receiver can measure frequency accurately on short pulse. However, the sensitivity of the receiver is only comparable with a crystal video

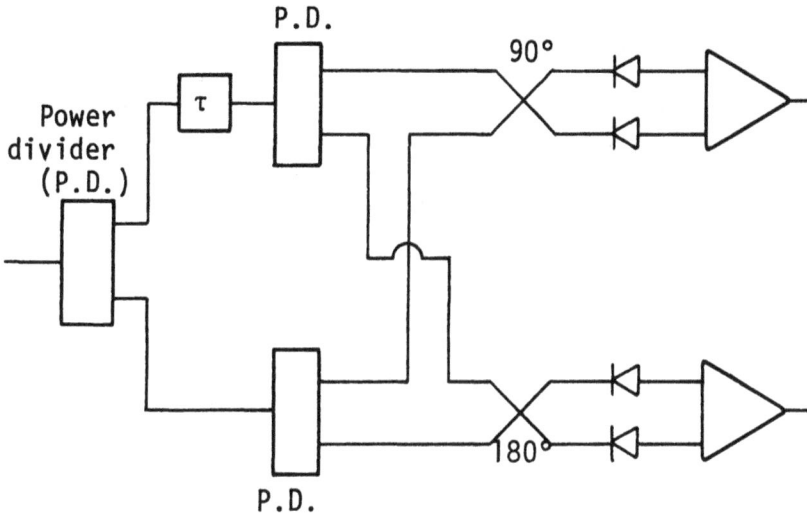

Figure 4.4. *Phase Shifter with Two Power Dividers, One 90 Degree and One 180 Degree Phase Shifters*

receiver. Eventually, if only one of the four diode detectors is considered, this receiver is a crystal video receiver.

4.4 LIMITING AMPLIFIER IN RF SIGNAL PATH

From Equations 4.33 and 4.34, it is noted that the outputs from the differential amplifiers not only produce frequency information ($\omega\tau$) but the amplitude of the incoming signal (A) is also retained. The output from the IFM receiver is proportional to the square of the signal amplitude. In the derivation of Equations 4.33 and 4.34, the amplitude fluctuations among the four signal paths to the detectors are neglected. In other words, the amplitude tracking among the four signal paths is assumed perfect. However, there is amplitude mismatch between the four paths in a practical receiver, and this mismatch will cause error in the frequency measurement. The mismatch can be reduced if the input signal is amplified to a fixed level and the video detectors always encounter the same input power level. Therefore, in modern IFM receivers the amplitude is measured independently from the frequency information. The amplitude information of the signal is measured by a separate circuit which contains a log video amplifier, sample and hold circuit, and D/A converter. The frequency of the input signal is measured by the IFM receiver.

The frequency measurement circuit measures only the frequency of the signal. The constant amplitude of the signal is derived by using a limiter (or limiting amplifier) in front of the frequency measurement circuit as shown in

Figure 4.5. The input signal is amplified through several RF amplifiers and the output is passed through a limiter which limits the signal at a fixed level. The total gain of the amplifier chain should be high enough to amplify the minimum desirable signal into the limiting level. Suppose the minimum signal to be detected is -60 dBm and the saturated output of the limiter is at 10 dBm, then a minimum gain of 70 dB is required. In this approach the diode detectors at the outputs of the phase discrimination will always see a constant input power. Therefore, the requirements on amplitude tracking among the four signal paths to the four detectors can be reduced.

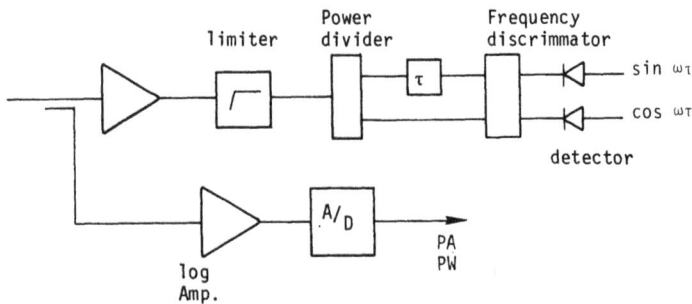

Figure 4.5. IFM Receiver with Amplifiers and Limiter in Front of the Frequency Measurement Circuits

Another advantage of using the limiter in front of the phase discriminator is to reduce the interference of simultaneous signals. An IFM receiver can generate only one frequency information at a time. If more than one signal arrives at the receiver simultaneously, only one signal frequency will be reported. The best that can be expected is for it to report one of the input signals (in general, the strongest one) correctly. The limiter described in Chapter 13 not only has the property to limit all the signals to a fixed level, but it will suppress the weak one even more.

Assume that two signals are present at the receiver simultaneously, a strong one and a weak one. If the strong signal is increased in amplitude, the limiter will not only limit the strong one but it will suppress the weak one (the capture effect of the limiter). For example, if one signal is 3 dB stronger than the other one at the input of the limiter, at the output of the limiter the strong one may be 6 dB higher than the weak one. With this capture effect of the limiter, a strong signal will get stronger when it arrives at the input of the phase discriminator. The receiver has a higher probability to report the frequency of the strong signal correctly because the weak one has less influence after the limiter. In general, an IFM receiver with an RF limiting amplifier can report one correct frequency constantly when the two signals are separated more than 3-6 dB in amplitude.

4.5 MULTIPLE DELAY LINES TO IMPROVE FREQUENCY RESOLUTION

In general, a receiver with one delay line of the maximum length without causing unambiguity frequency reading cannot generate the frequency accuracy and resolution desired. As discussed above, a receiver covering 2-4 GHz requires a maximum delay of 0.5 ns without causing frequency unambiguity. However, a delay line of 0.5 ns is not long enough to generate fine frequency resolution (say 1 MHz) because it must divide 2π into 2000 divisions to obtain the desired resolutions which is very difficult to accomplish if not impossible. The common approach to improving the frequency resolution is to use multiple delay lines. An example of such an arrangement is shown in Figure 4.6.a. The input signal is separated into four parallel paths. Following each of the four paths there is an IFM receiver and each IFM receiver has a different length delay line. The shortest delay line (say τ_1) is short enough not to produce frequency ambiguity, where $\tau_4 > \tau_3 > \tau_2 > \tau_1$ are used to generate desired frequency resolution. In general, τ_4, τ_3, and τ_2 are multiples of τ_1. Figure 4.6b shows the result of the output of $\omega\tau$ versus frequency for different lines. In this special case $\tau_4 = 2\tau_3 = 4\tau_2 = 8\tau_1$. It is obvious that $\omega\tau_1$ does not have any ambiguity problem in the frequency f_1 and f_2. However, the slope of the $\omega\tau_1$ is rather flat and it is difficult to measure the $\omega\tau$ angle and predicate the frequency accurately. Suppose the angle $\omega\tau$ can be measured in four divisions at intervals of $0 - 2\pi$ (Figure 4.6b). Then from delay line τ_1, the frequency can be divided only in four regions. Combining with the other three delay lines with the same angle ($\omega\tau$) resolution, the frequency can be divided into 32 regions which is an 8 time improvement in frequency resolution. In a practical receiver design, one delay line can generally be more than two bit (four frequency regions) of information. The second delay line can be four times rather than two times the length of the first delay line as discussed in the above example.

The longer the delay line, the finer the frequency resolution. The longest delay line that can be used in an IFM receiver depends on the minimum pulse handling capability. The delay line delay time must be shorter than the shortest pulse the receiver can handle, otherwise the delayed portion and the undelayed portion of the signal cannot overlap. Thus, the phase angle between them cannot be measured. A minimum overlap is required on the shortest pulse width, since the measurement circuit following the delay line has limited frequency bandwidth, and a steady-state must be reached before any measurement can start.

The length of the delay line is very critical to the frequency measurement accuracy. Special lines with a very small temperature coefficient have been used in the IFM receiver. Sometimes temperature controlled ovens surrounding the delay lines are used to maintain the length of the line at a constant value.

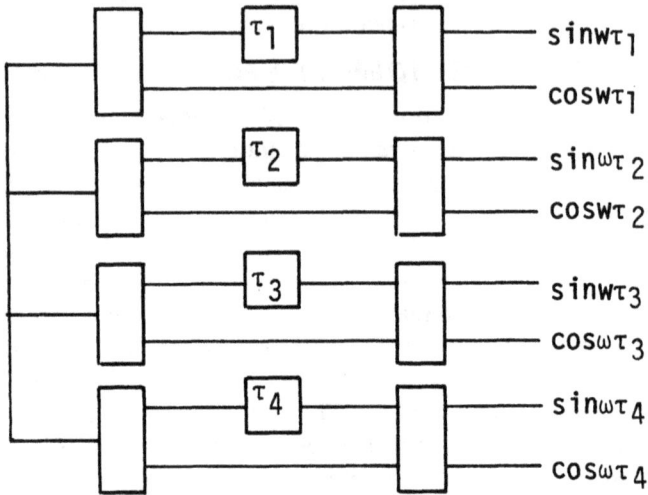

(a) Four Frequency Measurement Circuits

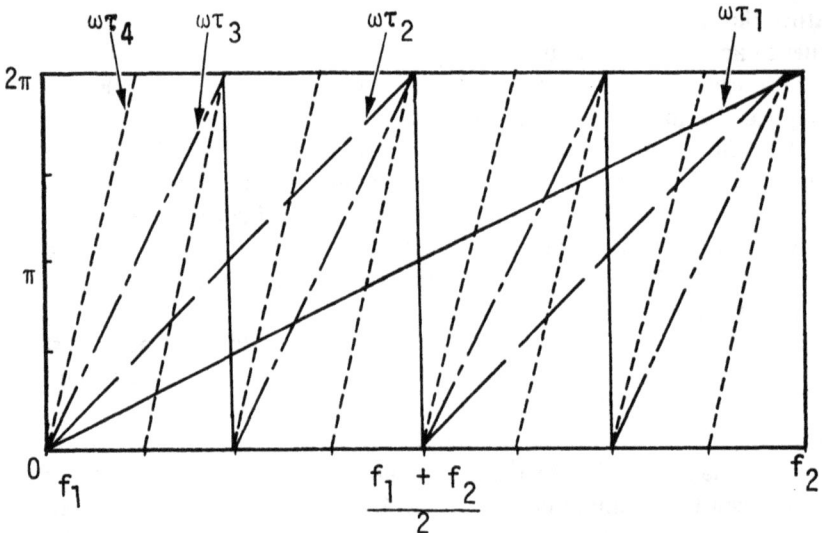

(b) $\omega\tau$ versus Frequency for Four Delay Lines

Figure 4.6. IFM Receiver with Multiple Delay Lines

4.6 DIGITIZING SCHEME

In digital IFM receivers, the time to sample the $\sin\omega\tau/\cos\omega\tau$ is very important. The general approach is to sample the outputs from the differential amplifiers as soon as the output signal reaches the steady-state. The sampling window is usually very narrow, approximately a few nanoseconds. If the incoming signal is a pulse, it takes some finite time for the output to reach steady-state because of the finite video bandwidth of the detectors and video amplifiers. Keeping the sampling window as close to the leading edge as possible can improve the receiver's capability to measure short pulse. At the same time the probability of having simultaneous signals from the leading edge of the pulse to the sampling time will be reduced. This will reduce the probability of generating erroneous frequency data caused by simultaneous signals. For example, if a signal is followed by a second one and the sampling window occurs before the arrival of the second signal, the second signal does not affect the frequency reading of the receiver even though the second pulse overlaps the first one. The receiver will report the frequency of the first signal; however, the frequency of the second pulse may not be recorded by the receiver.

There are various ways to digitize and encode the outputs from the differential amplifiers. One simple approach will be discussed here. Figure 4.7a shows the outputs $\sin\omega\tau$ and $\cos\omega\tau$ versus $\omega\tau$. Figure 4.7b lists the four bits output pattern. The output pattern should be a gray code (only one bit changes from one frequency bin to an adjacent one). If more than one bit changes from one frequency bin to another, they must change at exactly the same time which is difficult to accomplish. Otherwise, frequencies at the edge of the frequency bins may be encoded incorrectly. In order to generate the bit patterns in Figure 4.7b the conditions in Figure 4.7c must be fulfilled. For example, if the frequency is in bin 12, the following condition will occur (bit 4 = 0, bit 3 = 1, bit 2 = 0, bit 1 = 1) which implies

$$\sin\omega\tau < 0 \qquad (4.41)$$

$$\cos\omega\tau > 0 \qquad (4.42)$$

$$|\cos\omega\tau| - |\sin\omega\tau| < 0 \qquad (4.43)$$

$$||\cos\omega\tau| - |\sin\omega\tau|| > 0.54 \qquad (4.44)$$

To implement this kind of digitizing circuit, operational amplifiers and comparators can be used. The approach is straightforward for generating bits 4 and 3; the output signal $\sin\omega\tau/\cos\omega\tau$ is compared with zero reference voltage through a comparator. Diode circuits are used as full wave rectifiers to generate the absolute values of the $\sin\omega\tau$ and $\cos\omega\tau$ terms to produce bits 2 and 1. These terms are compared with comparators to generate the necessary digitizing information. For bit 1 the equation

$$||\cos\omega\tau| - |\sin\omega\tau|| > 0.54 \qquad (4.45)$$

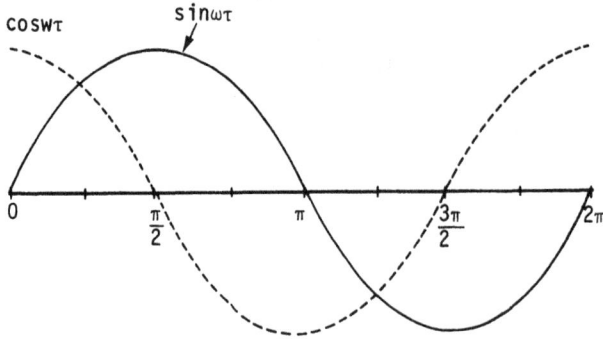

(a) Sin$\omega\tau$/cos$\omega\tau$ Outputs

(b) A Typical 4 bits Gray Code

bit	conditions to generate a logic "1"
4	$\sin \omega\tau > 0$
3	$\cos \omega\tau > 0$
2	$|\cos\omega\tau| - |\sin\omega\tau| > 0$
1	$||\cos\omega\tau| - |\sin\omega\tau|| > (\cos 22.5° - \sin 22.5°) = 0.54$

(c) Conditions to Generate the Bits in (b)

Figure 4.7. Typical Digitizing Scheme

can be considered as two equations.

$$|\cos\omega\tau| - (|\sin\omega\tau| + 0.54) > 0 \qquad (4.46)$$

or

$$(|\cos\omega\tau| + 0.54) - |\sin\omega\tau| > 0 \qquad (4.47)$$

which can be implemented through comparators.

Theoretically, the numerical value of 0.54 in Equations 4.46 and 4.47 can be represented by a constant voltage, if the input signal is well into saturation after the limiter circuit and the amplitudes of the $\sin\omega t$ and $\cos\omega t$ are always units. Otherwise, the numerical value of 0.54 can be obtained through a proper voltage divider at the output of the differential amplifiers which generate the $A\sin\omega\tau$ $A\cos\omega\tau$ outputs where A is a constant representing the amplitude change of the function. In this latter case, if the signal amplitude changes, the amplitude of the $A\sin\omega\tau/A\cos\omega\tau$ also changes, and proper voltage dividers will retain the relation $A(\cos 22.5° - \sin 22.5°) = 0.54A$.

Using the same ideas discussed above, more than 4 frequency bits can be generated. However, there is a practical limit on the number of bits that can be generated from one delay line. Although there is no definite rule to limit the maximum number of bits, it seems that 5-6 bits are the maximum. If the number of bits are increased, the digitizing circuit will become complicated and the frequency accuracy will deteriorate.

4.7 SIMULTANEOUS SIGNALS WITH TIME COINCIDENT LEADING EDGES[11]

The question always of interest to system engineers is that if two signals arrive at an IFM receiver simultaneously, what is the chance for the receiver to generate correct frequency information. Many laboratory tests have been carried out with two simultaneous signals on IFM receivers to produce some data. The general rule is that if the receiver can encode one frequency correctly, the performance is considered satisfactory. The general tested results are that when two signals arrive simultaneously and their amplitudes are within approximately 6 dB of each other, then the receiver will sometimes produce erroneous data. Sometimes the receiver will report one frequency correctly even though the two signals are equal in amplitude because when the two signals arrive at the input of phase discriminator, their amplitude may not be the same any more. The following paragraphs discuss the effect on the IFM receiver caused by different simultaneous signal conditions.

When simultaneous signals are discussed, it usually means the leading of the pulses are coincidental in time. Let us consider only two pulses of different frequencies and amplitudes. First, let us assume that there is no limiter at the input RF amplifier chain of the receiver and the output frequency is displayed in a polar coordinate on a scope. If one signal is applied, the output will display the frequency correctly. If the second signal is applied at a different frequency with smaller amplitude, the receiver will still read the strong signal. When the amplitude of the second signal increases, the output frequency from the receiver will move gradually from the first signal toward the second one. The receiver will display a frequency somewhere between the two input signals. This phenomena can be easily observed from an IFM receiver with polar display. From this experimental result one can see the response of the IFM receiver to simultaneous signals. Although there is no quantitative data obtained, the weak signal will start to effect the strong one when they are quite far apart in power.

However, with a limiter in front of the receiver and a digital output, the two signals start to interfere each other when they are approximately within 6 dB (which depends on the characteristic of the limiter and the gain flatness of the RF amplifier chain). The interfering range of the two signals is different for different IFM receivers. Figure 4.8 shows a typical probability of erroneous data produced by an IFM receiver versus the amplitude difference in signal strength. Even with two signals of equal amplitude (0 dB difference), the erroneous frequency data produced by the receiver is not close to 100 percent. Because the receiver does not have perfect amplitude response over the input frequency range, two signals of the same amplitude at input may not have the same amplitude at the output of the limiter. One signal can be distinguishably higher than the other one at the phase discriminator and get encoded correctly.

4.8 SIMULTANEOUS SIGNALS WITH NON-TIME COINCIDENT LEADING EDGES(11)

This time, let us consider that the leading edges of the two signals are separated slightly in time. If the first signal triggers the encoding circuit of an IFM receiver and the encoding is completed before the arrival of the second signal, then the second signal will never affect the performance of the receiver. In general, the receiver just misses the second signal but measures the first signal correctly. The discussion presented here will concentrate on the case that the first signal triggers the digitizing circuit, and a second signal arrives before the completion of the frequency measurement. If the second signal is much weaker than the first one, then the second signal will not disturb the measurement. If the second signal is stronger than the first one, then it will affect the measurement. Figure 4.9 shows a typical result of an IFM receiver measuring against two simultaneous signals with the second signal delayed close to the digitizing time. The probability of generating erroneous data is very high when the second signal is stronger than the first one. The reason for such bad performance of an IFM receiver is that when the second signal arrives, the output from the phase discriminator changes from a voltage correspond to the first signal of

another voltage corresponding to the second signal, and the sampling circuit is reading the frequency data at this transient period. A typical voltage

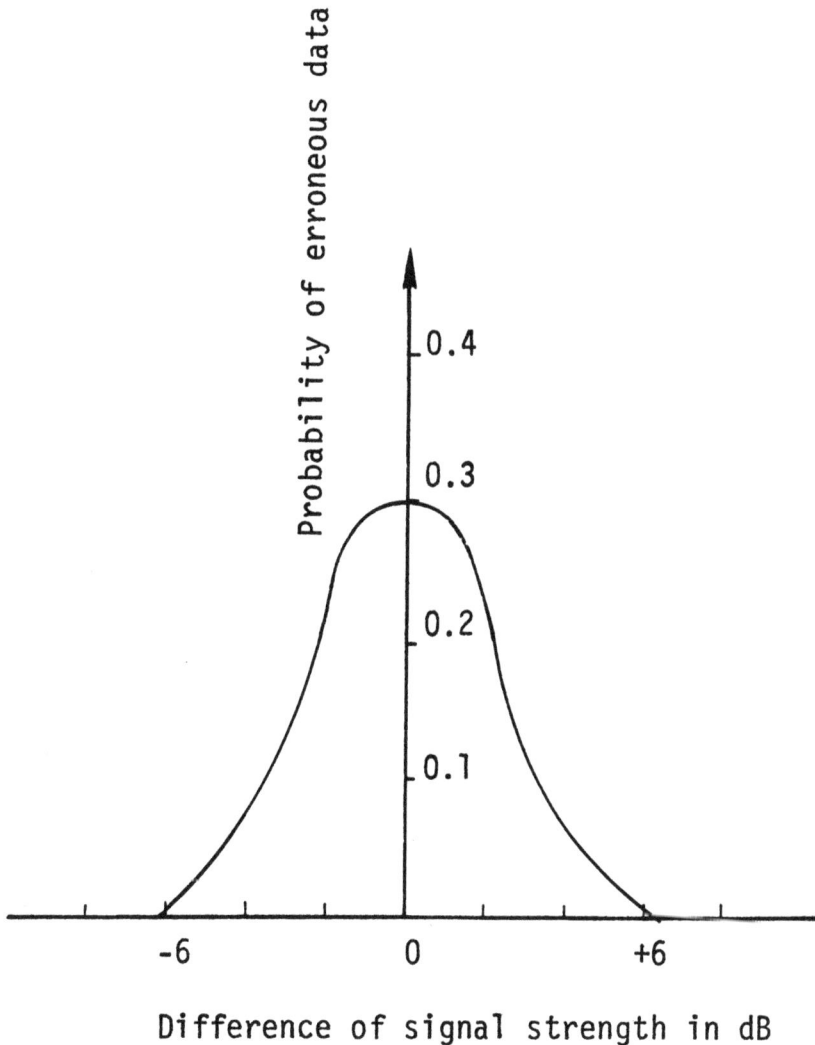

*Figure 4.8. Erroneous Frequency Data Produced by a Typical IFM
Receiver with Time Coincident Leading Edges Simultaneous
Signals*

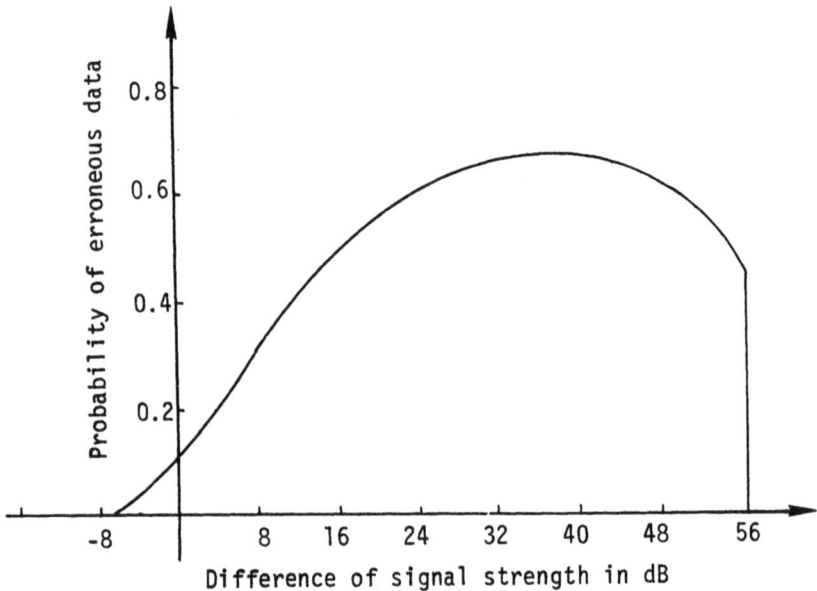

Figure 4.9. Erroneous Frequency Data Produced by a Typical IFM Receiver with Simultaneous Signals with Time Delay in Leading Edges Close to the Digitizing Time

output from the phase discriminator is shown in Figure 4.10 with the sample pulse shown. It is clearly seen in this figure that the sampling pulse is sampling at a voltage corresponding to neither signal. The worse case error probability can reach as high as approximately 80 percent in some IFM receivers. The amplitude of the second signal can be very strong even at the maximum power level the receiver can handle; and the receiver still does not read the frequency of the second signal, because of the interference of the first signal.

The time delay between the two signals is very critical in relation to the generation of erroneous data from an IFM receiver. When the leading edge of the second signal is very close to the first one, the situation is close to leading edges coincident case. When the delay time between the leading edges increases, the probability of producing erroneous data also increases. Further increasing of the delay time between the leading edges of the two signals will cause the probability of generating erroneous data to decrease slightly. When the second signal arrives after the sampling time, there is no longer erroneous data generated. Figure 4.11 shows a typical result of the "worst" probability as a function of the leading edge delay time.

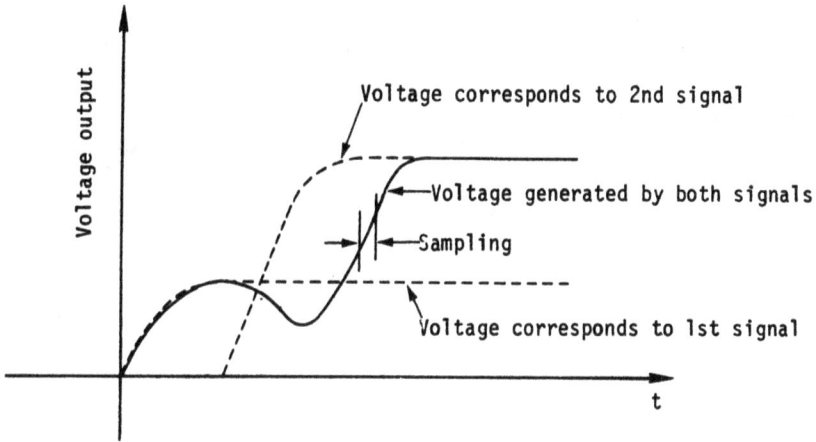

Figure 4.10. Phase Correlator Output from Two Signals with the Sampling Window

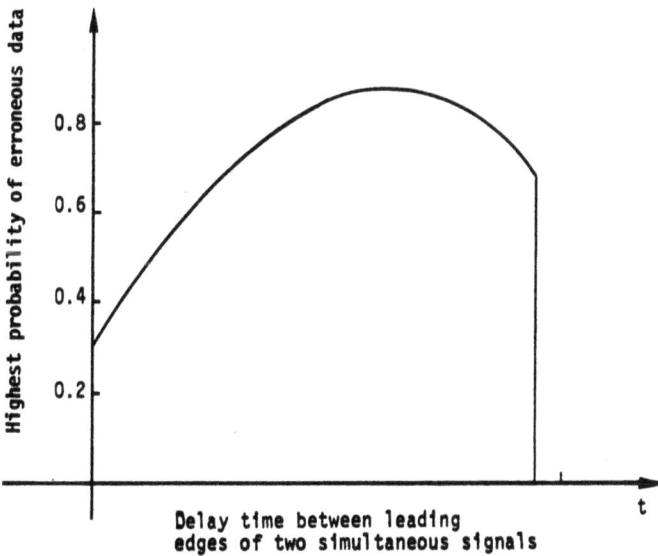

Figure 4.11. Highest Probability of Erroneous Data Generated by an IFM Receiver Versus the Leading Edge Delay Time

4.9 SUGGESTED SOLUTIONS TO SIMULTANEOUS SIGNAL PROBLEMS(12)

It is very desirable to detect the existence of simultaneous signals so that the erroneous frequency data reported by the IFM receiver can be neglected. The erroneous frequency data generated by the receiver can put a tremendous burden on the signal processor following the receiver because the processor will treat the erroneous data as a correct one and process them. Since these erroneous data do not represent a true signal frequency, the processor cannot make any sense of it. The processor may even treat them as exotic signals and wastes lots of valuable processing time on them. It is even helpful to generate a flag if simultaneous signals are present. Since the frequency data could be correct even under simultaneous signal conditions, the data will be processed as usual. However, when the data does not make any sense while the simultaneous signal flag is on at the same time, the data can be considered as erroneous and discarded.

The most common simultaneous signal detector uses either a mixer or a diode detector. If there is more than one signal arriving at the receiver, the mixer/diode will generate the beat frequency of the two signals. By detecting the existence of this beat frequency, one can determine that there are simultaneous signals. Figure 4.12 shows a simplified simultaneous signal detecting circuit detector. If two or more signals arrive at the diode simultaneously, then the square law characteristic of the diode will generate the beat frequency. The low-pass filter will stop the input signals and pass only the different frequency provided it is lower than the cut-off of the filter. The comparator used as a threshold detector makes the decision whether there are simultaneous signals.

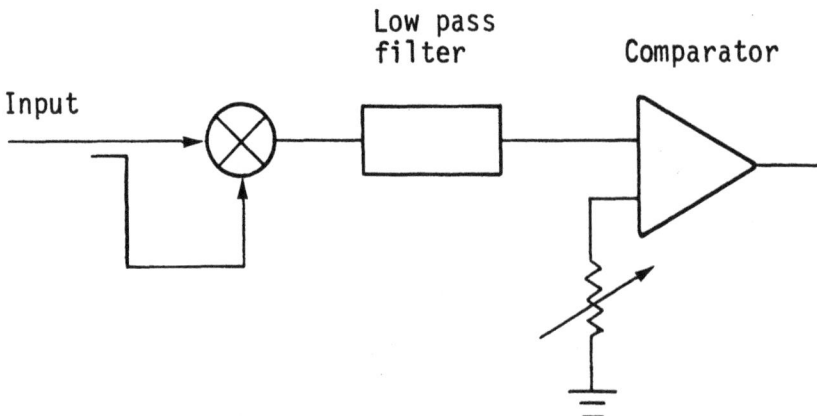

Figure 4.12. Simultaneous Signal Detector with Diode Detector

For example, if the input frequency range of an IFM receiver is 2-4 GHz, (in order to block the fundamental frequency to reach the comparator), the low-pass filter cut-off frequency must be less than 2 GHz. Therefore, when two signals are separated by approximately 2 GHz (one at 2 GHz, the other at 4 GHz), the detecting circuit cannot detect them. If the IFM receiver has a frequency range from 2-6 GHz, then any two simultaneous signals separated by more than 2 GHz cannot be detected.

In order for two signals to generate a difference frequency with large amplitude, the power levels of the two signals must be close together. Therefore, this detecting circuit limits itself to simultaneous signals of comparable amplitudes. Signals with nontime coincident leading edges which cause the IFM receiver to generate a high percentage of erroneous data cannot be effectively detected by such a detecting circuit either. There is also the possibility that the simultaneous signal detector detects the existence of simultaneous signals; however, the IFM receiver encodes one of the input signals correctly. Therefore, when simultaneous signals are detected, the frequency reported by an IFM receiver could still be correct. On the other hand when there is erroneous data caused by simultaneous signals, the detection circuit may not realize it. In addition, the detection circuit may generate a false alarm when there is only one signal. Thus, this circuit is far from an ideal case. An ideal simultaneous signal detector will produce an output only when the IFM receiver generates an erroneous frequency data caused by simultaneous signals. To devise such a detecting circuit, further development is required.

If there is a continuous wave (cw) signal at the input of an IFM receiver, the receiver always has a simultaneous signal condition when pulsed signals arrive. Thus, theoretically, one cw signal can put an IFM receiver out of work. The receiver can only read the cw signal and not the pulse signals. The usual solution to this problem is to put a notch filter in front of the IFM receiver to filter out the cw signal. The filter is controlled from the receiver output to block out the cw frequency.

Figure 4.13 shows the schematics of an approach to separate the cw and pulsed signals and encode both of them (Reference 12). The RF front end of the IFM receiver remains the same. At the output of the diode detectors, the signal paths are separated into two paths; an ac coupled and a dc coupled. The dc coupled path will contain the cw and the pulsed signal, while the ac coupled path contains only the pulsed signal. The clamping diodes in the ac paths will change the reference level of the pulsed signal to 0 voltage. Therefore, the difference outputs from the dc and ac coupled circuit leave only the dc components which correspond to the cw signal. Two digitizing (or display) circuits can encode the cw and pulsed signals separately. If more than one cw signal or pulsed signal is present, this circuit can only separate the cw signals and pulsed signals. It cannot separate one cw signal from another, nor can it separate two pulsed signals. The accuracy of the frequency readings on both the cw and pulsed signals may be slightly degraded when they are present simultaneously at the IFM receiver.

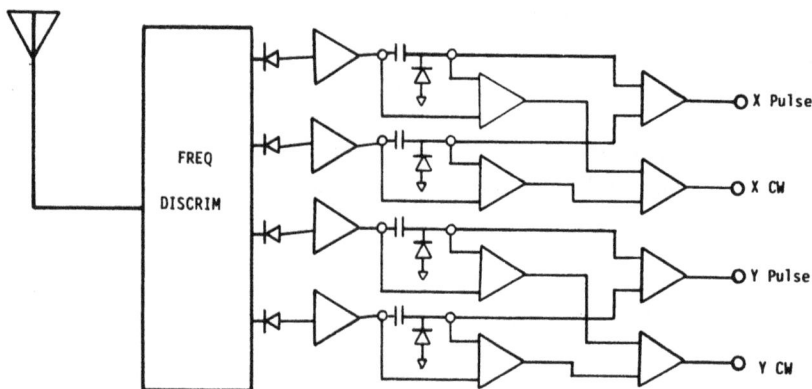

Figure 4.13. IFM Receiver with Capability to Separate Pulsed and cw Signals

4.10 APPLICATIONS OF IFM RECEIVERS

IFM receivers, in general, have the following characteristics:
- Moderate sensitivity
- Extremely wide RF input bandwidth over octave bandwidth
- Fine frequency measurement capability even on short pulses
- Simultaneous signals may cause erroneous frequency data
- Very wide dynamic range because the receiver can measure only one signal at a time

From these characteristics one can see that an IFM receiver is equivalent to a frequency counter with a wide input frequency range. Thus, this receiver can be used as an electronic warfare (EW) receiver because frequency measurement is one of the most important requirements in an EW receiver. The wide RF input bandwidth makes the probability of intercepting unknown signals very high. Using microwave integrated circuit (MIC) technology, the receiver can be very compact because of its relatively simple structure. The small size makes the IFM receiver suitable for airborne applications. However, the simultaneous signal problem is a major roadblock for the wide applications of IFM receiver. This problem can cause false alarm and slow down the operation of the digital processor following the receiver. This problem must be solved or reduced before the receiver can be widely adapted. The best that can be done today is to put a "flag" on a questionable frequency data. Other than EW applications, this receiver has little use in communication, because of its moderate sensitivity and poor selectivity.

REFERENCES

1. Earp, C.W., "Frequency indicating cathode ray oscilloscope," U.S. Patent 2434 914, January 27, 1948.
2. Cumming, R.C., Myers, G.A., "Performance of receivers and signal analyzers using broadband frequency-sensitive devices," Technical Report No. 1905-1, Stanford Electronic Laboratories, SU-SEL-66-125, Stanford University, March 1976.
3. Wilkens, M.W., Kincheloe, W.R., Jr., "Microwave realization of broadband phase and frequency discriminators," Technical Report No. 1962/1966-2, Stanford Electronics Laboratories, SU-SEL-68-057, November 1968.
4. Myers, G.A., Cumming, R.C., "Theoretical response of a polar-display instantaneous-frequency meter," IEEE Trans. Instrumentation and Measurement, Vol. IM-20, pp. 38-48, February 1971.
5. Heaton, D., "The systems engineer's primer on IFM receivers," Microwave Journal, p. 71, February 1980.
6. Gourse, S.J., Worrell, E.A., "Wide-band IFM flyable brassboard," Final Engineering Report, Litton Amecom, College Park, Maryland, November 1974.
7. Blachman, N.M., "The effect of noise polar-display instantaneous frequency measurement," IEEE Trans. Information and Measurement, Vol. IM-25, pp. 214-221. September 1976.
8. Grossbach, R., "Degradation of polar-discriminator performance by non-ideal components," Microwave Journal, p. 53, December 1974.
9. Saul, D.L., "Design a Ka-band polar frequency discriminator," Microwave, p. 74, April 1976.
10. Gysel, U.H., Watjen, J.P., "Wide-band frequency discriminator with high linearity." IEEE MTT-S. Int. Microwave Symp. Digest 1977.
11. Shaw, R.L., Tsui, J.B.Y., "IFM receiver test and evaluation," Technical Report AFAL-TR-79-1049, Air Force Avionics Laboratory, Wright-Patterson Air Force Base, April 1979.
12. Tsui, J.B.Y., Schrick, G.H., "Instantaneous frequency measurement (IFM) receiver with capability to separate cw and pulsed signals," United States Patent No. 4,194,206, March 18, 1980.

CHAPTER 5
CHANNELIZED RECEIVERS

5.1 INTRODUCTION[1-6]

In order to widen the radio frequency (RF) bandwidth to improve the probability of intercept, the most straightforward approach in concept is to use brute force to build a large number of parallel narrow band receivers with adjacent frequencies. A channelized receiver is a realization of this idea which uses a large number of contiguous filters to sort the input signal. An input signal with certain frequency will fall into a certain filter. By measuring the output of all the filters, the input signal frequency is determined. Although the idea is simple, this receiver is expensive to fabricate because of the large number of filters required. The size of this receiver will be bulky, and the maintenance of it will be difficult due to the large number of components used. That is why there has been little development in channelized receivers in the past. With the advances in microwave integrated circuits (MIC) and surface acoustic wave (SAW) filters, the development of channelized receivers has grown.

Due to the difficulty of fabricating filters in the high frequency range, it is common practice to down convert the high frequency input signals to a lower frequency before channelization. Thus, a channelized receiver can be considered as a large number of superheterodyne receivers operating in parallel. However, there is an essential difference. In channelized receiver, the local oscillator frequency is fixed and does not tune over a range as in a superheterodyne receiver. Although the operational principle of a channelized receiver is obvious, there are still many detailed design problems that need to be solved.

5.2 WIDE RADIO FREQUENCY (RF) BANDWIDTH CHANNELIZATION

The application of a channelized receiver will be very wide. In general, one can build a fine frequency receiver (either an instantaneous frequency measurement (IFM), a compressive, a channelized, or a Bragg cell receiver) which may have an instantaneous bandwidth from a few hundred megahertz to several gigahertz. However, the RF spectrum of interest could be over ten gigahertz bandwidth. The only way the fine frequency receiver mentioned above can be used to monitor the entire spectrum of interest is to divide the wide RF spectrum into many bands with bandwidth matches to that of the fine frequency receivers and then convert them to the proper frequency ranges. This approach is a broadband channelization. For example, a fine frequency receiver covers the frequency range of 2-4 GHz and the spectrum of interest covers 0.5 to 18 GHz. The 0.5 to 18 GHz will be divided into 2 GHz bands from 0.5-2, 2-4, 4-6,...16-18 GHz as shown in Figure 5.1. With the exception of the 2-4 GHz band, all the other bands will be converted to 2-4 GHz frequency range to match the input frequency of the receiver. The bands adjacent to the 2-4 GHz band usually

filter

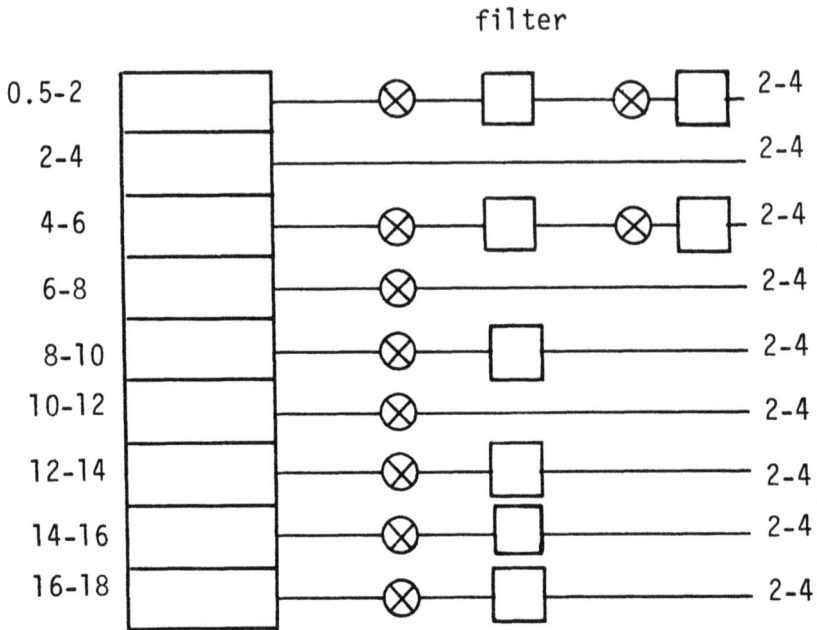

(a) General Channelization and Down Convert

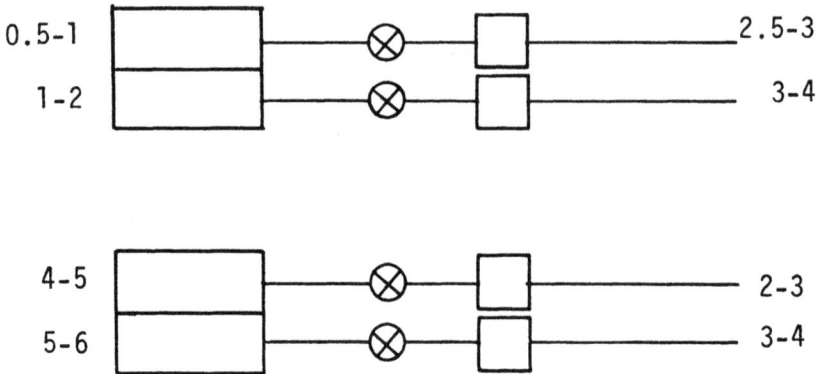

(b) Down Convert through Sub-channelization

Figure 5.1. RF Front End Channelization

are not converted directly to the 2-4 GHz range, but are converted up to some frequency then down converted to 2-4 GHz in order to avoid severe spurious products produced by the mixers. In order to change 4-6 GHz directly to 2-4 GHz, the local oscillator (LO) frequency is either at 2 GHz or 8 GHz. If the LO frequency is at 2 GHz, the LO frequency will "leak through" the mixer and appear at its output as a signal. Another problem with the direct conversion is that when the input signal is close to 4 GHz, the signal at the mixer output can appear at both 2 and 4 GHz. The 2 GHz signal is the desired signal from down conversion of the mixer, while the 4 GHz is fed directly through. Therefore, the input signal may be read as two input frequencies: one at 2 GHz, the other one at 4 GHz.

If the LO frequency is at 8 GHz, then the 6 GHz frequency is down converted to 2 GHz, where the 4 GHz frequency is down converted to 4 GHz. In this arrangement, the LO frequency will not leak through. However, the signal below 6 GHz will fold into the 2-4 GHz band through $2f_s - f_0$ where f_s is the signal frequency and f_0 is the LO frequency. This is the so called 2×1 spur which is rather strong in most mixers. Besides, the input signal at GHz will arrive at the output frequency 4 GHz through two paths: direct and converted. These two signals may interfere with each other and cause a measurement problem in the fine frequency receiver.

These problems can be reduced through double conversions as shown in Figure 5.1a. The other approach is to divide 4-6 GHz into 4-5 GHz and 5-6 GHz. The 4-5 GHz can be converted to 2-3 GHz while the 5-6 GHz can be converted to 3-4 GHz as shown in Figure 5.1b. However, both solutions mentioned above use two mixers and two LOs which doubles the amount of hardware required for the other channels. The 0.5-2 GHz band can be handled in a similar manner.

5.3 TRADE-OFFS IN A CHANNELIZED RECEIVER

In order to minimize confusion in the discussion of channelized receivers, the following terms are used.

1. Bands: The results of the first channelization. For example, the spectrum of interest (0.5-18 GHz) is divided into nine 2 GHz bands.
2. Channels: A band is subdivided into channels.
3. Slots: A channel is further subdivided into slots.

A. Frequency Down Conversion

If a receiver which covers a 2-4 GHz frequency band with 10 MHz frequency resolution is required, the obvious solution is to build a channelized receiver with a 200 10 MHz contiguous filter covering the 2-4 GHz band. However, it is impractical to build filters at GHz frequency with 10 MHz bandwidth, because the Q of the filter (defined as $Q = f_0/\Delta f$ where f_0 is the center frequency of the filter and Δf is the bandwidth) will be very high. A more reasonable approach is to down convert the GHz frequency range to a few hundred MHz frequency range, then channelize at this lower frequency to 10 MHz bandwidth slots. For example, the 2-4 GHz band can be divided into 10 channels, each consisting of 200 MHz. These 200 MHz channels should be less than an octave bandwidth. Therefore, the channel should be above the 200-400 MHz range (say 210-410 MHz) in order to keep outside the octave bandwidth. In this configuration, 20 fine frequency filters can cover the 210-410 MHz range, and each has a bandwidth of 10 MHz. This channel structure will be duplicated nine more times to cover the entire 2 GHz bandwidth. If one channel is designed and tested, the rest can be duplicated.

The SAW filters can be used for the slot assembly. Theoretically, the SAW filters can be mass produced, if the photo mask is designed. However, the following characteristics must be considered if SAW filters are used.

1. Insertion loss: The high insertion loss (10-25 dB) of the SAW filters must be recovered by additional amplifiers.
2. The out of band rejection level of the SAW filter limits the instantaneous dynamic range of the receiver. If a receiver dynamic range of 35 dB is required, then the out of band rejection of the filter must be approximately 50 dB (a signal-to-noise ratio of 15 dB is usually required to reduce false alarm rate).
3. Relative long time delay: it usually takes hundreds of nanoseconds to a few microseconds for a signal to pass the filter. This delay time should be properly compensated, and it can also benefit some special applications.
4. Time domain spurious responses will limit the throughput of the receiver. The receiver must wait until the time domain spurious responses die-off in order to process another signal. The most significant time domain response is the triple-transit effect. The response must be approximately 50 dB lower than the true signal in order to keep a 35 dB receiver dynamic range.

B. Hardware and Performance Trade-Off

In Section 5.3A, the configuration discussed is used to measure frequency only. Other information such as pulse width (PW), pulse amplitude (PA), and time of arrival (TOA) can also be measured at this slot level. However, the amount of circuitry involved will be high. The PW, PA, and TOA information can also be measured at channel or band level with less hardware. But if there are simultaneous signals arriving at the channel or band level, the information measured is ambiguous.

C. Short Pulse and Fine Frequency Resolution

If the short pulse handling capability is the prime requirement of a channelized receiver, then the fine frequency resolution and accuracy is limited. A rough approximation of the fine frequency resolution is $1/PW_{min}$, where PW_{min} is minimum pulse width. If the minimum pulse width to be measured is 1 μs, then the best frequency resolution that can be expected is 1 MHz. The PW and PA measurement is also limited by the same basic rule. The PW and PA cannot be measured accurately after a narrow filter. Their accuracy is limited by the impulse response of the filter.

D. Spurious Output from Mixer

Because many mixers are used in a channelized receiver, the spurious response generated must be carefully considered. The spurs table discussed in Chapter 14 is the guideline for choosing the input, output, and LO frequencies. Proper filtering is required both in front and after the mixer. One of the most serious spurs in the mixer is the 2×1 spur.

E. Common Local Oscillator (LO)

If the channelized receiver is properly designed, one LO can be shared among two or more mixers. This approach is illustrated by the example in Figure 5.1. In order to convert a 10-12 GHz band into a 2-4 range, a LO of either 8 or 14 GHz can be used. Theoretically, the 8 GHz LO can also mix with the 4-6 GHz to down convert it to 2-4 GHz. However, in practical design the 4-6 GHz is not converted to an adjacent band 2-4 GHz, but the 14 GHz can be used for both the 10-12 and 16-18 GHz band mixers.

5.4 FREQUENCY ENCODING SCHEMES FOR WIDE-BAND RECEIVERS

The idea of how to determine the input frequency from the output of the receiver is not only applicable to channelized receivers but also to other wide bandwidth receivers (i.e., compressive and Bragg cell receivers) as well. There are several approaches to handle this problem. Each has its advantages and disadvantages. This section will discuss this subject and the same ideas will be applicable to compressive and Bragg cell receivers.

Let us first consider an ideal case. The receiver consists of several banks of filters (these filters are referred to as slots) to provide the fine resolutions of the receiver. Each filter has an infinite sharp boundary and two adjacent slots share exactly the same common boundary as shown in Figure 5.2. A signal will fall in only one of the filters regardless of the signal strength. If the signal is very close to a boundary, shown in Figure 5.2 as S, it will fall either in B or C but never in B and C simultaneously. When two simultaneous signals arrive at the adjacent filters, both filters will have outputs. Under this ideal case, there is no problem to determine the input frequency; the maximum frequency error is half the filter bandwidth. Detectors are placed at the output of each slot filter. When there is an output from the detected output, there is a signal in the slot. However, in the actual fabrication of the receiver, sometimes there is a gap between two boundaries as shown in Figure 5.3a or sometimes an overlap in the boundaries as shown in Figure 5.3b, then an error will be generated. In Figure 5.3a a signal near the B, C boundaries may be missed totally which is an unacceptable situation. In Figure 5.3b, a signal near the boundaries may be detected in both B and C. Instead of one signal, the receiver will report this as two signals which has to be resolved later by a signal processor.

In addition to the problem mentioned above, it is impossible to have a filter with very sharp boundaries. A filter with sharp boundaries means many sections of filters are required. It is impractical to use large numbers of filters with many sections in a channelized receiver. Besides the high cost and large size of these filters, the transient time of such a filter with many sections is also rather long which will affect the receiver capability on handling short pulse.

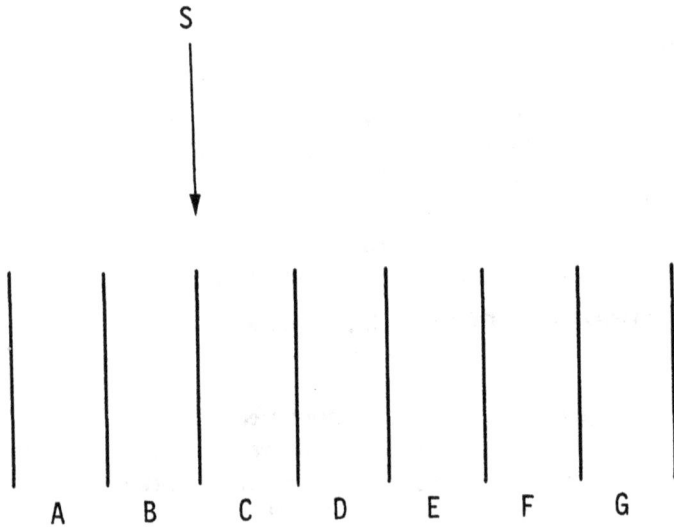

Figure 5.2. *Contiguous Filters with Sharp Boundaries*

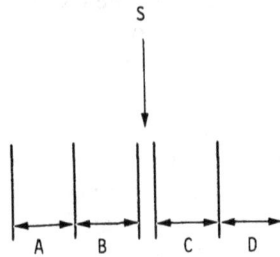

(a) Gap Between Boundaries

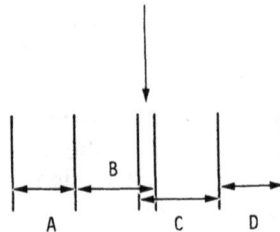

(b) Overlap Between Boundaries

Figure 5.3. *Filter Boundary Conditions*

One way to solve the above problem is through overlapping filters. Figure 5.4 shows nine continuous overlapping filters: A, B, C, D, E, F, G, H, and I. These filters overlap their adjacent filters by one-third of their bandwidth. The filters are so designed that one signal, regardless of strength, can trigger one output of a filter or a maximum of two outputs from two adjacent filters. The bottom part of Figure 5.4 shows frequency bins which are numbered as 1, 2, 3, etc. If filter B is the only bin and has an output, then the signal frequency is encoded as bin 3. If filters B and C are on, then the signal frequency is encoded as bin 4, the frequency bin below both filters B and C. The maximum frequency error is half a frequency bin. In Figure 5.4, there are 9 filters and 17 frequency bins. In general, the number of frequency bin n is related to the filter number N by $n = 2N - 1$. If all the filters have the same bandwidths, then the two end frequency bins are twice the size of the rest of the frequency bins as shown in Figure 5.4. The two end frequency bins can be reduced to equal the rest of the frequency bins by reducing the widths of the two filters to two-thirds of their original width. This encoding scheme will improve the frequency resolution. This is an advantage which will have saved hardware for the same frequency resolution as compared with the approach in Figure 5.2.

The shortcoming of the encoding scheme is the limitation to read two or more simultaneous signals. Suppose one signal is at S_1 and another signal is at S_2 as in Figure 5.4. It is noted that signal S_1 will trigger filters A and B, and signal S_2 will trigger C and D. Therefore, it is very difficult to determine the frequencies of the signal. Even the correct number of signals is difficult to determine. For instance, using the encoding rule for single signals as mentioned above, if two adjacent filters have outputs, the frequency bin under both filters will be assigned as the output. If filters A, B, C, and D all have outputs, the three frequency bins 2, 4, and 6 will be reported as the desired outputs while only two input signals are present. If the second input signal is at S_3 instead of S_2, then the correct frequency

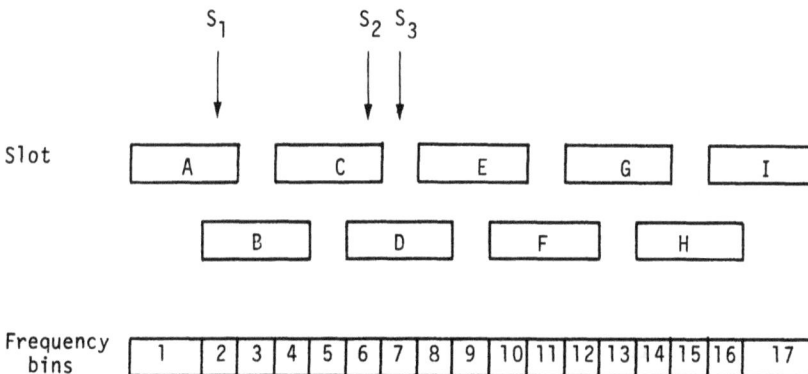

Figure 5.4. Slot Filters with One-Third Overlapping on Adjacent Boundaries

bins 2 and 7 will be reported. Therefore, this encoding approach can correctly encode simultaneous signals separated by five frequency bins or more. In this arrangement, when there is only one signal present, the encoding circuit always reports the correct frequency.

Another arrangement is shown in Figure 5.5. The slot filter boundaries are slightly overlapping each other to avoid signals on the boundaries being missed. In this arrangement the logic circuit following the slot filters can make one of two choices: (1) whenever this is output from a certain slot filter, there is an output from the corresponding frequency bin; (2) when two adjacent filters have outputs, only one of the frequency bins will provide an output. In the first approach, the deficiency is that when an input signal falls in the boundary region and activates two adjacent filters, they will be recorded as two frequencies. Two signals with frequencies in two adjacent filters will be encoded correctly. The second approach has the advantage of always encoding one frequency correctly. Even when the input signal is in two adjacent filters, only one of the frequency bins will have an output; and the adjacent one will be suppressed. The deficiency of this scheme is when two signals fall into two adjacent slot filters, only one will be reported; and the other one will be missed. The logic can be designed in such a manner that when signal S_1 and S_2 triggers filter A, B, and C, then frequency bins 1 and 3 will be reported. Therefore, when two signals are separated by more than one frequency bin, they can be encoded correctly. In both of the above arrangements, the two end filters have the same bandwidths as the rest of the filters.

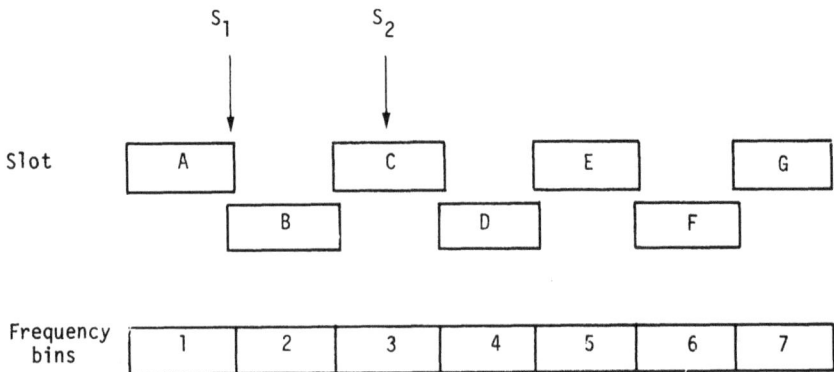

Figure 5.5. Slot Filters with Slight Overlapping on Adjacent Boundaries

The designer can choose any of the above approaches. Each scheme has its advantages and deficiencies. However, the arrangement in Figure 5.4 is often selected since it can provide finer frequency resolution for the same amount of hardware on a single signal. In all the above arrangements, the boundaries of the filters should be independent of the input power of the signals.

To make the frequency boundaries independent of the input signal strength, the most obvious approach is to amplify the input signal into a fixed power level. Amplifiers and limiters are used to accomplish this goal. However, the limiter is a nonlinear device and spurious signals will be generated when simultaneous signals are applied. To eliminate the spurious signals from the limiter, the limiter should be put after the slot filter, thus the simultaneous signals have less chance to reach the limiter because of the narrow bandwidth. It should be noted that simultaneous signals within the bandwidth of the slot filter cannot be recognized by the receiver anyway, and they will be encoded as one signal.

5.5 FINE FREQUENCY CHANNELIZATION[7]

The most crucial part of a channelized receiver is the frequency determining circuits. The circuits should provide sharp frequency boundaries as discussed in the last section for both continuous waves (cw) and pulsed input signals.

There are many possible ways to implement slot filter assemblies with sharp boundaries. However, all approaches have to consider one common problem, the transient phenomena of a pulsed signal going through a filter. It is well known that a pulsed signal in time domain can be represented in the frequency domain through Fourier transform. For example, a pulse of amplitude A and width T can be written as

$$S(t) = A \quad \text{for} \quad -\frac{T}{2} < t < \frac{T}{2} \qquad (5.1)$$
$$ = 0 \quad \text{elsewhere}$$

Its Fourier transform is

$$S(f) = \int_{-\frac{T}{2}}^{\frac{T}{2}} A\exp(-j\omega t) \, dt = AT \frac{\sin \pi fT}{\pi fT} \qquad (5.2)$$

where f represents frequency and $\omega = 2\pi f$. The power spectrum $|S(f)|^2$ plotted in dB for $A = T = 1$ is shown in Figure 5.6.

Therefore, if a pulsed signal with a power spectrum as shown in Figure 5.6 is at the input of the filter bank, its energy will spread in many channels. This effect can be demonstrated in Figure 5.7. A pulsed input signal at the center of the bank of filters will produce outputs not only from the center filter but also from the rest of the filters. The outputs from all the filters do not resemble the input signal which is flat in amplitude. The output signals from the filters have larger amplitudes at the leading and trailing edges of the pulse which is referred to as the transient phenomena. This is critical in a channelized receiver design. It should be noted that the frequencies

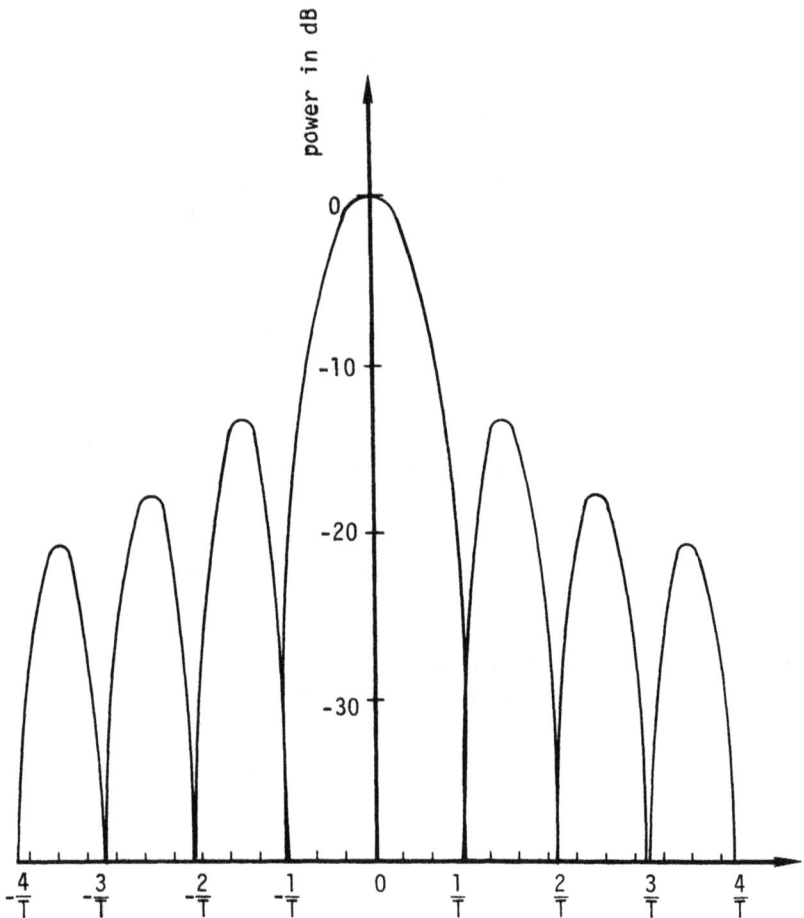

Figure 5.6. Power Spectrum for a Pulsed Signal

at the leading and trailing edges also shifted slightly. These frequencies will be shifted toward the center frequency of the filter. The output pulse shape can be predicted if the transfer function of the filter can be obtained (see Chapter 12).

Before discussing the frequency determination schemes in detail, it is appropriate to discuss the isolation filters and the amplifier-limiter chains first. Figure 5.8 shows this arrangement. The input signal is either frequency multiplexed or power divided into many parallel outputs. A frequency isolation filter is required at each output. Following the frequency isolation filter, there is an amplifier-limiter chain.

The frequency isolation filters are used to separate the input signals according to their frequencies. Their frequency response shape determines

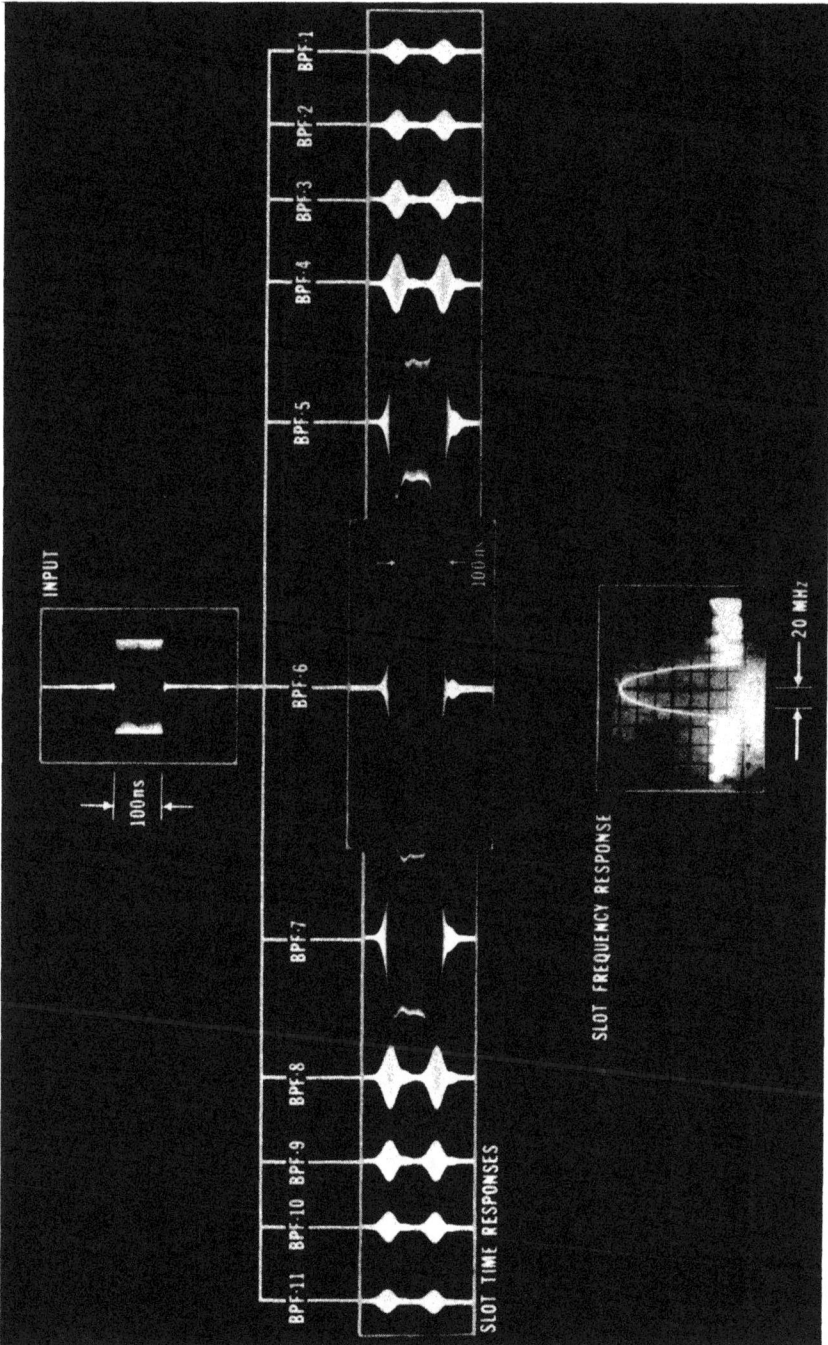

Figure 5.7. Output from a Filter Bank with One Input Signal (Courtesy of Texas Instruments)

the receiver's capability of handling simultaneous signals (see Chapter 2). In Figure 5.9 three contiguous isolation filters, F_1, F_2, and F_3, are shown with two input signals, S_1 and S_2.

Suppose S_2 is a stronger signal than S_1, and S_1 is at the center of filter F_2 while S_2 is on the skirt of filter F_2. If filter F_2 can provide enough isolation and reduce the strength of signal S_2 below that of S_1 at the output of filter F_2, S_1 will capture the amplifier-limiter chain. The frequency determining circuit following the limiter will recognize signal S_1 and encode it.

If filter F_2 does not provide enough isolation, at the output of the filter S_2 will be stronger than S_1, and S_2 will capture the amplifier-limiter chain. The frequency determining circuit will recognize S_2 as a outband signal, thus signal S_1 will not be received by the receiver. The ability of the channelized receiver to handle simultaneous signals with different frequency

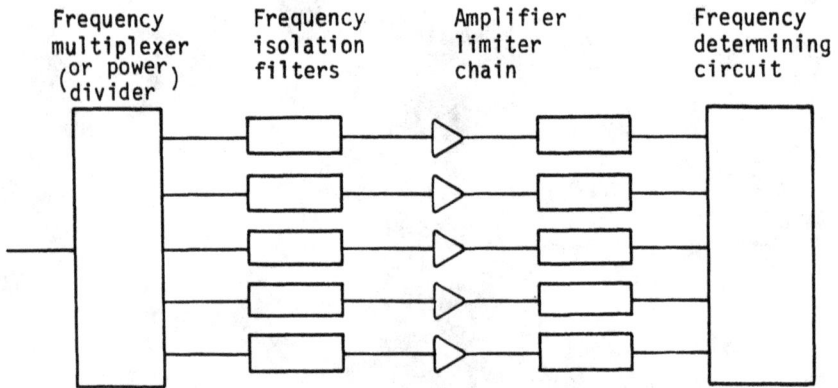

Figure 5.8. *Fine Frequency Channelization and Frequency Determining Circuits*

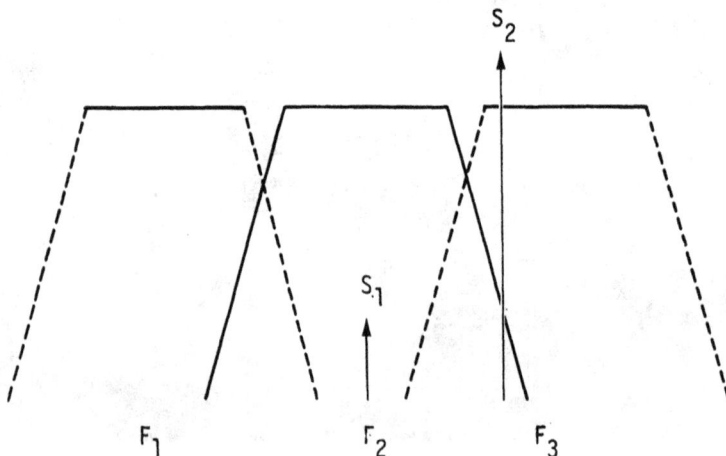

Figure 5.9. *Frequency Isolation Filters*

and amplitude is basically the frequency shape of the isolation filters. One can imagine that an isolation filter with a steep skirt (many sections) has better frequency isolation property; however, a steep skirt will have longer transient response in time domain which will affect the receiver's capability to receiver short pulses. The principal factor in choosing isolation filters is a compromise between simultaneous signal capability and minimum PW capability.

The amplifier-limiter chain amplifies the signal after the isolation filters to a constant power level, so that the frequency determining circuit will measure the fine frequency with minimum amplitude effect. Simultaneous signals are separated by the isolation filter before the amplifier-limiter chain, which further reduces the number of signals in the chain through capture effect. At the output of the amplifier-limiter chain, there is usually only one signal and the frequency measurement circuit only has to determine the frequency of that signal. The center frequency of the isolation filter is, in general, not required to be very accurate, but the center frequency of the filter following the isolation filter will be a carefully selected. This filter is used to determine the frequency of the input signal. The output of the amplifier-limiter chain is not totally amplitude independent. There is some change from one input to another input condition.

Four approaches to create a filter with sharp boundaries for both cw and pulsed signals will be discussed below:

1. Amplitude comparison scheme: The outputs from adjacent filters are compared in amplitude.
2. Dual detection scheme: The signal after the fine frequency filter and limiter is divided into two paths. They are detected and their amplitudes are compared.
3. Energy detection scheme: The signal after the fine frequency filter and limiter is detected to charge a capacitor. The voltage on the capacitor is compared with a reference voltage to determine whether the signal is in the filter pass band.
4. Peak-valley comparison scheme: The signal after the fine frequency filter is detected then separated into two paths. One path is delayed with respect to the other one, the leading edge peak is compared with the after leading edge valley to determine the frequency of the signal.

The detailed discussions of the four frequency encoding schemes to generate sharp boundaries will be discussed in the following sections.

5.6 AMPLITUDE COMPARISON SCHEME TO DETERMINE FREQUENCY

Since a short pulse will produce outputs from a number of contiguous parallel filters, it seems obvious that the filter with the strongest output will represent the frequency of the input signal. The intuitive approach is to compare the amplitude of a filter output with the outputs from its two adjacent filters to determine the center frequency of the signal. A simple

amplitude comparison scheme is shown in Figure 5.10a. The output from
a filter is compared with the two outputs of its two adjacent filters.
However, if there is no signal in the filter bank, the noise from these filters
will trigger the comparator and put them into a random state which is
undesirable. To improve this circuit, a threshold can be added to each
comparator as shown in Figure 5.10b. Assume the threshold is applied in

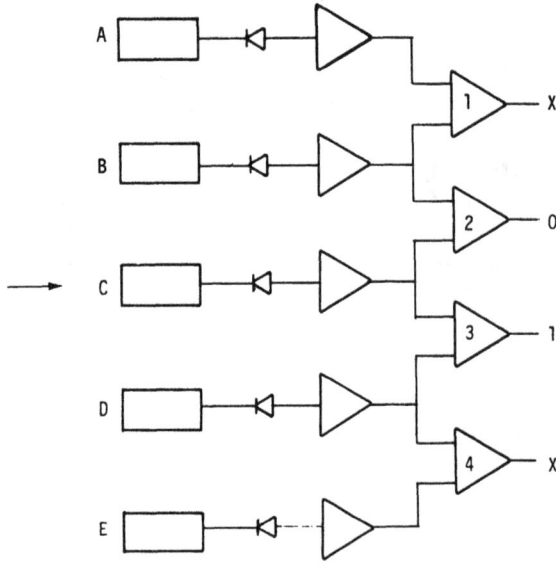

(a) A Simple Amplitude Comparison Scheme

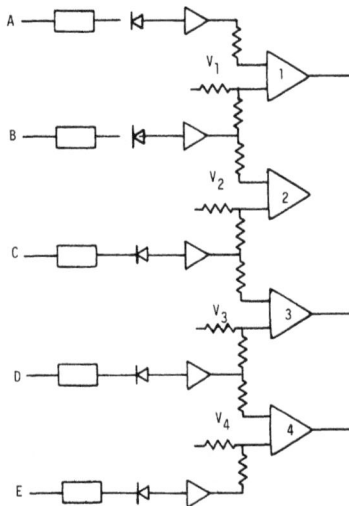

(b) Amplitude Comparison with Threshold

Figure 5.10. Amplitude Comparison Scheme

such a way that if there is no input in a certain channel, the lower input of the comparator is higher than the upper input and the output of the comparator is zero. Therefore, if there is no signal arriving at the receiver, all the outputs are zero. If the signal is in filter C, the output of comparator 3 will be 1 and the rest of the outputs are still zero. If the energy in the pulse is spilled into channels C and D, both the outputs 3 and 4 will be 1. Under this condition, the frequency of the input signal will appear as at the center of filter D. Therefore, inaccuracy in the frequency reading will occur. The frequency resolution of this approach approximately equals the bandwidth of the filters used in the receiver.

There are some approaches to improve the frequency resolution without narrowing the bandwidth of the filters. In other words, the frequency resolution can be improved without increasing the total number of filters. As shown in Figure 5.11a, the output of each detector is divided into two paths: the unattenuated ones labeled B, C, and D, and the attenuated ones labeled B', C', and D'. Instead of one comparator used between two adjacent filters, there are three comparators between them. The attenuation provided to slot C is shown in Figure 5.11b. Comparing the amplitude C' against B, D will further divide each filter into three equal frequency ranges. If the signal is in the center part of filter C, in addition to $C > B$ and $C > D$, the conditions $C' > B$ and $C' > D$ are also true. The outputs from the comparators for this condition are shown in Figure 5.11a.

Since the adjacent filters have the same bandwidth with center frequencies close together and the same number of poles, their transient responses should be quite similar. Therefore, the outputs from the comparator will stay at the same states whether the signal is in the transient or steady-state. However, whether the output of a filter will actually track during the transient state in comparison with its adjacent filter outputs must be carefully studied for the individual filter used in the receiver. In addition, the detectors and the amplifiers following the detectors must be tracked over the dynamic range of the receiver which might be difficult to accomplish. If poor tracking occurs in the transient of the filters, detectors, or amplifiers, erroneous frequency information may be generated.

In addition to the above difficulties, the alignment of this scheme is complicated because each signal must be compared with at least two adjacent outputs. If one output is adjusted it will affect its two neighboring slots. Although this amplitude comparison scheme seems to be a reasonable approach, it is not widely considered in channelized receiver designs.

5.7 DUAL DETECTION SCHEME TO DETERMINE FREQUENCY[8]

In this dual detection scheme (Reference 8), the signal after the amplifier-limiter chain is divided into two paths. Both paths have filters of the same center frequency but different bandwidths as shown in Figure 5.12. In Figure 5.12 filters A_1 and A_2 have the same center frequency. Let us assume that filter A_1 has a narrower bandwidth than filter A_2. An attenuator of proper

(a) Schematic Diagram

(b) Filter Outputs

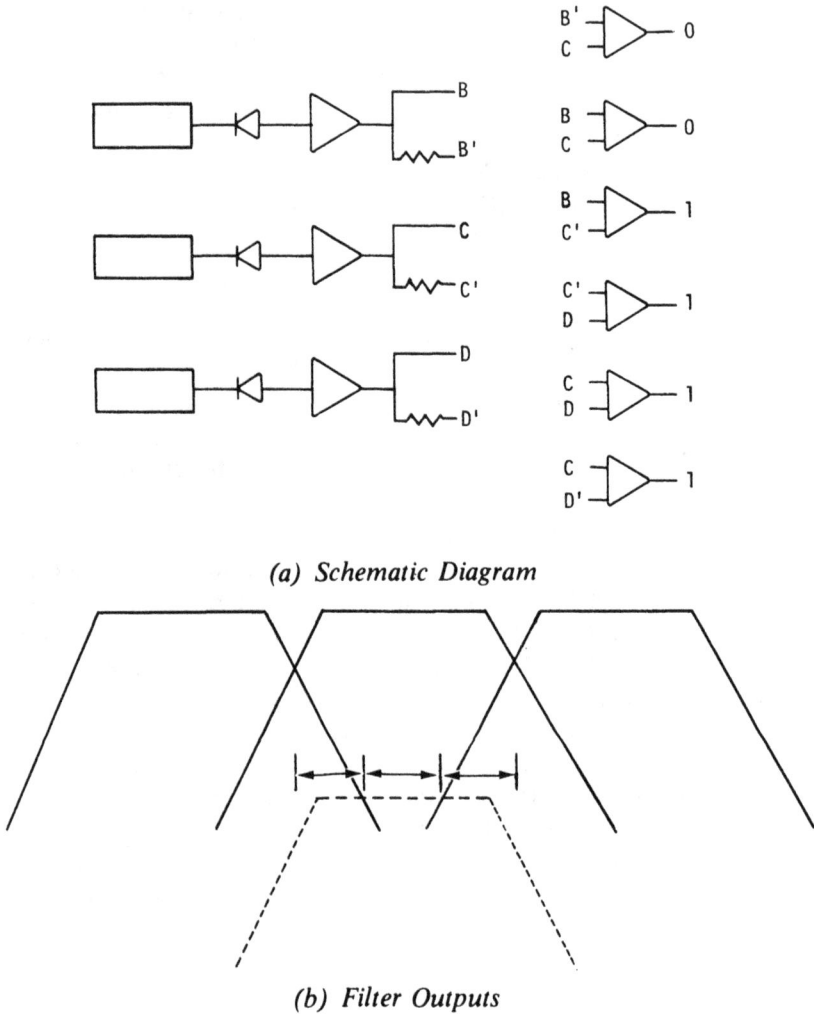

*Figure 5.11. Amplitude Comparison Scheme with Improved Frequency
     Resolution*

value is put in series with filter A_2 to reduce its output amplitude. In
general, the two filters have the same number of poles (or sections). The
outputs from filters A_1 and A_2 in the frequency domain are shown in Figure
5.13. When the input is in the frequency range f_{A_1} to f_{A_2}, the output from
the filter A_1 is stronger than that of filter A_2 and the output from comparator
A is positive. Under such conditions, the input signal is considered in slot A.
When the output of comparator A is 0, the signal is outside the frequency
range f_{A1} to f_{A2} and there is no signal in slot A. However, the problem is not
this simple. The transient phenomenon in the pair of filters are different. The

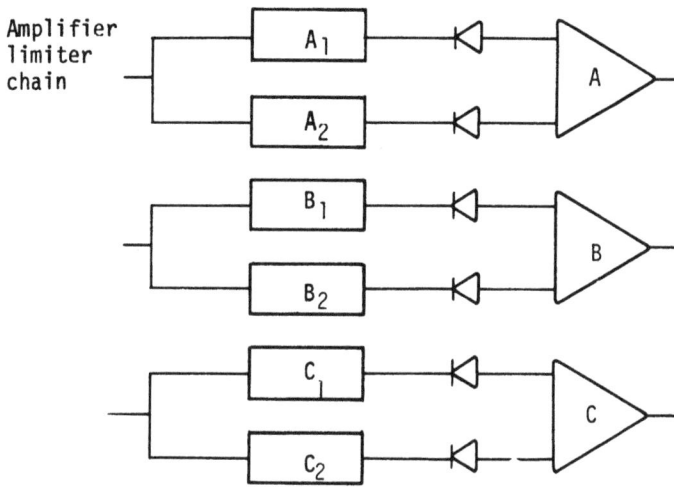

Figure 5.12. Dual Detection Scheme

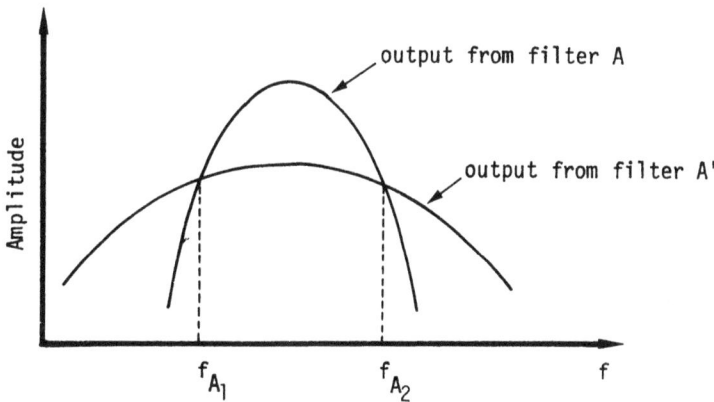

Figure 5.13. Outputs from Dual Detection Filters in Frequency Domain

wider the filter bandwidth, the shorter the transient time. In other words the rise time of a signal passing a wide-band filter is shorter than that of a narrow filter if the two filters have the same center frequency and the same number of poles. The outputs from the two filters in the time domain are shown in Figure 5.14a and b. In Figure 5.14a the signal is inside the slot A, while in Figure 5.14b the signal is outside of the slot A.

There is no problem in detecting the "in" slot signal as shown in Figure 5.14a. However, the signal outside the slot generates two outputs at the leading and trailing edges of the pulse which are usually referred to as ears. The detection circuit has to be able to ignore the ears, otherwise false output will be reported. The width of the ear is related to the filter bandwidth and

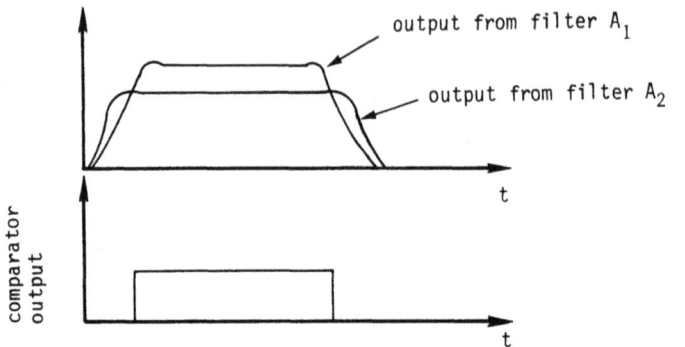

(a) *Signal at the Center*

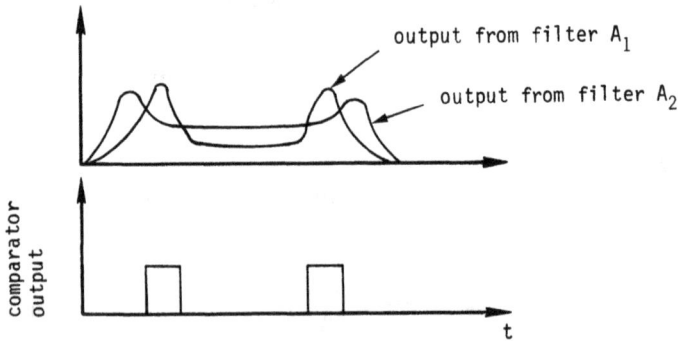

(b) *Signal Outside of Slot A*

Figure 5.14. Output from Slot Filters and Comparator A

the number of poles of the filter. The narrower the bandwidth, the larger number of poles and the wider the ears. The detection circuit must make a decision after the first ear but before the second ear. However, if the input signal pulse width decreases, the time between the two ears also decreases. When the two ears are very close together, there is not enough time for the detection circuit to make a decision between the two ears which is the minimum PW the dual detection scheme can handle. In actual receiver design, filters of few poles and proper bandwidth are used to minimize the transient effect and reduce the ear length.

The dual detection scheme can be simplified if the variation of the output level from the amplifier limiter chain is very small. The second filter used in each slot (as shown in Figure 5.12) can be eliminated, and the new circuit is shown in Figure 5.15. The output from the frequency determining filter is detected and compared with a constant voltage value. The output from the comparator will be similar to that of Figure 5.14a and b. The signal

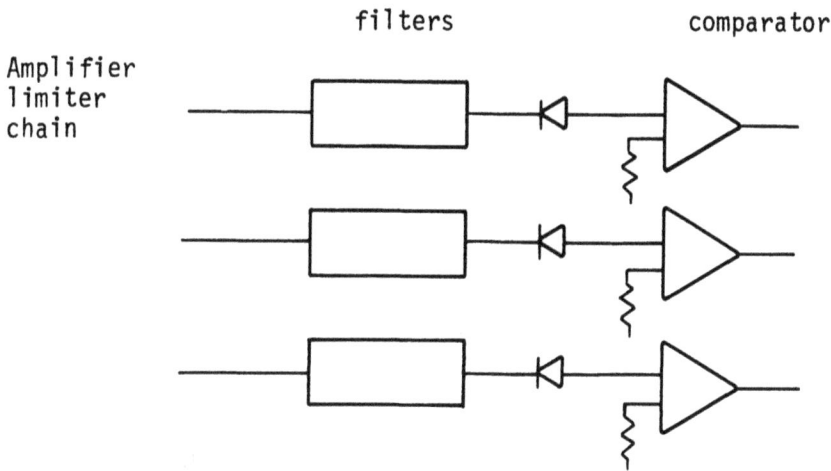

Figure 5.15. Simplified Dual Detection Scheme

outside the slot will sometimes generate ears. In this simplified detection scheme, if the output power from the amplifier-limiter chain changes, the output from the comparator may change and cause frequency inaccuracy. In the dual detection scheme, if the power from the amplifier-limiter chain varies, both detector output levels will change and the amplitude variation will be somewhat compensated. Thus, the change in amplitude in the amplifier-limiter chain has less effect on the dual detection scheme.

The dual detection scheme has proven that the frequency reading is practically amplitude independent. The equivalent filter skirt is very sharp. Slots with this frequency measurement arrangement are shown in Figure 5.16. Figure 5.16a shows a total of 16 slots: 8 in the top container, 8 in the bottom container. Figure 5.16b shows a total of eight slots with SAW filters to demonstrate the reduction in size. It should be noted that in Figure 5.16b the assembly does not include the isolation filters. Figure 5.17 shows the schematic of the amplifier-limiter chain with the isolation filters. It contains two isolation filters, three amplifiers, and two limiters. Figure 5.18 shows the dual detection circuit following the amplifier-limiter chain.

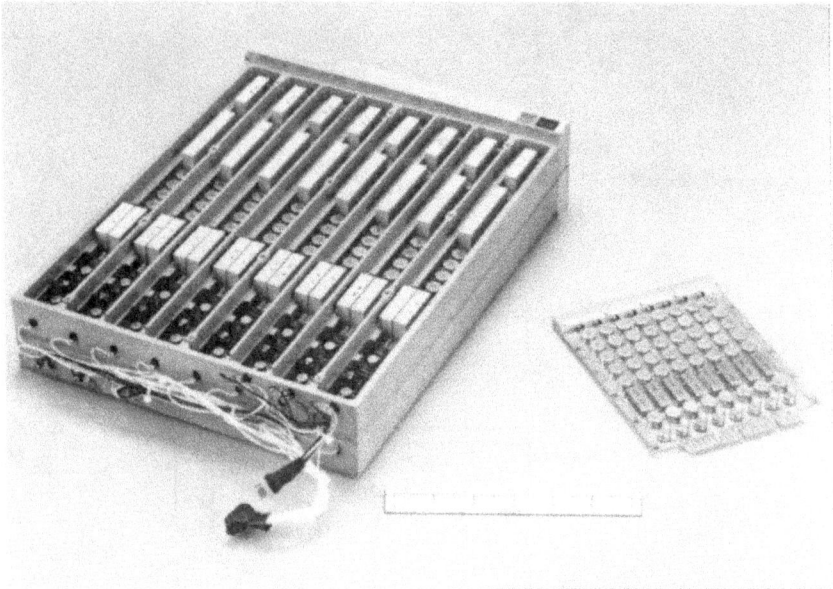

(a) With Conventional Filters   (b) With SAW Filters

Figure 5.16. Dual Detection Assemblies (Courtesy of Air Force Wright Aeronautical Laboratories)

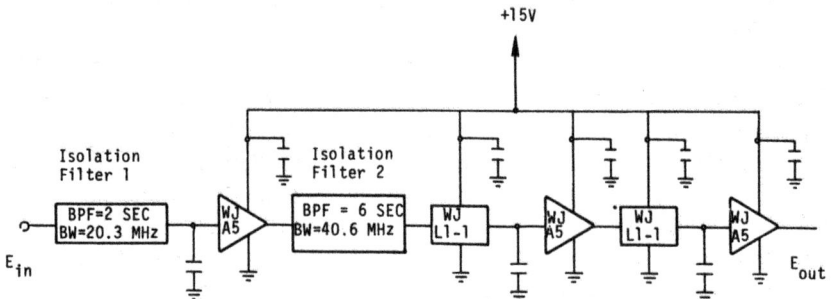

Figure 5.17. Isolation Filters and Amplifier-Limiter Chain

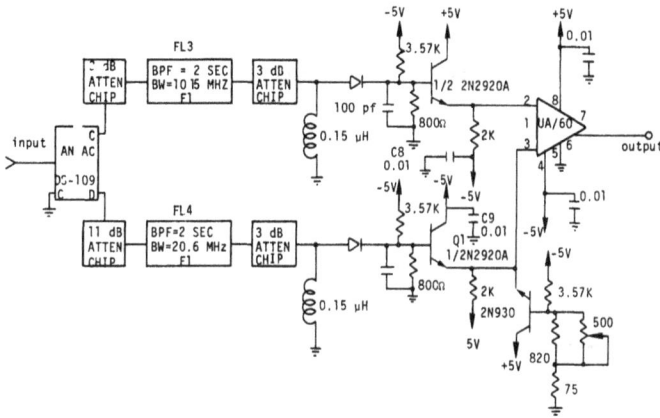

Figure 5.18. Dual Detection Scheme

5.8 ENERGY DETECTION SCHEME TO DETERMINE FREQUENCY(9)

In this detection scheme after the signal passes through the amplifier-limiter chain, it passes through another fine frequency filter. The input signal is then detected and compared with a fixed threshold. The output of the comparator is amplified and stored in a capacitor C (as shown in Figure 5.19). Output from the capacitor is compared again with another fixed threshold to determine whether it is an "in" slot or an "out" slot signal.

If the signal from the detector is above the first threshold, then the transistor Q will be biased off and capacitor C is charged. If the signal from the detector is below the threshold, then the transistor Q is switched on and the charge on the capacitor is drained through the transistor. The voltage

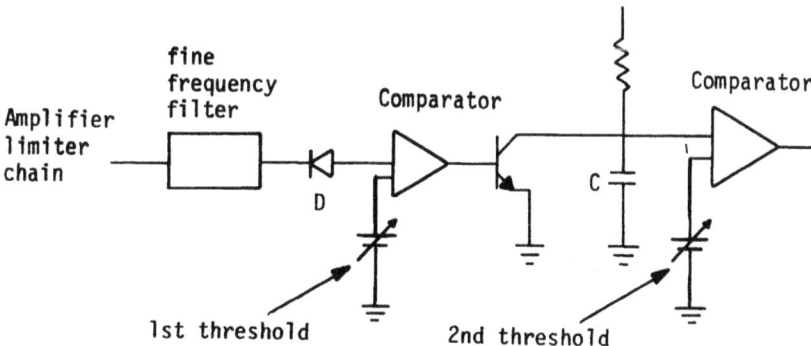

Figure 5.19. Frequency Determining Circuit for a Channelized Receiver by Energy Detection Scheme

on the capacitor is again compared with a fixed reference (second threshold). When the voltage on the capacitor is above the second threshold, the signal is considered in the slot. The ear generated at the leading edge of the pulse may start the charging of the capacitor, but as soon as the steady-state is below the first threshold, the charge is drained and the second threshold will not be crossed. Figure 5.20 shows the slot outputs and the

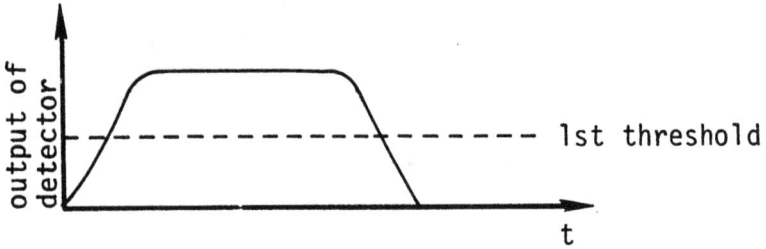

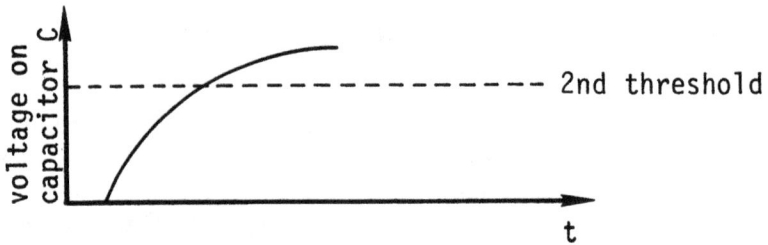

(a) *Output of Signal with Frequency in the Slot*

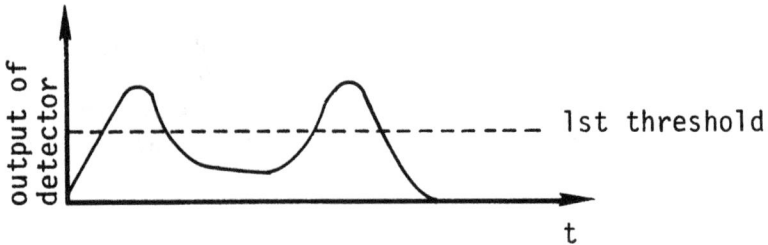

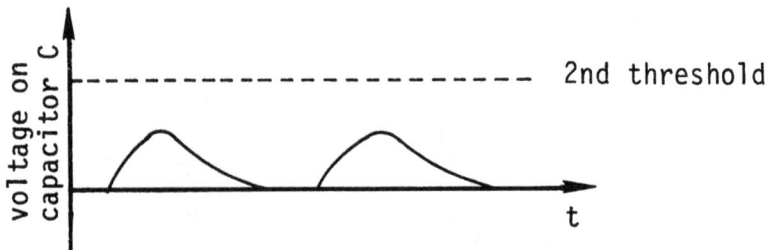

(b) *Output of Signal with Frequency Out of the Slot*

Figure 5.20. Slot Outputs and Voltages on Capacitor

voltages on the capacitor for both an "in" slot signal and an "out" slot signal. The advantage of this approach is to reduce the complexity of sampling the output of the comparator at the proper time.

In this approach, when the PW is reduced, the leading and trailing edges will be closer together and the capacitor will be charged above the second threshold, although the signal is an "out" slot signal. This effect limits the minimum PW the receiver can handle.

5.9 PEAK VALLEY COMPARISON SCHEME TO DETERMINE FREQUENCY (10)

In this detection scheme, after the signal passes through the amplifier-limiter chain, it passes a fine frequency filter. At the output of the fine frequency filter, the signal is detected and separated into two parallel paths. One is delayed a fixed time with respect to the other one as shown in Figure 5.21. The amplitude of the two signals are properly adjusted and fed to a comparator. The inputs and the outputs of the comparator are shown in Figure 5.22. When the signal is "in" slot as shown in Figure 5.22a, the valley of the undelayed signal is higher than the peak of the delayed signal and the comparator output is positive. When the signal is outside the frequency slot, the outputs are shown in Figure 5.22b. The peak of the delayed signal is higher than the valley of the undelayed one and the corresponding output from the comparator is shown in Figure 5.22b. The output from the comparator is positive, then goes to zero level and up to positive again. If a sampling pulse is properly positioned, it can determine the relative amplitudes of the peak and valley, then the signal frequency. When the PW decreases, this encoding scheme also generates false information as the other schemes mentioned above.

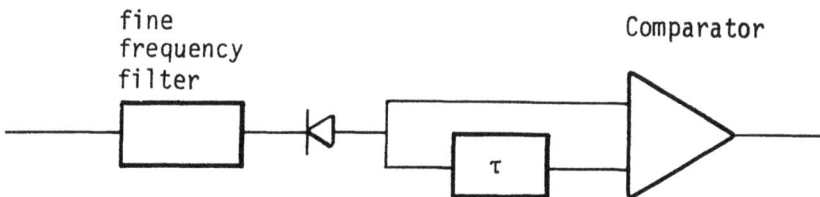

Figure 5.21. Peak-Valley Comparison Scheme for Channelized Receiver Frequency Determination

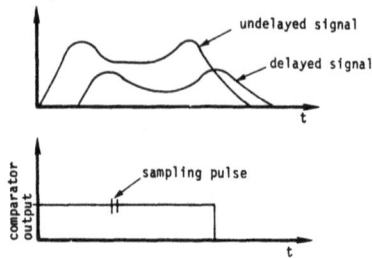

(a) Output of Signal with Frequency in the Slot

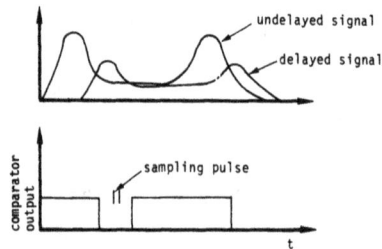

(b) Output of Signal with Frequency Out of the Slot

Figure 5.22. Inputs and Outputs of a Comparator for the Peak-Valley
Comparison Scheme

5.10 FINE FREQUENCY RESOLUTION ON SHORT PULSES

From the discussions above, one noted that the transient at the leading
and trailing edges affected the frequency measurement circuit. All the frequency measurement circuits mentioned above have a general limitation
that when the PW decreases, the frequency measuring circuit will generate
erroneous information. In other words, in a channelized receiver the
minimum PW the receiver can measure correctly depends on the width of
the frequency measurement filter. The narrower the filter bandwidth, the
longer the minimum PW. An approximate relation between the filter
bandwidth and the minimum PW [PW(min)] is

$$PW_{min} \cong \frac{1}{B} \tag{5.3}$$

where B is the filter bandwidth. If the minimum PW is the determining
factor, then the frequency accuracy/resolution of the channelized receiver
will be determined accordingly. The relation in Equation 5.3 is very close
to the experimental results obtained from actual channelized receiver
measurement. This is far from the results predicated in Equation 2.4.

In an ideal channelized receiver, the frequency resolution should be PW
dependent. In other words, the longer the PW the finer the frequency
resolution. Without excess hardware and complexity, this desired receiver

cannot be built. There is a trade-off, however, and one can obtain fine
frequency resolution to shorter PW. It has been discussed in Chapter 4 that
an IFM receiver using a delay line to measure frequency can measure fine
frequency on a short pulse. If a delay line discriminator is used to measure
frequency after the isolation filter, then finer resolution on short pulse can
be accomplished. One possible approach is shown in Figure 5.23.

Power
divider

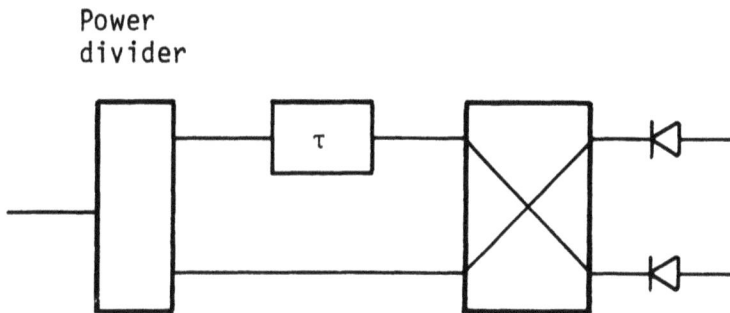

*Figure 5.23. Channelized Receiver with Delay Line Discriminator as Fine
Frequency Measurement Circuit*

In this arrangement the isolation filter cannot be very narrow. Equation
5.3 still needs to be observed, because the filter will actually shift the frequency in the signal as discussed in Section 5.5. If the minimum PW anticipated is 0.1 μs, then the narrowest filter that can be used is approximately
10 MHz; and the fine frequency can be measured by the frequency discriminator. This approach can be considered as a hybrid one between a channelized
and an IFM receiver. There are some other approaches using delay lines to
measure the fine frequency on short pulse.

5.11 IMPACT OF SURFACE ACOUSTIC WAVE (SAW) FILTERS ON CHANNELIZED RECEIVERS

At the beginning of this chapter, it was discussed that a SAW filter is
one of the practical ways to make a channelized receiver. The detailed performance of a SAW filter will be discussed in Chapter 12. The most significant effects of a SAW filter on a channelized receiver is the reduction in
size. Figure 5.16a and b clearly show the small size of the SAW filter
assembly compared with that of conventional filters. The high insertion
loss of the SAW filters must be compensated with additional gain. The
amplifier-limiter chain of the SAW assembly in Figure 5.16b is shown in

Figure 5.24. Comparing with the amplifier-limiter chain in Figure 5.17, two additional amplifiers are required to compensate the high insertion loss of the SAW filters. A total of five amplifiers and two limiters are used in the chain. The dual detection circuit is shown in Figure 5.25. It should be noted that the power divider to split the input signal to the two parallel filters is built as part of the SAW filters.

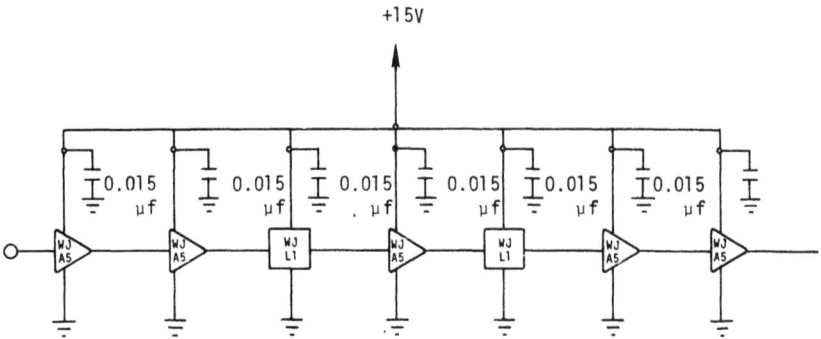

Figure 5.24. Amplifier-Limiter Chain for the SAW Channel Assembly

Other significant effects of SAW filters on channelized receivers are the limited out-of-band rejection level and triple-transit effects.

The effect of the limited out-of-band rejection is obvious. It limits the dynamic range of the receiver. The filter responses in frequency domain are shown in Figure 5.26. It shows a conventional filter and a SAW filter with the same center frequency. Outside the pass band of the conventional filter, the attenuation increases with increasing frequency separation. While outside the pass band of a SAW filter, the rejection is more or less a fixed value regardless of the signal frequency. Thus, a signal far from the center of the SAW filter can also leak into the filter with more or less a fixed insertion loss. The dynamic range of the receiver is less than that of the SAW filters by the signal to noise required to reduce false alarm rate. If the dynamic range of the SAW filter is 60 dB, then the dynamic range of the receiver will be 45 dB if 15 dB is the required signal-to-noise ratio. It should be noted that the signal after the amplifier-limiter chain practically has limited amplitude variation. Therefore, the limited dynamic range imposed by the SAW filter after the amplifier-limiter chain does not affect the dynamic range of the receiver. In other words, the isolation filter must have high dynamic range; however, the dynamic range of the fine frequency is not very critical.

The triple-transit of the SAW filter will affect the throughput rate of the receiver. The encoding circuit of the receiver has to wait until the triple-transit is over to process another signal. While the triple-transit of the isolation filter affects the receiver throughput rate, the triple-transit in the fine frequency filter does not affect the throughput rate, because the signal after the isolation filter has almost constant amplitude. Improving the performance of SAW filters can reduce these problems.

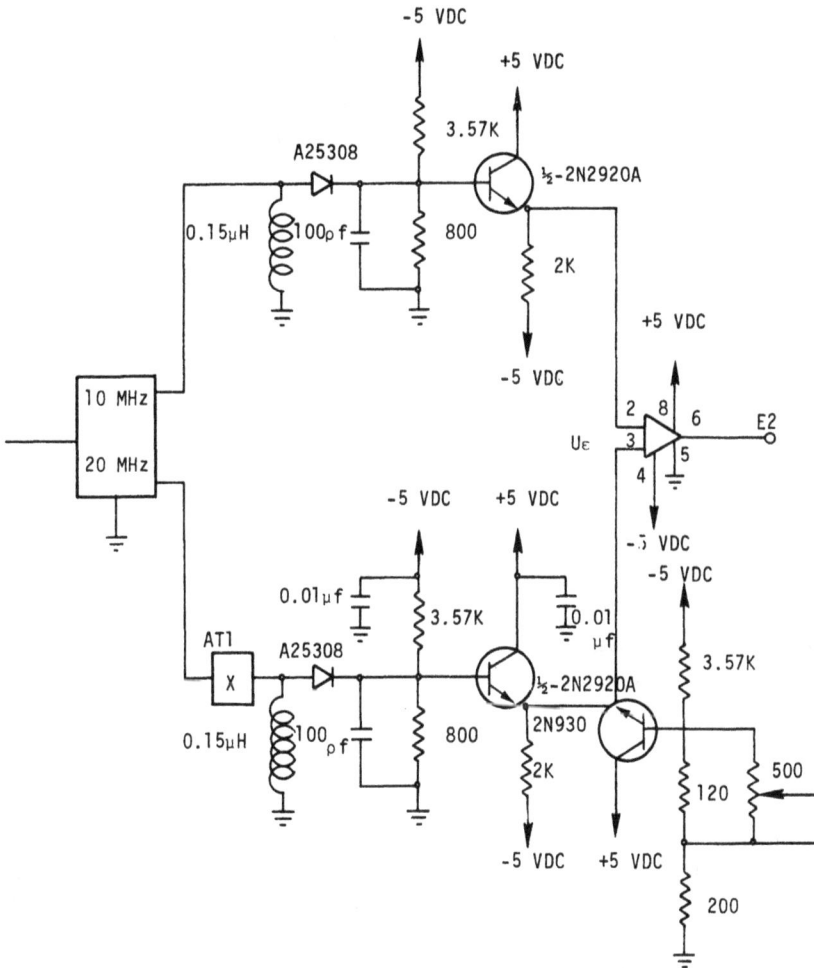

Figure 5.25. Dual Detection Circuit for the SAW Assembly

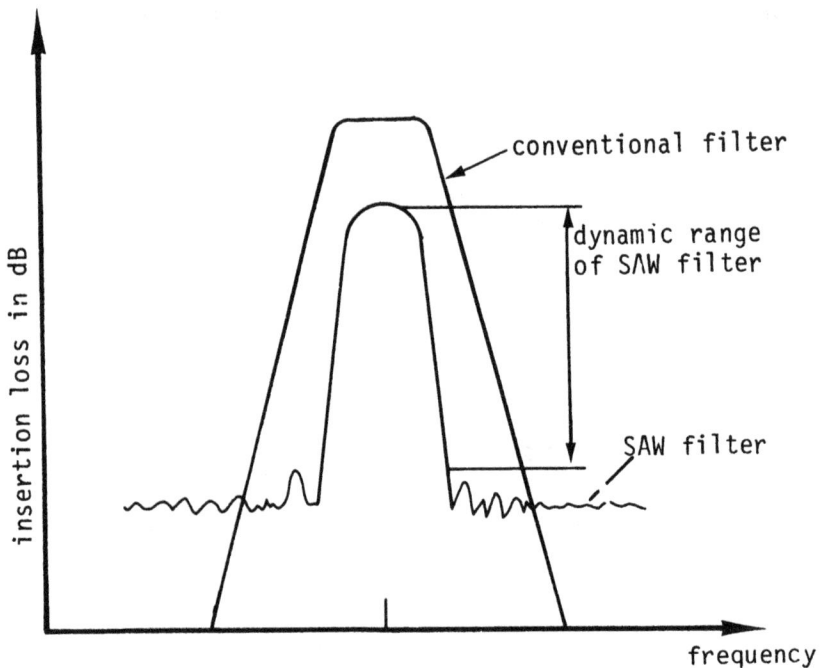

Figure 5.26. Frequency Reponse of SAW Filter and Conventional Filter

5.12 SUMMARY

A channelized receiver has an analogy to a number of parallel superhet receivers. It has all the good characteristics of a superhet receiver: high sensitivity, wide dynamic range, fine frequency resolution, and very limited simultaneous signal problem. In addition, a channelized receiver has wide instantaneous bandwidth which can provide high probability of intercept. However, the large number of parallel channels makes the receiver bulky and expensive.

The front end of an electronic warfare (EW) receiver usually uses a channelized approach. Other kinds of receivers can be used after the front end channelization. Thus, most EW receivers can be considered as a hybrid receiver between a channelized receiver and another kind of receiver.

The most critical part of a channelized receiver is the fine frequency encoding circuits. Four different approaches are discussed. Although SAW filters can reduce the size of the receiver, it also will reduce the performance of the channelized receiver slightly.

REFERENCES

1. Harper, T., "High probability of intercept receivers," Watkins Johnson Co., Tech-Notes, Vol. 2, No. 4, July/August 1975.
2. Harper, T., "New trends in EW receivers," Countermeasures, p. 34-38, December/January 1976.
3. "The channelized receiving systems," Staff Report, Microwave System News, p. 63, December/January 1976.
4. Hennessy, P., Quick, J.D., " The channelized receiver comes of age," Microwave System News, p. 36, July 1979.
5. Harper, T., "Hybridization of competitive receivers," Watkins Johnson Co., Tech-Notes, Vol. 7, No. 1, January/February 1980.
6. Hoffmann, C.B., Baron, A.R., "Wideband ESM receiving systems," Part I, p. 24, Microwave Journal, September 1980; Part II, p. 57, February 1981.
7. Tsui, J.B.Y., Adair, J.E., Hawkins, J.E., LaFleur, S.J., "Transient response of filters," Air Force Avionics Laboratory, Technical Report AFAL-TR-77-249, November 1977.
8. Hollis, R., "Double detection filter techniques test report," Watkins Johnson Co., Palo Alto, California, 1973.
9. Private Communication with Lundy, D., Grover, K., Motorola Co.
10. Private Communication with Daniels, W., Higgins, T., Texas Instruments Inc.

CHAPTER 6

COMPRESSIVE RECEIVERS (MICROSCAN RECEIVERS)

6.1. INTRODUCTION[1]

The name compressive receiver comes from the fact that a dispersive delay line (DDL) is used to compress the input radio frequency (RF) signal into a narrow pulse. It is also referred to as a microscan receiver because a fast sweeping local oscillator (LO) is used to convert the input signal into frequency modulated (FM) signals. The idea of using a DDL to measure frequency was patented in 1960 (Reference 1). The advances in surface acoustic wave (SAW) technology and logic circuits make the fabrication of a compressive receiver feasible.

A compressive receiver is a wide-band receiver. While the channelized receiver has many parallel filters to produce many parallel outputs, a compressive receiver uses one DDL and a sweeping LO to generate many series of output pulses. By measuring the positions of the output pulses in time domain, the frequency of the input signal can be obtained.

The key components in a compressive receiver are the DDL and sweeping LO. They will be discussed in separate chapters. Chapter 9 includes different kinds of delay lines while Chapter 15 concentrates on oscillators. This chapter will be devoted to the functional operation of the compressive receiver.

6.2 PRINCIPLES OF OPERATION[2-9]

A compressive receiver uses a DDL to compress an FM signal which is generated from the input signal into a pulse. Measuring the time of the compressed pulse emerging from the DDL, the frequency information of the input signal can be obtained. Suppose an FM signal changes frequency linearly from f_0 to f_1 in the time interval from t_0 to t_1 with a frequency versus time slope of m as shown in Figure 6.1a. If this signal enters a DDL with a frequency versus time slope of $-m$ as shown in Figure 6.1b, the output from the DDL will be a burst of energy at time $t_0 + t_1$. The leading edge of the signal with frequency f_0 entering the DDL at t_0 is delayed by t_1 while the trailing edge of the signal with frequency f_1 entering the delay line at t_1 is delayed by t_0; therefore, the entire input signal from t_0 to t_1 is compressed and output at time $t_0 + t_1$. Compressive receivers use this basic idea to compress the input signals into short pulses. The time difference between t_0 and t_1 is the differential delay time, while the frequency separation from f_0 to f_1 is the bandwidth of the delay line. The time bandwidth product of a DDL is the differential delay time times the bandwidth. The time bandwidth product is a very important factor which determines the processing gain of the compressive receiver. The differential delay time is related to the frequency resolution of the receiver, and the bandwidth of the DDL is related to the input bandwidth and the probability of intercept

119

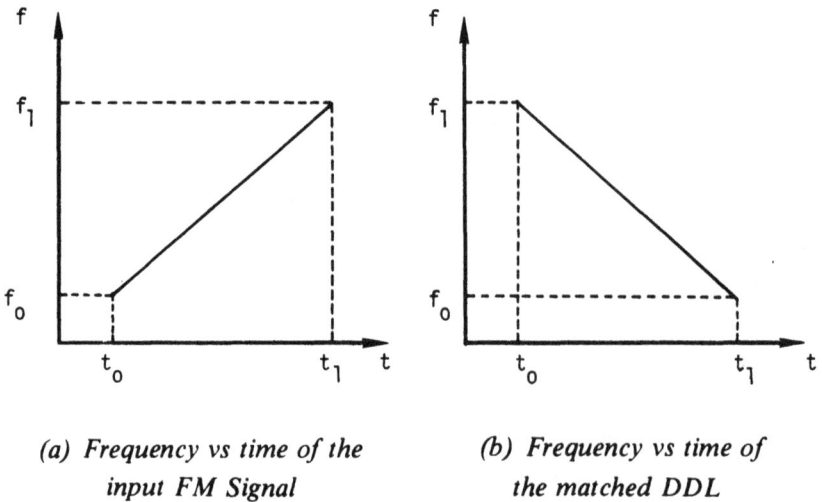

(a) Frequency vs time of the
input FM Signal

(b) Frequency vs time of
the matched DDL

Figure 6.1. Input FM Signal and Matched DDL

of the receiver. A compressive receiver block diagram is shown in Figure 6.2. The input signal is mixed with an FM signal generated by an LO. A weighting filter is following the mixer to modify the signal from the mixer. The purpose of the weighting filter is to make the output pulse from the DDL into a desirable shape. The output from the weighting filter passes through the DDL and is detected by the detector.

The FM signal to be compressed in the receiver is generated internally. The frequency versus time slope of the sweeping LO matches that of the DDL. Here, matching means that the slope of the sweeping LO and the slope of the DDL are equal in amplitude but opposite in sign. An input signal with constant frequency is converted to a linear FM signal at the output of the mixer. This signal is modified by the weighting filter and is compressed in the time domain into a pulse at the output of the delay line. Note that the RF signal (Figure 6.2) is drawn backwards in time domain which means the low frequency end of the RF signal enters the DDL first.

To demonstrate how the compressive receiver reads frequency, three continuous wave (cw) signals at the frequencies f_A, f_B, and f_C are shown in Figure 6.3. It is assumed that frequencies f_A and f_C are at the edge of the receiver input band, while f_B is at the center of the input band. It is also assumed that the RF input bandwidth (B_R) equals the intermediate frequency (IF) input bandwidth (B_I) and the sweeping LO has zero flyback time. The LO must sweep a frequency range which equals the RF bandwidth and the IF bandwidth together. This relation can be observed from Figure 6.3c. In order for B_I to intercept both $f_A + f_0$ and $f_C + f_0$, and fill the entire bandwidth B_I, the LO must sweep the bandwidth $B_R + B_I$. From Figure 6.3, it is seen that frequency f_A enters the IF bandwidth and the DDL first and is followed by f_B and f_C. At the output of the delay line f_A will come out first as a compressed pulse then followed by f_B and f_C. The relative positions of output pulses represent the frequency of the input signals.

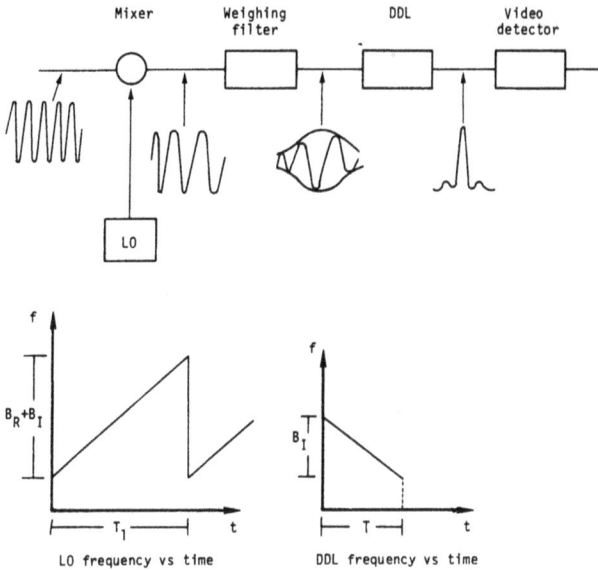

Figure 6.2 A Basic Compressive Receiver

Let us examine the signals in the DDL a little closer. In order to eliminate confusion, the period of the sweeping LO is represented by T_1 while the dispersive delay time of the delay line is called T. From Figure 6.3c one can see that the signals enter the DDL in the first half of T_1 while the signals emerge from the DDL on the second half of T_1. This phenomena can also be observed from Figure 6.3d; the pulses come out from the DDL in the period from $T_1/2$ to T_1, $3T_1/2$ to $2T_1$, and so on. The pulses coming out only at $T_1/2$ are a very important factor. This phenomenon is used in a compressive receiver with dual scan to improve the probability of intercept. It should be emphasized here that this phenomena happens only when the $B_R = B_I$.

It is also clearly shown in Figure 6.3 that the probability of intercepting the input signal is not 100 percent in the time domain. For example, if a pulsed signal of frequency f_A arrives at the receiver between time $T_1/2$ to T_1 (as shown by the heavy line), the mixer output frequency is outside the

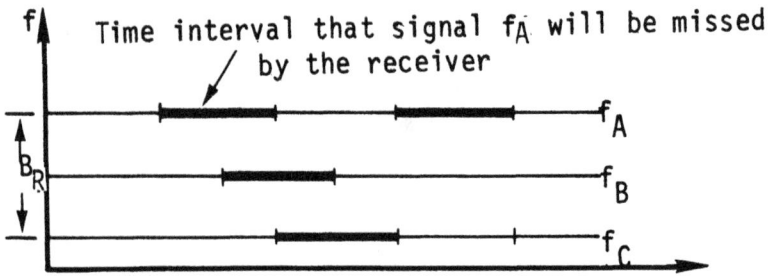

(a) Input Signal

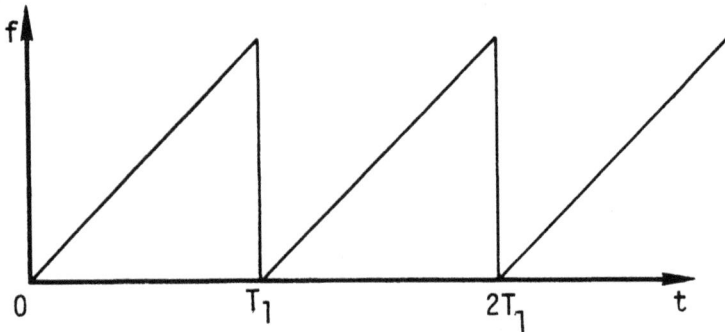

(b) Sweeping LO

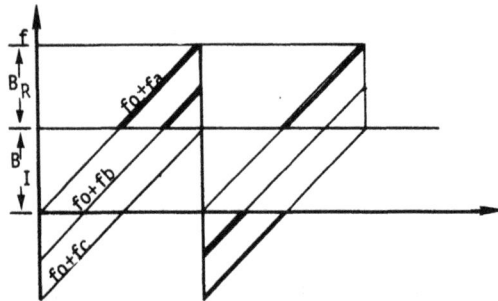

(c) Mixer outputs intercepted by IF band B_I

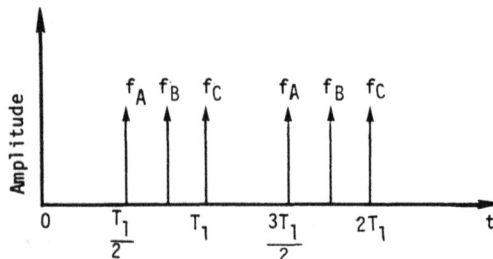

(d) Output pulses from DDL

Figure 6.3. Time and Frequency Relation of a Compressive Receiver with $B_R = B_I$

IF bandwidth and the receiver will miss the signal. A similar argument holds true from f_B between $3T_1/4$ to $5T_1/4$ and f_C between T_1 to $3T_1/2$. In general, a pulsed signal can arrive at the receiver in half the scan period time and cannot be detected by the receiver at all. This discussion indicates that a compressive receiver will sometimes miss short pulses. The probability of intercept will be discussed in more detail in Section 6.5.

In a compressive receiver the IF bandwidth and the RF bandwidth are not necessarily equal. They can have almost any relation. Figure 6.4 shows the time and frequency relationship of a compressive receiver with an IF bandwidth equal to half the RF bandwidth ($B_I = \frac{1}{2}B_R$) and the sweeping LO period $T_1 = 3T$. In this configuration, the time bandwidth product requirement of the delay time is not as stringent, but the probability of intercept is degraded also. Figure 6.5 shows the time frequency relationship of a compressive receiver with $B_R = \frac{1}{2}B_I$ and $T_1 = 3/2T$. This configuration improves the probability of intercept, but a large value of time bandwidth product is required on the delay line. Most practical compression receivers are designed with either $B_R = B_I$ or $B_R > B_I$.

The LO must sweep a frequency range that equals the sum of the RF and IF bandwidths. The output frequency from the mixer can either chirp up or down. The output frequencies from the examples in Figures 6.3 through 6.5 all chirp up. Theoretically, there is no difference in receiver design for a chirp up FM or a chirp down FM. However, in practical cases in the DDL, the longer the delay time, the higher the insertion loss. At higher frequency, the insertion loss is higher than that at lower frequency. To equalize the insertion loss over the frequency range in the delay line, the higher frequency usually has a shorter delay time while lower frequency has longer delay time. To match such a delay line, a chirp up FM signal is required at the input of the delay line. However, this does not mean the LO must chirp up; it only means that the output from the mixer must chirp up. Besides the desired frequency, other frequency components from the mixer do not match the frequency versus time slope of the delay line, and their energy is not compressed into a pulse in time domain. Therefore, the spur products from a mixer cause less problem in a compressive receiver.

6.3 MATHEMATIC ANALYSIS[8,9]

From the discussion in the preceding section, it appears that the output from the delay line is a pulse with zero width. This is not true. The compressed pulse has a certain pulse width and shape. This section will derive the mathematic model to: (1) further understand signal passing through the receiver; (2) predict the output pulse shape from the DDL; and (3) introduce some approaches to modify the output pulse shape for better detection. Suppose the DDL has a bandwidth B_I and dispersive delay time T, and a weighting filter is inserted before the delay line in Figure 6.2. The purpose of the weighting filters is to shape the output pulse.

As mentioned above, the output from the mixer will pass the weighting

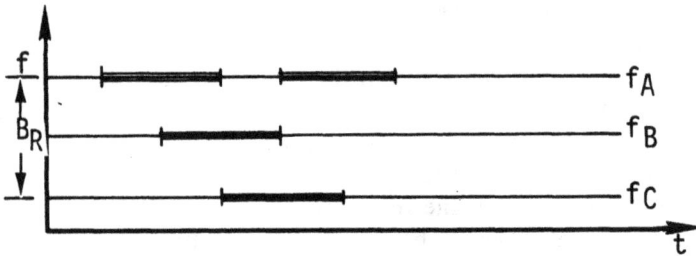

(a) Input Signal

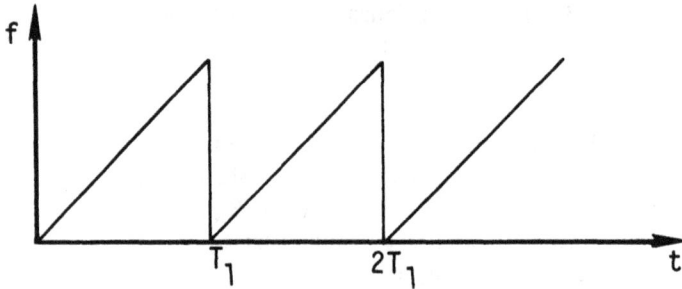

(b) Sweeping LO

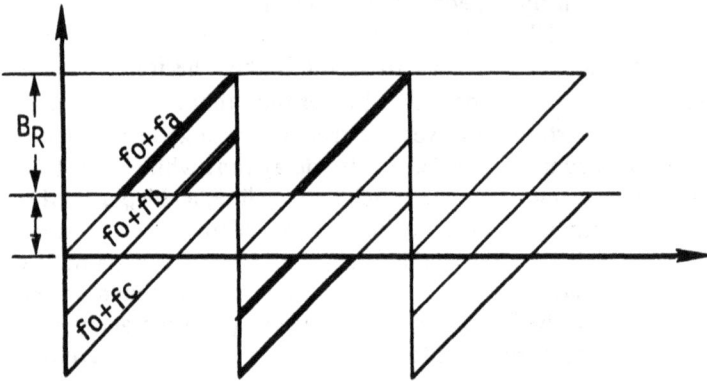

(c) Mixer output intercepted by IF band B_l

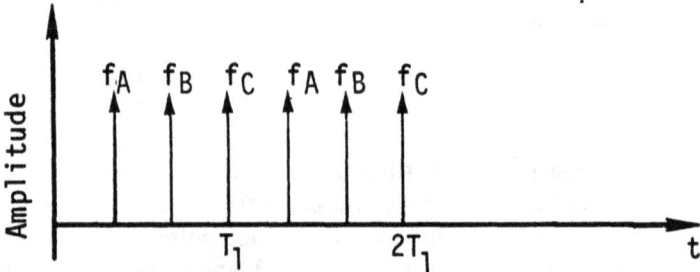

(d) Output pulses from DDL

Figure 6.4. Time and Frequency Relation of a Compressive Receiver with $B_R = 2B_l$

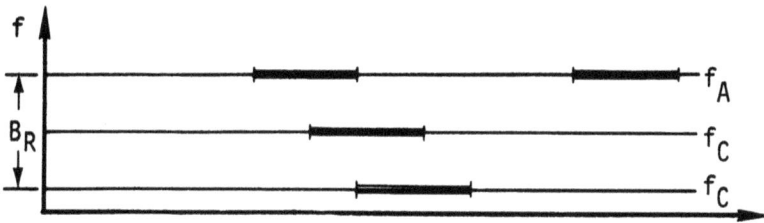

(a) Input Signal

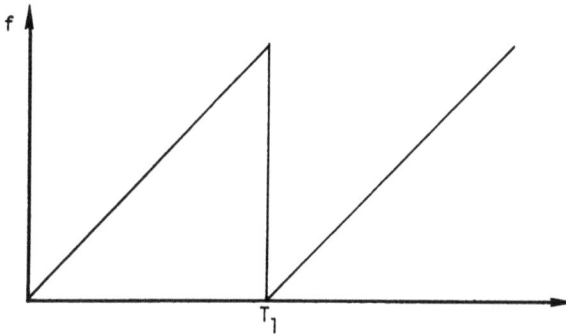

(b) Sweeping LO

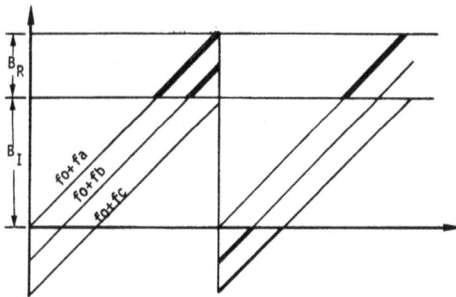

(c) Mixer outputs intercepted by IF band B₁

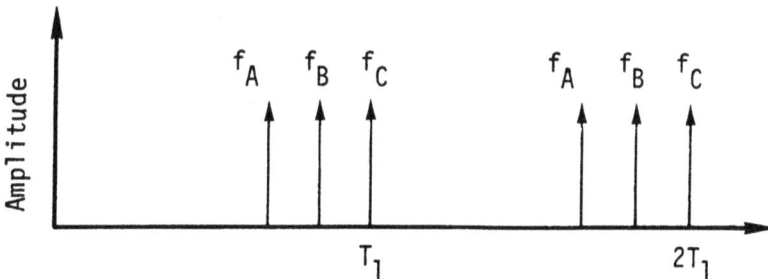

(d) Output pulses from DDL

Figure 6.5. Time and Frequency Relation of a Compressive Receiver with $B_R = \frac{1}{2}B_I$

function and enter the IF filter. Examining from the time domain, it is seen (Figures 6.3 through 6.6) that the only time the signal from the mixer enters the IF filter is in the time period T. Therefore, the FM pulsed signal, at the output of the DDL without the weighting filter, can be represented by (References 8 and 9)

$$S(t) = \cos(\omega_0 t + \frac{\mu t^2}{2}) \quad -\frac{T}{2} < t < \frac{T}{2} \tag{6.1}$$

$$= 0 \qquad\qquad\qquad \text{elsewhere}$$

where ω_0 is the angular frequency of the center of the DDL, t is time, and μ is the scan rate which can be expressed

$$\mu = \frac{2\pi B_I}{T} \tag{6.2}$$

Equation 6.1 can be written in exponential form as

$$S(t) = \exp[j(\omega_0 t + \frac{\mu t^2}{2})] \quad \text{for } -\frac{T}{2} < t < \frac{T}{2} \tag{6.3}$$

$$= 0 \qquad\qquad\qquad \text{elsewhere}$$

This form is easier to operate in equations with integrations.

The transfer function of the DDL can be expressed as (Reference 9)

$$H(\omega) = \exp\left[j\frac{(\omega - \omega_0)^2}{2\mu}\right] \tag{6.4}$$

The signal at the output of the delay line is given by

$$G(\omega) = H(\omega)\, S(\omega) \tag{6.5}$$

where $S(\omega)$ is the Fourier transform of $S(t)$ of Equation 6.3 and

$$S(\omega) = \int_{-\frac{T}{2}}^{\frac{T}{2}} S(t)\, \exp(-j\omega t)\, dt \tag{6.6}$$

where ω is the angular frequency.

To find the time domain output of the DDL, one needs to do an inverse Fourier transform of $G(\omega)$

$$g(t) = \frac{1}{2\pi} \int_{-\infty}^{\infty} G(\omega) \exp(j\omega t)\, d\omega \qquad (6.7)$$

by substituting Equations 6.3, 6.4, 6.5 and 6.6 into Equation 6.7, one obtains

$$g(t) = -\frac{1}{2\pi} \int_{-\infty}^{\infty} H(\omega) \left\{ \int_{-\frac{T}{2}}^{\frac{T}{2}} w(\tau) \exp\left[j(\omega_0\tau + \frac{\mu\tau^2}{2})\right] \right.$$

$$\left. \exp(-j\omega\tau)d\tau \right\} \exp(j\omega t)d\omega \qquad (6.8)$$

In Equation 6.8, $w(\tau)$ represents the weighting filter. The weighting filter should be a function of frequency. However, since after the mixer the output from a cw signal is a linear FM signal where the frequency is linearly proportional to time, the weighting filter can be written as a function of time. The order of integration of Equation 6.8 can be rearranged as

$$g(t) = \int_{-\frac{T}{2}}^{\frac{T}{2}} w(\tau) \exp\left[j(\omega_0\tau + \frac{\mu\tau^2}{2})\right]$$

$$\left\{ \int_{-\infty}^{\infty} H(\omega) \exp[j\omega(t - \tau)]d\omega \right\} d\tau \qquad (6.9)$$

The integration in the braces with the substitution of $H(\omega)$ can be written as

$$\int_{-\infty}^{\infty} H(\omega) \exp[j\omega(t - \tau)]d\omega$$

$$= \int_{-\infty}^{\infty} \exp\left[\frac{(\omega - \omega_0)^2}{2\mu}\right] \exp[j\omega(t - \tau)]d\omega$$

$$= \int_{-\infty}^{\infty} \exp\left\{ j\left[\frac{\omega^2}{2\mu} + \left(-\frac{\omega_0}{\mu} + t - \tau \right)\omega + \frac{\omega_0{}^2}{2\mu} \right] \right\} d\omega \quad (6.10)$$

Let

$$a^2 \equiv \frac{1}{2\mu} \qquad (6.11)$$

$$2ab \equiv \frac{-\omega_0 + \mu(t - \tau)}{\mu}$$

or

$$b \equiv \frac{-\omega_0 + \mu(t - \tau)}{2a\mu} = \frac{-\omega_0 + \mu(t - \tau)}{\sqrt{2\mu}} \qquad (6.12)$$

and

$$c \equiv \frac{\omega_0{}^2}{2\mu} \qquad (6.13)$$

After substituting Equations 6.11, 6.12, and 6.13 into Equation 6.10 and completing the square, Equation 6.10 can be rewritten as

$$\int_{-\infty}^{\infty} \exp\{j[(a\omega + b)^2 + c - b^2]\}d\omega$$

$$= \exp[j(b^2 - c)] \int_{-\infty}^{\infty} \exp[j(a\omega - b)^2]d\omega \qquad (6.14)$$

Using the relation

$$e^{jx} = \cos x + j\sin x \qquad (6.15)$$

and

$$\int_0^{\infty} \sin(a^2x^2)dx = \int_0^{\infty} \cos(a^2x^2)dx = \frac{\sqrt{\pi}}{2a\sqrt{2}} \qquad (6.16)$$

The integration in Equation 6.14 can be evaluated and the result is

$$\int_{-\infty}^{\infty} H(\omega) \exp[j\omega(t - \tau)]d\omega$$

$$= \sqrt{2\pi\mu} \exp\left\{ j[\omega_0(t - \tau) - \frac{\mu}{2}(t - \tau)^2 + \frac{\pi}{4}] \right\} \qquad (6.17)$$

After substituting Equation 6.17 into Equation 6.9, one can obtain

$$g(t) = \sqrt{\frac{\mu}{2\pi}}\exp[j(\omega_0 t - \frac{\mu t^2}{2} + \frac{\pi}{4})] \int_{-\frac{T}{2}}^{\frac{T}{2}} w(\tau)\exp(j\mu\tau t)d\tau \qquad (6.18)$$

where τ is a dummy variable.

The term $\exp[j(\omega_0 t - \mu t^2/2 + \mu/4)]$ represents the frequency of the output pulse. The center frequency of the output pulse from the DDL is at ω_0, the center frequency of the DDL, but the phase is shifted by $\pi/4$. The amplitude of the output signal can be obtained by integrating the function

$$R(t) = \int_{-\frac{T}{2}}^{\frac{T}{2}} w(\tau) \exp(j\mu t\tau)d\tau \qquad (6.19)$$

The constant $\sqrt{\mu/2\pi}$ is neglected for this discussion. Equation 6.19 states that the amplitude of the output pulse from the DDL is the Fourier transform of the weighting function. For the unweighted signal, $W(\tau) = 1$ and the integration of R(t) yields

$$R(t) = \frac{\sin\frac{\mu Tt}{2}}{\frac{\mu Tt}{2}}T \qquad (6.20)$$

This is the product of a sinc function (sinx/x) with the delay time T. If this output is detected by a log video detector, the output has many side lobes. The first side lobe is about 13 dB below the main lobe. The first side lobe can be found as

$$\frac{\sin x}{x} = 1 \qquad \text{for } x = 0 \qquad \text{the main lobe}$$

$$\frac{\sin x}{x} = \frac{-2}{3\pi} \qquad \text{for } x = \frac{3\pi}{2} \qquad \text{the first side lobe}$$

$$10 \log \frac{\left[\frac{\sin x}{x}\right]^2 \quad x = \frac{3\pi}{2}}{\left[\frac{\sin x}{x}\right]^2 \quad x = 0} = 10 \log \frac{4}{9\pi^2} = -13.5 \text{ dB} \qquad (6.21)$$

The rest of the side lobes decrease monotonically (see Figure 5.6 in Chapter 5). Therefore, for one input signal the compressive receiver will generate a string of pulses (or a pulse with many side lobes in the time domain) rather than a single pulse. To measure the frequency of the input signal, the position of the main lobe must be measured. The side lobes of the output pulse limit the two signal dynamic ranges of the compressive receiver, because they tend to mask signals whose amplitudes are less than the side lobe of another signal. For an unweighted DDL, the largest side lobe is 13 dB below the main lobe, which limits the dynamic range of the receiver to 13 dB.

The usual way of improving the dynamic range of a compressive receiver is to suppress the side lobes by adding a weighting filter before the DDL. In some cases (i.e., SAW DDLs), the delay line itself can be weighted. The generally used weighting is a cosine square on pedestal weighting filter which can be expressed as

$$W(t) = K + (1 - K)\cos^2\left(\frac{\pi t}{T}\right) \qquad (6.22)$$

where K is a constant between 0 and 1. Substituting W(t) in Equation 6.19 and carrying out the integration yields

$$R(t) = \frac{T \sin\frac{\mu T t}{2}}{2}\left[\frac{1 + K}{\frac{\mu T t}{2}} + \frac{(1 - K)\frac{\mu T t}{2}}{\pi^2 - \left(\frac{\mu T t}{2}\right)^2}\right] \qquad (6.23)$$

When K = 0.08, this weighting function is referred to as the Hamming weighting which will reduce the side lobes to 42.8 dB below the main lobe. The power spectrum output from a Hamming weighting filter is shown in Figure 6.6. Although the Hamming weighting suppresses the time side lobes, it broadens the main lobe. Broadening the main lobe decreases its measurement accuracy in the time domain which, in turn, decreases the frequency measurement accuracy of a compressive receiver. This is always a trade-off in compressive receiver design. There are many different weighting functions to reduce the side lobes of a receiver and improve the dynamic range.

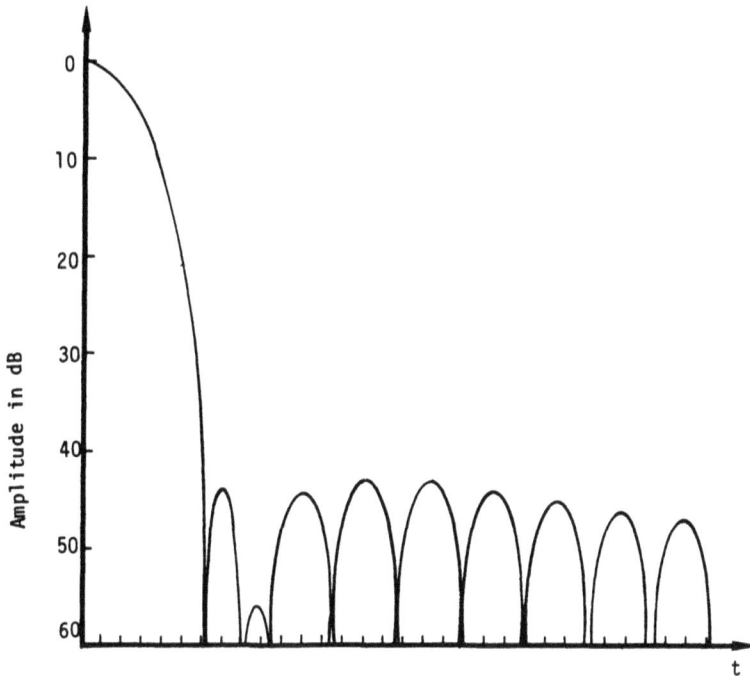

Figure 6.6. Output from a DDL with a Hamming Weighting Filter

6.4 BASIC EQUATIONS

In this section the relationship between the different parameters of the compressive receiver will be derived. It is helpful to have all these relationships listed together for quick reference. The scan rate and scan time will be discussed first.

If the bandwidth of the DDL in a compressive receiver is B_I and the dispersive delay time is T, then the sweeping LO must scan at a rate to match the frequency versus time slope of the DDL. The scan rate is

$$S = \frac{B_I}{T} = \frac{\mu}{2\pi} \qquad (6.24)$$

If the input frequency bandwidth of the receiver is B_R, as mentioned in Section 6.2, that LO must sweep

$$B_T = B_R + B_I \qquad (6.25)$$

Therefore, the total scan time is

$$T_1 = \frac{(B_R + B_I)T}{B_I} \qquad (6.26)$$

If the sweeping LO has a retrace time of T_R, then that total period is

$$T_1 = \frac{(B_R + B_I)T}{B_I} + T_R \qquad (6.27)$$

The frequency resolution of the receiver is determined by the width of the main lobe of the compressed pulse. From both Equations 6.20 and 6.23, the main lobe is proportional to $1/T$. In other words, the longer the delay time T, the narrower the main lobe; therefore, the finer the frequency resolution. One can simply write the frequency resolution (Δf) as

$$\Delta f = \frac{k}{T} \qquad (6.28)$$

where k is a proportionality constant. The value of k is usually greater than 1 and depends on the weighting function and the detection scheme.

The maximum number of output pulses from the DDL equals the total RF bandwidth divided by the frequency resolution. Thus,

$$N = \frac{B_R}{\Delta f} = \frac{B_R T}{k} \qquad (6.29)$$

which is proportional to the time bandwidth product of the DDL.

From Figures 6.3, 6.4, and 6.5, and the discussion in Section 6.2 that the compressed pulses come out of the DDL only part of the sweeping time T_1. If T_0 is the time period that the delay line can have output, then

$$T_0 = \frac{B_R}{B_I}T \qquad (6.30)$$

This relation is obvious by observing Figures 6.3, 6.4, and 6.5. The output pulse width from the delay line is equal to the time period T_0 divided by the total number of pulses N which can be expressed as

$$P = \frac{T_0}{N} \qquad (6.31)$$

Substituting Equations 6.29 and 6.30 into Equation 6.31

$$P = \frac{k}{B_I} \qquad (6.32)$$

Equation 6.32 reveals an interesting relationship; the wider the bandwidth of the DDL, the narrower the output pulse. The width of the pulse has an important impact on the video circuit following the detector. The video bandwidth must be wide enough to pass the narrow pulse. The relation in Equation 6.32 can also be derived from Equation 6.20. When $\mu Tt/2 = \pi$ or $t = 2\pi/\mu T$, $\sin(\mu Tt/2) = 0$ which labels the first zero of the output pulse in time domain. The width of the pulse can then be approximated as

$$P = \frac{2\pi}{\mu T}k \qquad (6.33)$$

Substituting Equation 6.24 into Equation 6.33

$$P = \frac{k}{B_I} \qquad (6.33)$$

The previous relationships are collected in Table 6.1.

Some important factors are included in these equations, and they will be discussed in the following paragraphs.

It is general practice to increase the time bandwidth product of a DDL to improve the receiver performance. Increasing dispersive delay time can improve the frequency resolution of the receiver. Wider bandwidth B_I will improve the probability of intercept which will be discussed in the next section. The dispersive delay time T shall be determined by the minimum pulse width that the receiver intended to receive. In other words, the frequency resolution of the receiver should be approximated by

$$\Delta f \cong \frac{1}{PW_{min}} \qquad (6.34)$$

TABLE 6.1. KEY PARAMETERS FOR COMPRESSIVE RECEIVER

RF bandwidth	B_R
Dispersive delay line bandwidth	B_I
Dispersive delay time	T
Oscillator Scan Rate	$S = \dfrac{B_I}{T}$
Oscillator Scan Width	$B_T = B_R + B_I$
Time to scan the total bandwidth (without retracing time)	$T_1 = \dfrac{B_R + B_I}{B_I} T$
Total scan period including retrace time T_R	$T_1 = \dfrac{B_R + B_I}{B_I} T + T_R$
Frequency resolution (k > 1)	$\Delta f = \dfrac{k}{T}$
Maximum number of pulses from delay line per scan	$N = \dfrac{B_R}{\Delta f} = \dfrac{B_R T}{k}$
Time period delay line has output per scan	$T_0 = \dfrac{B_R T}{B_I}$
Output pulse width from delay line	$P = \dfrac{k}{B_I}$

This relation is applicable to all receivers with a fine frequency channel to measure frequency. Superhet, channelized, compressive, and Bragg cell receivers belong in this category. From Equation 6.24 and 6.28, the delay time

$$T \cong PW_{min} \qquad (6.35)$$

This means that the minimum pulse width is proportional to the delay time.

However, if a long delay time T can be accomplished in the dispersive line, it may be helpful in the receiver design. This will be discussed in Section 6.9. A wider bandwidth B_I will require that the sweeping oscillator cover a wider frequency range. A wider B_I also generates very narrow output pulses from the delay line. For example, a delay line with 1 GHz bandwidth will generate approximately 1 ns pulses. These narrow pulses are very hard to handle by logic circuits. Any weighting filter added to widen the width of the pulse is equivalent to shortening the delay line delay time which, in turn, will degrade the frequency resolution of the receiver. Widening the compressed pulse is equivalent to increasing the value of k in Equations 6.28, 6.29, and 3.32. If the log video amplifier that followed the DDL does not have enough bandwidth, it will also increase the output pulse width.

In many cases, the bandwidth of the DDL is determined by how narrow a pulse the logic circuit can handle and the desired frequency resolution.

6.5 PROBABILITY OF INTERCEPT FOR COMPRESSIVE AND INTERLACE SCANNED RECEIVER

If a superhet receiver with a frequency resolution ΔF is used to scan over a wider RF bandwidth to improve the probability of intercept, the receiver must scan with a speed less than the impulse response time of the filter. As discussed in Chapter 3, the maximum scan rate equals $(\Delta F)^2$. Compared to a compressive receiver with the same frequency resolution, the scan rate of the compressive receiver is B_I/T, where T is approximately equal to $1/\Delta F$. Therefore, the compressive receiver can scan $B_I T$ times faster than the superhet receiver. In other words the compressive receiver has a probability of intercept better than a superhet receiver with the same frequency resolution by its time bandwidth product of the DDL. The actual probability of intercept of a compressive receiver is rather complicated to define. The probability of intercept is not only a function of the pulse width of the input signals but also depends on its relative time of arrival (TOA) with respect to the start of a scan. To discuss the probabilities of intercept, let us redraw part of Figure 6.3 with retrace time in Figure 6.7.

The probability of intercept for a short pulse is relatively easy to figure out. For a pulse with zero PW, the probability of intercept (POI) is determined by

$$POI = \frac{T}{T_1} = \frac{T\,B_I}{(B_R + B_I)\,T + T_R B_I} \tag{6.36}$$

In deriving at Equation 6.36, the relation in Equation 6.27 was used. The shorter the retrace time, the higher the probability of intercept. When $T_R = 0$, Equation 6.36 will be simplified as

$$POI = \frac{B_I}{B_R + B_I} \tag{6.37}$$

which means the wider the dispersive bandwidth, the better the probability of intercept for a fixed B_R. Although it does not make too much sense to

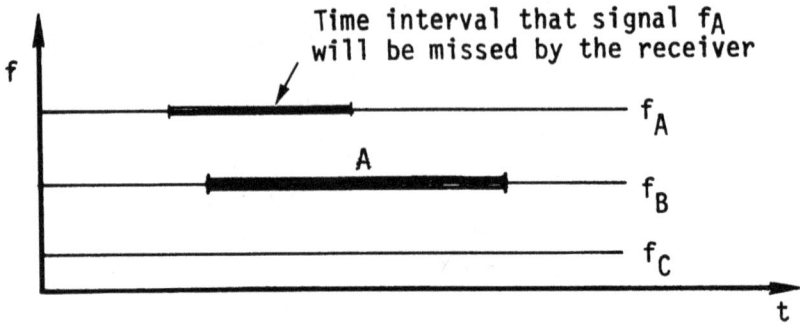

(a) Input signal

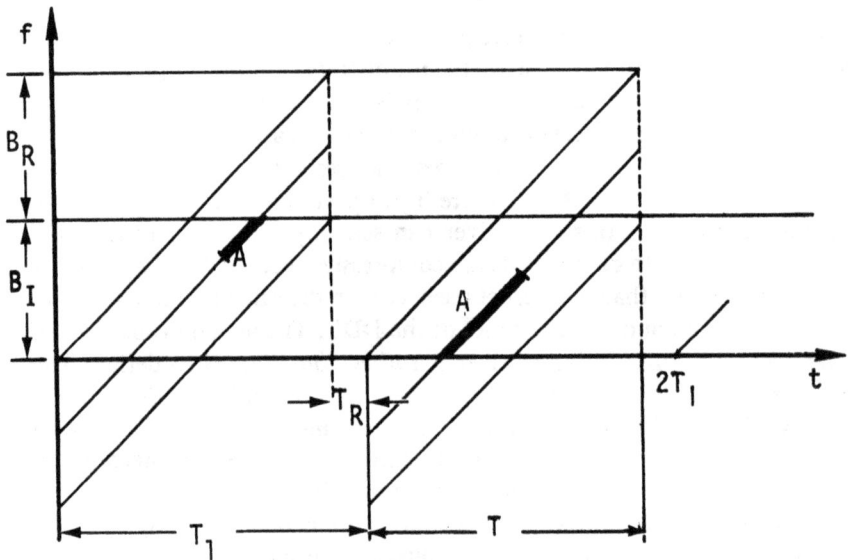

(b) Mixer outputs intercepted by IF band B_I

Figure 6.7. Time Frequency Relation of a Compressive Receiver with Retrace Time T_R

intercept a signal with zero pulse width, it gives the important result that increasing B_I will improve the probability of intercept.

As shown in Figure 6.7, pulsed signal A is intercepted by two consecutive scans. Since the signal does not exist in the dispersive line for a full time T (or as sometimes referred to, the signal that does not fill the line), the energy at the output of the delay line is decreased. Even though the signal is received, it is received with a lower sensitivity. If the signal is weak, it may not break the threshold of the receiver. If the signal only fills half the line, say T/2, the output energy is decreased by 3 dB. In addition to the loss of output energy, the energy in the output pulse is spread in time; therefore, the amplitude of the output pulse will be further decreased which may be equivalent to a total of 6 dB loss in sensitivity. From the above discussion, it is hard to define the probability of intercept. In order to clarify the definition, start with a proper definition. Assume that the signal has to fill the compressive line or the receiver will receive the signal with full sensitivity; then the signal is considered intercepted by the receiver. In order for a signal to fill at least one full scan under any input time interval, its minimum pulse width must be $T_1 + T$ (see Figure 6.7). Any pulse shorter than this value, the receiver may still intercept it, but with less sensitivity. In order to improve the probability of intercept on short pulses, a DDL with short delay time should be used. However, this will degrade the frequency resolution and the sensitivity. Therefore, a careful compromise must be reached.

From Figure 6.3, it has been noted that the delay line has output only over half of the scan period. The other half never has outputs. If another scan period is added between the first and the second one, the probability of intercept will be improved and the DDL will have outputs during the full scan period. This design is usually only applicable when $B_I = B_R$. This additional scan is referred to as the interlacing scan. The configuration of an interlacing scan is shown in Figure 6.8a. An additional set of LO and mixer are required, and their scan time must be properly controlled. In order to fill up the dispersive line, the minimum pulse width is $2T + T_R/2$ (Figure 6.8b) compared with $3T + T_R$ as required to fill up a line without interlacing scan. If the retrace time is zero for a signal to fill the DDL, the minimum pulse equals to $2T$. However, there is another point of view. A compressive receiver with interlace scan will always intercept the entire pulse. The pulse may be divided into consecutive scans. In this sense the compressive receiver can be considered with 100 percent probability of intercept. Although the interlacing scan improves the probability of intercept of a compressive receiver, more hardware is required and the number of output pulses is also doubled as shown in Figure 6.8. At the input, the signal is divided into two parallel paths so more amplification is required. The logic circuits following the detector must handle double the number of output pulses. These are some of the disadvantages of a compressive receiver with interlace scan.

6.6 DISPERSIVE DELAY LINE (COMPRESSIVE LINE)

The first component to choose in a compressive receiver is the DDL. The line has to have the desired IF bandwidth and the dispersive delay time. It can be an electromagnetic DDL (often referred to as the meander line) or a surface acoustic wave DDL as discussed in Chapter 9. The insertion loss of the dispersive line should be uniform across the band. However, due to a higher insertion loss at higher frequency, the higher frequency end has a relatively short delay time compared to that of the lower

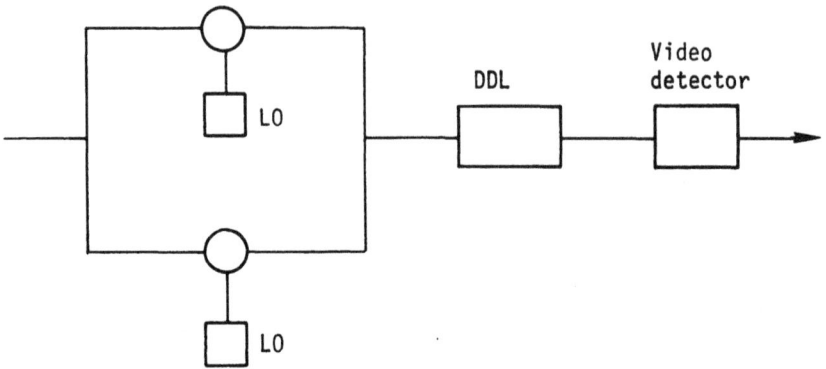

(a) Configuration of an Interlacing Scan

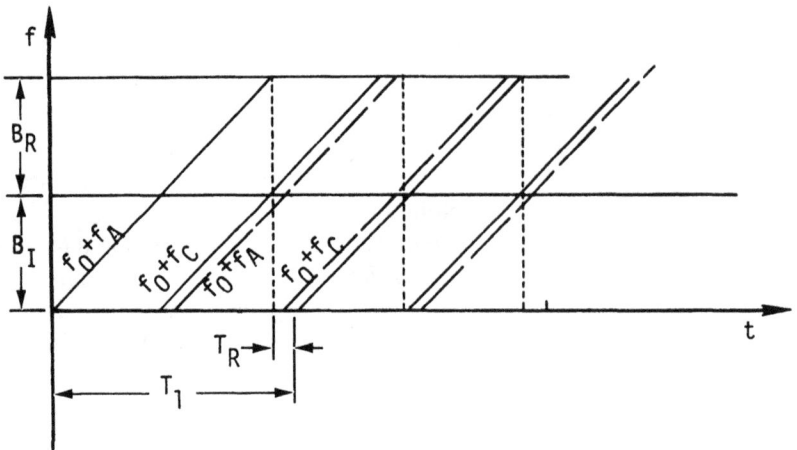

(b) Mixer outputs intercepted by IF band B_I

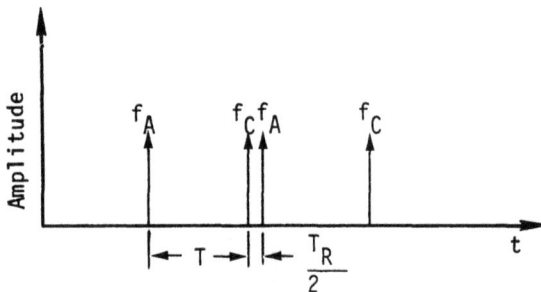

(c) Output pulses from DDL

Figure 6.8. Compressive Receiver with Interlace Scan (Retrace Time is
Included)

frequency end. Sometimes an equalizer is required to induce additional insertion loss at the lower frequency end to obtain the desired uniform insertion loss across the band. The dispersive delay time is chosen to generate the desired frequency resolution. The bandwidth of the line is chosen by the desired probability of intercept of the receiver, the RF bandwidth, and how narrow an output pulse the detection circuit can handle. The center frequency of the line can be arbitrarily chosen. In general, the dispersive line is less than an octave bandwidth due to manufacturing difficulties. If the dispersive line is more than an octave bandwidth, the second harmonic of the low frequency can enter the line. However, the second harmonic from the sweeping LO is sweeping at a frequency versus time rate double the desired rate; the output frequency from the dispersive line does not have the compressive effect. And the second harmonic is low compared to the desired output pulse. The frequency versus time slope of the dispersive line should be as linear as possible. Its performance can be measured through a network analyzer.

6.7 SWEEPING LOCAL OSCILLATOR (LO) [10]

The sweeping LO must sweep the range that equals the RF and IF bandwidths with the frequency versus time slope matching the dispersive line. However, there is no known easy technique to measure the performance of the LO. One of the possible schemes to evaluate the LO is to measure the output of the compressive line. The output of the compressive line with a perfectly matched sweeping LO can be theoretically calculated. Comparing the measured output pulse shape with the calculated one can provide some general idea of how well the sweeping LO matches the dispersive line.

The most common way to build a sweeping oscillator is by applying a ramp voltage on a voltage controlled oscillator (VCO). Usually the frequency versus voltage curve of a VCO is not linear. Therefore, applying a linear voltage on the VCO will not generate a linear FM signal. Linearizing circuits are designed to generate the proper voltage which in turn will generate a linear FM signal. One scheme to design a linearizing circuit is to measure the steady-state frequency versus voltage of the VCO. Then divide the frequency axis into many equal sections as shown in Figure 6.9. If the voltage applied to the VCO changes from V_1 to V_2, V_2 to V_3, etc. in equal time, the frequency output from the VCO will be nonlinear with respect to time t. A typical voltage versus time curve is shown in Figure 6.10, the voltage applied does not change linear with respect to time. In order to change the sweeping rate, the time interval in Figure 6.10 has to change accordingly. Theoretically, the voltage versus time curve can be stored in the memory circuit to generate the desired pattern. However, due to the short time required through the entire frequency band, sometimes resistor-capacitor charge/discharge curves (to match sweeping up/sweeping down) have to be used.

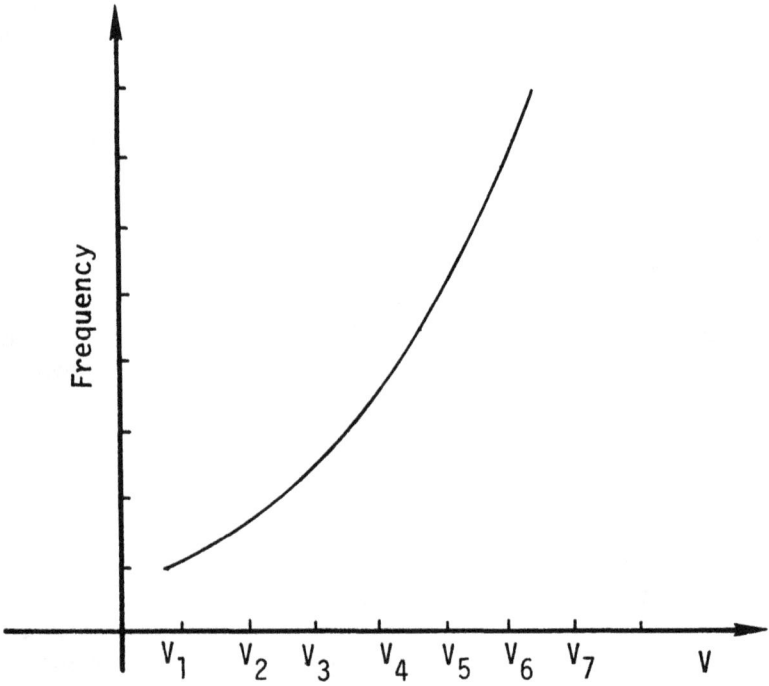

Figure 6.9. Frequency versus Voltage of a VCO

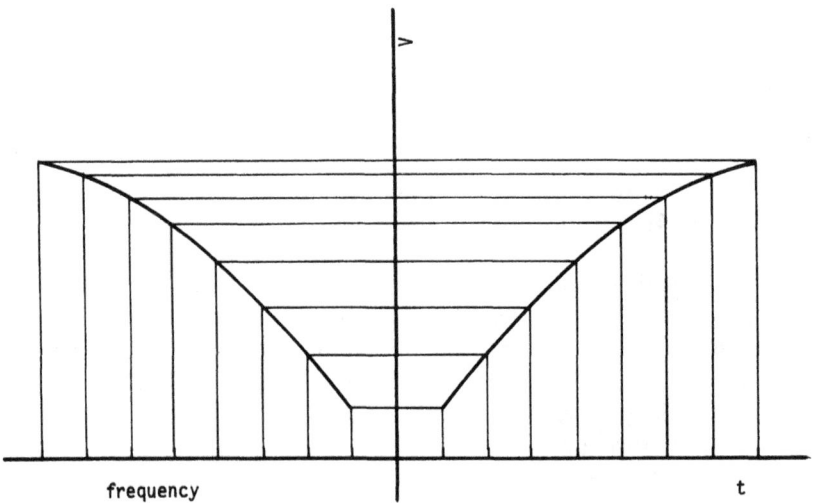

Figure 6.10. VCO Applied Voltage versus Time and Frequency

Another approach to produce the sweeping frequency (Reference 10) is to pulse a DDL. At the input of the DDL, an RF pulse with frequency at the center of the delay line is applied. Since the frequency domain representation of a short pulse is a wide frequency, the DDL will delay one end of the frequency with respect to the other end, and a sweeping frequency is generated. One practical problem with such an aproach is that the energy in the pulsed signal is relatively low and the FM signal generated at the output of the delay line is very weak. If the DDL of the same performance are used for both the compressive line and sweeping oscillator, the bandwidth of the sweeping LO must be doubled, because the oscillator scan widths are equal to the sum of the RF and IF bandwidths. One way to increase the sweeping oscillator bandwidth is to use two identical delay lines (Reference 2) as shown in Figure 6.11. Input signal with frequency (f_{IF}) at the center of the delay line is sent through two time gates which gate the input signal into narrow pulses. These two time gates gate out the pulses alternately at a time interval equal to the dispersive time of the individual line. At the outputs of the dispersive lines, frequency sweeps from f_1 and f_2 in time T intervals alternately as shown in Figure 6.11. Mixers M_1 and M_2 shift the frequencies f_1 and f_2 to the desired range. Frequency f_A and f_B are separated by $\Delta f = f_2 - f_1$, or $f_B - f_A = f_2 - f_1$. The output from the summing circuit is an FM signal chirp from $f_A + f_1$ to $f_B + f_2$ (or $f_A - f_1 + 2f_2$). The frequency sweeps a total bandwidth of $2(f_2 - f_1)$ in time 2T. Therefore, the chirp rate equals $(f_2 - f_1)/T$ and the

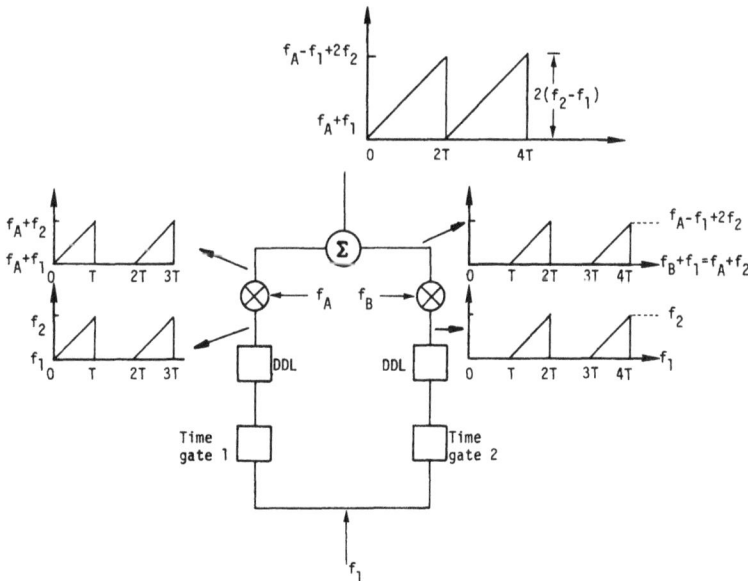

Figure 6.11. Sweeping Local Oscillator

bandwidth equals $2(f_2 - f_1)$ as required for the sweeping LO. In this arrangement there is no flyback time in the sweeping LO. However, at the transition time T, 3T, etc., it is very difficult to maintain a continuous phase. Another approach is to use a delay line of the same bandwidth of the dispersive line but with a dispersive delay time double that of the compressive line and use a frequency doubler to double the frequency bandwidth to the desired value. The phase discontinually can then be avoided in this scheme. Of course, there is always the straight forward approach which is using different DDLs for the LO and the compressive line, one with a bandwidth double the other one.

6.8 LOGARITHMIC AMPLIFIER AND VIDEO DETECTION

In order to measure the amplitude of the signals, a wide dynamic range is required. As mentioned in Chapter 13, a logarithmic amplifier using amplifiers and a detector to expand the dynamic range is a logical choice. If the signal can fill up the compressive line, then the amplitude of the output from the DDL represents the amplitude of the input signal. Since the outputs from the compressive lines are short pulses, the response time of the logarithmic amplifier has to be extremely fast. If the dispersive bandwidth is 500 MHz, the output pulse is about 2 ns in width. The logarithmic amplifier must have a comparable video bandwidth of 500 MHz, which will cause a severe design problem.

To determine the frequency of the input signal, as mentioned earlier in this chapter, the position of the output pulse from the compressive line must be measured. One of the simplest approaches is to compare it with a fixed threshold as shown in Figure 6.12. The shortcoming of this approach is that when the amplitude of the input signal increases, the base of the pulse becomes wider even though the side lobes of the output pulse in this arrangement are assumed to be properly suppressed. Therefore, for a

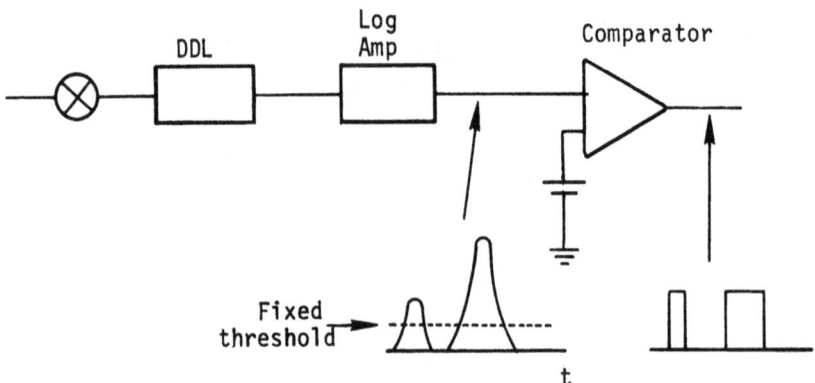

Figure 6.12. Simple Detection Circuit of a Compressive Receiver

strong signal, the position of the compressed pulse is difficult to locate accurately. Another approach is to compare the output pulse with many comparators with different thresholds as shown in Figure 6.13. The weak signal only crosses the threshold V_1 and generates a narrow pulse, while the strong one crosses all three thresholds and generates a narrow pulse from output of comparator 3. The logic circuit following the comparators should be designed to measure the pulse position from the highest threshold the pulse crossed to determine the input signal frequency. These comparators can also be used as a digitizer to generate digitized information for the signal amplitude.

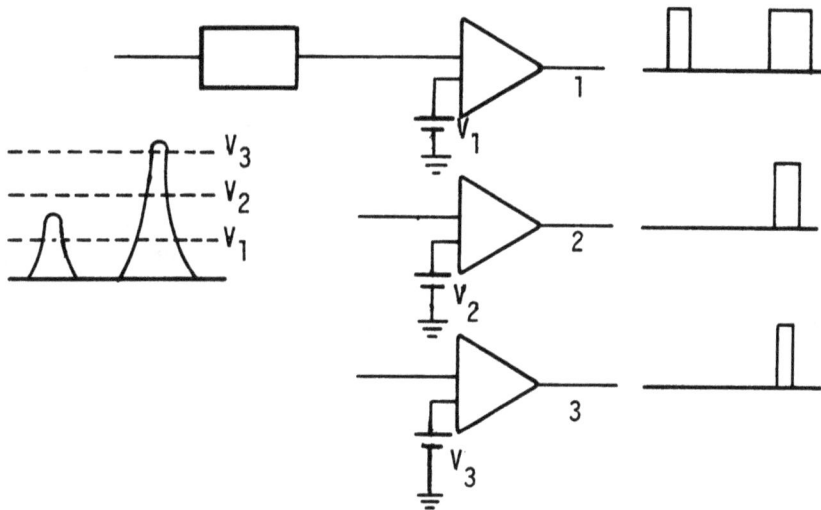

Figure 6.13. Compressive Receiver Detection Circuit with Multiple Thresholds

6.9 TRADING OFF TIME BANDWIDTH PRODUCT FOR WIDER COMPRESSED PULSE AND DUAL DETECTION SCHEME(11)

One of the primary difficulties in designing a compressive receiver is to process the narrow pulses emerging from the DDL. In order to cover a wide RF bandwidth with high probability of intercept, B_I must be wide which implies short compressed pulses from the DDL. One way to widen the compressed pulse but keep the B_I constant is to use a longer dispersive delay time. The best way to explain this approach is through an example. If the desired B_I is 1000 MHz and the frequency resolution is 10 MHz, a DDL with B_I = 1000 MHz and T = $1/\Delta f$ = 100 ns ($B_I T$ = 100) is needed. All these relations can be found in Table 6.1. The compressive pulse width

$P = 1$ ns and the maximum number of pulses per scan is 100. In this conventional approach, the video bandwidth will be 1000 MHz to pass the 1 ns pulse.

A different approach is to keep the same $B_I = 1000$ MHz and increase the bandwidth $T = 1000$ ns ($B_I T = 1000$). Theoretically, this approach can generate 1 MHz frequency resolution with 1 ns compressed pulse. The maximum number of outputs from the DDL per scan is 1000. If the video bandwidth after the detector is 100 MHz, then the compressed pulse will be expanded to 10 ns. Therefore, the maximum number of pulses per scan is 100 rather than 1000. The frequency resolution is 10 MHz rather than 1 MHz. The advantage of this approach is that the output pulse from the DDL has wider pulse width. The disadvantage is that the DDL with a high time bandwidth product is difficult to fabricate. The TOA resolution measured by receiver will be degraded also which will be discussed in the next section.

In order to differentiate the main lobe from the side lobes of the output of the DDL, a dual detection scheme similar to that used in channelized receivers (Section 5.7) can be used. In this approach the weighting filter can be put after the DDL. Since both the DDL and the weighting filter are both linear components, they can be interchanged without changing the performance of the circuit. Instead of one weighting filter, two parallel weighting filters with different weighting factors can be used as shown in Figure 6.14a. For the purpose of discussion, let us assume that both weighting filters are cosine square on pedestal (Equation 6.22) with different k values. For instance, $k = 0.06$ for the #1 filter and $k = 0.14$ for the #2 filter. The outputs from the two weighting filters, as shown in Figure 6.14b, are detected and fed into a comparator. The output from the comparator is shown in Figure 6.14c. Theoretically, this approach does not generate any side lobes from the output of the comparator which should simplify the following processing circuits. The disadvantages of this scheme is that two weighting filters and detection circuits are required, and the PW of the comparator output is wider than the main lobe thereby reducing the frequency resolution.

6.10 OTHER SIGNAL PARAMETER MEASUREMENT CAPABILITIES

Besides the frequency information, the compressive receiver can be used to measure other parameters of the input signals (i.e., pulse amplitude, pulse width, TOA, and angle of arrival). As discussed in former sections, the amplitude information of the signal is still retained after the DDL. Therefore, by measuring the amplitude of the compressed pulse, pulse amplitude information can be obtained. If the pulse to be measured is partially intercepted by the receiver, an error in pulse amplitude will occur. The measured value will be less than the true pulse amplitude. The pulse width and TOA resolutions are equal to one scan period. Although a

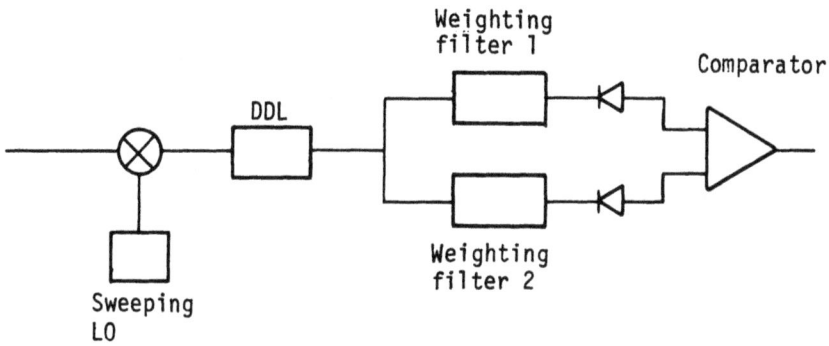

(a) Two Parallel Weighting Filters

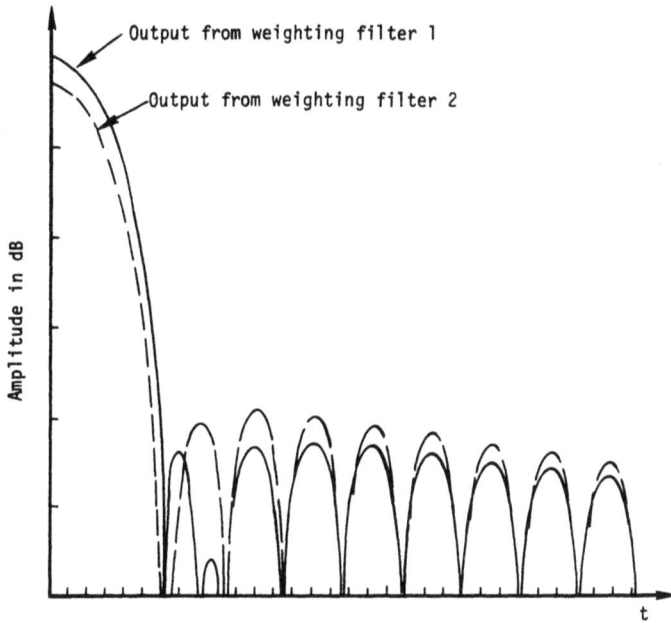

(b) Output from Weighting Filters

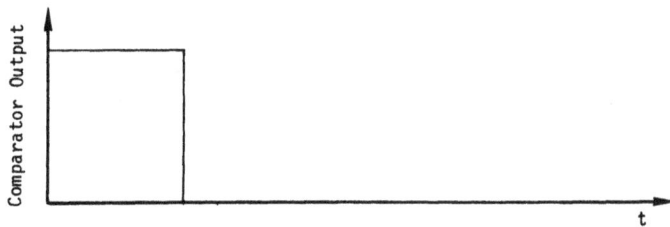

(c) Output from Comparator

Figure 6.14. Dual Detection Scheme for Compressive Receiver

longer delay time can generate fine frequency resolution, it will provide relatively poor time resolution. Theoretically, the outputs from two compressive receivers can be used to measure angle of arrival information of the input signal. From the discussion in Section 6.3, it is shown that not only is the amplitude information of the input signal included in the compressed pulse, but the phase is also retained. Thus, compressive receivers can be used either in an amplitude comparison or a phase comparison system. Because the output pulse from the DDL is extremely short, further research will be needed to build an angle of arrival measurement system by using compressive receivers.

We have already known that high-speed logic circuits are required to process the output of a DDL. The data comes out of the receiver at a very high rate. To make this problem even worse, if an input signal is intercepted by a compressive receiver in N consecutive scans, the same frequency and pulse amplitude information will be reported N consecutive times. These are redundant data which should be reduced. One approach is to compare the frequency information of one scan to the next. If they are the same, they can be considered as one signal and reported accordingly.

6.11 SUMMARY

A compressive receiver can be considered a channelized receiver with all information output in series rather than in parallel. The input bandwidth of the receiver can be very wide. To calculate the sensitivity of the receiver, the RF bandwidth used in the calculation is equal to the frequency resolution. Thus, this kind of receiver can provide high sensitivity. The dynamic range is limited by the side lobes of the compressed pulse from the DDL. Proper weighting should be able to surpress the side lobes more than 40 dB below the main lobe. Therefore, the dynamic range of the receiver should reach 40 dB. Interlace scan can improve the probability of intercept. The advance in SAW technology has a strong impact on the development of compressive receivers. A digital processor following the detector to process the narrow pulse requires developing work. The pulse width and TOA resolution of the receiver are rather coarse and equal to one scan period.

Compressive receivers are mainly in the developing stage. Presently, their primary application is to intercept communication signals in a dense signal environment. These receivers have relatively narrow RF bandwidth but very fine frequency resolutions. If a compressive receiver with a wider RF bandwidth and coarse resolution are developed, their applications should be able to extend to electronic warfare areas.

REFERENCES

1. White, W.D., "Signal translation apparatus utilizing dispersive network and the like, for panoramic reception, amplitude-controlling frequency response, signal frequency gating, frequency-time domain conversion, etc.," U.S. Patent 2, 954, 465, September 27, 1960.

2. Kincheloe, W.R., "The measurement of frequency with scanning spectrum analyzers," Systems Techniques Laboratory, Stanford Electronics Labs., Stanford, CA, Technical Report No. 557-2. Air Force Contract AF30(602)-2398, October 1962.

3. White, W.D., Saffitz, I.M., "Compressive receivers," Airborne Instruments Lab, A Division of Cutler-Hammer, Inc., Deer Park, L.I., New York, "Topics in Electronics," Vol. 3, 1962.

4. Sweet, R.G., Kincheloe, W.R., Jr., "A real-time scanning spectrum analyzer using a tapped sonic-delay line filter," Technical Report No. 1967-1, SU-SEL-64-058, Systems Techniques Laboratory, Stanford Electronics Laboratories, Stanford University, California, June 1964.

5. Sweet, R.G., Hewith, H., Kincheloe, W.R., Jr., "A high-resolution rapid-scan receiver and signal recorder for communications frequencies," Technical Report No. 1967-2, SU-SEL-66-111, Systems Techniques Laboratory, Stanford Electronics Laboratories, Stanford University, California, December 1966.

6. Harper, T., "New trends in EW receivers," Countermeasures, p. 33, December/January 1975, 1976.

7. Hoffman, C.B., Baron, A.R., "Wide-band ESM receiving systems," Microwave Journal, Part I, p. 24, September 1980; Part II, p. 57, February 1981.

8. Bernfeld, M., Cook, C.E, Paolillo, J., Palmieri, C.A., "Matched filtering, pulse compression and waveform design," Microwave Journal, Part I, pp. 57-64, October 1964; Part II, pp. 81-90, November 1964; Part III, pp. 70-76, December 1964; Part IV, pp. 73-81, January 1965.

9. Cook, C.E., Bernfeld, M., "Radar signals an introduction to theory and applications," Chapter 7, Academic Press, 1967.

10. Harrington, J.B., Nelson, R.B., "Compressive intercept receiver uses SAW devices," Microwave Journal, p. 57, September 1974.

11. Tsui, J.B.Y., "Dual detection scheme for compressive receivers," U.S. Patent 4, 200, 840, 29 April 1980.

CHAPTER 7
BRAGG CELL RECEIVERS (OPTICAL PROCESSORS)

7.1 INTRODUCTION[1-7, 28]

Optical processing, in general, means using light to analyze signals and perform special signal processing functions (i.e., correlation). A Bragg cell receiver is one kind of optical processor which performs a Fourier transform and obtains the frequency information on the input signals. Thus, the discussions in this chapter are limited to the operation and performance of a Bragg cell as a microwave receiver.

In a Bragg cell receiver, input electrical radio frequency RF signals are first transformed into spatial patterns which modulate a light beam. While many modulation techniques can be used in optical processors (e.g., thermo-plastic deformation and electro-optic modulation), Bragg cell receivers use acousto-optic modulation. In this case, the electric signal is converted into an acoustic wave which propagates through an optically transparent material (Bragg cell). Through the elasto-optic effect, the acoustic wave produces a spatial modulation of the refractive index in the Bragg cell. When a light wave is passed through the Bragg cell, the refractive index modulation (and hence the electric signal waveform) is impressed onto the optical wavefront as a spatial phase modulation. A suitable optical lens system is used to Fourier transform the modulated optical wavefront and the transformed signal is converted back to electrical form using a set of optical detectors. The photo detector output is the result of the optical processing.

Although the principles discussed here have been known for many years, it was only fairly recently that the requisite technologies have become sufficiently developed to make Bragg cell optical processors feasible. Of particular importance is the development of the laser since high brightness in the optical wave is essential to achieve satisfactory performance. The spatial coherence of lasers can be used to generate some special detection schemes to improve the dynamic range of the Bragg cell receiver. However, these special detections will not be discussed in this chapter. Other significant technological advances over the last decade include achievement of large (> 100) time-bandwidth products in Bragg cells and development of large one- and two-dimensional photo detector arrays. While much work remains to be done, especially in the areas of dynamic range and output rate of detector arrays, these recent developments in optical processor technology have encouraged the exploration of Bragg cell processors for microwave receiver applications.

The most attractive aspect of using the Bragg cell as a microwave receiver is its potentially extremely small size and low cost. Theoretically, a Bragg cell receiver can perform as a channelized receiver without the hundreds of filters required in a channelized receiver. The Bragg cell receiver can have a maximum time-bandwidth product of approximately 1000 which is equivalent to 1000 outputs in a channelized receiver. Furthermore, the development of integrated optical circuits (IOCs) makes the integration of the laser source, the Bragg cell transducer, output detector arrays, and the optical lens system on a single chip possible. An integrated optical Bragg cell receiver could have a volume as small as $0.1 \times 2 \times 6$ cm^3 (Reference 28).

In addition to the potential of mass production of Bragg receivers in IOC form, conventional Bragg cell receivers have been packed in compact form of about 10 in^3. These developments make the Bragg cell approach the most attractive electronic warfare (EW) receivers for airborne applications.

7.2 OPTICAL FOURIER TRANSFORM[5-13]

The operational theory of the Bragg cell is based on the Fourier transform property of an optical lens. As shown in Figure 7.1, a collimated light beam passes through an input plane with modulation mark $t(x, y)$ and then through a lens located at distance d behind the input plane. The light distribution in the focal plane represents the Fourier transform of $t(x, y)$. This relationship can be expressed by (Reference 9)

$$U(x_f, y_f) = \frac{A \exp\left[j\frac{k}{2F} (1 - \frac{d}{F})(x_f^2 + y_f^2) \right]}{j\lambda F}$$

$$\cdot \int \int_{-\infty}^{\infty} t(x, y) \exp\left[-j \frac{2\pi}{\lambda F} (xx_f + yy_f) \right] dxdy$$

(7.1)

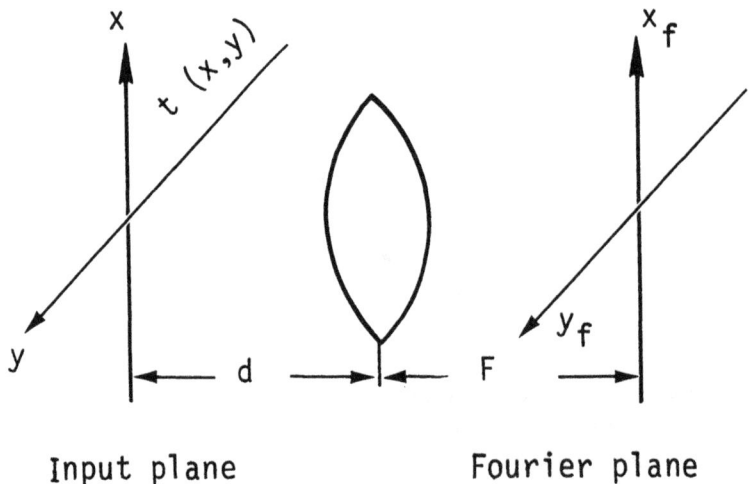

Figure 7.1 Basic Optical Fourier Transform

where x, y are coordinates on the input plane and x_f, y_f are coordinates on the Fourier plane, k is the wave number $= 2\pi/\lambda$, λ is the wavelength of the incident light, d is the distance between the object plane and the lens, F is the distance between the lens and the focal plane, A is the input light amplitude, t(x, y) is the modulation of the input plane, and $U(x_f, y_f)$ is the complex optical wave amplitude in the Fourier plane.

In Equation 7.1 the transformation differs from a true Fourier transform by the phase factor preceding the integral. But when F = d, the phase factor disappears and Equation 7.1 represents the exact Fourier transform. However, in the Bragg cell receiver application, only the power spectrum I = UU* (where U* is the complex conjugate of U) of the output is of interest; and the phase factor can be neglected. The power spectrum can be calculated as

$$I(x_f, y_f) = \frac{A^2}{\lambda^2 F^2} \left| \int \int_{-\infty}^{\infty} t(x, y) \exp\left[-j \frac{2\pi}{\lambda F} (xx_f + yy_f) \, dxdy \right] \right|^2$$

(7.2)

If standard photo detection techniques are used, it is I (x_f, y_f) which is detected in the Fourier plane.

If only RF frequency and amplitude are to be obtained in the Bragg cell receiver, then a one-dimensional implementation is appropriate. Taking the amplitude to be unity, the analog to Equation 7.1 for one-dimensional systems can be written as

$$U(x_f) = \int_{-\infty}^{\infty} t(x) \exp(-j2\pi f_x x) \, dx$$

(7.3)

where $f_x = x_f/\lambda F$.

For a sinusoidal amplitude grating of length l, t(x) can be expressed as

$$t(x) = \left(\frac{1}{2} + \frac{m}{2} \cos 2\pi fx \right) \text{rect} \frac{x}{l}$$

(7.4)

where m is the modulation depth, f is the spatial frequency of the grating (equal to the inverse of the grating spacing periodicity), and rect is the rectangle function defined as

$$\text{rect} \frac{x}{l} = 1 \qquad\qquad \text{when } |x| < \frac{l}{2}$$

$$= 0 \qquad\qquad \text{otherwise}$$

Substituting t(x) in Equation 7.3

$$U(f_x) = \int_{-\infty}^{\infty} \left(\frac{1}{2} + \frac{m}{2} \cos 2\pi fx \right) \text{rect}\frac{x}{l} \exp(-j2\pi f_x x)\, dx \qquad (7.5)$$

Equation 7.5 can be evaluated expediently.

Using the Fourier transform convolution theorem, which states that the Fourier transform of a product (ab) is equal to the convolution of the Fourier transforms "a" and "b". Defining F(a) as the Fourier transform of a, we have

$$F(ab) = A * B, \qquad (7.6)$$

where A and B are the Fourier transform of a, and b, respectively, and "*" represents convolution.

It is straight forward to show that

$$A = F\left\{ \frac{1}{2} + \frac{m}{2} \cos 2\pi fx \right\} = \frac{1}{2}\delta(f_x) + \frac{m}{4}\delta(f_x - f)$$
$$+ \frac{m}{4}\delta(f_x + f) \qquad (7.7)$$

where $\delta(fx)$ is the Dirac delta function [$\delta(x - x_0) = 0$ for $x \neq x_0$ and $\int g(x) \delta(x - x_0)\, dx = g(x_0)$], and that

$$B = F\left\{ \text{rect}\frac{x}{l} \right\} = l \text{ sinc}(lf_x) \qquad (7.8)$$

where

$$\text{sinc } lf_x = \frac{\sin \pi l f_x}{\pi l f_x}$$

The convolution of A and B is then

$$A * B = \frac{l}{2}\left[\text{sinc } lf_x + \frac{m}{2}\text{sinc } l(f_x - f) + \frac{m}{2}\text{sinc } l(f_x + f) \right] \qquad (7.9)$$

Therefore, the output in the Fourier plane with arbitrary amplitude k is

$$U(x_f) = K\left[\text{sinc}\frac{\ell x_f}{\lambda F} + \frac{m}{2}\text{sinc }\ell\left(\frac{x_f}{\lambda F} - f\right) + \frac{m}{2}\text{sinc }\ell\left(\frac{x_f}{\lambda F} + f\right)\right]$$

$$(7.10)$$

The output optical wave amplitude consists of three sinc functions as depicted in Figure 7.2.

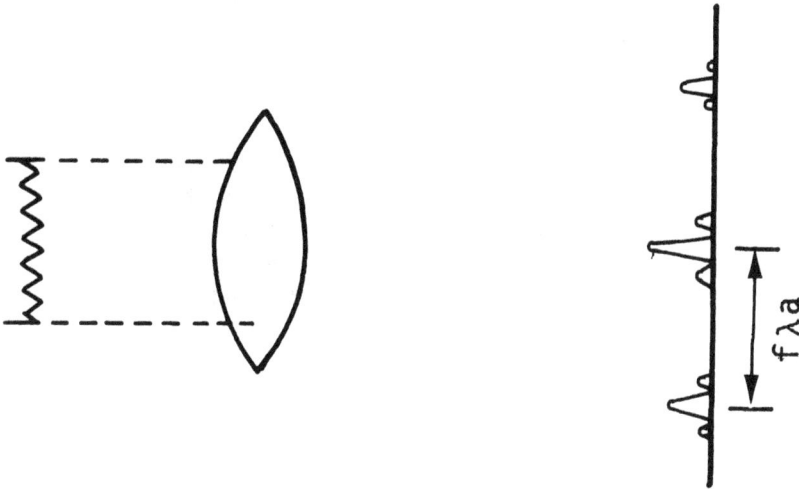

Figure 7.2. Fourier Transform of a Rectangular Window with Sinusoidal Grating

The light intensity on the detector is proportional to square of $U(x_f)$ or

$$I(x_f) = |U(x_f)|^2 \qquad (7.11)$$

At the center of the Fourier plane is the zero order output, while the first order outputs are located at positions

$$x_f = \pm f\lambda F \qquad (7.12)$$

The position of the first order outputs are proportional to the input frequency f when λ and a are fixed. By measuring the light spot position, the input frequency f can be obtained. The shape of the light spot is the square of a sinc function which is the square of the Fourier transform of the input window.

7.3 BRAGG DIFFRACTION

The results of Section 7.2 are valid for thin sinusoidal amplitude and phase gratings. However, in most Bragg cell application, the acoustic wave (and hence the grating) extends over a significant distance along the light propagation path. In this case, the modulation mask must be regarded as a thick grating. An important consequence of this fact is that the incidence angle of the optical wave onto the acoustic wavefronts must be appropriately adjusted to achieve maximum energy transfer to the first order.

The important parameters for describing the Bragg cell operation are illustrated in Figure 7.3. A sinusoidal acoustic wave of frequency f and corresponding acoustic wavelength Λ propagates with velocity V_s along the X direction in an optically isotrapic Bragg cell of refractive index n. A collimated optical wave of free space wavelength λ impinges on the sound field at angle θ_i from the z-axis, as measured in the Bragg cell. The width of the acoustic wave along the z-axis, and hence the resultant elasto-optically formed phase grating thickness, is L. Analysis of Maxwell's equations shows that two optical waves emerge from the cell, one being the transmitted incident beam and the other being a first order diffracted beam at angle θ_d from the z-axis as measured in the Bragg cell. In order for energy transfer to the first order beam to be maximum, it is necessary that

$$\theta_i = \sin^{-1}\left(\frac{\lambda f}{2nV_s}\right) \tag{7.13}$$

and that

$$\theta_d = \theta_i \tag{7.14}$$

By analogy to X-ray diffraction in crystals, Equations 7.13 and 7.14 are called the Bragg conditions, and the incidence angle θ_i satisfying Equation 7.13 is called the Bragg angle. In the case of Bragg cell optical processors, $\theta_i \ll 0.1$ rad and Equation 7.14 can be approximated as

$$\theta_i = \frac{\lambda f}{2nV_s} \tag{7.15}$$

In terms of the external incidence angle, θ_i', one can write

$$\theta_i' = \frac{\lambda f}{2V_s} \tag{7.16}$$

and

$$\theta_d' = \frac{\lambda f}{2V_s} \tag{7.17}$$

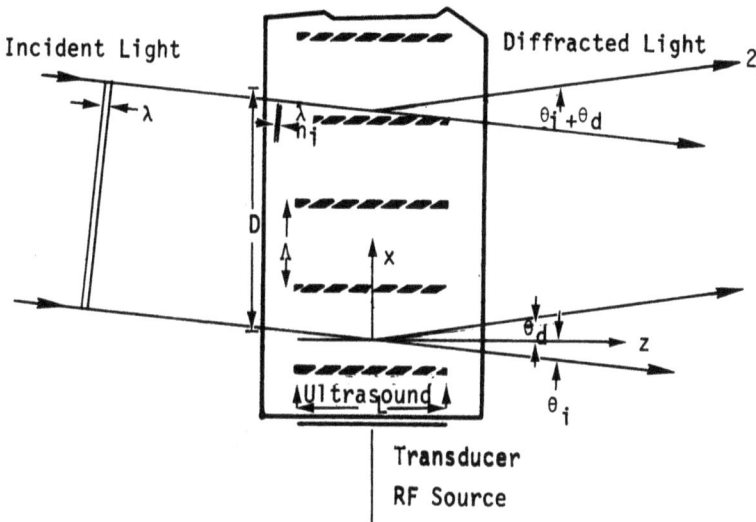

Figure 7.3. Traveling Wave Acousto-Optic Modulator (Courtesy of Air Force Wright Aeronautical Laboratories)

The optical wave exiting the Bragg cell is composed of two parts, one whose phase is identical with that of the incident wave and one whose phase is linearly modulated over the X-axis extent of the Bragg cell. The Fourier transform lens then yields an optical amplitude distribution in the focal plane consisting of two sinc functions. The displacement of the first order spot from the undeflected spot is given by

$$X_f = F(\theta_i' + \theta_d) = \frac{F\lambda f}{V_s} \tag{7.18}$$

and is seen to be proportional to the input acoustic frequency f and focal length F.

The Bragg conditions can be obtained without using Maxwell's equations by the following argument. The acoustic wave is represented as a set of equally spaced partially reflecting mirrors as depicted in Figure 7.4. In this case θ_i will equal θ_d and the deflected wave will be a superposition of waves reflected from the series of mirrors. The deflected wave will have maximum strength when the waves reflected from each mirror are all in phase. This will occur when the optical path difference between waves reflected from successive mirrors is one optical wavelength, i.e., $2\Lambda \sin \theta_i = \lambda$. Using $\Lambda = V_s/f$, we find $\theta_i = \sin^{-1}(\frac{\lambda f}{2V_s})$ as before.

The undeflected light from the Bragg cell is called the zero order output and does not contribute any information to the optical processor, but the light scattered from the zero order will increase the background noise of the Bragg receiver and degrade its sensitivity. The zero order scattered light is usually properly absorbed.

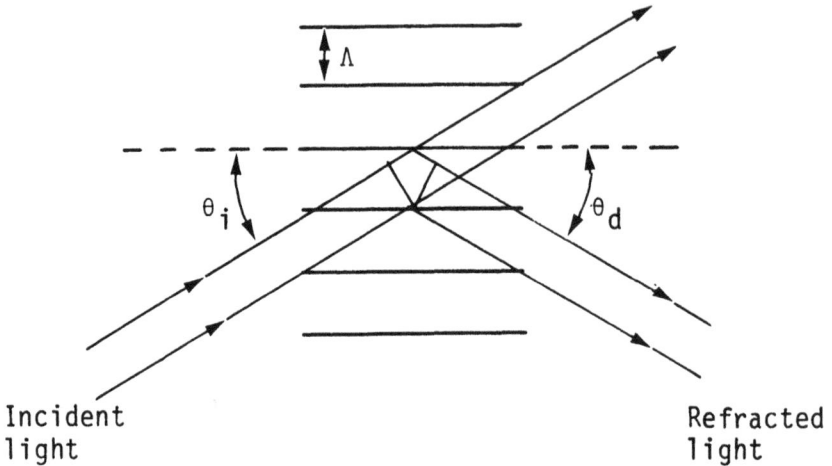

Figure 7.4. Reconstruction Geometry

Equations 7.13 and 7.14 show that if the acoustic frequency f changes, both the incident and the deflected angles should change accordingly. However, this situation is undesirable in a practical Bragg cell receiver, because it is not possible to appropriately steer the input optical beam without a prior knowledge of the RF frequency.

In practice, the incident optical angle is fixed at the angle which fulfills Equation 7.13 at the center of the operational band. In this case the deflected beam will change angle according to the acoustic frequency, while the incident angle stays the same. This is true because natural diffraction spreading of the acoustic waves automatically provides a range of incidence angles for a fixed optical beam. This is conveniently illustrated using the following conservation of momentum argument. The momentum analog of the Bragg conditions is that the incident optical wave vector $\vec{k_i}$, the deflected optical wave vector $\vec{k_d}$ and the acoustic wave vector $\vec{k_a}$ are all related according to

$$\vec{k_i} + \vec{k_a} = \vec{k_d} \qquad (7.19)$$

Since the thickness of the Bragg cell is limited (dimension L in Figure 7.3), the acoustic wave spreads and $\vec{k_a}$ has a distribution of values as shown in Figure 7.5. Therefore, there is a range of $\vec{k_d}$ values that can fulfill

Equation 7.19. In other words, when θ_i is fixed, θ_d changes with the acoustic frequency f over the range of $\vec{k}_a$ values in the acoustic wave. The limit to the range of $\vec{k}_a$ is populated imposes a frequency bandwidth limitation on a Bragg cell receiver.

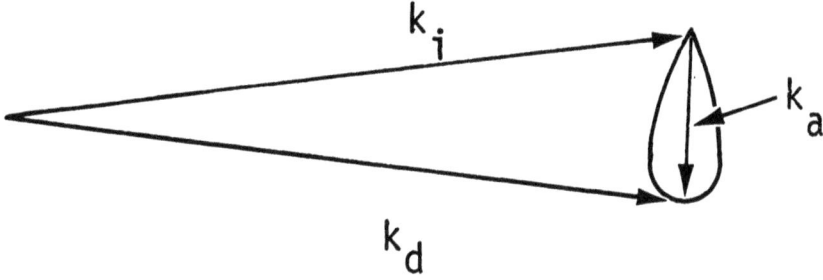

Figure 7.5. Wave Vectors of the Incident, Acoustic, and Deflected Wave

Before leaving this discussion, it should be noted that if two (or more) RF frequencies are input to the Bragg cell, there will be two (or more) deflected beams with each deflection as given in Equation 7.18.

7.4 BASIC BRAGG CELLS(14-19, 32, 33)

A Bragg cell is made of a light transmitting crystal with an input transducer as shown in Figure 7.6. The input transducer changes with the input RF signal into an acoustic wave. The acoustic wave in the crystal, in turn, changes the index of refraction of the crystal which deflects the laser beam. The thickness of the Bragg cell along the direction of the light propagation is about 10 mm. The transducer is a piece of piezoelectric material which is sandwiched between two conductors and then bonded on one end of the Bragg cell. The input signal is applied to the two conductors through an impedance matching network. The thickness of the transducer is dependent on the operating frequency of the cell. For an input frequency of 1 GHz, the thickness is around 1 μm. The electric field applied across the piezoelectric material will create a mechanical vibration which will couple to the crystal. At the opposite end of the Bragg cell crystal, there is absorbing material which will prevent the acoustic wave from being reflected back. Thus, in the Bragg cell, there is traveling acoustic wave instead of a standing wave.

The piezoelectric material used in the transducer can be any one of several materials. The most commonly used ones are lithium niobate (LiNbO$_3$) and zinc oxide (ZnO). The conducting material which is deposited on the piezoelectric material can be either gold or aluminum. The input impedance of the transducer is rather low, and the matching network is usually used to match the impedance to 50Ω for maximum power transfer efficiency. Spoiling the reflection at the end of the Bragg cell can be accomplished by just roughing the material itself or cutting the end of the Bragg cell at an angle so that the reflected acoustic wave will not reflect in the same direction. A Bragg cell with matching network is shown in Figure 7.7.

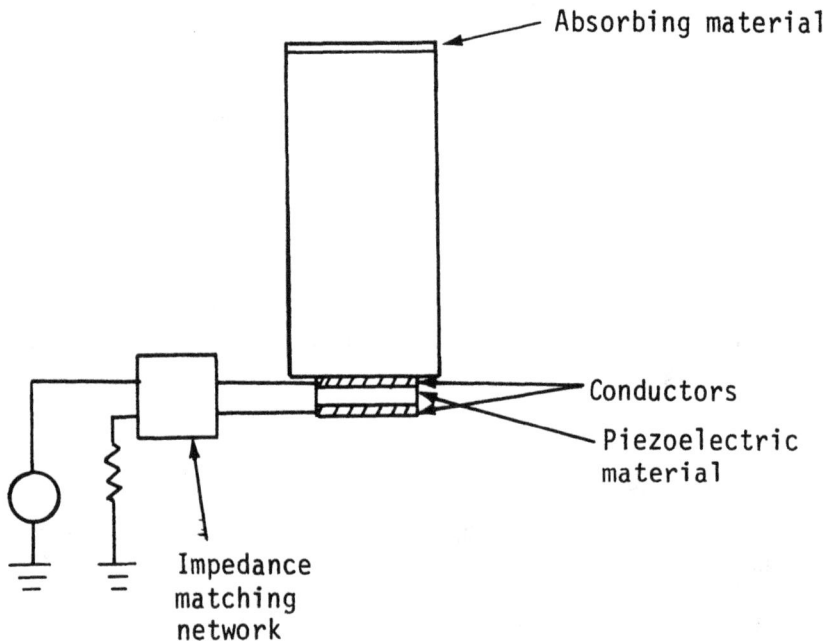

Figure 7.6. Bragg Cell Configuration

One basic physical requirement necessary for the Bragg cell to work is that the thickness of the Bragg cell L (along the light direction), the acoustic wavelength Λ, the light wavelength λ, and the index of refraction n are related approximately by (References 18, 32, and 33)

$$L \geq \frac{2n\Lambda^2}{\lambda} \qquad (7.20)$$

The acoustic velocity in solids is typically about 6×10^3 m/sec. Then for a 1 GHz input signal the acoustic wavelength is approximately 6 μm with light wavelength of 0.6328 μm (HeNe laser) and a value of n = 3.31 (GaP material), the thickness of the Bragg cell should be greater than 0.38 mm.

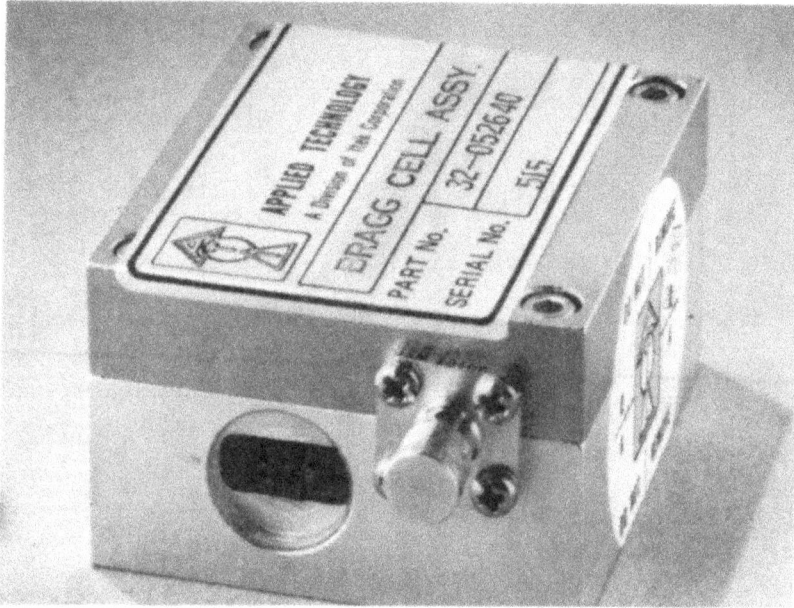

Figure 7.7. Bragg Cell (Courtesy of Applied Technology, Division of Itek)

The time-bandwidth products of some materials used to make Bragg cells are shown in Figure 7.8. The frequency bandwidth is shown on the vertical scale and the time delay τ is shown on the horizontal axis. The time delay is related to the frequency resolution through

$$\Delta f = \frac{k}{\tau} \tag{7.21}$$

where k is a constant and has a value of $1.2 \sim 2$.

The maximum time bandwidth product of the Bragg cell is generated by reading the bandwidth on the vertical axis and the time delay on the horizontal axis. The time bandwidth product at low frequency is limited by the size of the crystal which determines the transient time of the acoustic wave. At high frequency, the time bandwidth product is limited by the acoustic loss in the crystal.

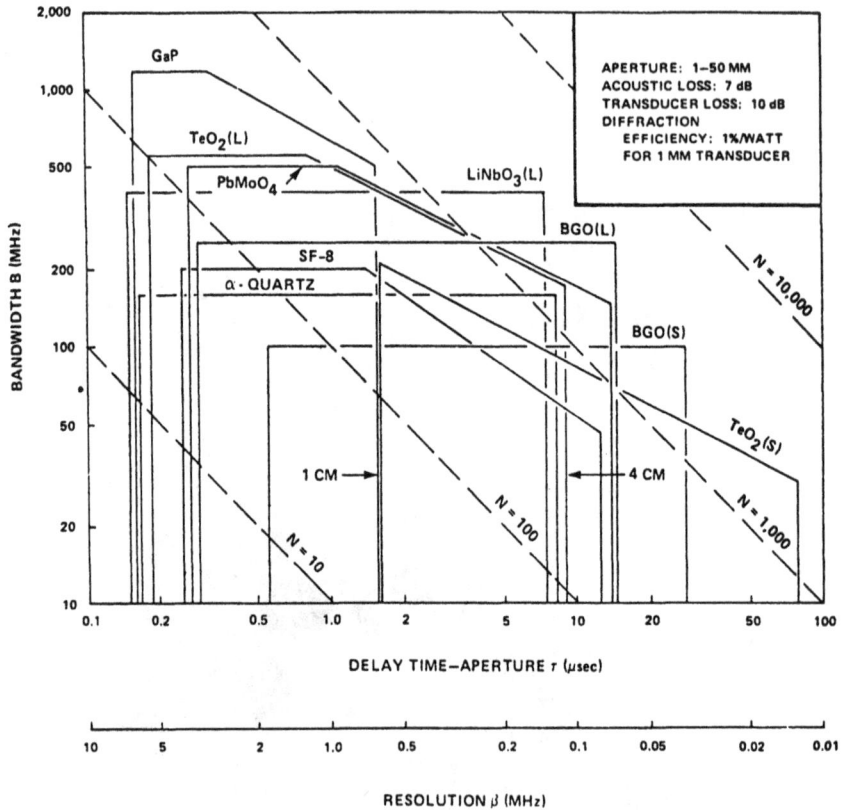

Figure 7.8. Bandwidth-Resolution Contours (Courtesy of Applied Technology, Division of Itek)

The material constants shown in Figure 7.8 are also functions of the light wavelength. For example, GaP has a very wide time-bandwidth product and can work at high RF frequencies with HeNe and solid-state diode lasers. It does not work with an Argon laser, because GaP does not transmit blue light. The surfaces of the Bragg cell should be properly coated to reduce the mismatch reflection losses at the surfaces. Mismatch at the surfaces will increase the light transmission loss through the Bragg cell and sometimes may even cause multiple reflection problems which may produce spurious outputs.

A. Bragg Cell Efficiency

Besides the time-bandwidth product, the efficiency of a Bragg cell is also a very important parameter. The efficiency is defined as the ratio of the diffracted light power to the input light power. The diffracted light power is not only dependent on the Bragg parameters cell but also on the

RF input power. Therefore, the common way of defining the efficiency of a Bragg cell is percentage per watt. For example, an efficiency of 30 percent per watt means that the Bragg cell will diffract 30 percent of the input light beam with a 1 watt RF input power. An efficiency of 200 percent per watt means that the Bragg cell will diffract all of the light with 0.5 watt RF input power. The Bragg cell output is linear in RF power only for actual operating efficiencies of about 10 percent or less. Above that the diffraction process becomes highly nonlinear. This limits the linear dynamic range of the cell. In general, the higher the efficiency of the cell, the lower the dynamic range. Compromise between the efficiency and the dynamic range of a Bragg cell has to be carefully evaluated.

B. Bandwidth of Bragg Cell[10]

As discussed in Section 7.3, the Bragg angle cannot be exactly matched for all frequencies over a wide-band. Efficiency is determined for each frequency according to the ultrasonic transducer angular radiation pattern intensity (W) in the direction required for the Bragg condition. This limits interaction length and corresponding efficiency for a given bandwidth. For a simple uniform transducer, the angular spectrum is the $(sinc)^2$ function. Then for normal diffraction

$$W(F,F_m) = sinc^2 \left[\frac{1}{2}\left(\frac{L}{L_O}\right)(FF_m - F^2) \right] \qquad (7.22)$$

where F is the frequency normalized to midband frequency and F_m is the frequency at which the Bragg angle is matched. L_O is the acousto-optic characteristic length at the midband frequency:

$$L_O = \frac{\Lambda^2 n}{\lambda \cos\theta_i} \qquad (7.23)$$

Normal diffraction bandshapes are shown in Figure 7.9.

7.5 WEIGHTING EFFECT IN THE BRAGG CELL[10]

The position of the light spot is directly proportional to the RF input of the Bragg cell as shown in Equation 7.18. One can expect that using a one-dimensional detector array to measure the position of the light spot permits determination of the input signal frequency. However, by examining the Fourier transform carefully, one can see that the light spot on the Fourier plane is not a simple spot. It is the Fourier transform of the Bragg cell shape as shown in Equation 7.3. If the window is a one-dimensional rectangular function, then the output at the Fourier plane is a sinc function. This means that the light spot will have a main lobe and many side lobes. If only one signal is measured by the Bragg cell receiver, the main lobe or the center of the many side lobes can be used to represent the position of

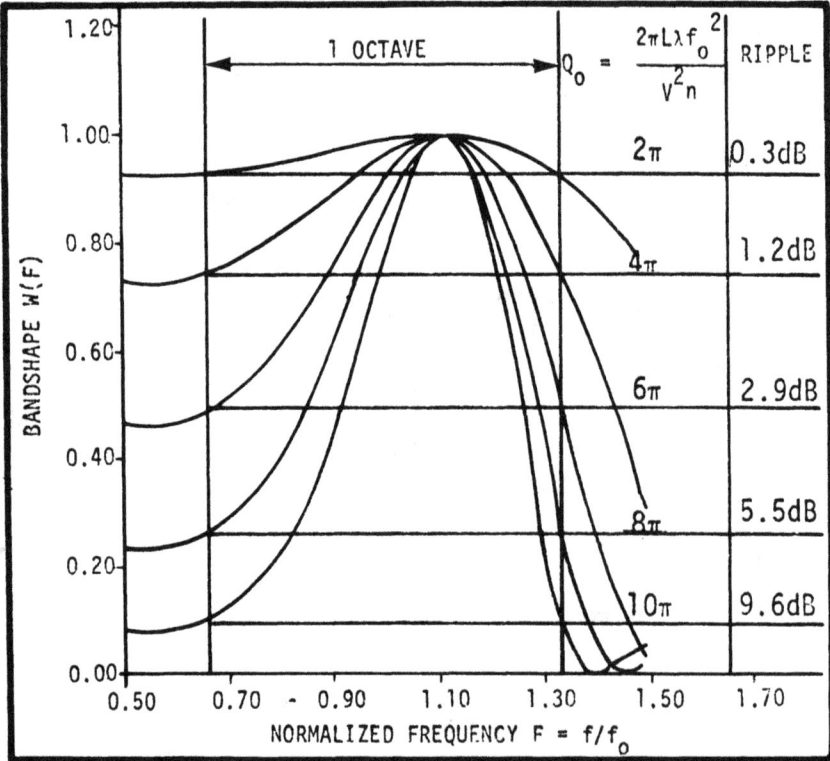

Figure 7.9. Acousto-Optic Band Shapes for Normal Matching with Octave Bandwidth (Based on Hecht, Reference 10)

the light spot. If the receiver has to receive multiple signals, it is necessary to distinguish a side lobe from another signal. The complexity of the problem is increased when high dynamic range is required from the receiver. The same physical problem happens in a channelized receiver and compressive receivers as discussed in Chapters 5 and 6. Of course, there are many solutions to this problem. The simplest solution seems to be adding a proper weighting function to the Fourier transform to reduce the side lobes. As mentioned in Chapter 6, weighting filters can be added in a compressive receiver to reduce the side lobes. The weighting filter in a Bragg cell receiver is spatial and has been referred to as apodization. It can be an absorbing material with a desirable spatial attenuation distribution installed in front or behind the Bragg cell. This arrangement changes the spatial light intensity of the laser beam which can be used to convolve with the Bragg cell window to reduce the side lobes at the Fourier plane. In addition, there are a number of naturally occurring weighting functions. To see how they affect the spot size and its side lobe, the following discussion is presented.

Equation 7.3 is rewritten here as

$$U(x_f) = \int_{-\infty}^{\infty} t(x) \exp(-j2\pi f_x x)\, dx \qquad (7.3)$$

where $t(x)$ represents the composite natural weighting function composed of the window function, the acoustic amplitude attenuation function and the optical beam profile. The "window" function $t_1(x)$ can be rewritten as

$$t_1(x) = \text{rect}\left(\frac{x}{D} - \frac{1}{2}\right) \qquad (7.24)$$

where D is the height of the Bragg cell (Figure 7.6). The window is extended from $X = 0$ to $X = D$. The acoustic amplitude attentuation function $t_2(x)$ is

$$t_2(x) = \exp(-\alpha x) \qquad (7.25)$$

where α is a frequency dependent loss factor in nepers/seconds. The optical beam amplitude profile $t_3(x, T)$

$$t_3(x, T) = \exp\left[-4T^2\left(\frac{x}{D} - \frac{1}{2}\right)^2\right] \qquad (7.26)$$

where $T = D/2W_0$, and W_0 is the half width of the laser beam at $1/e^2$ of the center intensity. $T = 0$ represents uniform beam. The combined weighting function is the product $t_1 t_2 t_3$:

$$t(x) = \exp\left[-\alpha x - 4T^2\left(\frac{x}{D} - \frac{1}{2}\right)^2\right]\text{rect}\left[\frac{x}{D} - \frac{1}{2}\right] \qquad (7.27)$$

Substituting Equation 7.27 into 7.3, the Fourier transform can be obtained from numerical integration through a computer. A typical spectrum is shown in Figure 7.10 where $T = 1.5$ and $\alpha = 0$. As expected, the side lobes are reduced in amplitude, and the main lobe is broadened. All side lobes are lower than about 34 dB relative to the central peak, and the side lobe peaks drop-off at the expected rate of 20 dB/decade (or 6 dB/octave). The slight rise of side lobes at high frequency (>8) shown in Figure 7.10 is due to the computation and is not a physical effect.

7.6 PHOTO DETECTORS[20-24]

A. Size and Number of Photo Detector Requirements

In a Bragg cell receiver, the photo detectors should be small in size, have

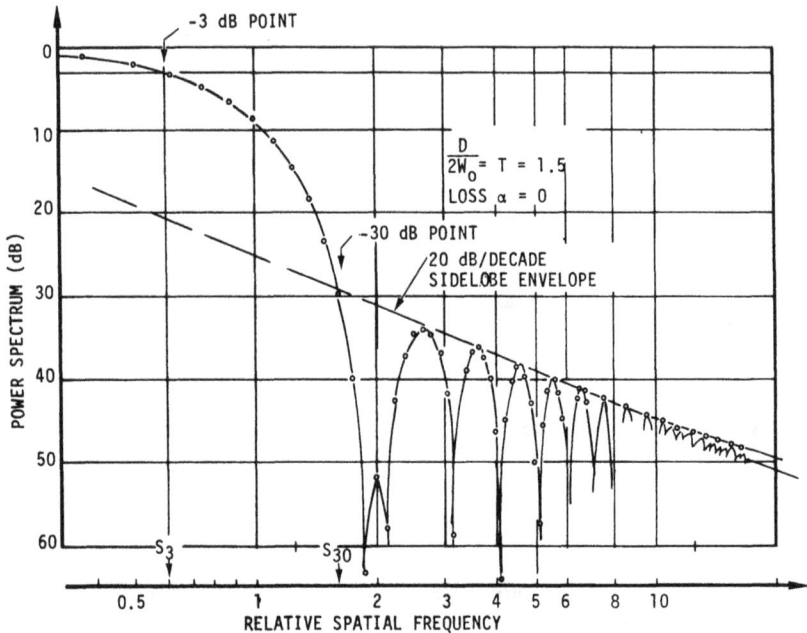

Figure 7.10. Spectrum of Truncated Gaussian (Based on Hecht, Reference 10)

a high dynamic range, and have fast readout speed. The small size requirement is necessary so that a large number of detectors can be positioned in the Fourier transform plane to detect the position of the light spots. The elements in a photo detector array can be packed very close together. For example, a Reticon RL1024C linear array contains 1024 elements with approximately 1 mil center to center spacing. The total number of detectors required in the linear array should be double the number of resolution cells of the Bragg cell. For example, a Bragg cell of 100 MHz width with 1 MHz frequency resolution can produce 100 resolution spots. Considering the possibility that a light spot can shine on two detectors, 200 photo detectors are required to generate 100 resolutions. When two adjacent detectors have outputs simultaneously, they can be considered as one single frequency output. Two simultaneous signals must be separated by a minimum of two detectors (1 MHz) to be detected as two separate signals.

B. Dynamic Range of a Photo Detector

Photo detectors can be viewed as square law detectors in that their output current is proportional to the input optical power. This can lead to some confusion in describing their dynamic range since the dynamic range referred to input optical power is half that referred to output electrical power. For example, a photo detector with 100 dB output dynamic range only covers a 50 dB dynamic range of the input power range. In Bragg cell receiver applications, the detector input optical power is proportional to the Bragg cell input RF power.

The dynamic range of photo detector is often defined as extending from the noise level to saturation. The noise floor of a detector is specified in terms of the noise equivalent power (NEP), defined as the root mean square (rms) incident optical power required to give an output rms signal current (or voltage) equal to the rms noise current (or voltage) (References 20 and 21). If the photo detector responsivity R is defined as

$$R = \frac{\text{output current (or voltage)},}{\text{input optical power}} \qquad (7.28)$$

the NEP can be written as

$$NEP = \frac{N}{R} \qquad (7.29)$$

where N is the rms noise. For silicon photo detectors, R is approximately 0.4 amp/watt.

With this definition, the NEP is given in watts and the NEP describes a specific detector under specific operating conditions. It will generally be a function of detector bias, illumination (optical) wavelength, illumination frequency and electrical (video) frequency bandwidth, because the noise and responsivity depend on these parameters. In order to permit comparisons of detectors, manufacturers will generally specify the NEP at some low frequency (e.g., 1 KHz) with 1 Hz bandwidth and at the peak response wavelength. Usually the NEP will be independent of frequency from approximately 1 KHz to some upper frequency limit (typically 10-100 MHz).

Another frequently encountered definition of the noise equivalent power is

$$NEP' = \frac{N}{\sqrt{B}\,R} \qquad (7.30)$$

Where B is the video bandwidth. In this case the NEP' has the units of watt/$\sqrt{\text{Hz}}$. The NEP' is numerically equivalent to NEP if the video bandwidth in the NEP specification is 1 Hz.

To determine the electrical noise floor of a detector when the NEP is specified for the specific detector operating conditions, one simply uses Equation 7.29, that is $N = NEP \cdot R$. However, if the NEP is specified at 1 Hz bandwidth or if NEP' is given, it is necessary to make the assumption that the noise spectral density (mean square noise per unit bandwidth) is frequency independent. In this case, the electrical noise floor is computed as

$$N = NEP' \cdot R \cdot \sqrt{B} \qquad (7.31)$$

Then the optical noise floor (NEP) can be computed from Equation 7.29.

For example, the RCS C38016 silicon PIN photo diode/preamplifier module is specified to have NEP' of 3×10^{-12} w/$\sqrt{Hz}$ and responsivity of 10^4 V/W at 900 nm wavelength. If this detector is operated at 1 MHz bandwidth, the rms electrical noise floor is 3×10^{-5}V. To obtain the optical noise floor (i.e., NEP) at 633 nm (HeNe laser line) the responsivity at 633 μm is required. Taking R(633nm) to be 60 percent of R(900nm), Equation 7.29 yields an NEP of 5×10^{-9} watts or -53 dBm.

If the maximum illumination where the detector is linear is 0.5 mw (-3 dBm), a 50 dB linear dynamic range linear results. Of course, in an operational Bragg cell receiver, threshold must be set above the noise floor to reduce false alarm rate.

In integrating photo detectors where charge transfer techniques are used for readout, the NEP is not generally used as a figure of merit because the dominant noise sources are different and the output is proportional to energy rather than power. In such a device, the incident radiant flux generates charge carriers which are stored in capacitors during the integration period. After the integration period, the stored carriers are clocked out. One way to clock them out is through using charge coupled device (CCD) shift registers. For such devices, noise is generally described in terms of noise electrons which are produced in the photo detection readout process. One way to characterize the noise floor of a CCD detector is to compute the number of noise electrons N_e which would be associated with a given detector element. An equivalent to the NEP can be defined as the average optical power incident to the element over integration time τ which would result in a stored charge of N_e electrons. Assuming the photon-to-charge-carrier conversion factor (quantum efficiency) is η, then P_0, the equivalent to NEP, can be written as

$$P_0 = \frac{N_e hf}{\eta \tau} \text{ watts} \qquad (7.32)$$

where h is the Plank's constant $= 6.625 \times 10^{-34}$ Joul-sec, and f is the frequency of the incident light.

If the number of noise electrons N_e is given, the value of P_0 can be computed for use in system performance calculations. For example, if $N_e = 1000$ and f $= 4.74 \times 10^{14}$ Hz ($\lambda = 633$ nm) for 1 ms integration time with $\eta = 0.5$, the P_0 can be calculated as

$$P_0 = \frac{1000 \times 6.625 \times 10^{-34} \times 4.74 \times 10^{14}}{0.5 \times 10^{-3}} = 6.28 \times 10^{-13} \text{ watts}$$

$$= 6.28 \times 10^{-10} \text{mw} = -93 \text{ dBm}$$

C. Output Rate of the Photo Detectors

Photo detectors used in Bragg cell receivers can be divided into two groups: discrete photo detectors and detector arrays. Popular discrete photo detectors are silicon planar photo diodes and avalanche diodes. Avalanche diodes, requiring high bias voltage (in hundreds of volts), make the design complicated. The discrete diodes usually have high dynamic range and very fast response time. For example, the Hewlett Packard 4205 PIN photo diodes package of 1.5 mm diameter with an active area diameter of 0.25 mm have over 100 dB dynamic range with less than 1 ns response time. If these detectors are arranged in a linear array, about 16 diodes can be packed in approximately 1 inch (25 mm) (approximately the total Fourier transform plane space of a normal Bragg cell receiver). Usually this arrangement, with only 16 frequency resolution cells, does not provide the desired frequency resolution. To achieve high frequency resolution using such detectors, fiber optics in a density package must be placed in the Fourier plane to couple the light to individual diodes. Each photo diode in such a scheme would be followed by an individual electron circuit to further process the signal. This design using discrete photo diodes would be very complicated; however, the parallel outputs can provide very good time of arrival (TOA) information. Photo detectors in array form are available; however, the outputs are in parallel and the complexity of the following circuit still exists.

There are CCDs in array form which can be used as photo detectors in a Bragg cell receiver. They are available in one-dimensional and two-dimensional arrays and are fabricated on a single substate. For example, a Reticon RL 1024 linear array contains 1024 active elements with 1 mil centers. Figure 7.11 shows an equivalent circuit of a linear detector array in a Bragg cell receiver. All the outputs are readout in series. When the commutating switch is connected to a certain output, the charge on that output port is readout, while all the other output ports are collecting information from the Bragg cell receiver. It takes a rather long time to read the entire array. If it takes 1 μs to switch from one port to the next one, 1024 μs would be required to read the 1024 element array. If the input signal is a short pulse, then the detector will integrate noise most of the time. The time that the detector collects useful information is equal to the pulse width or the Bragg cell window transient time, whichever is longer. Besides the long integration time, the switching noise in the linear array is relatively high. Fortunately, the noise is of fixed pattern; therefore, it may be compensated. Since the detector array has only one output, the electronic circuit following the detector is much simpler than that required for parallel outputs. But this kind of detector is not very sensitive to detecting short pulses. In addition, the TOA resolution generated from each output is usually too large to be useful for signal sorting. The two-dimensional detector arrays can be used for two-dimensional optical signal processing.

Other linear photo detector arrays have outputs that can be readout individually through external control. They are usually referred to as random access arrays, and their applications to a Bragg cell receiver are still to be explored. Another kind of detector array has parallel outputs in which each output reads the signal serially from many detectors rather than from one detector. It is referred to as a photo detector array with series/parallel outputs. This array has some benefits of both the series and the parallel detectors. The properly designed one with the desired number of detectors and output ports has the potential to provide a near term solution to a Bragg cell receiver.

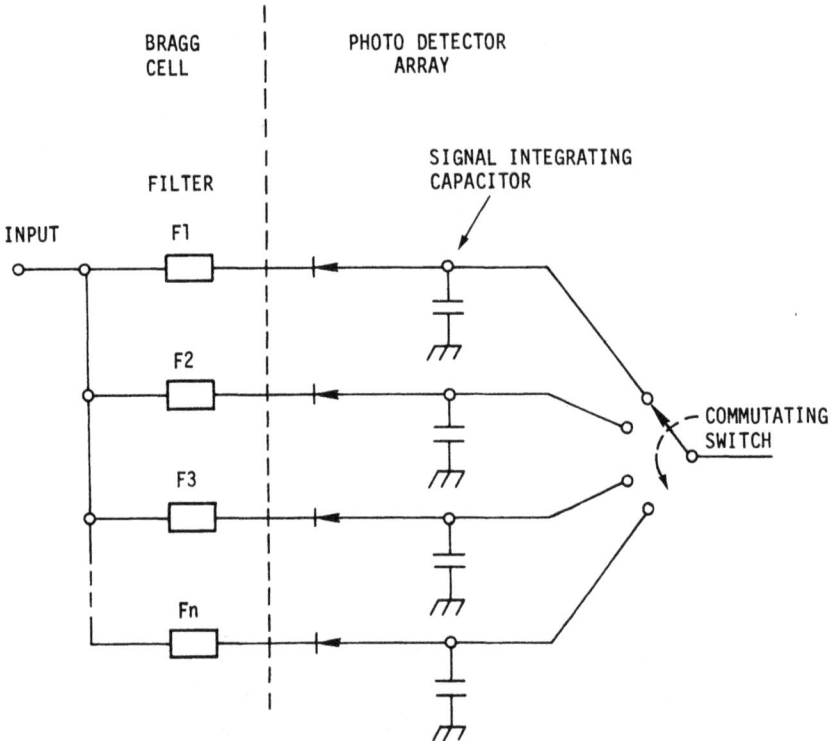

Figure 7.11. Electrical Equivalent of Optical Detectors

7.7 BRAGG CELL RECEIVERS

The key components necessay to build a Bragg cell receiver are the laser, Bragg cell, optical lenses, and detector array. The simplest arrangement is a one-dimensional Bragg cell receiver which is commonly used to measure the signal frequency, and amplitude as shown in Figure 7.12.

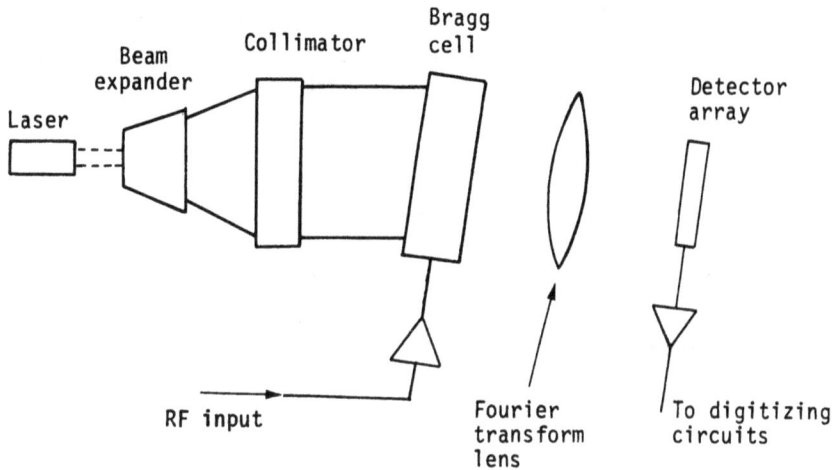

Figure 7.12 Bragg Cell Receiver

The laser beam is converted to a collimated sheet beam by the beam expander/collimator. This beam is incident on the Bragg cell at the Bragg angle. The output is focused on the detector array through the Fourier transform lens. The outputs from the detector array can be readout in series, parallel or series/parallel, followed by a log amplifier. In general, an RF amplifier is required at the input of the Bragg cell to boost the input signal to the level required by the Bragg cell. The output power from the amplifier is in the order of hundreds of milliwatts to a watt. Improving the efficiency of the cell can reduce the output power requirement of the amplifier.

The dynamic range of a Bragg cell receiver can be estimated through the following procedure. The lower limit (the sensitivity) of the receiver is determined by the noise floor of the detector. The upper end is limited by the laser power or the detector. Let us assume that a solid-state laser of 10 dBm output power is used and that the Bragg cell receiver has a total loss of 20 dB from the laser to the detector including the efficiency of the Bragg cell. Then the signal power at the detector is -10 dBm. If the detector is not in saturation at this level and its noise floor is assumed at -60

dBm, the dynamic range is 50 dB. However, at the noise floor the false alarm rate is rather high. If 10 dB above the noise floor is the required threshold setting, the useful dynamic range is 40 dB.

In order to keep the second harmonics out of the receiver, its bandwidth is usually limited to less than an octave as in other wide-band receivers. If two or more strong signals are present in the receiver, it may happen that the strong signal suppresses the weak one; also, it degrades the accuracy of the amplitude measurement. The signals can also generate intermodulation products. These effects may take place in the RF amplifier as well as in the Bragg cell. As long as the amplifier works in the linear region, the capture effect and the intermodulation products can be neglected. As for the Bragg cell, these effects can be minimized by limiting the operation of the Bragg cell at low diffraction efficiency (a few percent or less).

From all practical points of view, the Bragg cell receiver is equivalent to a channelized receiver. The maximum number of possible parrallel output channels equals the time-bandwidth product of the Bragg cell. If discrete photo detectors are used as output detectors, then the performance of the Bragg cell is comparable to a channelized receiver.

7.8 DYNAMIC RANGE FOR SHORT PULSES

One very important phenomenon is that most of the detectors used in an optical processor are energy detectors rather than peak power detectors. When the pulse width of the incoming signal is longer than the integration time which represents the time between successive readouts of the output detector, the full designed sensitivity and dynamic range will be realized. However, if the pulse width is small the processor performance will suffer. If the pulse width is shorter than the integration time but longer than the window transient time of the Bragg cell, the energy on the detector will decrease linearly with the pulse width. If the pulse width is shorter than the Bragg cell window time, the spectrum spreading effect should also be considered. Suppose in a Bragg cell receiver, the Bragg cell window time is τ (the receiver frequency resolution is approximately $1/\tau$), and the detector integration time (or TOA resolution) is T_S where $T_S > \tau$. The three incoming signal conditions are:

$$PW > T_S, \quad \tau < PW < T_S, \quad \text{and} \quad PW < \tau$$

The power on the detector as a function of time is shown in Figure 7.13 for signals of the same pulse amplitude. (Note that the $PW = \tau$ condition is also shown in Figure 7.13.) The sensitivity of the receiver decreases as the total received energy (areas in Figure 7.13). The sensitivity (expressed in dBm) can be approximately represented as

(a) $PW > T_s$, full sensitivity S   (7.33)

(b) $\tau < \mathrm{PW} < \mathrm{T_s}$, $\mathrm{S} + 10 \log \dfrac{\mathrm{T_s}}{\mathrm{PW}}$ \hfill (7.34)

(c) $\mathrm{PW} < \tau$, $\mathrm{S} + 10 \log \dfrac{\mathrm{T_s}}{\tau} + 20 \log \dfrac{\tau}{\mathrm{PW}}$ \hfill (7.35)

(a) $PW > T_S$

(b) $\tau < PW < T_S$

(c) $PW = \tau$

(d) $PW < \tau$

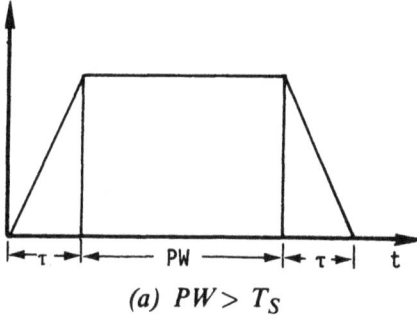

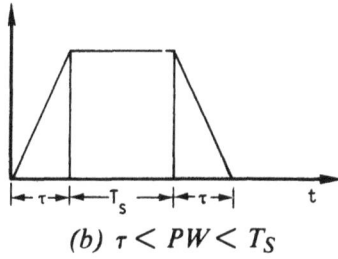

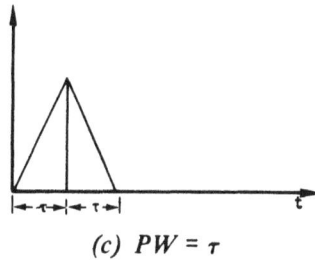

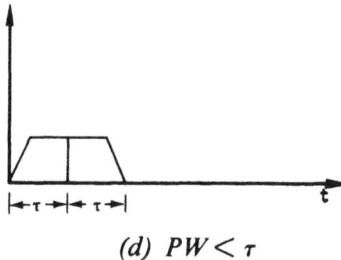

Figure 7.13. Energy on a Detector for Different Signal Pulse Width

The factor 20 in Equation 7.35 is due to energy reduction as well as spectrum spreading. The results are shown in Figure 7.14.

An example to demonstrate this point is discussed here. Suppose the Bragg cell time delay is 1 μs, the frequency resolution of the receiver is approximately 1 MHz, the time resolution on the detector is 10 μs, and the sensitivity of the receiver is -60 dBm. Then the receiver can measure pulses longer than 10 μs with the full sensitivity (-60 dBm). If the pulse width is 5 μs, the receiver can detect the pulse at -57 dBm ($-60 + 10 \log 10/5$). For a pulse of 100 ns, the receiver can detect it at -30 dBm ($-60 + 10 \log 10/1 + 20 \log 1/0.1 = -60 + 10 + 20$). If a peak detector is used in the Bragg cell receiver, the integration effect can be avoided, but the spectrum spreading effect still exists. The spectrum spreading effect also happens on the leading and trailing edges of the pulse, because when the pulse enters and leaves the Bragg cell it appears as a short pulse. This phenomenon will cause "rabbit ears" outputs from adjacent photo detectors just as discussed in channelized receivers (Chapter 5). Detection circuits following the photo detectors must be able to handle these outputs in order to avoid the generation of spurious outputs.

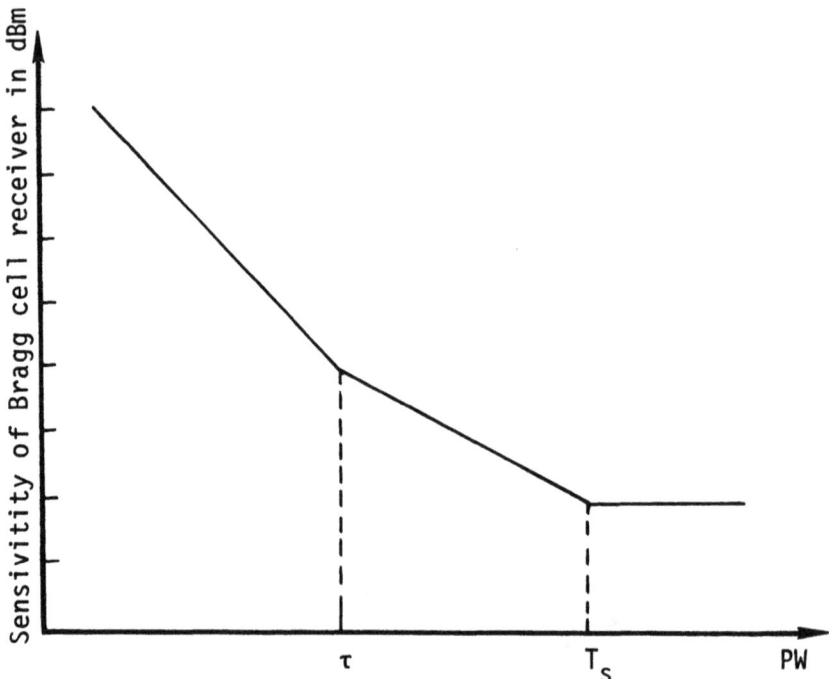

Figure 7.14. Sensitivity of Bragg Cell Receiver Versus Pulse Width

7.9 TWO-DIMENSIONAL OPTICAL PROCESSING(11)

Since the Bragg cell is a two-dimensional device, one dimension can be used for frequency reading and the other dimension can be used to measure some other parameter of the input signal. The most popular use of the second dimension is for angle of arrival (AOA) information. Its theory of application is identical to the interferometry of antenna as discussed in Chapter 2. Actually one can consider it as a special output scheme for the antenna interferometry. Its schematic is shown in Figure 7.15. In this case the Bragg cell has four input transducers; each one is connected to an antenna. Each transducer generates its individual acoustic wave in the Bragg cell. Because of the time delay in the four input antennas, the four acoustic waves in the Bragg cell have specific phase relationship among them.

A laser input is at the Bragg angle for the frequency plane; its output is registered on a two-dimensional array. The ordinate represents the frequency information and the abscissa direction represents the angular information. The angle information measured from this scheme is obtained from the phase relation of the signal. Therefore, the angle measured in this arrangement depends on the signal frequency. The higher the frequency, the more the light deflection along the abscissa, even though the incident angle is the same. The frequency AOA output shown in Figure 7.15b has the keystone effect. The AOA information should be calibrated accordingly.

The input transducers on the Bragg cell are not necessarily equally spaced. Generally, in phase interferometry, the closest antenna space is less than one half of the shortest wavelength to be measured in order to eliminate ambiguity problems. Larger antenna spacings are required to generate fine angle information. Thus, a linear array with nonuniform antenna spacings is a common practice in phase interferometry. If the antennas used on the Bragg cell optical receiver are nonuniformly spaced, the input transducers on the Bragg cell must also be spaced accordingly; the spacings of the input transducers and the spacings of the input antennas must have the same ratio.

7.10 INTEGRATED OPTICAL BRAGG CELL RECEIVER(25-31)

The Bragg cell receiver discussed in this chapter has been implemented with an integrated optics approach. The advantages of using integrated optics for the Bragg cell receiver are the extremely small size and potentially low production cost. The entire Bragg cell receiver including laser source, Bragg cell, acoustic transducer, detector arrays, and the optical lens system could all be fabricated on one chip. The disadvantage of this approach is that only a one-dimensional processing can be implemented. Thus only the frequency of the incoming signal can be measured. The operational frequency of an integrated optical Bragg cell receiver is usually lower than a bulk wave Bragg cell receiver because: (1) the surface acoustic

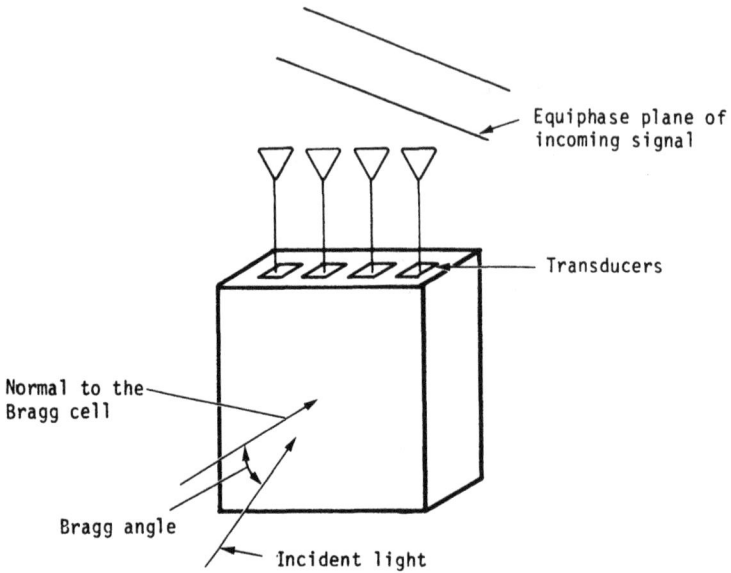

(a) *Two-Dimensional Bragg Cell Signal Processing*

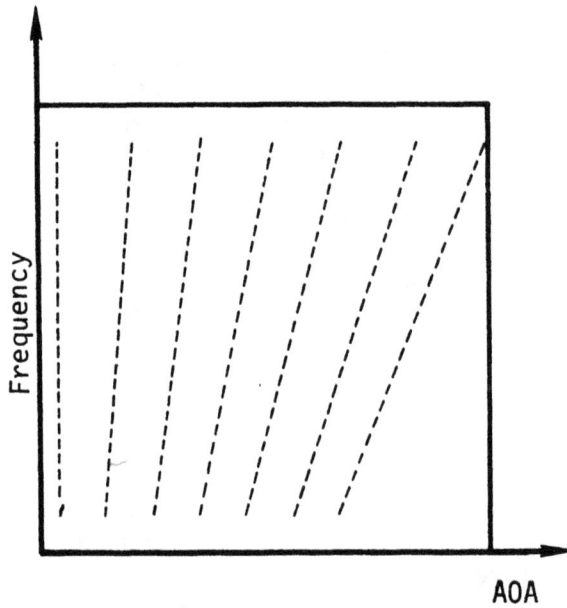

(b) *Two-Dimensional Display*

Figure 7.15. Bragg Cell Used for Frequency and Angle of Arrival Measurement

wave is usually slightly slower then the bulk wave; (2) the propagation loss is higher in surface wave than in bulk wave; (3) the interaction between the surface acoustic wave (SAW) and the light in the waveguide is less at higher frequencies; and (4) it is difficult to manufacture surface wave transducers at high frequency. Since the operational frequency is lower for integrated optics, the RF operational bandwidth is often narrower, because the bandwidth of a receiver is often less than an octave. The integrated optical Bragg cell receiver is still in the research stage; however, the feasibility has been demonstrated.

A. The Basic Operation

The operational principle of the integrated optics approach is exactly the same as the Bragg cell receiver mentioned in Section 1 (Reference 30). A (GaAl)As double-heterostructure injection laser is used as the light source because of its small size and efficiency. The light is transmitted in a waveguide rather than in free space. The light waveguide is a thin layer of material with higher index of refraction than the substrate, in which the light is trapped by total internal reflection at the interface. The Bragg cell uses SAW at the surface of the substate to interact with the light in the waveguide. A detector array is used to detect the deflected light spot which, in turn, determines the incoming signal frequency. Two lenses, one for beam expansion and collimation, the other for the Fourier transform, are both fabricated on the substate. A basic integrated optics Bragg cell receiver is shown in Figure 7.16. It is almost a duplicate of the bulk wave Bragg cell receiver.

B. Substrate Materials

There are three basic materials on which the integrated optics can potentially be fabricated. Each material has its own advantages and shortcomings. The three materials are: gallium-aluminum arsenide (GaAl)As, lithium niobate ($LiNbO_3$), and silicon (Si_2).

The (GaAl)As has the potential to be used to fabricate the laser source, the lenses, the Bragg cell [(GaAl)As itself is a piezoelectric material], and the detector array. In other words, there is the potential that the entire Bragg cell receiver can be made in monolithic form on a single (GaAl)As chip. However, a different material would have to be used for the light waveguide. In addition, the piezoelectric effect in GaAs is very weak; a piezoelectric material overlay on the substrate may be required to provide the means to generate a SAW. Extensive research on this subject is required in order to make the Bragg cell receiver in monolithic form.

$LiNbO_3$ has superior piezoelectric and electro-optic coupling characteristics and it is one of the prime materials used in SAW device fabrication. Optical waveguides on $LiNbO_3$ can be produced by diffusion of metals, i.e., titanium (T_i), in place of some of the niobium (Nb). The indiffused waveguide is several micrometers (μm) thick and exhibits insertion loss of about 0.3-1.0 dB/cm. An electro-acoustic transducer can be deposited on the substrate directly because of the strong piezoelectric coupling, and a frequency bandwidth of 1 GHz may be possible. However, the laser and detector array cannot be fabricated on this substrate. They would have to be hybrid coupled to the waveguide in a compact fashion. Butt end coupling techniques are being developed with approximately 50 percent efficiency.

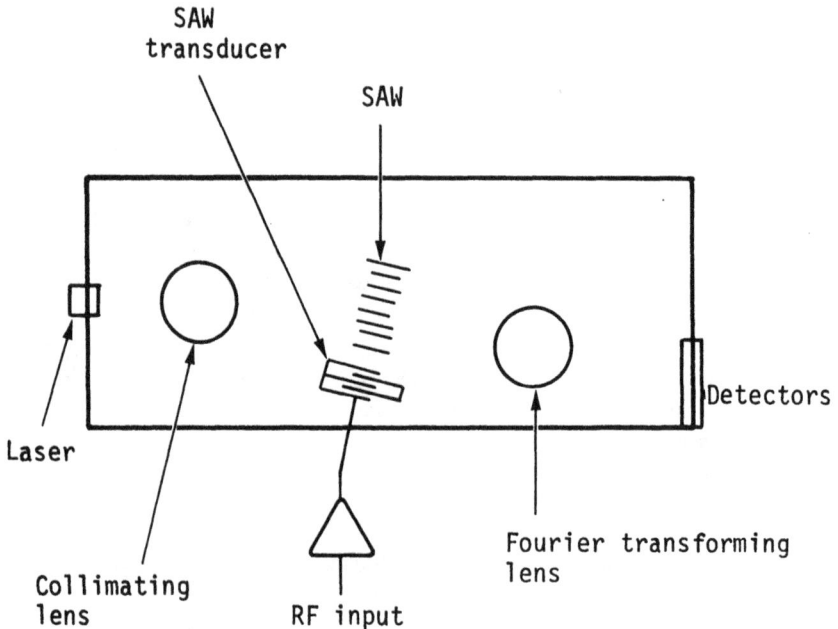

Figure 7.16. Basic Integrated Optics Bragg Cell Receiver on a Chip

Silicon is attractive because Si wafer processing is a well established technology in integrated circuit design. Another main advantage of using Si as the substrate is that the detectors can be fabricated directly on the substrate. However, the laser would have to be coupled to the waveguide. Silicon absorbs visible light; therefore, the optical waveguide cannot be fabricated directly on the chip. A silicon dioxide (SiO_2) layer on top of the substrate to isolate the optical waveguide is required. A glass layer can be laid on top of the SiO_2 as waveguide. Corning 7059 glass is a typical material having insertion loss of 0.2-0.6 dB/cm with an index of reflection of 1.57. Since Si, SiO_2, and glass are not good piezoelectric materials, zinc oxide (ZnO) is generally used as the transducer material.

C. Transducers

Transducers are used to convert electric energy to acoustic waves. Since the light is concentrated in the optical waveguide on the surface of the substrate, surface acoustic waves rather than bulk acoustic waves are used to deflect the light. Metal fingers on piezoelectric materials are used to generate acoustic waves. To increase the RF bandwidth, multiple transducers, each tuned to a specific RF band are often used. Each one can be fabricated at a slightly different angle with respect to the optical beam to have the proper Bragg angle at different frequencies. This arrangement is referred to as stagger-tilted transducer array as shown in Figure 7.17.

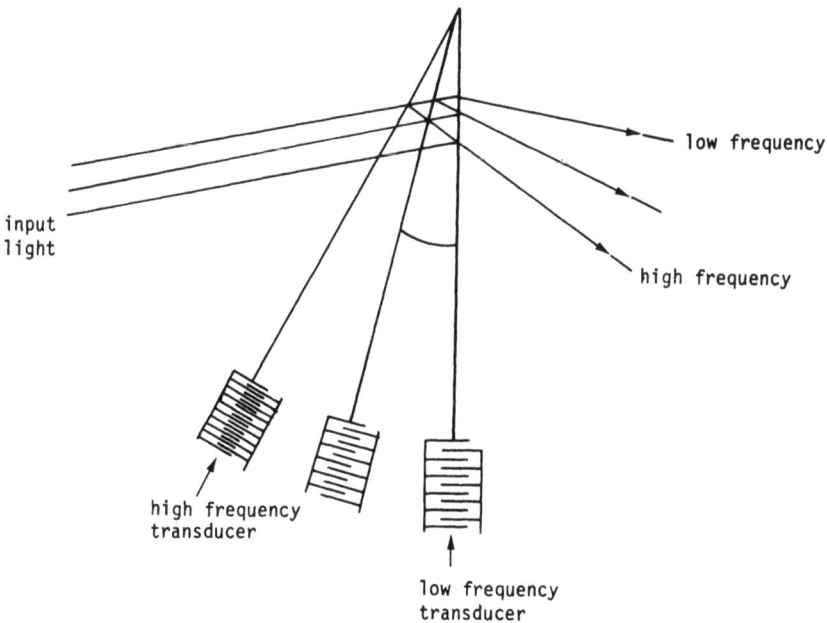

Figure 7.17. Angular Stagger-Tilted Transducer Array

D. Lenses

Two lenses are required for the integrated optics Bragg cell receiver: one is used as a collimating lens which refracts the light from a point source (solid-state diode laser) into a collimated beam. The other one is the Fourier transform lens which focuses the light diffracted by the Bragg cell onto the detectors. There are two basic structures for the lenses: Luneburg lens and geodesic lens.

The Luneburg lens is formed by depositing a material with an index of refraction higher than that of the waveguide material on top of the waveguide. Light inside the waveguide is refracted by this material and changes the propogation direction. For the Si substrate with a glass waveguide, Luneburg lenses are made possible by using Ta_2O_5 (n = 2.1) and Nb_2O_5 (n = 2.2) thin films. However, for (GaAl)As and $LiNbO_3$ substrates, since the index of refraction of the waveguide is rather high, it is not presently practical to use Luneburg lenses.

A geodesic lens physically changes the light path length to achieve a focusing effect. Since this approach is independent of index of refraction, it can be used on any substrate. A geodesic lens is formed by grinding a quasispherical depression in the substrate as shown in Figure 7.18. The light entering the side of the lens travels a shorter distance than the light entering the center of the lens. This differential length of light path focuses the incoming parallel light into a point. The grinding and polishing of this kind of lens is tedious and time consuming with the existing technology.

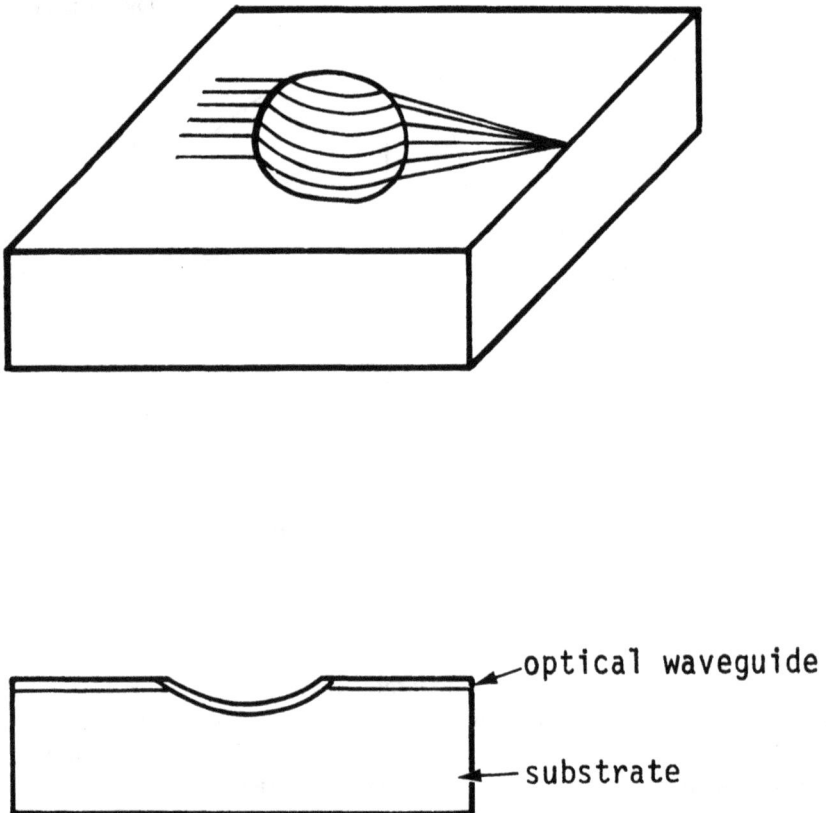

Figure 7.18. Geodesic Lens

E. Detector Array

The light passing through the Fourier lens is focused onto the detectors. The detector output can be made either in parallel or in series forms. Detectors can be fabricated on the chip if a Si substrate is used; however the detectors must be butt-coupled to the waveguide in for a $LiNbO_3$ substrate. Most of the discussions on detectors in Section 5 can be applied here also.

F. Recent Development

Most of the individual components discussed in the above sections for integrated optics Bragg cell receiver applications have been demonstrated already. A chip containing all these components to perform as an RF receiver has also been demonstrated. The achieved bandwidth was 400 MHz centered at 600 MHz. The frequency resolution achieved is better than 4 MHz. Series detectors were used in this feasibility demonstration model. The success of fabricating a Bragg cell receiver on a chip really opens the possibility of mass production of low-cost, small sized microwave receivers.

7.11 SUMMARY

A Bragg cell receiver is a wide-band receiver with fine frequency resolution. It can be considered as a channelized receiver where the channelization is through optical Fourier transform. The fine frequency resolution implies that the receiver will have high sensitivity. The dynamic range of the receiver is not fully explored yet. At the present time, the dynamic range is limited by the photo detectors and the output power of the laser. If the photo detector is sensitive to the energy rather than power, the sensitivity and dynamic range of the receiver are pulse width dependent. The receiver can measure the pulse amplitude and pulse width. The pulse width measurement is limited by the window time of the Bragg cell which is closely related to frequency resolution.

A Bragg cell receiver has the potential performance very desirable for EW applications. Although the feasibility of a Bragg cell receiver has been demonstrated, extensive research and development are still required to realize the full capability of the receiver.

REFERENCES

.1 Hopkins, H.H., "On the diffraction theory of optical images," Proc. Royal Society A., Vol. 217, p. 408, 1953.

2. O'Neill, E.L., "Spatial filtering in optics," IRE Trans. Information Theory, Vol. IT-2, pp. 56-65, June 1956.

3. Lugt, A.V., "Coherent optical processing," Proc. IEEE, Vol. 62, pp. 1300-1319, October 1974.

4. Petiotis, S., "Acousto-optics light the path to broadband ESM receiver design," Microwaves, p. 54, September 1977.

5. Chang, I.C., "Acousto-optic devices and applications," IEEE Trans. on Sonics and Ultrasonics, Vol. 23, pp. 2-22, January 1976.
6. Hecht, D.L., "Multifrequency acousto-optic diffraction," IEEE Trans. on Sonics and Ultrasonics, Vol. SU-24, pp. 7-18, January 1977.
7. Coppock, R.A., Croce, R.F., Regier, W.L., "Bragg cell RF signal processing," Microwave Journal, p. 62, September 1978.
8. Hecht, D.L., "Acousto-optic device techniques -400-2300 MHZ," Ultra-sonics Symposium Proceedings, IEEE Cat. #77CH1264-ISU, 1977.
9. Goodman, J.W., *Introduction to Fourier optics,* McGraw-Hill Book Co., 1968.
10. Hecht, D.L., "Spectrum analysis using acousto-optic devices," Optical Engineering, Vol. 16, pp. 461-466, September/October 1977.
11. Tippet, J.T. (Ed.), *Optical and electro-optical information processing,* Chapter 38, Cambridge MIT Press, 1965.
12. Thomas, C.E., "Optical spectrum analysis of large space bandwidth signals," Applied Optics, Vol. 5, pp. 1782-1790, November 1966.
13. Psaltis, D., Vijaya Kumar, B.V.K., "Acousto-optic spectral estimation: a statistical analysis," Applied Optics, pp. 601-605, 15 February 1981.
14. Meitzler, A.H., "Piezoelectric transducer materials and techniques for ultrasonic devices operating above 100 MHz," Ultrasonic Transducer Materials, Mathial, O.E. (Ed.), Plenum, New York, 1971.
15. Larson, J.D., III, Wilson, D.K., "Ultrasonically welded piezoelectric transducers," IEEE Trans. Sonics Ultrasonics, Vol. SU-18, pp. 142-152, July 1971.
16. Sittig, E.K., "Design and technology of piezoelectric transducers for frequencies above 100 MHz," in Physical Acoustics, Vol. IX, Mason, W.P., Thurston, R.N. (Eds.), Academic Press, New York, 1972.
17. Pinnow, D.A., "Guided lines for the selection of acousto-optic materials," IEEE J. Quantum Electron, Vol. QE-6, pp. 223-238, April 1970.
18. Uchida, N., Niizeki, N., "Acousto-optic deflection materials and techniques," Proc. IEEE, Vol. 61, pp. 1073-1092, August 1973.
19. Chang, I.C., Hecht, D.L., "Doubling acousto-optic deflector resolution utilizing second-order birefringement diffraction," Applied Physics Letters, Vol. 27, pp. 517-518, November 1975.
20. "RCA electro-optics handbook," Chapter 10, Technical Series EOH-11 RCA/Commercial Engineering/Harrison, New Jersey, 1974.
21. Wolfe, W.L., Zissis, G.J. (Eds.), "The infrared handbook," Chapter 11, Office of Naval Research, Dept. of the Navy, Washington, D.C.
22. Sheridon, N.K., Berkovitz, M.A., "Optical processing with the Ruticon," SPIE, Vol. 83, Optical Information Processing, pp. 68-75, 1976.

23. Lee, J.P.Y., "Acousto-optic spectrum analysis of radar signals using an integrating photo detector array," Applied Optics, Vol. 20, pp. 595-600, February 1981.
24. Borsuk, G.M., "Photo detectors for acousto-optic signal processors," Proc. IEEE, Vol. 69, pp. 100-118, January 1981.
25. Dakss, M.L., Kuhn, L., Heidrich, P.F., Scott, B.A., "Grating coupler for efficient excitation of optical guided waves in thin films," Applied Physics Letters, Vol. 16, pp. 523-525, June 1970.
26. Anderson, D.B., Davis, R.L., Boyd, J.T., August, R.R., "Comparison of optical-waveguide lens technologies," IEEE J. Quantum Electronics, Vol. QE-13, pp. 275-282, April 1977.
27. Boyd, J.T., Chen, C.L., "An integrated optical waveguide and charge-coupled device image array," IEEE J. Quantum Electronics, Vol. QE-13, pp. 282-287, April 1977.
28. Hamilton, M.C., Wille, D.A., Miceli, W.J., "An integrated optical RF Spectrum Analyzer," IEEE Ultrasonics Symposium, October 1976.
29. Anderson, D.B., Boyd, J.T., Hamilton, M.C., August, R.R., "An integrated-optical approach to the Fourier Transform," IEEE J. Quantum Electronics, Vol. QE-13, pp. 268-275, April 1977.
30. Anderson, D.B., "Integrated optical spectrum analyzer: an imminent 'chip'," IEEE spectrum, pp. 22-29, December 1978.
31. Hamilton, M.C., "Wide-band acousto-optic receiver technology for electronic warfare systems." J. of Electronic Defense, pp. 50-55, Jan/Feb. 1981.
32. Klein, W.R., and Cook, B.D., "Unified approach to ultrasonic light diffraction," IEEE Trans. on Sonics and Ultrasonics, Vol. SU-14, pp. 123-134, July 1967.
33. Young, E.D., Jr., and Yao, S.K., "Design considerations for acousto-optic devices," Proc. of IEEE, Vol. 69, pp. 54-64, January 1981.

CHAPTER 8
TRANSMISSION LINES

8.1 INTRODUCTION

Conventional ways to transmit microwave energy are by free space, transmission lines, and waveguides. Electromagnetic waves of all frequencies can theoretically travel in free space if proper antennas can be built to launch the waves. Transmission lines with two conductors in coaxial form can be used from dc up to 18 GHz frequency range. Above 18 GHz the insertion loss of the line increases rapidly and limits its application. Waveguides of rectangular and cylindrical shape made of hollow metallic tubes are used to guide electromagnetic waves at higher frequences from a few GHz up to several hundred. Rectangular shaped waveguides are more popular because the well defined electromagnetic field orientation made the component's design easier than that for the cylindrical waveguides. Since the sixties, advances in dielectric materials and integrated circuits technology have made the use of strip lines and microstrip lines very attractive in many components and systems.

In microwave receiver systems, the components only need to handle power in the milliwatts to a few watts range. Therefore, for components and circuitry in a receiver, power handling capability is not the prime design factor to be considered. Instead, size and weight are the most important considerations especially in an airborne receiver system. In modern receiver development and fabrication, almost all of the microwave circuitry is made of strip lines or microstrip. The use of waveguides and coaxial cables is not very prevalent in recent microwave receiver development work especially for frequency below 18 GHz although they have been used as the primary transmission lines in microwave receivers in the past. Waveguide and coaxial transmission lines have been discussed in many microwave or electromagnetic field theory books. Thus, this chapter will concentrate only on strip line, microstrips, slotlines, and millimeter (mm) wave lines.

Although there are no clear definitions for strip lines and microstrip lines it is commonly agreed by scientists and engineers, in the microwave field, that a conductor printed on a dielectric substance and sandwiched by another homogenous dielectric parallel planes is called strip line (or balanced strip line) as shown in Figure 8.1a. Conducting plates are on both sides of the dielectric plates. If the printed line is on one side of a dielectric sheet with a shield conductor on the other side as shown in Figure 8.1b, it is referred to as microstrip line (sometimes called open strip line). The microstrip lines are enclosed in metal containers.

The microstrip line was originally abandoned by microwave designers in favor of the strip line because of the radiative nature of the open strip line. The use of thin and high dielectric materials greatly reduces the radiation

183

from the open strip, and the microstrip has been used in almost all microwave integrated circuits (MIC). Another disadvantage of the microstrip line is that it cannot support a transverse electromagnetic (TEM) mode which means that the characteristic impedance of the line is not defined. However, when high dielectric substrate materials are employed and the dimensions of the line are small compared to a wavelength, the impedance and relative velocity can be approximated as a TEM mode.

Coupled microstrips and slotlines will improve the versatility of circuit designs. The fin lines and dielectric transmission lines (Sections 8.8 and 8.9) are used in mm wave applications.

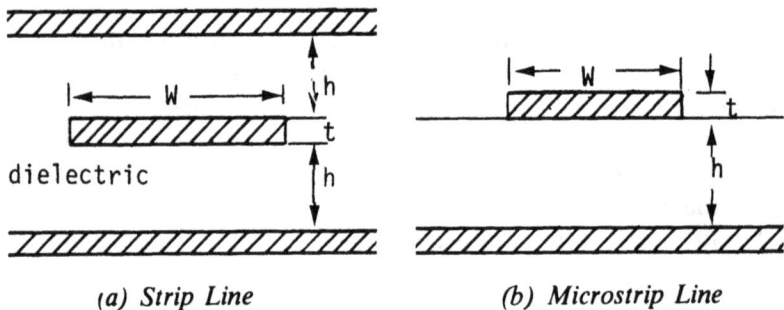

(a) Strip Line   (b) Microstrip Line

Figure 8.1. Strip Line and Microstrip Line

8.2 STRIP LINES[1-5]

Strip lines were used for microwave circuits because the fabrication procedure is simple in comparison with waveguides and the line is well shielded. Since the strip lines can be considered totally enclosed, they work in a TEM mode, where the electric (E) and magnetic (H) fields are perpendicular to the direction of propagation. In other words, there are no E and H components in the direction of propagation. Besides, the E and H fields are always perpendicular to each other and the charactistic impedance is defined as E/H. Wheeler (References 1 and 3) analyzed the thin strip lines and obtained the following empirical equations for analysis and synthesis. He introduced a normalized impedance for convenience

$$z = \frac{\sqrt{\varepsilon_r}Z}{377} \tag{8.1}$$

$$Z = \frac{377z}{\sqrt{\varepsilon_r}} \tag{8.2}$$

$$\frac{W'}{h} = \frac{16}{\pi} \frac{\sqrt{(\exp 4\pi z - 1) + 1.568}}{(\exp 4\pi z - 1)} \tag{8.3}$$

From the value of z the following relation can be obtained

$$Z = \frac{30}{\sqrt{\varepsilon_r}}$$

$$\bullet\, \ln \left\{ 1 + \frac{1}{2}\left(\frac{16h}{\pi W'}\right)\left[\left(\frac{16h}{\pi W'}\right) + \sqrt{\left(\frac{16h}{\pi W'}\right)^2 + 6.27}\,\right] \right\}$$

The relative error in Z is < 0.005 for $\dfrac{W'}{h} < 20$ \hfill (8.4)

where z is the normalized impedance, ε_r is the relative dielectric constant,

Z is the impedance of the strip line, $\sqrt{\dfrac{\mu_0}{\varepsilon_0}} = 377\Omega$ impedance of free

space, W' is the width of an equivalent thin strip, and h is the height (separation) of the strip line from each ground plane as shown in Figure 8.1a and $\ln$ is the natural logarithm operator.

The width W' is related to the actual conductor width W by the following equations. The equivalent width W' is caused by the fringing effect at the end of the conductor

$$W' - W = \frac{t}{\pi} \ln \frac{e}{\sqrt{\left[\dfrac{1}{\dfrac{4h}{t} + 1}\right]^2 + \left[\dfrac{\dfrac{1}{4\pi}}{\dfrac{W}{t} + 1.10}\right]^m}} \qquad (8.5)$$

or

$$W' - W = \frac{t}{\pi} \ln \frac{e}{\sqrt{\left[\dfrac{1}{\dfrac{4h}{t} + 1}\right]^2 + \left[\dfrac{\dfrac{1}{4\pi}}{\dfrac{W'}{t} - 0.26}\right]^m}} \qquad (8.6)$$

where

$$m = \frac{6}{3 + \dfrac{t}{h}} \qquad (8.7)$$

and t is the thickness of the strip conductor in Figure 8.1a and e is the base of natural logarithm = 2.718.

When the thickness t is approaching zero, the equivalent width is approaching the physical width W. Equation 8.3 is used for synthesis while

Equation 8.4 is used for analysis. If the impedance Z of a strip line is given, the normalized impedance z can be obtained from Equation 8.1 provided the dielectric material is selected. The $\dfrac{W'}{h}$ ratio can be obtained from Equation 8.3, and the true width can be found in Equations 8.5 and 8.7. The results are also shown in Figure 8.2, for relative dielectric constants of 1, 2, 4, 8, and 16. As expected, the higher the dielectric constants, the lower the characteristic impedance for the same physical dimensions. For the same dielectric constant, the narrower the strip line width, the higher the characteristic impedance.

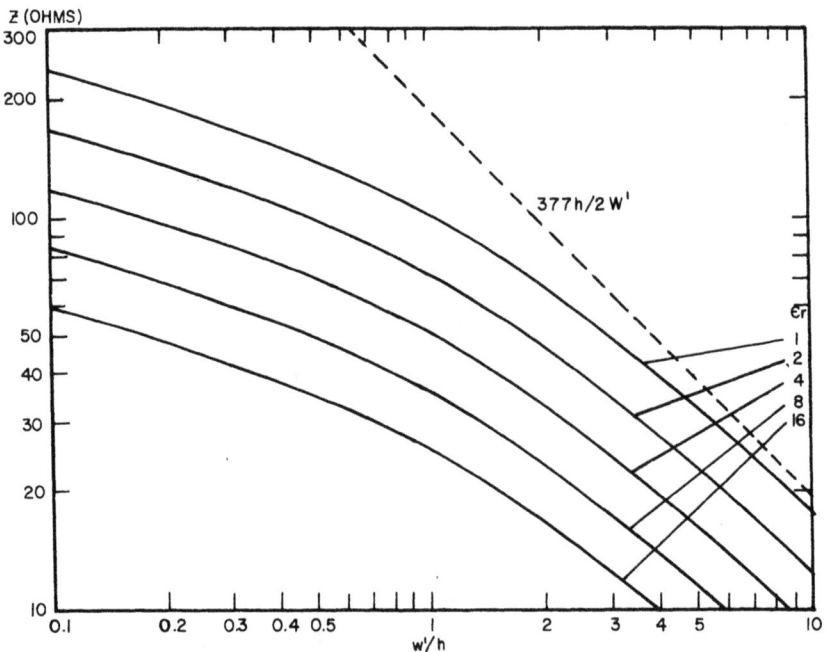

Figure 8.2. The Characteristic Impedance of a Thin Strip in Dielectric Between Parallel Planes (Based on Wheeler, Reference 3)

Strip line circuits were used before microstrip lines became popular. Since strip lines are large and complex to fabricate (compared with microstrip lines), they are replaced by microstrip lines in microwave receivers.

8.3 CHARACTERISTIC IMPEDANCE OF MICROSTRIP LINES[6-11]

There are many studies on microstrip lines, but no exact solutions to determine the E and H fields for this configuration. In general, the approaches fall into three categories: (1) conformal mapping (Reference 6), (2) relaxation method (References 7 and 8), and (3) variational method (References 9 and 10). These approaches use numerical calculations and involve digital computers, and their results are represented in figure and table forms. It should be noted that a microstrip cannot support a TEM wave mode. The characteristic impedance of such a transmission line is not defined. When the dielectric constant is high and the dielectric sheet is thin, the energy of the wave will concentrate in the dielectric. An approximation of TEM mode can be applied in the dielectric. Wheeler (Reference 6) used empirical formulas to analyze and synthesize thin metal strips on dielectric substrates. For analysis, the characteristic impedance of the microstrip lines can be obtained if the physical dimensions and dielectric constant of the substrate is known. These results are quite similar to the strip lines discussed in the last section. The characteristic impedance Z can be expressed as

$$Z = \frac{42.4}{\sqrt{\varepsilon_r + 1}} \ln \left\{ 1 + \left(\frac{4h}{W'}\right) \left[\left(\frac{14 + \frac{8}{\varepsilon_r}}{11}\right)\left(\frac{4h}{W'}\right) \right. \right. \tag{8.8}$$

$$\left. \left. + \sqrt{\left(\frac{14 + \frac{8}{\varepsilon_r}}{11}\right)^2 \left(\frac{4h}{W'}\right)^2 + \frac{1 + \frac{1}{\varepsilon_r}}{2} \pi^2} \right] \right\}$$

where ε_r is the relative dielectric constant, h is the thickness of the dielectric sheet, and W ' is the effective width of the microstrip which is related to the microstrip width W through

$$\frac{W' - W}{t} = \frac{1}{\pi} \ln \frac{4e}{\sqrt{\left(\frac{t}{h}\right)^2 + \left(\frac{\frac{1}{\pi}}{\frac{W}{t} + 1.10}\right)^2}} \tag{8.9}$$

or

$$\frac{W' - W}{t} = \frac{1}{\pi} \ln \frac{4e}{\sqrt{\left(\frac{t}{h}\right)^2 + \left(\frac{\frac{1}{\pi}}{\frac{W}{t} - 0.26}\right)^2}} \tag{8.10}$$

where t is the thickness of the microstrip line and e is the base of natural logarithms (2.718). It should be noted that when $t \to 0$, W ' equals W. The error induced by Equation 8.8 should be less than .02 Z.

For synthesis, if the dielectric substrate and the desired characteristic impedance are given, then the equivalent width of the line can be calculated as

$$\frac{W'}{h} = 8 \frac{\sqrt{\left[\exp\left(\frac{Z}{42.4}\sqrt{\varepsilon_r + 1}\right) - 1\right]\frac{7 + \frac{4}{\varepsilon_r}}{11} + \frac{1 + \frac{1}{\varepsilon_r}}{0.81}}}{\left[\exp\left(\frac{Z}{42.4}\sqrt{\varepsilon_r + 1}\right) - 1\right]} \tag{8.11}$$

The actual width of the line can be calculated by Equation 8.10. Some of the results are shown in Figure 8.3. The results are similar to Figure 8.2. However, for the same dielectric constant and width-to-height ratio $(\frac{W'}{h})$, microstrip lines have higher characteristic impedance than that of strip lines. In Figure 8.3, the effective filling factor q of the dielectric material of $\varepsilon_r = 3$ is also plotted. Since the microstrip line is partially surrounded by air and by dielectric material, the effective dielectric constant ε_e is a single value representing the overall effect. In other words, if the strip line is surrounded entirely by a uniform dielectric material of ε_e, the same effect will be obtained as a microstrip line.

The effective dielectric constant ε_e is related to q, the filling factor, through

$$\varepsilon_e = 1 + q(\varepsilon_r - 1) \tag{8.12}$$

From Equation 8.12 and q values in Figure 8.3, the value of ε_e can be found and the wave speed ratio $\frac{1}{\sqrt{\varepsilon_e}}$ can then be determined.

8.4 LOSSES IN MICROSTRIP[12-19]

Attenuation constant is another important constant in microstrip lines. The attenuation constant has three parts: (1) dielectric loss, (2) ohmic loss, and 3) radiation loss. The dielectric loss and the ohmic loss are the major losses in the microstrip and the radiation loss can be neglected if the microstrip is properly fabricated. The discussion here basically follows Reference 12. The voltage and current on the transmission line can be expressed in terms of an attenuator factor α as

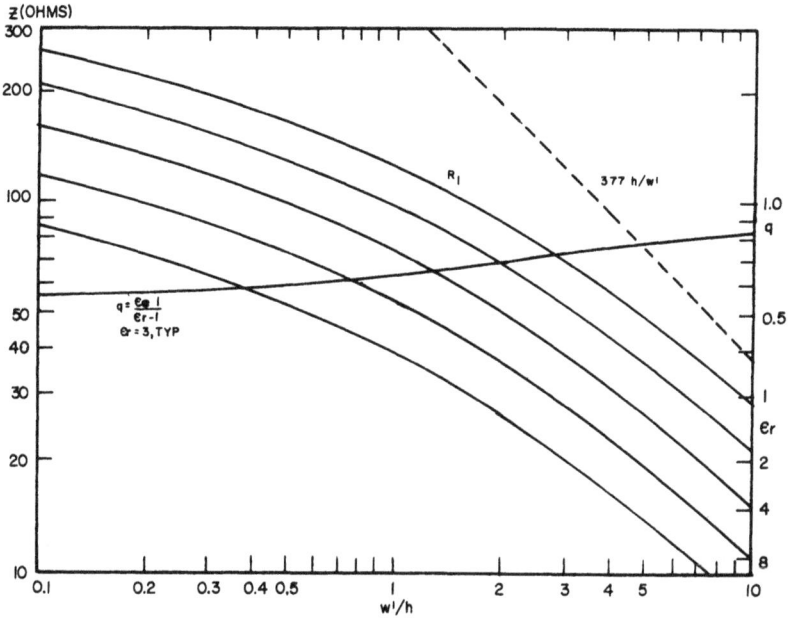

Figure 8.3. The Characteristic Impedance of a Thin Strip on a Dielectric Sheet on a Plane (Based on Wheeler, Reference 6)

$$V(z) = V_0 e^{-\alpha z}$$

$$I(z) = I_0 e^{-\alpha z} \tag{8.13}$$

where $V(z)$ and $I(z)$ are the voltage and current at distance z from the source.

V_0 and I_0 are the source voltage and current at $z = 0$. z is the distance along the microstrip line. The transmitted power along the z axis is

$$P(z) = P_0 e^{-2\alpha z} \tag{8.14}$$

where P_0 is the power at $z = 0$. Letting $\alpha = \alpha_d + \alpha_c$, the sum of the dielectric attenuation factor α_d and ohmic attenuation factor α_c, then

$$\alpha = \alpha_d + \alpha_c = -\frac{\dfrac{dP(z)}{dz}}{2P(z)} \cong \frac{P_d + P_c}{2P(z)} \frac{\text{nepers}}{\text{cm}} \tag{8.15}$$

or

$$\alpha_d \cong \frac{P_d}{2P(z)} \frac{\text{nepers}}{\text{cm}} \tag{8.16}$$

or

$$\alpha_c \cong \frac{P_c}{2P(z)} \, \frac{\text{nepers}}{\text{cm}} \qquad (8.17)$$

where P_d and P_c are the average dielectric power loss and the average conductor loss per unit length.

The dielectric attentuation constant α_d can be expressed as

$$\alpha_d \cong 27.3 \left(\frac{q\varepsilon_r}{\varepsilon_e} \right) \frac{\tan \delta}{\lambda_g} \, \frac{\text{dB}}{\text{cm}} \qquad (8.18)$$

or

$$\cong 4.43 \, \frac{q}{\sqrt{\varepsilon_e}} \sqrt{\frac{\mu_0}{\varepsilon_0}} \, \sigma \, \frac{\text{dB}}{\text{cm}} \qquad (8.19)$$

where q is the filling factor discussed in the last section, ε_r is the relative dielectric constant, ε_e is the effective dielectric constant which is related to the filling factor q by Equation 8.12, tan δ is the loss tangent of the dielectric material, λ_g is the wavelength of the guided wave, μ_0 is the free space permeability $= 4\pi \times 10^{-7}$ Henry/m, ε_0 is the free space permittivity $= 1/36\pi \times 10^{-9}$ Farad/m, and σ is the conductivity of the dielectric material.

It should be noted that the unit of α_d changes from $\frac{\text{neper}}{\text{cm}}$ in Equations 8.16 and 8.17 to $\frac{\text{dB}}{\text{cm}}$ in Equations 8.18 and 8.19. The conversion factor between neper and dB is 20 log e = 8.69.

The guided wavelength is related to the free wavelength by

$$\lambda_g = \frac{\lambda_0}{\sqrt{\varepsilon_e}} \qquad (8.20)$$

Calculating the filling factor q, (References 1 and 11), is rather tedious.

Pucel et. al (Reference 12) plotted the quantity $(\frac{q\varepsilon_r}{\varepsilon_e})$ versus $\frac{w}{h}$ in limited range as a function of dielectric constant ε_r as shown in Figure 8.4 The wider the microstrip, the higher the ε_r value, the more energy is concentrated between the strip line and the ground conductor, thus the higher the ratio $\frac{q\varepsilon_r}{\varepsilon_e}$.

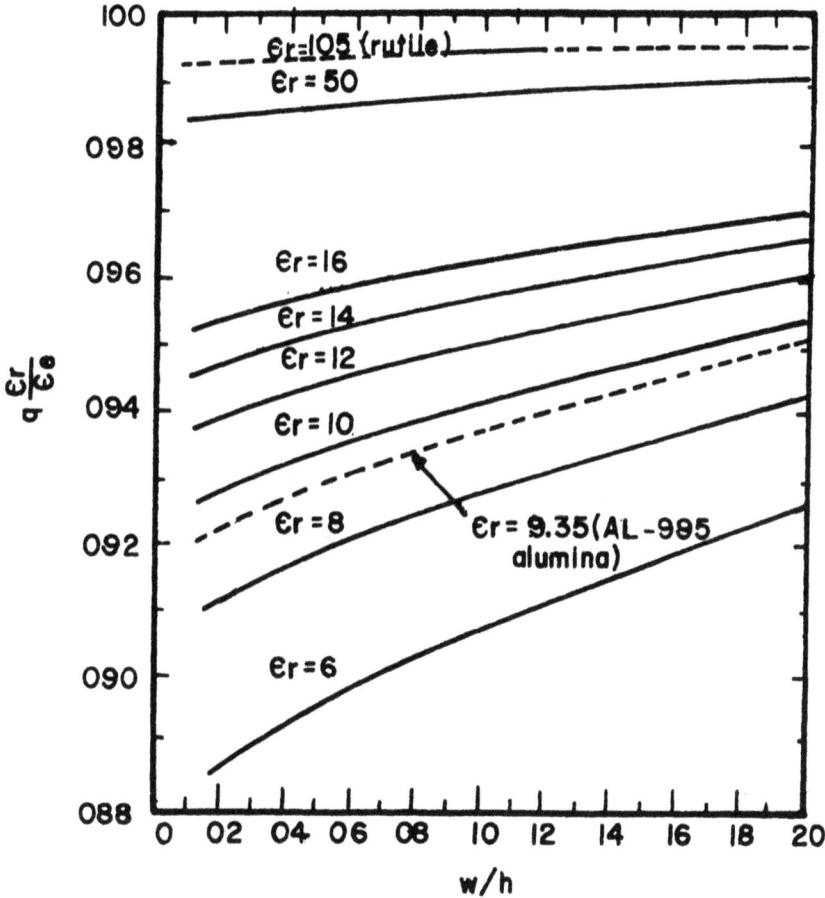

Figure 8.4. *Filling Factor for Loss Tangent of Microstrip Substrate as a Function of* $\frac{W}{H}$ *(Based on Pucel, Masse and Hartwig, Reference 12)*

Equation 8.18 is more convenient for nonconductive substrates whereas Equation 8.19 is appropriate for substrates in which the conduction loss is the predominant component of loss. However, the conductivity and the loss tangent are related as

$$\tan \delta = \frac{\sigma}{\omega \varepsilon} \qquad (8.21)$$

where ω is $2\pi f$ the angular frequency of the electromagnetic wave.

8.5 OHMIC LOSSES

If a low-loss dielectric substrate is used for the microstrip line, the predominant source of loss at microwave frequencies are the ohmic loss.

The high frequency will produce a skin effect on the conductor. The current density in the conductor is concentrated in a sheet a skin thickness deep on the surfaces exposed to the electric field. The skin depth can be expressed as

$$\delta = \frac{1}{\sqrt{\pi f \mu_0 \sigma}} \tag{8.22}$$

where f is the frequency, σ is the conductivity of the conductor, and μ is the permeability of the conductor which usually equals the free space value μ_0.

Equation 8.22 indicates the higher the frequency, the higher the conductivity, the shallower the skin depth. If the current distributions were known, the ohmic attenuation factor directly uses the expression (Reference 18).

$$\alpha_c = \frac{R_{s_1}}{2Z_0} \int_{c_1} \frac{|J_1|^2 \, dx}{|I|^2} + \frac{R_{s_2}}{2Z_0} \int_{c_2} \frac{|J_2|^2 \, dx}{|I|^2} \tag{8.23}$$

where R_{s_1} and R_{s_2} are the surface skin resistivity for the upper and lower conductor

$$R_{s_1} = \sqrt{\frac{\pi f \mu_1}{\sigma_1}} \tag{8.24}$$

$$R_{s_2} = \sqrt{\frac{\pi f \mu_2}{\sigma_2}} \tag{8.25}$$

μ_1, μ_2 are the permeability of top and bottom conductors, in general, $\mu_1 = \mu_2 = \mu_0$; σ_1 and σ_2 are the conductivity of top and bottom conductors, respectively; Z_0 is the characteristic impedance of the microstrip line; J_1, J_2 are the current density of the top and bottom conductors, respectively; I is the total current per conductor; c_1 and c_2 are the cross section of the top and bottom conductors, respectively.

If the width of both conductors are W and a uniform current density $\frac{I}{W}$ is assumed, then Equation 8.23 can be simplified as

$$\alpha_c = \frac{R_s}{Z_0 W} \frac{neper}{cm} = 20 \log e \frac{R_s}{Z_0 W} = 8.68 \frac{R_s}{Z_0 W} \frac{dB}{cm} \qquad (8.26)$$

where $R_{s_1} = R_{s_2} = R_s$. This simple expression is valid only for arbitrarily large strip width $\frac{W}{h} \rightarrow \infty$. Pucel et. al., (Reference 12) pointed out that when $\frac{W}{h} < 2$, Equation 8.26 may overestimate the skin loss by 80 percent, because the current density for narrow strips are far from uniform. The ohmic losses in the microstrip derived in Reference 13 based on Wheeler's analysis (References 1 and 19) are presented here.

For $\frac{W}{h} < \frac{1}{2\pi}$

$$\frac{\alpha_c Z_0 h}{R_s} = \frac{8.68}{2\pi} \left[1 - \left(\frac{W'}{4h}\right)^2 \right]$$

$$\cdot \left[1 + \frac{h}{W'} + \frac{h}{\pi W'} \left(\ln \frac{4\pi W'}{t} + \frac{t}{W'} \right) \right] \qquad (8.27)$$

For $\frac{1}{2\pi} < \frac{W}{h} < 2$

$$\frac{\alpha_c Z_0 h}{R_s} = \frac{8.68}{2\pi} \left[1 - \left(\frac{W'}{4h}\right)^2 \right]$$

$$\cdot \left[1 + \frac{h}{W'} + \frac{h}{\pi W'} \left(\ln \frac{2h}{t} - \frac{t}{h} \right) \right] \qquad (8.28)$$

For $\frac{W}{h} > 2$

$$\frac{\alpha_c Z_0 h}{R_s} = \frac{8.86}{\left\{ \frac{W'}{h} + \frac{2}{\pi} \ln \left[2\pi e \left(\frac{W'}{2h} + 0.94 \right) \right] \right\}^2}$$

$$\cdot \left[\frac{W'}{h} + \frac{\frac{W'}{\pi h}}{\frac{W'}{2h} + 0.94} \right] \left[1 + \frac{h}{W'} + \frac{h}{\pi W'} \left(\ln \frac{2h}{t} - \frac{t}{h} \right) \right]$$

$$(8.29)$$

where

$$W' = W + \frac{t}{\pi}\left(\ln\frac{4\pi W}{t} + 1\right), \quad \frac{W}{h} < \frac{1}{2\pi}$$

$$\left(\frac{2t}{h} < \frac{w}{h}, \frac{1}{2\pi}\right) \tag{8.30}$$

$$W' = W + \frac{t}{\pi}\left(\ln\frac{2h}{t} + 1\right), \qquad \frac{W}{h} > \frac{1}{2\pi} \tag{8.31}$$

These results are also shown in Figure 8.5. The loss factor $\frac{\alpha_c Z_0 h}{R_s}$ is

plotted against $\frac{w}{h}$ for different $\frac{t}{h}$ ratio.

8.6 COUPLED MICROSTRIP LINES[20-30]

Coupled microstrip lines constitute a major building block of microwave integrated circuits. Data such as characteristic impedance, capacitance, and velocity of propagation are greatly needed for design of filters, directional couplers, and phase shifters. There are many studies on the coupling effect. Bryant and Weiss, (Reference 20) treated the problem in a regionous manner (vacuum and the dielectrics) and introduced a "dielectric Green's function" which expresses the discontinuity of the fields at the dielectric coupling effect. Ramadan and Westgate (Reference 22) solved Laplace's equation by assuming an approximate boundary condition at the dielectric vacuum interface. Akhtarzad et. al., (Reference 21) solved the problem through conformal mapping. Judd et. al., (Reference 29) claimed that using finite difference techniques to solve microstrip lines takes a lot of computer time, so they solved the Laplace equation analytically through boundary matching. As expected, there are no simple solutions to this problem. Most of the results are presented in graphical form. Figure 8.6 shows the configuration of coupled pair microstrip lines. The width of the line is W. They are separated by distance S and height H above a conducting place. The lines are printed on a dielectric sheet of ϵ. In the coupled microstrip lines, there are two modes—even and odd.

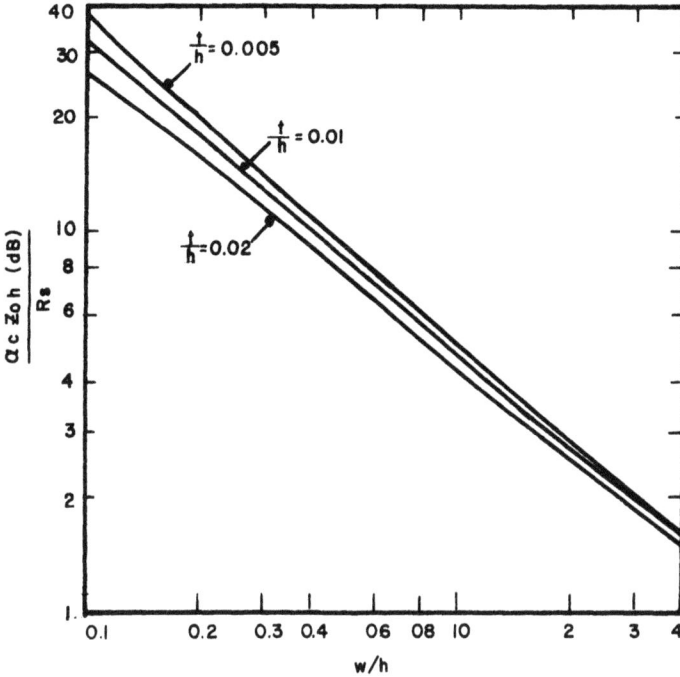

Figure 8.5. Theoretical Conductor Attenuation Factor of Microstrip as a
Function of $\dfrac{w}{h}$ (Based on Pucel, Reference 13)

The even mode can be considered as an electrostatic case with both conductors having the same sign and amount of charges. In the odd mode, the two conductors have the same amount of charges but are opposite in sign. Actually, an even mode corresponds to voltages and currents of equal and the same polarity on the conductors. An odd mode corresponds to voltages and currents of equal and opposite polarity on the conductors. The following boundary conditions will be helpful in understanding the even and odd modes (see Figure 8.6).

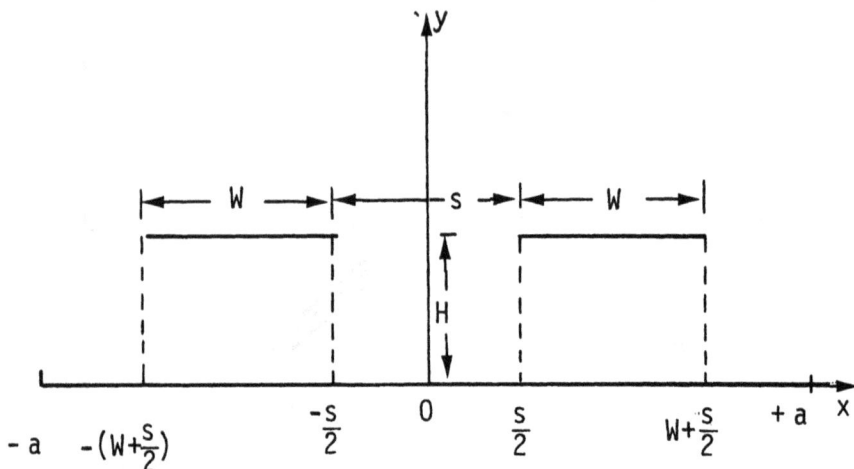

Figure 8.6. Transverse Cross-Section of Coupled Microstrip Lines

Even Mode

$$\phi(x, 0) = 0 \qquad\qquad (8.32)$$

$$\phi(x, \infty) = 0 \qquad\qquad (8.33)$$

$$\phi(\pm\infty, y) = 0 \qquad\qquad (8.34)$$

$$\frac{\partial\phi}{\partial x}(0, y) = 0 \qquad\qquad (8.35)$$

Odd Mode

$$\phi(x, 0) = 0 \qquad\qquad (8.36)$$

$$\phi(x, \infty) = 0 \qquad\qquad (8.37)$$

$$\phi(\pm\infty, y) = 0 \qquad\qquad (8.38)$$

$$\phi(0, Y) = 0 \qquad\qquad (8.39)$$

where ϕ is the potential function which satisfies the Laplace equation

$$\nabla^2\phi(x, y) = 0 \qquad\qquad (8.40)$$

Since a TEM mode is assumed on the microstrip line.

Akhtarzad et. al., (Reference 21) provided equations and figures for both analysis and synthesis. Their results are presented here. In the synthesis procedure, Z_{0_e} and Z_{0_0} which correspond to the even and odd mode characteristic impedances of the coupled lines are given, it is required to find the ratio $\frac{W}{H}$ and $\frac{S}{H}$ (see Figure 8.6 for W, H, and S). It takes two steps to find $\frac{W}{H}$ and $\frac{S}{H}$. The first step is to find $(\frac{W}{H})_s$ which represents the shape ratio for the equivalent single line in general. To be more specific, the first step is to find the two single-line, shape ratios $(\frac{W}{H})_{se}$ and $(\frac{W}{H})_{so}$ which correspond to the impedance $\frac{Z_{0e}}{2}$ and $\frac{Z_{0o}}{2}$, respectively.

$$(\frac{W}{H})_s = \frac{2}{\pi}(d-1) - \frac{2}{\pi}\ell n\,(2d-1)$$

$$+ \frac{\varepsilon_r-1}{\pi\varepsilon_r}\left[\ell n\,(d-1) + 0.293 - \frac{0.517}{\varepsilon_r}\right] \qquad (8.41)$$

where

$$d = \frac{60\pi^2}{Z_0(\varepsilon_r)^{1/2}} \qquad (8.42)$$

For the even mode, $Z_0 = \frac{Z_{0e}}{2}$ is used in Equation 8.42, and the value of $(\frac{W}{H})_s$ calculated in Equation 8.41 corresponds to $(\frac{W}{H})_{se}$. For the odd mode, $Z_0 = \frac{Z_{0o}}{2}$ is used in Equation 8.42 and the value of $(\frac{W}{H})_s$ obtained from Equation 8.41 corresponds to $(\frac{W}{H})_{so}$. The results of $(\frac{W}{H})_s$ can also be obtained from Figure 8.7. When Z_0 is replaced by $\frac{Z_{0e}}{2}$ or $\frac{Z_{0o}}{2}$, the corresponding $(\frac{W}{H})_s$ can be read directly from the abscissa of Figure 8.7.

The second step is to find $\frac{W}{H}$ and $\frac{S}{H}$ from $(\frac{W}{H})_{oe}$ and $(\frac{W}{H})_{oo}$. The $\frac{W}{H}$ and $\frac{S}{H}$ for the coupled lines can be found by the simultaneous solution of the following formulas.

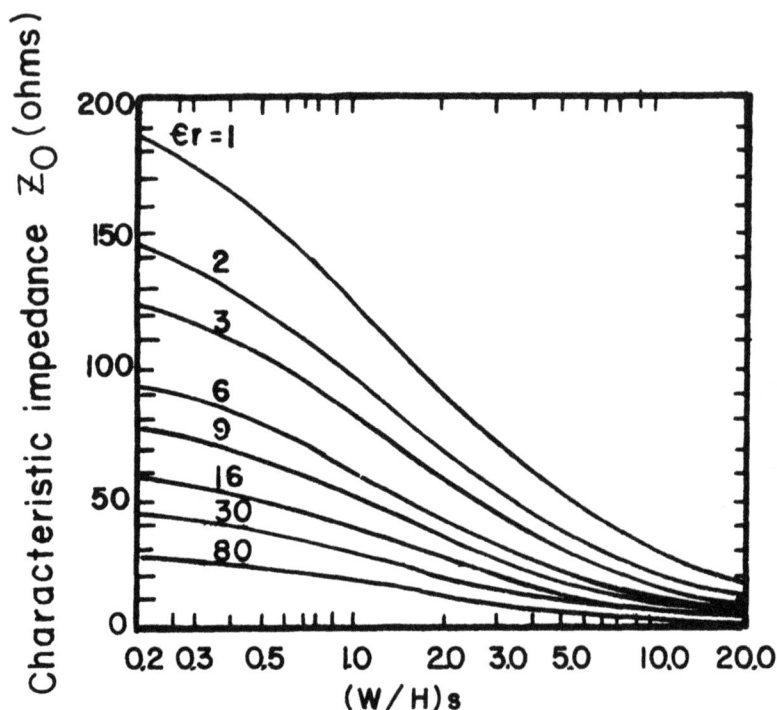

Figure 8.7. Single Microstrip Line Characteristic Impedance (Based on Akhtarzad et. al., Reference 21)

$$\left(\frac{W}{H}\right)_{se} = \frac{2}{\pi}\cosh^{-1}\left(\frac{2h - g + 1}{g + 1}\right) \qquad (8.43)$$

$$\left(\frac{W}{H}\right)_{so} = \frac{2}{\pi}\cosh^{-1}\left(\frac{2h - g - 1}{g - 1}\right) \qquad (8.44a)$$

$$+ \frac{4}{\pi\left(1 + \frac{\varepsilon_r}{2}\right)}\cosh^{-1}\left(1 + 2\frac{W/H}{S/H}\right), \ \varepsilon_r < 6$$

$$\left(\frac{W}{H}\right)_{so} = \frac{2}{\pi}\cosh^{-1}\left(\frac{2h - g - 1}{g - 1}\right)$$

$$+ \frac{1}{\pi}\cosh^{-1}\left(1 + 2\frac{W/H}{S/H}\right), \ \varepsilon_r > 6 \qquad (8.44b)$$

where

$$g = \cosh\left[\frac{\pi}{2}\left(\frac{S}{H}\right)\right] \qquad (8.45)$$

and

$$h = \cosh\left[\pi\left(\frac{W}{H}\right) + \frac{\pi}{2}\left(\frac{S}{H}\right)\right] \qquad (8.46)$$

Equations 8.43 and 8.44 are given in graphical form in Figure 8.8 where the curves are plotted for a fixed value of permittivity $\varepsilon_r = 6$. The relationship to a single microstrip line cannot be made independent of the permittivity, but the formula with $\varepsilon_r = 6$ gives accurate results for substrates with permittivity $\varepsilon_r = 6$ and above. For permittivities down to $\varepsilon_r = 2$, the formula for $\varepsilon_r = 6$ (Figure 8.8) has errors up to above 10 percent for the worst $\frac{S}{H}$ and $\frac{W}{H}$ values. For greater accuracy at low permittivities and certainly for values of permittivity less than $\varepsilon_r = 2$, the equation for $\varepsilon_r < 6$ should be used, and in this case it will be noticed that ε_r appears as a variable.

The solution of simultaneous Equations 8.43 and 8.44 is greatly eased by ignoring the second term in Equation 8.44. A value of $\frac{S}{H}$ is then given directly by

$$\left(\frac{S}{H}\right) = \frac{2}{\pi}$$

$$\cdot \cosh^{-1}\left\{\frac{\cosh\left[\frac{\pi}{2}\left(\frac{W}{H}\right)_{se}\right] + \cosh\left[\frac{\pi}{2}\left(\frac{W}{H}\right)_{so}\right] - 2}{\cosh\left[\frac{\pi}{2}\left(\frac{W}{H}\right)_{so}\right] - \cosh\left[\frac{\pi}{2}\left(\frac{W}{H}\right)_{se}\right]}\right\} \qquad (8.47)$$

In most cases Equation 8.47 is sufficiently accurate. Equation 8.47 is a starting point for an optimization process in the solution of Equation 8.43 and 8.44 if more accuracy is desired (particularly at lower values of ε_r).

In the analysis procedure, $\frac{W}{H}$ and $\frac{S}{H}$ for the coupled lines are known and it is required to find Z_{0_e} and Z_{0_o}. The first step is to find the two single-line shape ratio $\left(\frac{S}{H}\right)_{se}$ and $\left(\frac{W}{H}\right)_{so}$ corresponding to $\frac{W}{H}$ and $\frac{S}{H}$ for the even and odd modes, respectively. They can be obtained from Equations 8.43 and 8.44 or Figure 8.8. The characteristic impedance (Z_0) for a single strip corresponding to the shape ratio $(\frac{W}{H})_{se}$ and $(\frac{W}{H})_{so}$ can be read directly from Figure 8.7 or calculated by

$$Z_0 = \frac{120\pi \left(\dfrac{1}{\varepsilon_r}\right)^{\frac{1}{2}}}{(\dfrac{W}{H})_s + 0.882 + K + \left(\dfrac{\varepsilon_r - 1}{\varepsilon_r^2}\right)(0.164)} \qquad (8.48a)$$

where

$$K = \left(\frac{\varepsilon_r + 1}{\pi\varepsilon_r}\right)\left\{ ln\left[(\frac{W}{H})_s + 1.88\right] + 0.758\right\} \qquad (8.48b)$$

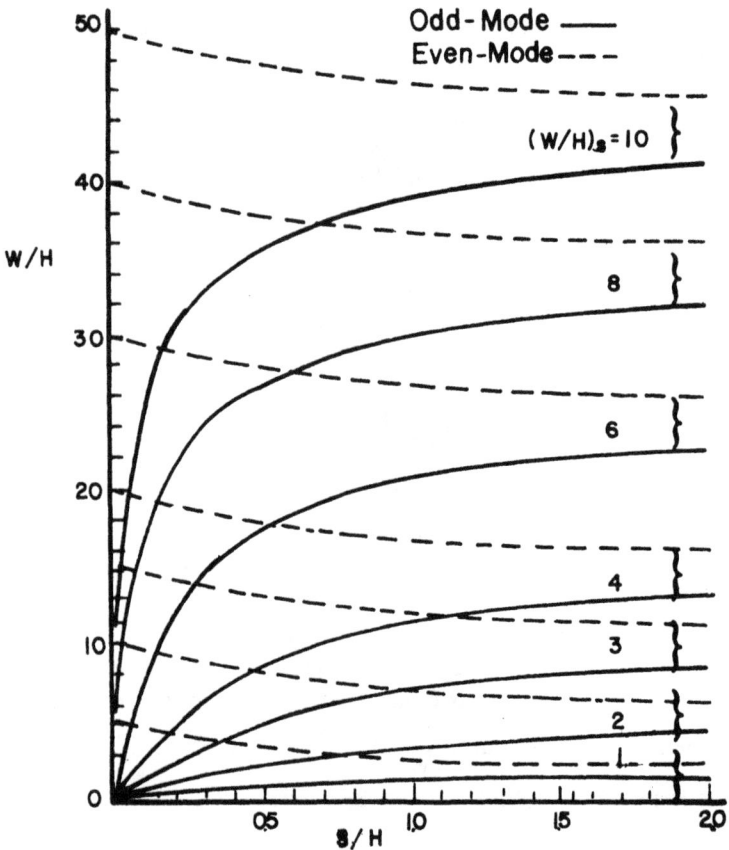

Figure 8.8. Synthesis Curves for Coupled Microstrip Lines (Based on Akhtarzad et. al., Reference 21)

The even and odd mode impedance are then given by

$$Z_{0_e} = 2 Z_0 \quad \text{for shape ratio } (\frac{W}{H})_{se}$$

$$Z_{0_o} = 2 Z_0 \quad \text{for shape ratio } (\frac{W}{H})_{so}$$

The synthesis and analysis approaches mentioned above are straightforward. The theoretical results agree within 12 percent with Reference 29 and 14 percent with Reference 20, over a wide range of $\frac{S}{H} = 0.1\text{-}2.0$ and $\frac{W}{H} = 0.1\text{-}2.0$. The results obtained from the above approaches also agree with experimental results.

8.7 SLOTLINES[31-34]

There is another kind of microwave transmission line which is fabricated on dielectric substrate. This transmission line is called slotline which was discussed by Cohn (Reference 31) in 1969. The slotline is formed by etching a narrow slot in the metallization on one side of the surface of the substrate, as shown in Figure 8.9. There is no conducting plate on the opposite side of the substrate. The energy is carried in both the dielectric substrate and the free space above it. A voltage difference exists between the slot edges. The electric field extends across the slot and the magnetic field is perpendicular to the slot as shown in Figure 8.10a. The longitudinal view in Figure 8.10b shows that in the air regions the magnetic field lines curve and return to the slot at half-wavelength intervals. The current paths on the conducting surface are shown in Figure 8.10c. The surface current density is greatest at the edges of the slot and decreases rapidly farther from the slot. Since the dielectric ε_r is greater than 1, more energy is carried through the dielectric substrate than in the air. The availability of slotlines can improve the flexibility of microwave designs.

In order to make a slotline as practical as a transmission line, radiation must be minimized. This is accomplished through the use of a high dielectric constant substrate, which causes the slot-mode wavelength λ' to be small compared to free space wavelength λ and thereby results in the fields being closely confined to the slot with negligible radiation loss. As a first order approximation, the propagation constant along the slot can be expressed into the following simple formula for the relative wavelength.

$$\frac{\lambda'}{\lambda} = \sqrt{\frac{2}{\varepsilon_r + 1}} \tag{8.49}$$

Since wavelength is inversely proportional to the square root of dielectric constant, the effective dielectric constant of a uniform medium replacing the two different dielectric half spaces may be defined as

$$\varepsilon_e = \sqrt{\frac{\lambda}{\lambda'}} = \frac{\varepsilon_r + 1}{2} \tag{8.50}$$

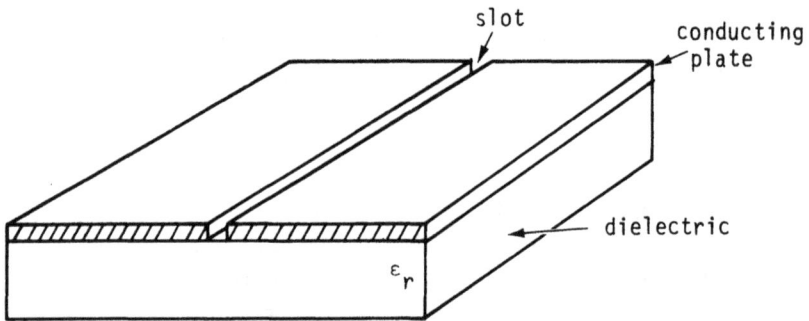

Figure 8.9. Slotline Configuration

Equation 8.49 provides a fair approximation for slotline, yielding values about 10 percent in typical cases, comparing with resolutions taking second order into consideration.

To fulfill the boundary condition, TEM cannot be supported by the slotline. Because of the non-TEM nature of the slotline mode, the characteristic impedance cannot be defined uniquely and is not a constant, but varies with frequency at a rather slow rate per octave. This behavior contrasts with quasi-TEM microstrip line, whose first-order approximated characteristic impedance Z_0 is independent of frequency. TEM mode is commonly assumed for analysis purposes. The most common definition of characteristic impedance Z is through the power relation.

$$Z = \frac{V^2}{2P} \tag{8.51}$$

where V is the peak voltage across the slot and P is the average power carried by the line.

The characteristic impedance of the slotline has been derived in Reference 31. The results will not be presented here because of their complexity. Instead, the characteristic impedance of a TEM mode will be presented. This result can be considered as a first-order approximation.

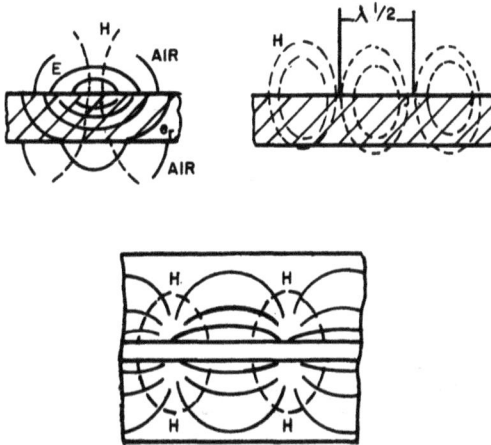

Figure 8.10. *Field and Current Distribution   a. Field Distribution in Cross Section   b. H field in Longitudinal Section c. Current Distribution on Metal Surface (Based on Cohn, Reference 31)*

$$Z_0 = \frac{591.7}{\sqrt{\varepsilon_e}\, \ln\left(\dfrac{8b}{\pi W}\right)}$$

$$= \frac{591.7\left(\dfrac{\lambda'}{\lambda}\right)}{\ln\left(\dfrac{8b}{\pi W}\right)} \quad \text{ohm} \quad \frac{b}{W} > 3 \quad \frac{b}{\lambda'} \to 0 \qquad (8.52)$$

where ε_e is the effective dielectric constant in Equation 8.50, b is the width of the slotline including the slot gap (Figure 8.9), and W is the width of the slot. For small value of b, Equation 8.52 can provide excellent results in comparison with results obtained by second-order solutions.

Electric signals can be coupled to and from the slotline through the coaxial cable. As shown in Figure 8.11, the coaxial cable is placed perpendicular to the slot, the inner conductor of the coaxial cable is connected to one side of the slot while the outer conductor is connected to the opposite of the slot. The slotline can also be fed through the microstrip line which is on the opposite of the substrate as shown in Figure 8.12a. The slotline and the microstrip line are perpendicular to each other and each one of the lines will overhang the other one by $\frac{\lambda}{4}$ as shown in Figure 8.12b. A magic

Tee in MIC form was fabricated through the use of microstrip lines and slotlines (Reference 32). (This will be discussed in Chapter 10.)

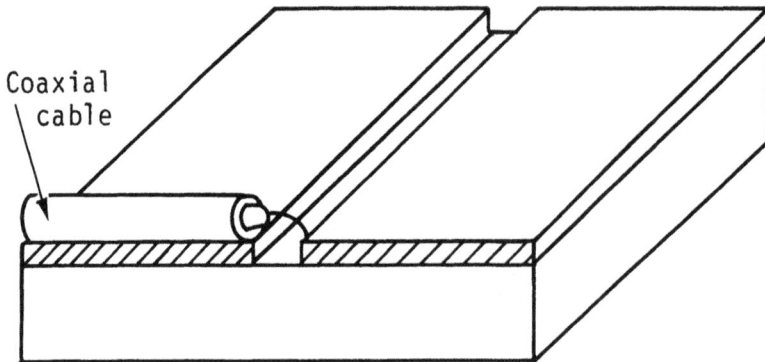

Figure 8.11. Slotline Feed Through Coaxial Cable

8.8 FIN LINES AND SUSPENDED STRIP LINES[35-41]

Fin lines and suspended strip lines are used at high frequencies, usually above 20 GHz and up to a few hundred gigahertz. They are transmission lines printed on substrate and enclosed in a rectangular shaped metallic waveguide. When the operating frequency is higher, the fundamental problems (i.e., radiation loss, dispersion, and higher mode) of ordinary microstrip lines will become more severe. Although these problems can be controlled by choosing thinner substrates at a higher operating frequency, this approach will degrade the Q-factor, compound tolerance problems, and restrict the range over which the characteristic impedance can be varied. Fin lines and suspended microstrips can be used at the frequency that is too high for ordinary microstrips. Many of the components (i.e., mixers) at higher frequencies have used fin line and suspended line structure. These kinds of transmission lines are not popular at lower frequencies because a conducting waveguide is required which complicates the fabrication. It is mostly used in the mm wave region.

Figure 8.13a shows an insulated fin line. The fins are printed on a dielectric substrate which does not have a conducting plate on the opposite side. This is referred to as the insulated fin line, because both the upper and lower fins are insulated from the metallic housing (waveguide). Bias may be introduced for active components mounted on the substrate. RF continuity between the fins and the waveguide wall is obtained by choosing

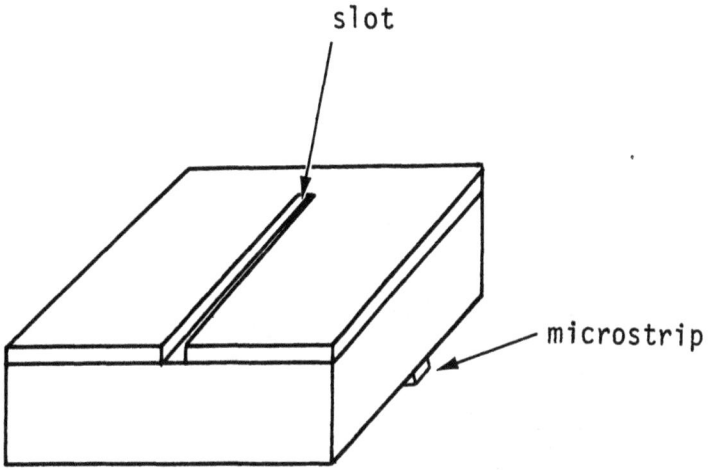

(a) Perspective View

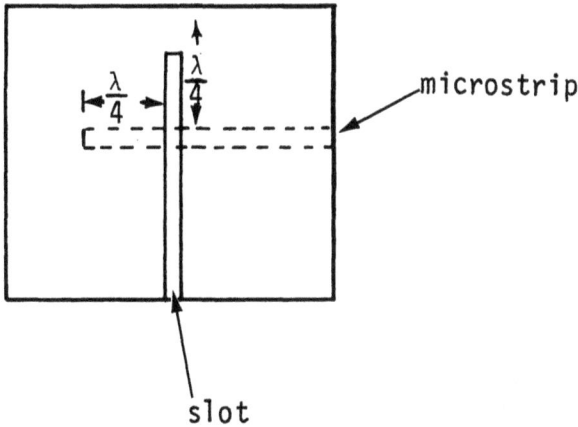

(b) Top View

Figure 8.12. Slotline Feed Through Microstrip Line

the thickness of the broad walls to be a quarter-wavelength in the dielectric medium and by selecting c ≪ a, where c is the dielectric thickness and a is the major inner dimension of the waveguide (Figure 8.13a).

A grounded fin configuration is shown in Figure 8.13b. The fins in this configuration are printed on both sides of a single dielectric substrate.

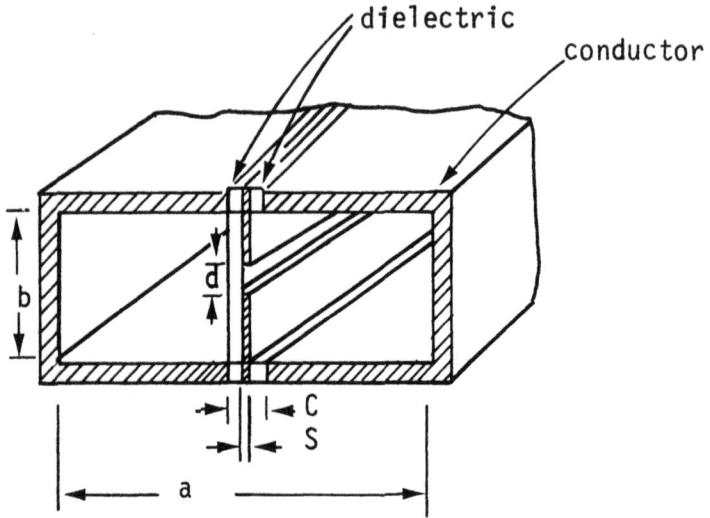

(a) Insulated Fins

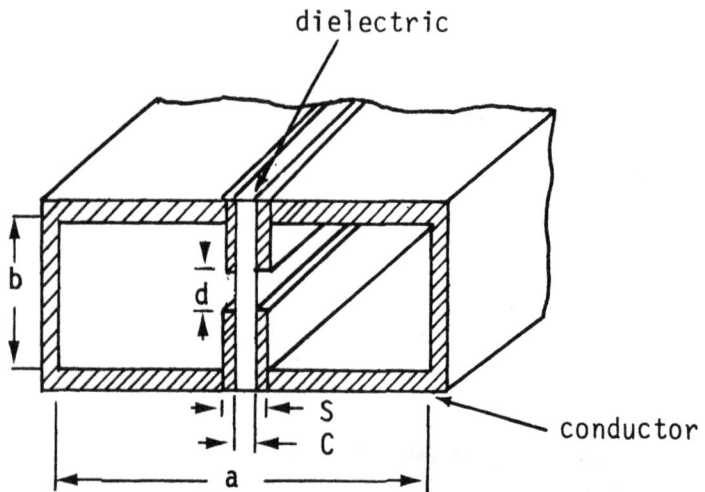

(b) Grounded Fins

Figure 8.13. Fin Line Cavity Configuration

Since the fins are directly grounded to the metal waveguide walls, it is impractical to apply dc bias to active components on the substrate. This configuration is applicable only to passive devices but it has lower loss than that of Figure 8.13a.

An important property of the fin line is the guided wavelength variation which can be calculated from the simple equation (Reference 35)

$$\lambda_g = \frac{\lambda_0}{\left[\varepsilon_e - \left(\frac{\lambda_0}{\lambda_c}\right)^2\right]^{1/2}} \tag{8.53}$$

where ε_e is the equivalent dielectric constant (see Section 8.2), λ_0 is the free space wavelength and λ_c is the cut-off wavelength for an air-filled ridge-loaded waveguide of the same dimensions (Reference 41). The value of ε_e may be experimentally determined for a given fin line configuration. It has been shown that the value of ε_e varies little with respect to d/b, when the fins are grounded directly to the waveguide. For example in Figure 8.13b, if the relative dielectric constant of the substrate is 2.5, $\frac{s}{a} = 0.07$ (where s is the equivalent ridge thickness, and d/b = 0.1), then $\varepsilon_e = 1.50$. For $\frac{d}{b} = 1.0$ and the other conditions are the same, then $\varepsilon_e = 1.25$, the change of ε_e is 0.25. If the value of ε_e is determined, the guided wavelength can then be obtained.

A suspended strip line is shown in Figure 8.14a and b. The circuit is printed on a dielectric substrate which is sandwiched in the conducting waveguide. Figure 8.14a is a single registry type while Figure 8.14b is a double registry type. Because these lines are doubly shielded, they have a potentially lower loss than microstrip. The effective dielectric constant of these lines is only slightly greater than unity, because most of the cross section is air. The transmission mode is very near TEM, because the configuration is similar to a coaxial transmission line. In Figure 8.14b both the printed conductors have the same potential and can be treated as one conductor. One material to fabricate this circuit has a commercial name of Duroid with a dielectric constant of approximately 2.2.

8.9 DIELECTRIC TRANSMISSION LINES(42-45)

Like the fin lines, the dielectric waveguides are used at a very high frequency often in the mm wave and optical frequency region. The concept of the dielectric waveguide was discussed in the fifties. They are becoming more popular in recent developments in mm wave integrated circuits. The dielectric transmission lines will only be briefly discussed here since they have not been widely adapted in microwave receivers yet. A simple dielectric waveguide is a piece of dielectric material with high dielectric

dielectric

conductor

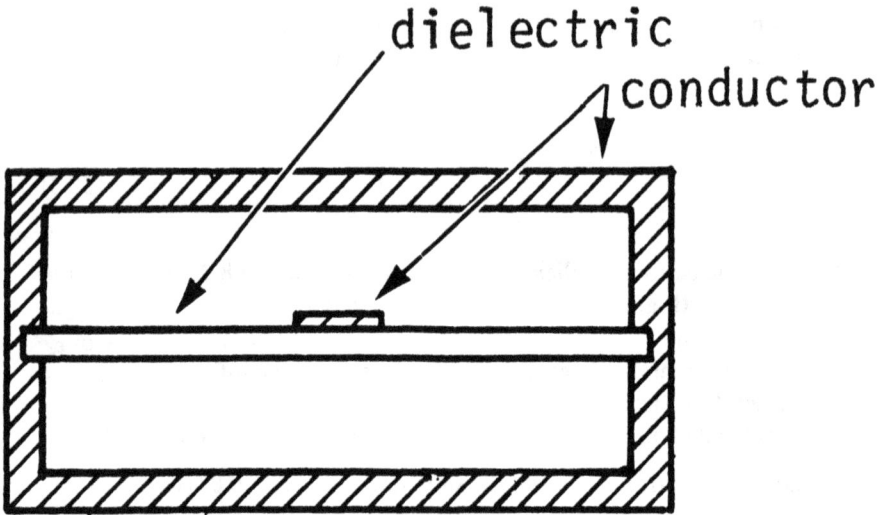

(a) Single Registry Type

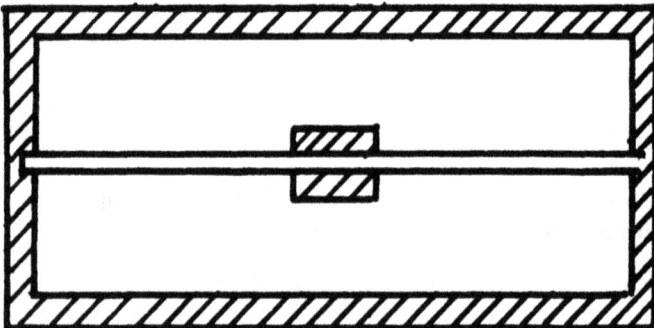

(b) Double Registry Type

Figure 8.14. Suspended Microstrip

constant and low loss. It can be either in rectangular or cylindrical form as shown in Figure 8.15. The high dielectric constant will retain the electromagnetic energy inside the dielectric material. There are many different configurations of a dielectric waveguide. A few of them will be presented here. A dielectric image line is a rectangular dielectric waveguide formed on a conductor as shown in Figure 8.16. Because of the conductor loss, the dielectric image line has higher loss than a dielectric waveguide. An insular dielectric waveguide is shown in Figure 8.17. It is a dielectric waveguide that is separated from the conductor by another dielectric sheet. This approach is

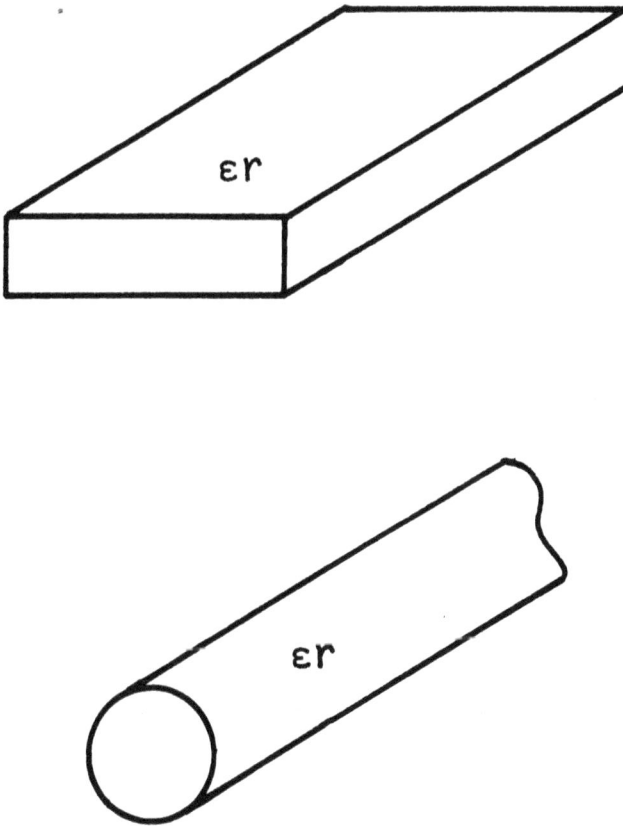

Figure 8.15. Dielectric Waveguide

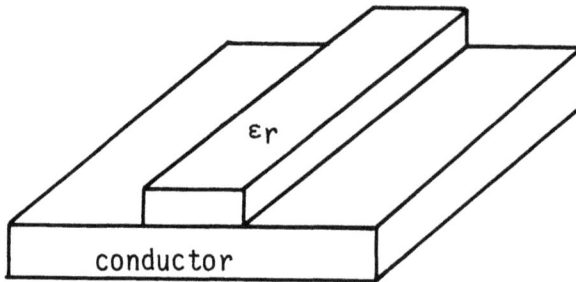

Figure 8.16. Dielectric Image Line

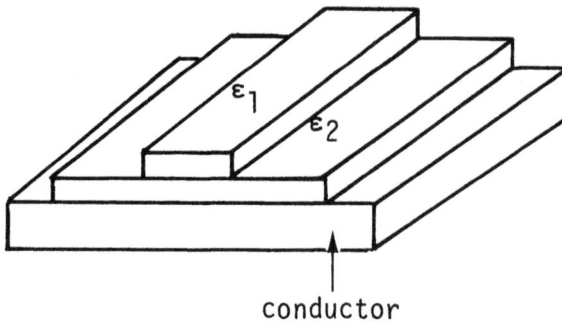

Figure 8.17. Insular Dielectric Waveguide

intended to reduce the loss of the image line. In general, $\varepsilon_1 > \varepsilon_2$ and the electromagnetic energy is most restricted in the waveguide. The conductor on the image line will provide a grounding plane for the possibility of mounting active devices in the circuit.

8.10 TRANSMISSION LINE SUBSTRATE MATERIALS[46-49]

There are many types of substrates which are useful for building strip line circuits (Reference 45). Copper conductor is usually laminated to one or both sides of the substrate. Some of these substrate materials are listed in Table 8.1.

TABLE 8.1. STRIP LINE SUBSTRATE MATERIALS

Substrate Materials	ε_r	Dissipation Factor
Glass Reinforced Teflon (Woven)	2.4 ∿ 2.62	.0015 ∿ .002
Glass Reinforced Teflon (Microfiber)	2.32 ∿ 2.40	.0004 ∿ .0008
Polyolefin	2.32	.0003
Cross-Linked Polystyrene	2.53	.00025 ∿ .00066
Glass Reinforced Polystyrene	2.62	.0004 ∿ .002
Polyphenelene Oxide (PPO)	2.55	.0016
Ceramic Filled Resins	1.7 ∿ 25	.0005 ∿ .005
Quartz Loaded Teflon	2.47	.0006

Among these materials, the glass reinforced teflon is one of the most commonly used materials. With the exception of the ceramic filled resins, the other materials listed have relative dielectric constant ranges from 2.3 to 2.6. Since the guided wave wavelength λ_g is related to the free space wavelength λ_0 by

$$\lambda g = \frac{\lambda_0}{\sqrt{\varepsilon_r}}$$ (8.54)

the guided wave wavelength is approximately .64 of that of free space wavelength.

The general requirements of a microstrip substrate materials are: (1) high dielectric constants (nine or higher), (2) low dissipation factor, (3) dielectric constants should remain constant over frequency and temperature range of interest, (4) high resistivity and dielectric strength and (5) high thermal conductivity. A few substrate materials, which are popular for microstrip applications, are listed in Table 8.2.

TABLE 8.2. MICROSTRIP SUBSTRATE MATERIALS

Substrate Materials	Loss Tangent at 10 GHz ($\times 10^{-4}$)	ε_r	Thermal Conductivity K(W/cm °C)
Alumina	2	10	0.3
Berryllia	1	6	2.5
GaAs	16	13	0.03
Rutile	4	100	0.02
Sapphire	1	9.3-11.7	0.4

Alumina (Al_2O_3) is the most commonly used substrate material and GaAs is used for high frequency circuits and monolithic MIC. If heat dissipation is the primary concern of the microwave integrated circuits, berryllia should be considered because of its good thermal conductivity.

The conductors on the substrate should be at least four times the skin depth to have a low electric resistance. The conductor materials commonly used on the substrates are (References 47 and 48) gold (Au), silver (Ag), copper (Cu), and aluminum (Al). With the exception of aluminum, the other three conductors have very poor adhesion to the substrates. It is possible to obtain good adhesion with high conductivity materials by using a thin film of relatively poor conductivity but good adhesion between the substrate and the good conductor. Some typical combinations are Cr-Au, Cr-Cu, and Ta-Cu.

REFERENCES

1. Wheeler, H.A., "Transmission line properties of parallel strips separated by dielectric sheet," IEEE trans. on Microwave Theory and Techniques, Vol. MTT-13, p. 172-185, March 1965.

2. Caulton, M., and Sobol, H., "Microwave integrated-circuit technology—a survey," IEEE J. Solid State Circuits SC-5, p. 292-303, December 1970.

3. Wheeler, H.A., "Transmission-line properties of a strip line between parallel plates," IEEE Trans. on Microwave Theory and Techniques, Vol. MTT-26, p. 866-876, November 1978.

4. Gruner, L., "Estimating rectangular coax cutoff," Microwave Journal, p. 88, April 1979.

5. Bahl, I.J., and Garg, R., "A designer's guide to strip line circuits," Microwaves, p. 90, January 1980.

6. Wheeler, H.A., "Transmission-line properties of a strip on a dielectric sheet on a plane," IEEE Trans. on Microwave Theory and Techniques, Vol. MTT-25, p. 631-647, August 1977.

7. Stinehelfer, H.E., "An accurate calculation of uniform microstrip transmission lines," IEEE Trans. on Microwave Theory and Technique, Vol. MTT-16, p. 439-444, July 1968.

8. Walters, K.C., and Clar, P.L., "Microstrip transmission lines on high dielectric constant substrates for hybrid microwave integrated circuits," presented at the 1967 International Microwave Symp., Boston, MA, May 1967, Session V-2.

9. Yamashita E., and Mittra, R., "Variational method for the analysis of microstrip lines," IEEE Trans. on Microwave Theory and Techniques,

Vol. MTT-16, p. 251-255, April 1968.

10. Yamashita, E., "Variational method for the analysis of microstrip-like transmission lines," IEEE Trans. on Microwave Theory and Techniques, Vol. MTT-16, p. 529-535, August 1968.

11. Garg, H., "Strip-line like microstrip configuration," Microwave Journal, p. 103, April 1979.

12. Pucel, R.A., Masse, D.J., and Hartwig, C.P., "Losses in microstrip," IEEE Trans. on Microwave Theory and Techniques, Vol. MTT-16, p. 342-350, June 1968.

13. Pucel, R.A., Masse, D.J., and Hartwig, C.P., "Correction to losses in microstrip," IEEE Trans. on Microwave Theory and Techniques, Vol. MTT-16, p. 1064, December 1968.

14. Schneider, M.V., "Dielectric loss in integrated microwave circuits," Bell System Technical Journal, Vol. 48, p. 2325-2332, 1969.

15. Solbach, K., "The measurement of the radiation losses in dielectric image line bends and the calculation of a minimum acceptable curvature radius," IEEE Trans. on Microwave Theory and Techniques, Vol. MTT-27, p. 51-53, 1979.

16. Kajfez, D., and Tew, M.D., "Pocket calculator program for analysis of glossy microstrip," Microwave Journal, p. 39, December 1980.

17. Wheeler, H.A., "Transmission—line properties of parallel wide strips by a conformal-mapping approximation," IEEE Trans. on Microwave Theory and Techniques, Vol. MTT-12, p. 280-289, May 1964.

18. Ramo, S., and Whinnery, J.R., "Fields and waves in modern radio," New York: Wiley, 1944, Chapter 8.

19. Wheeler, H.A., "Formulas for the skin effect," Proc. IRE, Vol. 30, p. 412-424, September 1942.

20. Bryant, T.G., and Weiss, J.A., "Parameters of microstrip transmission lines and of coupled pairs of microstrip lines," IEEE Trans. on Microwave Theory and Techniques, Vol. MTT-16, p. 1021-1064, December 1968.

21. Akhtarzad, S., Rowbotham, T.R., and Johns, P.B., "The design of coupled microstrip lines," IEEE Trans. on Microwave Theory and Techniques, Vol. MTT-23, p. 486-492, June 1975.

22. Ramadan, M., and Westgate, W.F., "Impedance of coupled microstrip transmission lines," Microwave Journal, Vol. 14, p. 30-35, July 1971.

23. Krage, M.K., and Haddad, G.I., "Characteristics of coupled microstrip transmission lines—I. Coupled-mode formulation of inhomogeneous lines," IEEE Trans. on Microwave Theory and Techniques, Vol. MTT-18, p. 217-222, April 1970.

24. Gauthier, F., and Besse, M., "Graphical design of coupled microstrip lines," Microwave Journal, p. 36, February 1974.

25. Chen, W.H., "Even and odd mode impedance of coupled pairs of microstrip lines," IEEE Trans. on Microwave Theory and Techniques,

Vol. MTT-18, p. 55-57, January 1970.

26. Wahi, P.K., "Parameters of coupled pair of strip line-like microstrip lines," Microwave Journal, p. 47-49, December 1979.

27. Chambers, D.R., "Take the hassle out of microstrip coupling," Microwaves, p. 48, July 1974.

28. Zehentner, J., "Analysis and synthesis of coupled microstrip lines by polynomials," Microwave Journal, p. 95, May 1980.

29. Judd, S.V., Whiteley, I., Clowes, R.J., and Rickard, D.C., "An analytical method for calculating microstrip transmission line parameters," IEEE Trans. on Microwave Theory and Techniques, Vol. MTT-18, p. 78-87, February 1970.

30. Liao, S.Y., Microwave Devices and Circuits, Prentice-Hall, June 1980.

31. Cohn, S.B., "Slotline on a dielectric substrate," IEEE Trans. on Microwave Theory and Techniques, Vol. MTT-17, p. 768-788, October 1969.

32. Aikawa, M., and Ogawa, H., "A new MIC magic-T using coupled slotlines," IEEE Trans. on Microwave Theory and Techniques, Vol. MTT-28, p. 523-528, June 1980.

33. Gupta, K.C., Garg, R., and Bahl, I.J., Microstrip lines and slotlines, Artech House, Inc., 1979.

34. Vogel, R., "Microstrip-slotline components for microwave ICs," Microwave Journal, p. 83, May 1980.

35. Meier, P.J., "Integrated fin line millimeter components," IEEE Trans. on Microwave Theory and Techniques, Vol. MTT-22, p. 1209-1216, December 1974.

36. Cohen, L.D., and Meier, P.J., "E-plane mm wave circuits," Microwave Journal, p. 63, August 1978.

37. Saad, A.M.K., and Schunemann, K., "A simple method for analyzing fin line structures," IEEE Trans. on Microwave Theory and Technique, Vol. MTT-26, p. 1002-1007, December 1978.

38. Bates, R.N., Coleman, M.D., Nightingale, S.J., and Davies, R., "E-planes drop millimeter costs," Microwave System News, p. 74, December 1980.

39. Harris, D.J., "Waveguiding difficult at near-millimeters," Microwave System News, p. 62, December 1980.

40. Rubin, D., Hislop, A.R., "Millimeter-wave coupled line filters—design techniques for suspended substrate and microstrip," Microwave Journal, p. 67, October 1980.

41. Hopfer, S., "The design of ridged waveguides," IRE Trans. Microwave Theory and Techniques Vol. MTT-3, pp. 20-29, October 1955.

42. King, D.D., "Dielectric image line," Journal Applied Physics, Vol. 23, p. 699, June 1952.

43. Knox, R., "Dielectric waveguide microwave integrated circuits—an overview," IEEE Trans. on Microwave Theory and Techniques, Vol.

MTT-24, p. 806-814, November 1976.

44. Itoh, T., "Inverted strip dielectric waveguide for millimeters wave integrated circuits," IEEE Trans. on Microwave Theory and Techniques, Vol. MTT-24, p. 821-827, November 1976.

45. Salbach, K., and Wolff, I., "The electromagnetic fields and the phase constants of dielectric image lines," IEEE Trans. on Microwave Theory and Techniques, Vol. MTT-26, p. 266-274, 1978.

46. Howe, H., *Strip line circuit design,* Chapter 1, Artech House, Inc., 1974.

47. Keister, F.Z., "An evaluation of materials and processor for integrated microwave circuits," IEEE Trans. on Microwave Theory and Techniques, Vol. MTT-16, p. 469-475, July 1968.

48. Sobol, H., "Applications of integrated circuit technology to microwave frequencies," Proc. IEEE, Vol. 59, p. 1200-1211, August 1971.

49. Ch'en, D.R., and Decker, D.R., "MMICs, the next generation of microwave components," Microwave Journal, p. 67, May 1980.

CHAPTER 9
DELAY LINES

9.1 INTRODUCTION[1-4]

A delay line is a passive component. The output signal is delayed by a specific amount of time with respect to the input signal. There are video delay lines and radio frequency (RF) delay lines. The video delay line is usually used after the detector or log video amplifier where the RF in the signal is filtered out. The delay line can delay the video signal for various signal processing purposes (i.e., the delay line canceler for moving-target-indicator (MTI) radars) (Reference 1). The video delay line must have enough bandwidth to pass the signal and retain the waveform.

An RF delay line delays signals with the frequency ranging from a few megahertz to over ten gigahertz. It is used before the detector in the RF or intermediate frequency (IF) circuits. Sometimes the video delay and RF delay lines have similar performance requirements. For example, a delay line (i.e., electromagnetic type) can be used either for the video signal or the RF signal; therefore, only the RF delay lines will be discussed here.

There are two kinds of RF delay lines: the nondispersive and the dispersive types. The nondispersive delay line delays the input signal by a fixed time. It is desirable to have the signal from the output of the delay line as a replica of the input signal. There are many kinds of nondispersive delay lines. The most commonly used are electromagnetic lines, surface acoustic wave (SAW), and solid-state bulk delay lines. The bandwidth of the delay line can be determined either by the insertion loss of the line (i.e., 3dB point) or the deviation of delay time.

In a dispersive delay line (DDL), the delay time is linearly related to the frequency. The DDL is a key component used in the compressive receiver (see Chapter 6) and in the pulse compression radar as in References 3 and 4. The characteristics, structures, and applications of the two delay lines will be discussed in the following paragraphs.

9.2 APPLICATIONS OF NONDISPERSIVE DELAY LINE TO A MICROWAVE RECEIVER

A typical use of a nondispersive delay line in a microwave receiver is to delay the signal for some measuring devices to get ready to further process the input signals. An example is shown in Figure 9.1. At the input of a receiver, the signal is channeled into a subband ΔB, but in order to reduce hardware, there is only one fine frequency encoder that covers the bandwidth of ΔB. If detectors are present to all the subbands, a signal will be detected in a certain subband, and the fine frequency encoder can be switched to that band to measure detailed information. However, it takes time to detect the signal, make logical decisions, and activate the switch. Delay

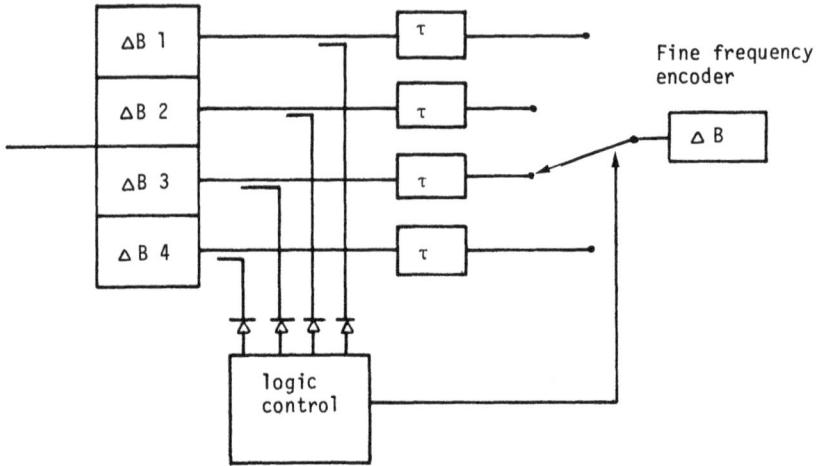

Figure 9.1. Delay Lines Used in Switchable Receiver

lines must be used in this design to switch the fine frequency encoder before the signal reaching the switch. When signals arrive at two subbands simultaneously, one of them cannot be measured by the fine frequency encoder. The nondispersive delay line can also be used as a signal processing element as in the instantaneous frequency measurement (IFM) receiver (see Chapter 4). In these lines, the delay time does not need to be very long, usually up to the ten nanoseconds range.

Another application of the nondispersive delay line is for the microwave memory. If the input signal is a single pulse, occasionally it is desirable to store the signal in a delay line for further processing. A simple memory arrangement is shown in Figure 9.2. The input signal is amplified and fed back to the input again after it is delayed time τ. In this arrangement, the gain of the amplifier and the insertion loss of the delay line must be properly balanced. If the input pulse width is wider than delay time τ, the amplitude of the pulse after time τ may change, because the undelayed and the delayed pulse will overlap and interfere with each other after time τ. In order to avoid this problem, the signal should be kept shorter than τ. In some designs, the longer input pulse is switched off after time τ. Another important factor in this arrangement is that the noise in the feedback loop may keep building up and finally, the noise will saturate the loop, causing the signal to be lost in the noise.

Another application of the nondispersive delay line is to use it as a resonator in an oscillator. The schematic circuit is quite similar to that of Figure 9.2, except there is no input. The noise from the output of the amplifier is fed back through the delay line to its own input. If the gain of the amplifier is higher than the insertion loss of the delay line, oscillation will start. However, the output of such an oscillator has multiple spectrum whenever the delay time $\tau = n\lambda$, where n is an integer and λ is the oscillation wavelength. Therefore, the oscillator is usually referred to as a comb generator, because the oscillator spectrum displayed in the frequency domain looks like a comb. If a single frequency is of interest, a filter with a sharp skirt must be used to isolate the desired frequency.

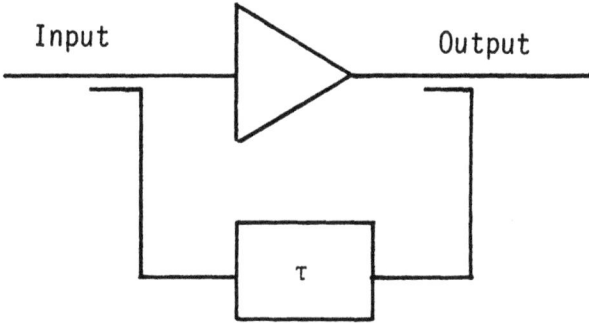

Figure 9.2. Delay Line Used for Microwave Memory

9.3 ELECTROMAGNETIC DELAY LINES (5,6)

An electromagnetic (EM) delay line is generally a coaxial cable (Reference 5 and 6). The delay time is proportional to the length of the line. The velocity of wave propagation for transverse electromagnetic (TEM) is

$$V = \frac{C}{\sqrt{\varepsilon_r}} \qquad (9.1)$$

where C is the velocity of light and ε_r is the relative dielectric constant of the cable.

If the length of the line is l, then the delay time

$$\tau = \frac{l}{V} = \frac{l\sqrt{\varepsilon_r}}{C} \qquad (9.2)$$

which states that the longer the delay line or the higher the dielectric constant, the longer the delay time. For a coaxial cable operating in TEM mode, the capacitance per unit length C and inductance per unit length L are (Reference 5)

$$C = \frac{2\pi\,\varepsilon_0\varepsilon_r}{\ln\dfrac{b}{a}} \text{ farad/m} \qquad (9.3)$$

$$L = \frac{\mu_0}{2\pi} \ln \frac{b}{a} \text{ henry/m} \qquad (9.4)$$

and the characteristic impedance is

$$Z = \sqrt{\frac{L}{C}} = \frac{1}{2\pi} \sqrt{\frac{\mu_0}{\varepsilon_0 \varepsilon_r}} \ln \frac{b}{a} = \frac{60}{\sqrt{\varepsilon_r}} \ln \frac{b}{a}$$

$$= \frac{138}{\sqrt{\varepsilon_r}} \ln \frac{b}{a} \text{ ohms} \qquad (9.5)$$

where ε_0 and μ_0 are the permittivity and permeability of free space, "b" is the radius of the inner wall of the outer conductor, and "a" is the radius of the center conductor.

The impedances of coaxial cable are usually 50Ω or 75Ω. Data on transmission lines can be found in Reference 6.

The impedances at both ends of the line should match the characteristic impedance of the line, otherwise multiple reflections will occur which will be discussed in detail for SAW delay line. In most applications, the size of the delay line will be minimal. For a fixed-delay time, this means coaxial cable of small diameters should be used to build delay lines. On the other hand, a delay line should have minimal insertion loss. But, in general, a coaxial cable of small diameter has higher loss per unit delay time. Therefore, in building coaxial cable delay lines, the volume of the delay line and the insertion loss must be carefully considered. In general, a delay time of a couple hundred nanoseconds requiring approximately 100 feet of cable might be the upper limit of EM delay lines. The insertion loss of the coaxial cable varies with frequency. The higher the frequency, the higher the loss. A delay line covering a certain bandwidth sometimes requires an equalizer to provide a constant insertion loss over the frequency band. A properly designed high-pass filter can be used as an equalizing circuit where the insertion loss decreases as the frequency increases.

When the nondispersive delay line is used in an IFM receiver, the accuracy of the delay time is very important, otherwise errors will be induced. Under this condition, the temperature coefficient of the line should be carefully compensated. In modern IFM receiver design, microstrip lines are often used as the delay lines. Substrate materials with dielectric constant independent of temperature must be selected for this application.

The ideal solution to an EM delay line is to improve the properties of dielectric materials. The higher the dielectric constant, the shorter the line length required for the same amount of time delay. Therefore, a dielectric material with high dielectric constant, low-loss, and low temperature coefficient is desirable for EM delay lines.

9.4 SURFACE ACOUSTIC WAVE (SAW) DELAY LINES[7, 8, 9, 10]

In this kind of delay line, the electric signal is transformed into a surface acoustic wave which will travel along the surface of a piezoelectric material; and finally, the acoustic wave is transformed back into electric signal at the output of the delay line. The operational principle of a SAW delay line is the same as a SAW filter (see Chapter 12).

Since the SAW travels at approximately $3 \sim 4 \times 10^3$ m/sec. (i.e., the surface wave velocity of standard, cutted quartz is 3158 m/sec and 3488 m/sec for Y-cutted LINbO3), a delay line of a few centimeters can provide microseconds of delay. At the input of the SAW delay line, there is a transducer which changes an electric signal into a mechanical signal. The most common transducer is of interdigital form as shown in Figure 9.3.

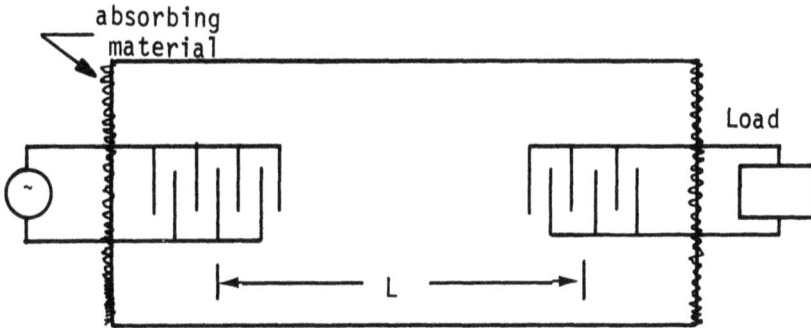

Figure 9.3. Simplified Configuration of a SAW Delay Line

The electric field applied between the interdigital fingers will deform the surface of the material underneath them and generate acoustic wave, because the substrate is a piezoelectric material in which an electric field will induce a corresponding mechanical displacement. The acoustic wave will travel in both directions along the axis of the transducer. The wave that travels to the left will be absorbed through properly added absorbing material to prevent the wave from being reflected from the edge and arrives at the output transducer as a spurious signal. The wave that travels to the right will be transformed into electric signal at the output transducer. The operation frequency of SAW delay line is usually under 2 GHz. If the input is a single pulsed signal an ideal SAW delay line will produce only one output pulse. However, in reality, if a pulsed signal is applied at the input, many pulses may appear at the output of the delay line as shown in Figure 9.4a. These outputs can be explained in the timing order as follows:

1. Direct Coupling. This is the first output which is caused by capacitive coupling between the input and output transducers. The shorter

the delay time or the closer the two transducers, the tighter the coupling; therefore, the stronger the direct coupling output. When the delay time is in the range from ten to a few hundred nanoseconds, this direct coupling can be very strong. In general, metal shielding between the input and output transducers is provided to reduce the coupling between them.

2. Desired Signal. The desired signal arrives at the output with a delay of time τ. This is the strongest signal and is used as the reference. All the other output signals are measured in dB below this level.

3. Multiple Transit Responses. If there are mismatches at the transducers, which are usually the case, part of the signal will reflect back. As shown in Figure 9.4b, part of the signal at the output transducer is reflected back and then reflected again at the input transducer and finally the signal reaches the output at time 3τ with respect to the input signal. This signal is referred to as the triple-transit effect as shown in Figure 9.4b. A similar effect will continue and signals will come out the delay line at 5τ, 7τ.... They are referred to as the 5th and 7th transits and so on. However, the amplitudes of these outputs decrease with time. In general, the triple-transit effect is of primary concern in many receiver designs. For a narrow bandwidth (less than 10 percent), the triple-transit can be made 50-60 dB below the desired signal. But for a wide bandwidth delay line (60 percent), the triple-transit is relatively high and is 20-30 dB below the desired signal. There are some approaches to reduce the magnitude of the triple-transit effect which will be discussed in the SAW filter section (Chapter 12).

4. Spurious Time Response. There are also outputs from the delay line which are not delayed by 3τ, 5τ, 7τ... from the input signal. These outputs can be ascribed to reflections from the edges of the substrate. The reflections can come from both the end edges and the top/ bottom edges. For example, if part of the input signal which travels to the left is not totally absorbed at the edge, it can be reflected and come out at the output of the delay line at some time later. If the surface wave travels at an angle rather than parallel to the direction of the input and output transducers, the signal can be reflected at the top and bottom edges of the substrate and comes out from the delay line. In general, improving the absorption at the edges of the substrate can reduce these spurious time responses.

It is desirable to reduce all these time domain responses except the desired signal. The effect of these spurious responses to the receiver is to reduce the signal processing speed. For example, a pulsed signal arrives at the receiver as shown in Figure 9.1 and the signal is channelized into subband 1. If the delay line is a SAW device, there will be more outputs than the desired one at the delay line output. All the spurious signals must be below a certain level before another signal in the same subband 1 can be processed. If a second signal arrives before the end of the spurious response, it cannot be properly processed. The total time from the direct coupling output to the last spurious output can be considered as shadow time caused by the delay line.

The insertion loss of the SAW delay line over the operation frequency range should be minimal and flat. The measured result (Reference 10) of a SAW delay line with 1.1 μs delay time with approximately 650 MHz bandwidth centered at 1350 MHz is shown in Figure 9.5. The insertion loss of this line is 27 dB at the center frequency with flatness of ± 0.6 dB across the band. The triple-transit measured in the time domain is about 23 dB. This delay line was designed for a channelized receiver. The picture of this delay line is shown in Figure 9.6.

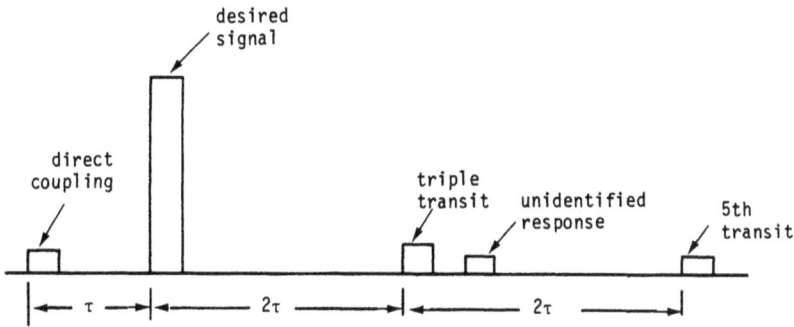

(a) *Outputs from a SAW Delay Line*

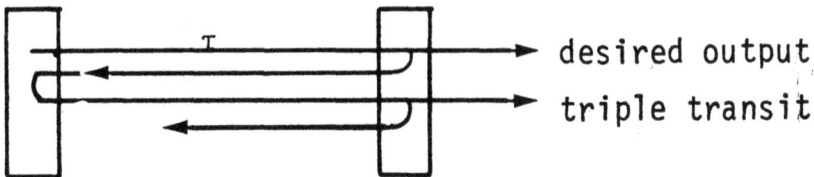

(b) *The Cause of Triple-Transit in a SAW Delay Line*

Figure 9.4. *Outputs from a SAW Delay Line*

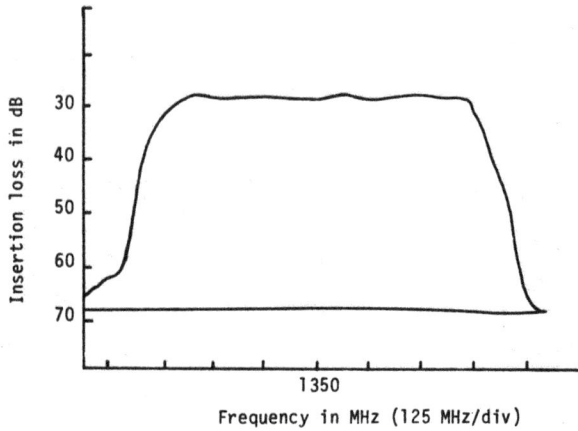

Figure 9.5. Test Result of a 1.1 μs SAW Delay Line f_o = 1350 MHz, Δf = 125 MHz/div. Insertion Loss 27 dB at Center of Band Amplitude 10 dB/div (Courtesy of Texas Instruments)

Figure 9.6. Picture of Delay Line (Courtesy of Texas Instruments)

9.5 SOLID-STATE BULK WAVE DELAY LINES (11, 12, 13)

The principle operation of this bulk wave delay line is quite similar to the SAW delay line. The input electric signal is changed into an acoustic wave through a transducer, the acoustic wave travels in the solid and is changed back into electric signal at the output of the delay line. The transducers are commonly made of piezoelectric (i.e., ZnO) film sandwiched between two metal films (as shown in Figure 9.7). The transducers are bonded on well polished parallel surfaces of the delay line. Microwave matching networks are usually provided to transform the impedance of the transducers to approximate 50Ω to match the characteristic impedance

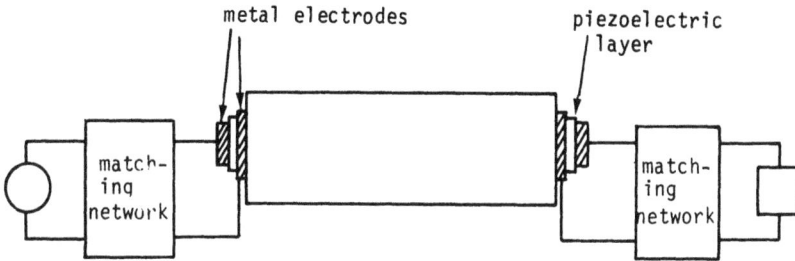

Figure 9.7. Bulk Wave Delay Line

of the transmission lines used in the system. The bulk wave velocity varies from several hundred meters per second to thousands of meters per second, i.e., in TeO_2 along $S[110]V = 620$ m/sec, while in diamond along $L[100]V = 17,500$ m/sec (Reference 11). In the most commonly used quartz, the sound velocity is 5970 m/sec (References 5 and 12). In general, the attenuation per unit length is less than that of the surface wave on the same material. The attenuation constant (dB/μs) of bulk wave is, in general, smaller than that of its surface wave at the same frequency. Using $LiNbO_3$ as an example, the attenuation of bulk wave is 0.1 dB/μs at 1 GHz (Reference 11) while the attenuation of surface wave is 1.07 dB/μs at the 1 GHz (Reference 9).

Thus, the bulk wave delay line can operate in a higher frequency range in comparison with SAW delay line. Moore et. al., (Reference 12) claimed that with low-loss acoustic media such as sapphire and yttrium aluminum garnet (YAG) and highly efficient transducers from ZnO films, bulk acoustic delay lines are available through X-band (8-12 GHz). They predicated that a delay line with 1 μs delay time at 10 GHz has 45 dB attenuation. They also claimed 20 percent bandwidth has been achieved and 40 percent bandwidth can be accomplished.

The spurious responses in the time domain similar to that discussed in SAW delay line such as direct coupling, triple-transit, and spurious outputs reflected from the surfaces are also present in the bulk wave delay lines. When they are used in a receiver, enough time must be allowed for the spurious responses to be under a certain specified value in order to process the next arriving signal.

9.6 MAGNETOSTATIC SURFACE WAVE DELAY LINES (14-20)

Another possible delay line used is magnetostatic wave delay line. In this kind of delay line, the electric signal is transformed into a magnetostatic wave in a magnetic material, traveling a fixed distance and coupled out in electric form again at the output. Yttrium iron garnet (YIG) is the most popular delay line material. It is usually in a single crystal form (Reference 13) in a magnetic field, and the wave travels on the surface of

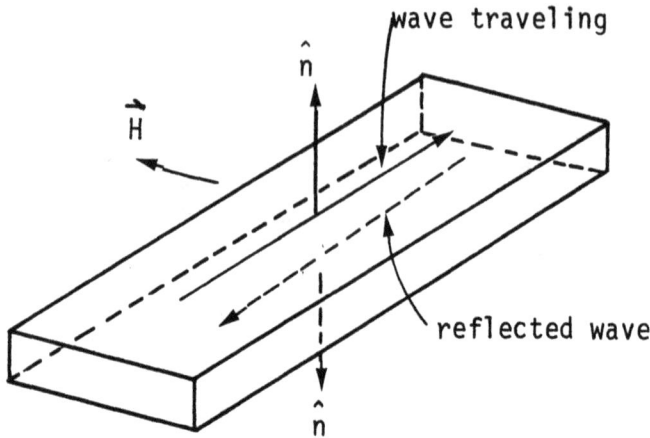

Figure 9.8. Magnetostatic Wave Delay Line

the material. Therefore, this kind of delay line is sometimes referred to as the magnetostatic surface wave delay line. The wave travels along the surface in the $\vec{H} \times \hat{n}$ (cross product) direction, where $\vec{H}$ is the direction of the applied magnetic field, and $\hat{n}$ is a unit vector outwardly directed normal to the slab surface as shown in Figure 9.8. Thus, a signal travels along the top surface of the slab, is reflected, and returns on the lower surface. The attenuation versus frequency for magnetostatic surface waves is shown in Figure 9.9. The attenuation is almost linearly increased with increase of operation frequency. Another form of the magnetostatic wave delay line is made of YIG film expitaxially grown on gadolinium-gallium garnet (GGG) substrate (Reference 15). The YIG film thickness can range from 1 μm to 100 μm. A YIG thin film delay line is shown in Figure 9.10. The input and output transducers of the magnetostatic can be a microstrip line which couples the electromagnetic field into the YIG layer. Different configurations of transducers can be fabricated. Figure 9.11 shows three kinds of transducers: a meander, an interdigital, and a parallel-bar type. In all three types the input to the transducer is a microstrip line.

 The performance of a magnetostatic delay line is quite similar to that of a SAW delay line or a bulk wave solid-state delay line. The direct coupling and the multiple time transit responses are also present in the magnetostatic delay line. The operation frequency of the magnetostatic delay line ranges from 1 GHz to approximately 20 GHz. The delay time can be varied by changing the applied magnetic field intensity as shown in Figure 9.12. An important factor that must be emphasized in Figure 9.12 is that for a fixed delay line length the delay time increases for higher frequency. Therefore, strictly speaking, the magnetostatic delay line is dispersive. However, the delay time versus frequency is not a linear relation which is desirable in a

DDL. Thus, the bandwidth of the magnetostatic wave delay line is limited by how much delay time deviation one can tolerate as well as the insertion loss over the band. The magnetostatic delay lines usually are used for delay times from 10-1000 ns. For a longer delay line, the insertion loss will be very high. The main attractive feature of a magnetostatic delay line is its potential to operate at a frequency range up to 20 GHz, where the SAW delay line does not operate and the EM delay line cannot provide enough delay time (usually less than 100 ns) limited by the high insertion loss at

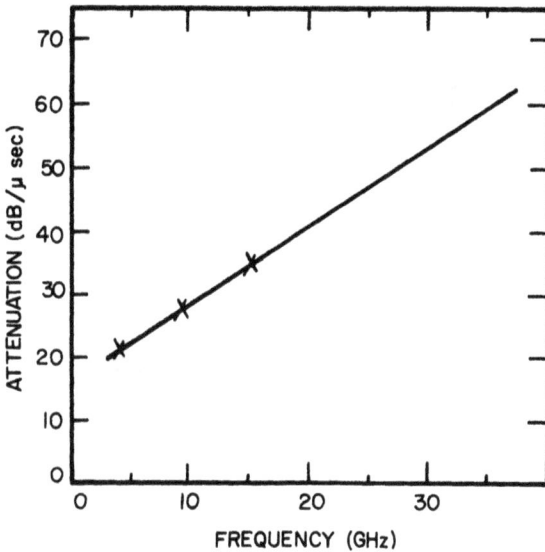

Figure 9.9. *Attenuation Versus Frequency for Magnetostatic Surface Waves (Based on Merry and Sethares, Reference 14)*

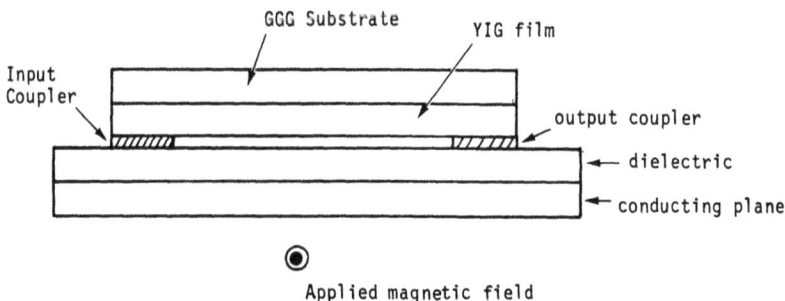

Figure 9.10. *A Thin Film Magnetostatic Wave Delay Line*

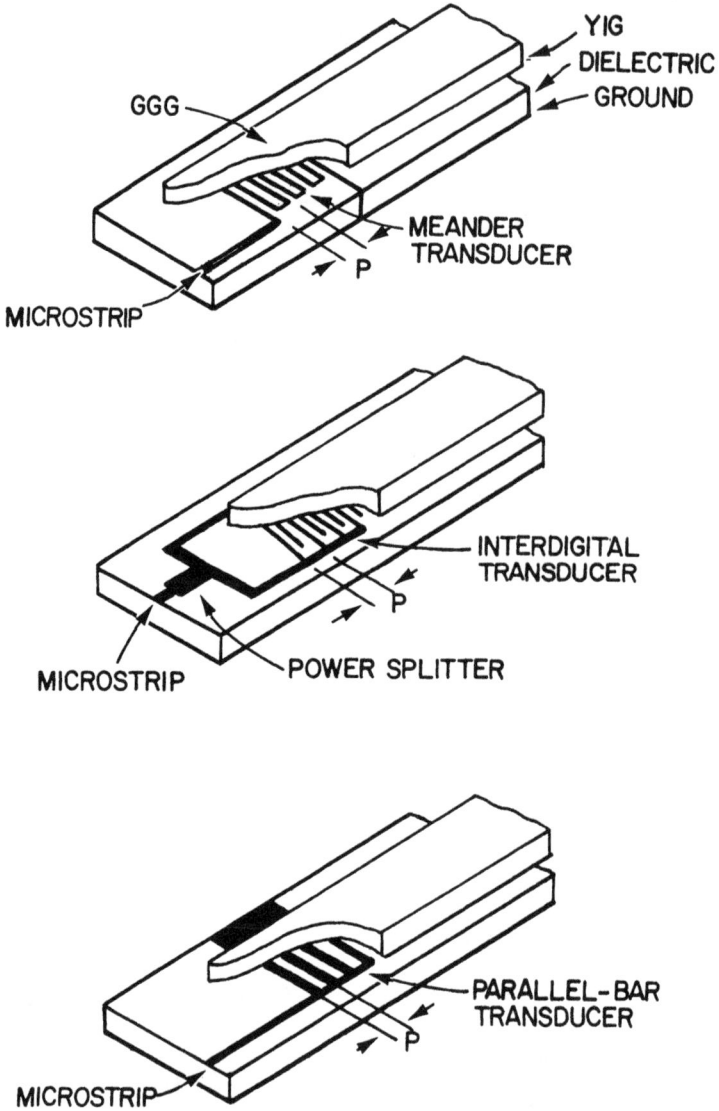

Figure 9.11. Three Differnt Transducers for Magnetostatic Delay Lines
(Adam et. al, Reference 19. Reprint from Electronics, May 8,
1980. Copyright© McGraw-Hill 1980 all rights reserved)

high frequency. Although the magnetostatic delay lines have the potential to be very useful, further development to reduce the insertion loss and lower the spurious time responses are required in order to be applicable to microwave receivers.

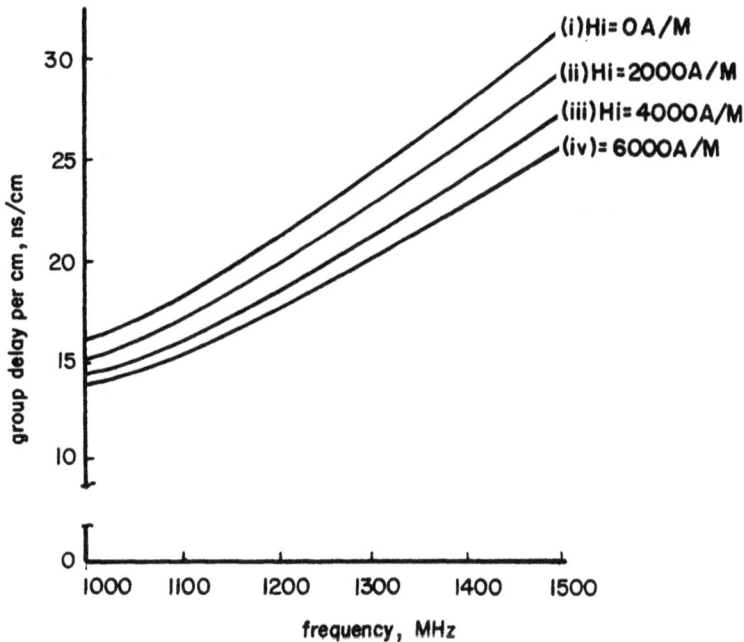

Figure 9.12. Theoretical Plot of Group Delay Against Frequency (Based on Adam et. al., Reference 15)

9.7 GENERAL CHARACTERISTICS OF A DISPERSIVE DELAY LINE (DDL)

A DDL is a RF delay line where the delay time varies linearly with frequency. The delay time can be either monotonously increased with increasing frequency or monotonously decreased with increasing frequency as shown in Figure 9.13. In general, the insertion loss in a delay line is high at higher frequency range. In order to keep the insertion loss uniform across the operation band of the delay line, the higher frequency range usually has shorter delay time. In other words, Figure 9.13b represents the commonly used DDL. The bandwidth of the delay line can be either determined by the insertion loss of the line (i.e., the 3 dB bandwidth) or the deviation from the linear delay time versus frequency relation. The term "time bandwidth product" is commonly used to describe the basic characteristic of a DDL. This quantity represents the processing gain of the device. The bandwidth is referred to as the operational bandwidth while the time is referred to as the differential delay time at the band edges. For example, DDL operates in the frequency range of 800 MHz to 1000 MHz with a delay time of 1.2 μs at 800 MHz and 1.0 μs at 1000 MHz, the time bandwidth product of the line is $(1000-800) \times 10^6 \times (1.2-1.0) \times 10^{-6} = 40$.

The DDL is used as a primary component in a compressive receiver (Chapter 6). It compresses a frequency modulated (FM) signal into a pulsed signal to obtain the frequency information. In an FM radar, the DDL performs the same function as in a compressive receiver to compress the FM signal into a pulse to improve range resolution. DDLs can be used in a compressive receiver to generate the FM signals which are used as the local oscillator of the mixer. The basic approach is to decompress a pulsed signal into a chirp signal. There are many kinds of DDLs, however, many of these DDLs are seldom used in modern receivers (i.e., the dispersive helix Reference 21) because of their limited time bandwidth product and poor frequency versus delay linearity. The DDLs presently used in receivers and other DDLs under development are discussed here. These DDLs can be divided into four groups. They are:

1. Folded Tape Meander Lines
2. SAW Devices
3. Magnetostatic Wave Devices
4. Crimped Coaxial Cables

Although acoustic bulk wave solid state devices can also be designed as DDL, they are in the early development phase and no useful devices have been reported at this time. The four groups of delay lines are discussed in the following sections.

9.8 FOLDED TAPE MEANDER LINES (21, 22, 23)

The folded tape meander line is an electromagnetic DDL. It was first reported by Dunn (Reference 21) in 1962. The meander line is a well established technology. Delay lines of time bandwidth product around 1,000 can be fabricated with this method. The operation frequency can approach 4 GHz. This kind of delay line is fabricated one at a time; a mass production method has not been developed. Compared to the other three kinds of DDL, the meander line is bulky and high in cost.

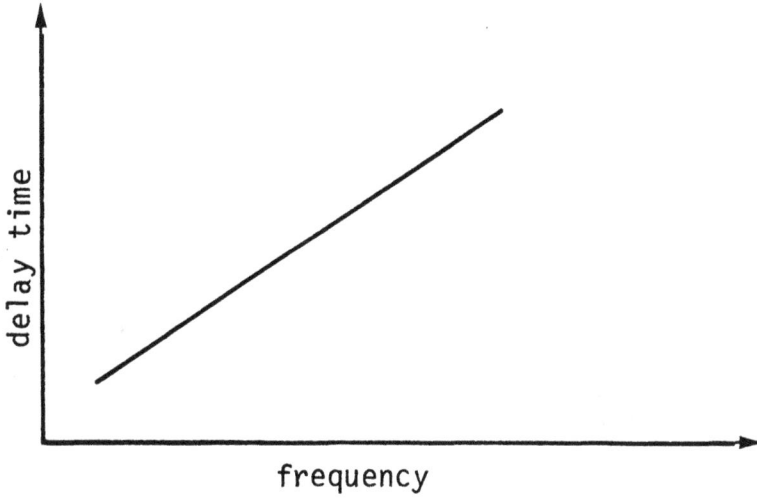

(a) Delay Time Increases with Increasing Frequency

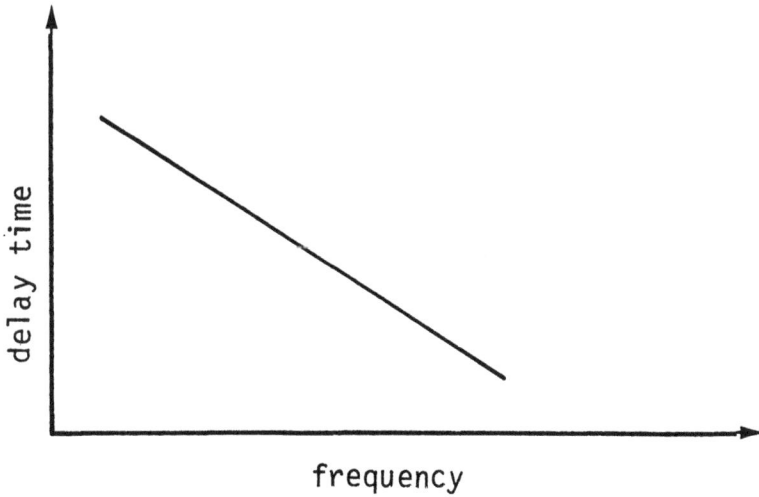

(b) Delay Time Decreases with Increasing Frequency
Figure 9.13. Characteristics of Dispersive Delay Lines

The physical configuration of a folded tape meander line is shown in Figure 9.14. The meander line is totally immersed in dielectrics. In practice, this is accomplished by two sheets of dielectric material of thickness D separating the meander line from the ground planes above and below, and individual pieces of dielectric material of proper size separating the meander line turns from each other. Dielectric strips are also placed at the four sides of the delay line.

The phase shift for each turn of the meander line (a turn is defined as a single piece of the conducting tape of width L within the periodic structures) is given implicitly by (Reference 23)

$$\cot^2\left(\frac{\omega L}{2C}\right) = \frac{\gamma - \cos\theta - \gamma'\cos 2\theta + \ldots}{\gamma + \cos\theta - \gamma'\cos 2\theta + \ldots}\cot^2\left(\frac{\theta}{2}\right) \qquad (9.6)$$

where ω is the angular frequency; L is the width of meander line (L = $\frac{\lambda_0}{4}$) C is the speed of light in the dielectric material; γ, γ' are coupling parameters determined by dimensions S, B, W, and D (as shown in Figure 9.14); and θ is the phase shift at frequency ω. The time delay of a single turn can be obtained by differentiating Equation 9.6 with respect to θ.

$$\tau = \frac{d\theta}{d\omega} \qquad (9.7)$$

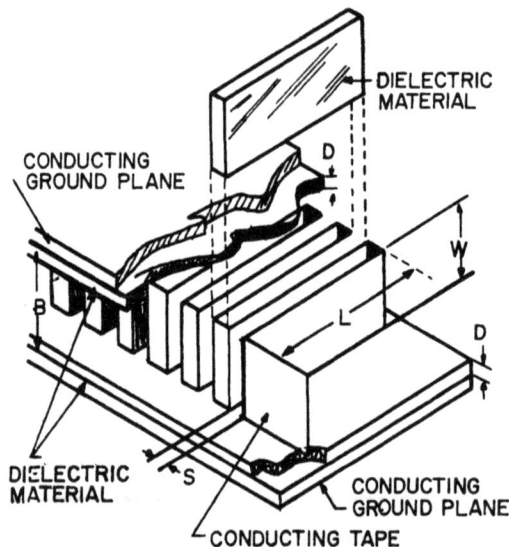

Figure 9.14. Folded Tape Meander Line Physical Configuration (Based on Hewitt, Reference 23)

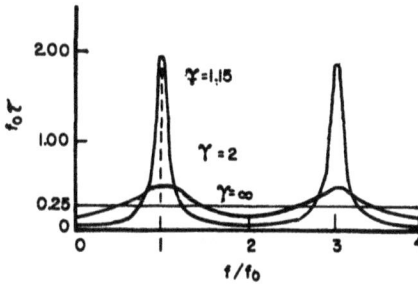

Figure 9.15. *Delay Versus Frequency for a Single Meander Line (Based on Hewitt, Reference 23)*

The delay time versus frequency is plotted in Figure 9.15 with different values of γ. The smaller the value of γ, the tighter the coupling. The values of γ and γ' can be expressed as (Reference 23):

$$\gamma = \frac{Y_a + Y_b + 2Y_c}{2(Y_a - Y_b)} \qquad (9.8)$$

$$\gamma' = \frac{2Y_c - Y_b - Y_a}{2(Y_a - Y_b)} \qquad (9.9)$$

Where λ_a, λ_b, and λ_c are the admittances for each of the three modes (a), (b), and (c) shown in Figure 9.16. They can be approximated as

$$Y_a \cong \frac{\sqrt{\varepsilon_r}}{30\pi} \left[\frac{W}{S} + \frac{2}{\pi} \ln 2 + \frac{1}{\pi} \ln \frac{1}{2 - 2\exp(-2\pi D/S)} \right] \qquad (9.10)$$

where $\dfrac{W}{S} > \dfrac{1}{2}$, $\dfrac{D}{S} > \dfrac{1}{4}$.

$$Y_b \cong \frac{\sqrt{\varepsilon_r}}{30\pi} \left[\frac{2D}{S} + \frac{2}{\pi} \ln 2 + \frac{1}{\pi} \ln \frac{1}{1 - 2\exp(-2\pi W/S)} \right]^{-1} \qquad (9.11)$$

where $\dfrac{W}{S} > \dfrac{1}{2}$, $\dfrac{D}{S} > \dfrac{1}{4}$.

$$Y_c \cong \frac{\sqrt{\varepsilon_r}}{30\pi} \left[\frac{W}{2S} + \frac{2}{\pi} \ln 2 + \frac{1}{\pi} \ln \frac{1}{1 - 2\exp(-2\pi D/S)} \right] \qquad (9.12)$$

where $\dfrac{W}{S} > 1$, $\dfrac{D}{S} > \dfrac{1}{2}$.

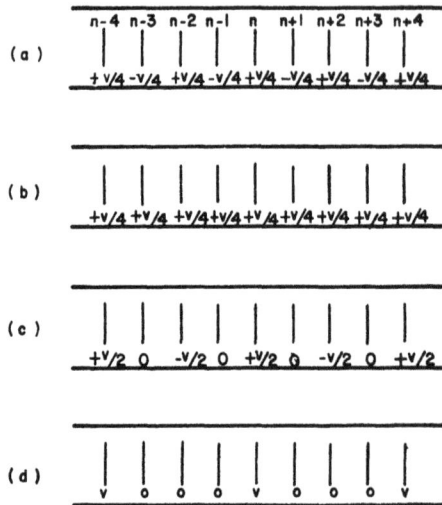

Figure 9.16. Cross Section of FTML Array Showing Superposition of Modes. Modes (a), (b), and (c) add to give (d). (Based on Hewitt, Reference 23)

The relationships between γ and γ' and the ratio $\dfrac{W}{S}$ are plotted in Figure 9.17 and 9.18 for four values of $\dfrac{D}{S}$. It is possible to construct a meander line with a delay accuracy of 1 percent by using the values of γ and γ' in Figures 9.17 and 9.18. More accurate results can be obtained from Equations 9.6 through 9.9. The total delay of a meander line at a given frequency is the sum of all the component delays of each turn at that frequency. Figure 9.19 shows that the delay time are varying with the three sections of a meander at frequencies f_1, f_2, and f_3, respectively. The three sections added will give the overall delay characteristics. The total delay time of the line is the sum of $\tau_1 + \tau_2 + \tau_3$. In general, to design a meander line, the delay of τ_k of any single section made of n turns for which $f = f_k$ will be directly proportional to n. Therefore, the synthesis of a desired delay function is accomplished by manipulating the number of sections, the center frequency of each section, the number of turns in each section, and the values of γ, γ' for each section.

A delay line from 2.2 to 3.8 GHz with a time bandwidth product of 160 was fabricated by using the above method. It has an insertion loss of approximately 25 dB. Some fine linearity adjustment on the meander line can be made through moving the dielectric materials around it on a trial and error basis. This procedure increases the complexity of manufacturing folded tape meander lines.

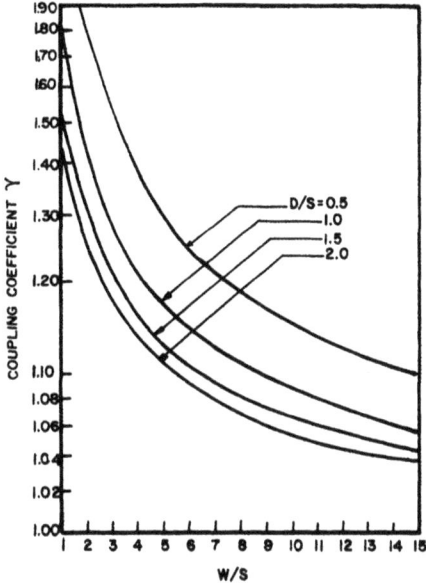

Figure 9.17. γ *versus* $\dfrac{W}{S}$ *and* $\dfrac{D}{S}$ *(Based on Hewitt, Reference 23)*

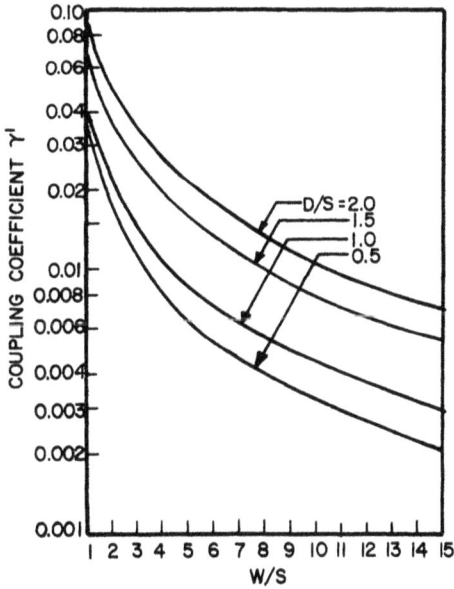

Figure 9.18. γ *versus* $\dfrac{W}{S}$ *and* $\dfrac{D}{S}$ *(Based on Hewitt, Reference 23)*

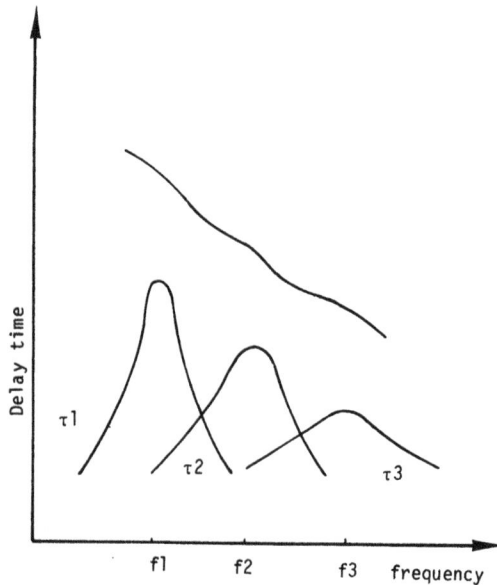

Figure 9.19. Linear Delay Synthesis Using Three Meander Line Sections

9.9 SURFACE ACOUSTIC WAVE (SAW) DISPERSIVE DELAY LINES (DDL)[13, 14, 24-26]

Just like the SAW nondispersive delay lines discussed earlier in this chapter and SAW filters discussed in Chapter 12, the electric signal is changed into an acoustic signal in a SAW DDL. The acoustic signal is delayed dispersively and then transformed back into electric signal at the output. Since the surface wave velocity is approximately five orders less than that of the EM wave, the size of the SAW delay line should be very small in comparison with an EM line. The common materials used for the SAW devices are $LiNbO_3$, $LiTiO_3$, and quartz. A time bandwidth product of 1,000 has been achieved (Reference 14). Higher time bandwidth can be achieved by cascading the DDLs. The operation frequency ranges from ten megahertz to a few gigahertz. To implement the SAW DDL, the two transducers must be properly designed to cover the frequency range. The distance between the two transducers must also be designed properly to produce the desired delay time. For a linear FM chirp, the electrodes of the transducer are shown in Figure 9.20. The locations of the electrode can be determined by (Reference 24):

$$X_n = Vt_n \qquad (9.13)$$

where V is the average surface wave velocity. The surface wave has different

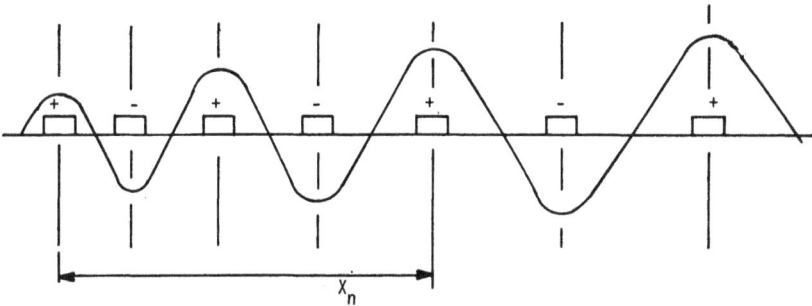

Figure 9.20. Electrodes of a Transducer that Produces an FM Chirp

velocities with the metal finger deposition on the surface. If the metal electrode width to the center-to-center spacing is equal everywhere to a constant β, and V_m and V_o are the surface wave velocities under the electrode and without the electrode, respectively, then the average velocity can be written as

$$V = \frac{V_o V_m}{V_m + \beta (V_o - V_m)} \tag{9.14}$$

The t_n in Equation 9.13 can be expressed as

$$t_n = \frac{T(f_o + \frac{B}{2})}{2B} \left[1 - \sqrt{1 - \frac{2Bn}{T(f_o + \frac{B}{2})^2}} \right] \tag{9.15}$$

where T is the total differential delay time; f_o is the center frequency; B is the bandwidth, and n is an integer. To make a down chirped waveform, the transducers are positioned with the high frequency regions close together as shown in Figure 9.21.

A different approach to fabricate the SAW DDL is through reflectors (Reference 13). As shown in Figure 9.22, the input and output transducers are slanted dispersive structures. Facing the transducers are dispersive gratings. The gratings can either be metal strips or grooves etched in the surface of the substrate. The high frequency path is shorter than the low frequency one, and the delay time is linearly related to the operation frequency. In this kind of design, there is one more dimension introduced comparing with that of Figure 9.21. Theoretically, it should be easier to design a DDL with high time bandwidth product by using the reflective structure, because the additional degree of freedom is available in this kind of design.

dispersive grating

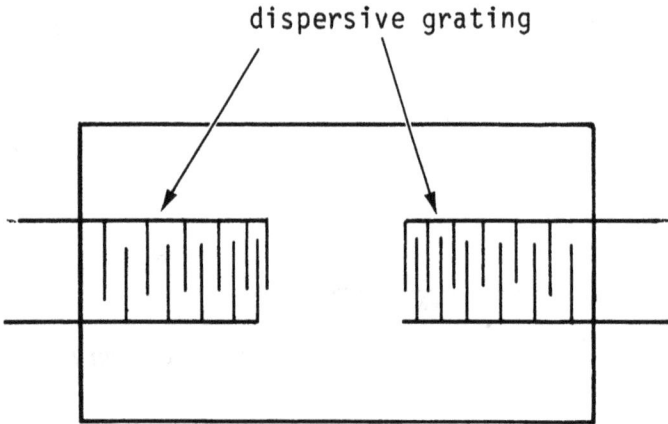

Figure 9.21. Down-Chirp Dispersive Delay Line Transducer Geometry

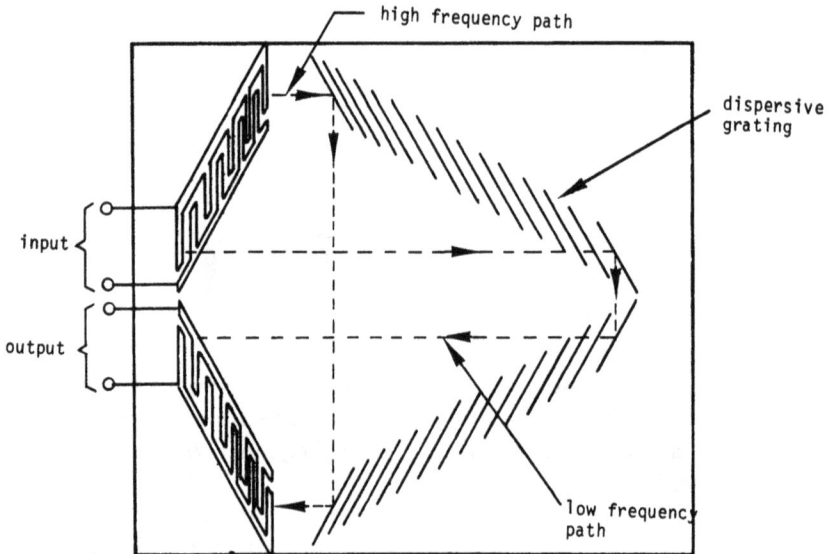

Figure 9.22. SAW DDL with Reflective Structure

9.10 MAGNETOSTATIC WAVE DISPERSIVE DELAY LINES (DDL)[27]

The magnetostatic wave DDL was reported by Uurtamo (Reference 27). The device was referred to as the time-prism filter. The time-prism filter could operate beyond 4 GHz with bandwidths approaching 1,000 MHz.

The specific magnetostatic DDL reported has the following characteristics:

Frequency Band of Operation   = 2.0 – 3.0 GHz
Nondispersive Delay (at 2.0 GHz)   = 480 ns
Dispersive Delay   = 340 ns/GHz
Insertion Loss   = 34 ± 2 dB

A magnetostatic DDL is a bulk YIG device whose propagation delay increases as the input frequency increases. The device is shown in Figure 9.23 and its exploded view in Figure 9.24. The input and output are coupled from the same end of the YIG crystal through loop antennas. The isolation between the input and the output of the device depends on the coupling factor of the loop antennas. Therefore, the isolation between the input and output is relatively poor. Typically, the isolation is about 6 dB which is a major deficiency of this device. Magnetic field with various strength is applied along the axis of the YIG rod. The required magnetic field is very critical, and specially designed pole pieces are used to generate the field.

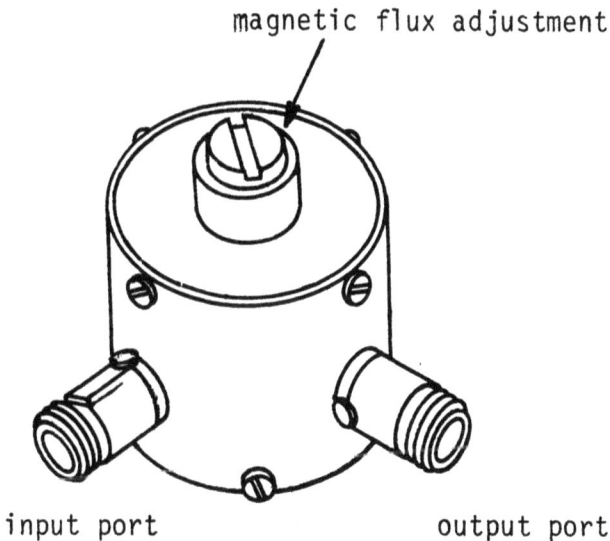

Figure 9.23. Completely Integrated and Packaged Time Prism Filter Assembly (Based on Uurtamo, Reference 27)

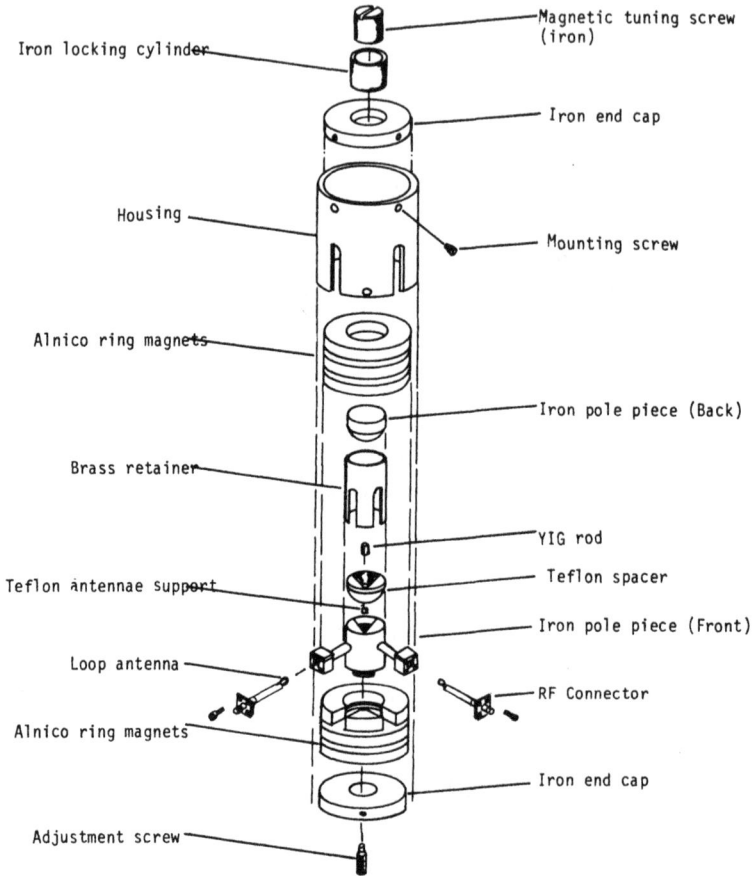

Figure 9.24. Exploded View of Integrated Time Prism Filter (Based on Uurtamo, Reference 27)

The operation of the magnetostatic DDL can be explained with the help of Figure 9.25. The electromagnetic energy coupled to the YIG rod initiates a backward magnetic wave, so called, because the wave phase and group velocities have respectively opposite senses. As the backward magnetic wave penetrates the YIG material along the z-axis, its group velocity diminishes to a point where a spin wave of opposite sense is formed. This spin wave starts to retrace the path of the backward magnetic wave from the turning point of Figure 9.25. The spin wave is initiated with a low velocity which continues to decay.

After traveling a very short distance in the YIG rod, typically less than 10 μm, the spin wave is converted to shear elastic wave at the cross-over point in Figure 9.25. This wave travels to the front face of the YIG rod where it is coupled to the output coupling loop. The delays caused by the spin wave and elastic wave are frequency dependent. The total delay T can be expressed as

$$T = D(f - \tfrac{1}{2}f_o) \qquad (9.16)$$

where f is the operation frequency in MHz and

$$f_o = |\lambda|H_o \qquad (9.17)$$

where $|\lambda|$ is the elastic wavelength, and H_o is the magnetic field intensity at z = o.

$$D = \frac{372}{H'(\dfrac{Oe}{mm})} \frac{nsec}{MHz} \qquad (9.18)$$

where H' is the magnetic field intensity gradient along z direction.

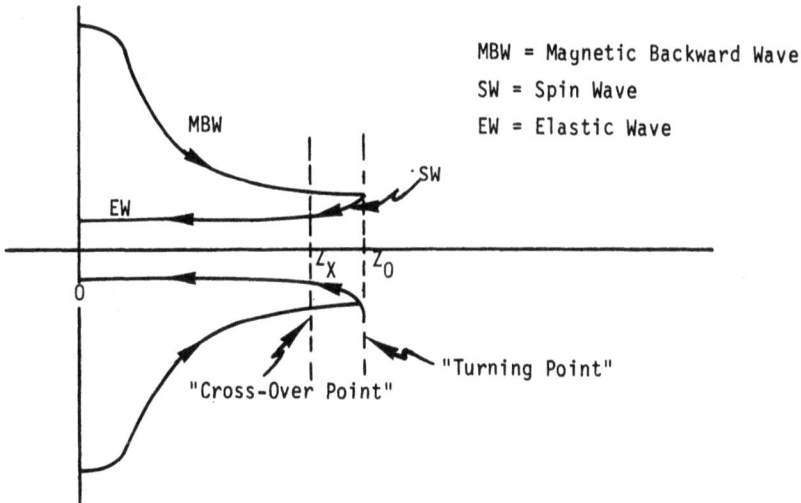

Figure 9.25. Schematic Presentation of the Propagation Modes Within the Crystal (Based on Uurtamo, Reference 27)

Equation 9.16 shows that the delay time is linearly proportional to the input frequency f. The proportional constant D is a function of the gradient of the applied magnetic field. In this DDL, the higher the input frequency, the longer the delay time. Equation 9.18 shows that a dispersion factor of D = 1 nsec/MHz requires a dc field gradient of 372 $\frac{Oe}{mm}$.

The frequency dependent time delay (dispersion) of the YIG crystal is normally quite nonlinear. This nonlinearity, however, can be minimized by carefully shaping the magnetic field boundary conditions within the crystal. This is accomplished by using a specially developed field synthesis techniques, which is the primary design concern. This kind of magnetostatic DDL is still in the development stage.

9.11 CRIMPED COAXIAL DISPERSIVE DELAY LINES (28)

This approach was first discussed by Fuller (Reference 28) in 1980. This delay line is a one-port device and both the input and output are at one end of the line. Therefore, a circulator must be used at the input of the line to isolate the input from the output. This delay line is fabricated by introducing discontinuities (crimps) on a semirigid microwave coaxial cable. The crimps are introduced by squeezing the outer conductor of the semirigid cable either by a tube cutter or by crimpers. The degree of crimp is very important. It should be deep enough to reflect the input signal, but not crack the outer conductor to interrupt the current flow. Cracking the conductor will introduce very high insertion loss. These crimps change the capacitance of the conductor and thus the impedance at those points.

As shown in Figure 9.26, if a signal of single frequency travels along the coaxial cable, energy will be reflected back at each discontinuity of the cable. These reflected signals will interface constructively at the input port after some time delay. If the positions of the discontinuities are properly located, the delay time introduced will vary linearly with the input signal frequency. The recurrence relationship used in calculating crimp positions is (Reference 28)

$$X_{i+1} = X_i + \frac{\sqrt{\varepsilon_r}\, C}{2f_L + 2\dfrac{(f_H - f_L)}{L}} \qquad (9.19)$$

where X_i is the length variable along the delay line where the crimp is introduced; L is the total length of the line; f_H and f_L are the high and low frequency of the DDL, respectively; C is the speed of light; and ε_r is the relative dielectric constant of the cable.

This kind of DDL is relatively easy to fabricate and has 50Ω characteristic impedance automatically. This technique is suited to generate delay line of wide bandwidth but relatively short dispersive delay time because long delay time will have high insertion loss. It was claimed that delay lines of 1-2 GHz bandwidth with 100 ns dispersive delay time could be fabricated by this scheme. The main disadvantage of this DDL is that this is a single port device. Input and output port isolation must be accomplished through a circulator.

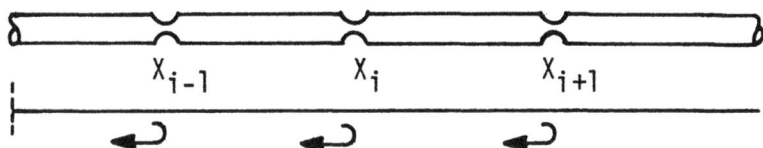

Figure 9.26. A Crimped Coaxial Cable Dispersive Delay Line

REFERENCES

1. Skolnik, M.I., *Introduction to radar systems,* McGraw-Hill Company, p. 119-140 and 493, 1962.
2. Jacobson, L., *Specifying delay lines,* Allen Avionics, Inc.
3. Skolnik, M.I., Editor, *Radar Handbook*, McGraw-Hill Company, 1970.
4. Cook, C.E. and Bernfeld, M., *Radar signals an introduction to theory and application,* Chapter 6, Academic Press, 1967.
5. Hayt, W.H., *Engineering electromagnetics,* 2nd ed., McGraw-Hill Company, p. 283, 1967.
6. *Reference data for radio engineers,* 5th ed., Chapter 22, Howard W. Sams Company, Inc., 1972.
7. Slobodnik, A.J. Jr., and Silva, J.H., "Ultra-flat UHF delay line modules," in-house report RADC-TR-77-257, Rome Air Development Center, Air Force Systems Command, Griffiss Air Force Base, New York, July 1977.
8. Ecklund, N.C., "Surface acoustic wave devices for electronic countermeasures (ECM)," Vol. 1, Technical Report AFAL-TR-77-240, Air Force Avionics Laboratory, Air Force Wright Aeronautical Laboratories, Air Force Systems Command, Wright-Patterson Air Force Base, Ohio, January 1978.
9. Slobodnik, A.J., *Acoustic surface design data,* printed by Air Force, Cambridge Research Laboratories.
10. Daniels, W., Texas Instruments, Inc., Private Communication.
11. Chang, I.C., "Acoustoptic devices and applications," IEEE Trans. on Sonics and Ultrasonics, Vol. SU-23, p. 2-22, January 1976.
12. Moore, R.A., Sundelin, R.N., Borsuk, G., Lane, J., Huber, C., and Lieberman, S., "Broadband matched and low triple-transit microwave delay line," Ultrasonics Symposium Proc. IEEE Cat #78CH-1344-ISU 1978.
13. Sundelin, R.N., Moore, R.A., McAvoy, B.R., and Lieberman, S., "Gigahertz bandwidth submicrosecond low spurious microwave bulk acoustic delay line," Ultrasonic Symposium Proceedings, p. 161-164, September 1979.
14. Merry, J.B. and Sethares, J.C., "Low-loss magnetostatic surface waves at frequencies up to 15 GHz," IEEE Trans. of Magnetics, Vol. MAG-9, p. 527-529, September, 1973.

15. Adam, J.D., Collins, J.H., and Owens, J.M., "Magnetostatic wave group-delay equilizer," Electronics Letters, p. 557-558, November 1973.
16. Brundle L.K., and Freedman, N.J., "Magnetostatic surface waves on a Y.I.G. slab," Electronics Letters, Vol. 4, p. 132-134, 1968.
17. Colliver, D., Hilsum, C., Joyce, B.D., Morgan, J.R., Rees, H.D., and Knight, J.R., "Experimental observation of magnetostatic modes in a Y.I.G. slab," Electronics Letters, Vol. 6, p. 434-436, 1970.
18. Adam, J.D., "Delay of magnetostatic surface waves in Y.I.G.," Electronics Letters, Vol. 6, p. 718-720, 1970.
19. Adam, J.D., Daniel, M.R., and Schroder, D.K., "Magnetostatic wave devices move microwave design into gigahertz realm," Electronics, p. 123-128, May 8, 1980.
20. Sethares, J.C., "Magnetostatic wave time delays for phased array technology," presented at Naval Research Laboratory Phased Array Technology Workshop, September 9-10, 1980.
21. Dunn, A.E., A pulse compression filter employing a microwave helix, Stanford Electronics Lab., Stanford, California Tech Report 557-1, October 1960.
22. Dunn, A.E., Realization of microwave pulse compression filters by means of folded-tape meander lines, Stanford Electronics Labs, Stanford, California, SEL-62-113 (TR-557-3), October 1962.
23. Hewitt, H.S., "A computer designed, 720 to 1 microwave compression filter," IEEE Trans. on Microwave Theory and Techniques, Vol. MTT-15, pp. 687-694, December 1967.
24. Smith, W.R., Gerard, H., and Snow, P.B., "Highly dispersive acoustic filter study final report, Research and Development Technical Report ECOM-0046-F. U.S. Army Electronics Command, Fort Monmouth, New Jersey, September 1973.
25. MacDonald, D.B., and Mellon, D.W., "Surface Acoustic Wave Devices for Electronic Countermeasures (ECM)," Vol. II, Acoustic Wave Compressive Delay Line, AFAL-TR-77-240, Vol. II, November 1977.
26. Hunsinger, B.J., and Datta, S., "Termination of surface acoustic wave velocity and impedance differences between metal strips and free surface regions of metallic gratings," Interim Report RADC-TR-81-4, Rome Air Development Center, Air Force Systems Command, Griffiss Air Force Base, February 1981.
27. Uurtamo, S.J., "Evaluation of the time-prism filter, a unique, broadband, microwave frequency dispersive, time delay device," NRL Memorandum Report 3598, Naval Research Laboratory, September 1977.
28. Fuller, H.W., "Broadband MW pulse compression using crimped coax delay lines," Microwave Journal, p. 52, April, 1980.

CHAPTER 10
DIRECTIONAL COUPLERS, HYBRID COUPLERS, AND POWER DIVIDERS

10.1 GENERAL CHARACTERISTICS OF DIRECTIONAL COUPLER[1-3]

A directional coupler is a passive component which couples out part of the transmission power by a known amount through another port. The device has four ports (input, transmitted, coupled, and isolated) as shown in Figure 10.1. Usually the isolated port is terminated with a matched load under normal usage. The signal coupled out through the directional coupler can be used to obtain the information (i.e., frequency and power level) on the signal without interrupting the main power flow in the system. The maximum power coupled out by a directional coupler is half the input power (or 3 dB below the input power level). Under this condition, the power on the main transmission line is also 3 dB below the input power and equals the coupled power. These couplers, with half the power coupled out, are sometimes referred to as 3 dB hybrids (or 3 dB couplers). The two outputs from a 3 dB hybrid are out-of-phase with each other. If the phase between them is 90 degrees, it is called quadrature (90 degrees) hybrid. If the phase difference is 180 degrees, it is called 180 degrees hybrid.

The important common properties desired by all directional couplers are: (1) wide operational frequency bandwidth, and (2) good input impedance matching at any arm when the other arms are terminated by matched loads. These performance characteristics of hybrid couplers and directional couplers are self-explanatory. Some other characteristics will be discussed here.

A. Isolation

Isolation of a directional coupler can be defined as the difference in signal levels in dB between the input port and the isolated port when the two output ports are terminated by matched loads. Isolation can also be defined between the two output ports. The latter defined isolation can be measured as follows: One of the output ports is used as the input; the other is considered the output port while the other two ports (input and isolated) are terminated by matched loads. The isolation between the input and the isolated ports may be different from the isolation between the two output ports. For example (Figure 10.1), the isolation between ports 1 and 4 can be 30 dB while the isolation between ports 2 and 3 can be a different value such as 25 dB. The isolation should be as high as possible.

B. Amplitude Balance

This terminology defines the power difference in dB between the two output ports of a 3 dB hybrid. In an ideal hybrid circuit, the difference should be 0 dB. However, in a practical device the amplitude balance is frequency dependent.

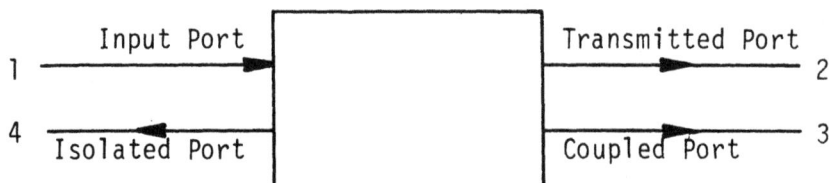

Figure 10.1. *A Hybrid or Directional Coupler with Arbitrarily Assigned Ports*

C. Phase Balance

The phase difference between the two output ports of a 3 dB hybrid should be 90 degrees or 180 degrees. However, the phase difference is sensitive to the input frequency and a typical phase balance can vary a few degrees.

D. Coupling Factor and Directivity

The coupling factor and directivity are defined as (Figure 10.1):

$$\text{Coupling factor (dB)} = 10 \log \frac{P_1}{P_3} \qquad (10.1)$$

$$\text{Directivity (dB)} = 10 \log \frac{P_3}{P_4} \qquad (10.2)$$

where P_1 is the input power at port 1; P_3 is the output power from the coupled port; and P_4 is the power output from the isolated port.

In measuring the coupling factor and directivity, all ports must be terminated with matched loads. The coupling factor represents the property of a certain directional coupler while the directivity should be as high as possible.

E. Scattering Matrix

This terminology expresses the characteristics of some microwave components. It can be explained through a simple two-port device as shown in Figure 10.2. If E_{i1} and E_{i2} represent the electric field of the input signals at ports 1 and 2, respectively, and E_{r1} and E_{r2} represent the electric field of the reflected signal at ports 1 and 2, respectively, then they can be related by

$$E_{r1} = S_{11} E_{i1} + S_{12} E_{i2} \qquad (10.3)$$

$$E_{r2} = S_{21} E_{i1} + S_{22} E_{i2} \qquad (10.4)$$

where S_{11}, S_{12}, S_{21}, and S_{22} are proportional constants. Generally, these are complex quantities. Equations 10.3 and 10.4 can be written in matrix form as

$$\begin{bmatrix} E_{r1} \\ E_{r2} \end{bmatrix} = \begin{bmatrix} S_{11}\ S_{12} \\ S_{21}\ S_{22} \end{bmatrix} \begin{bmatrix} E_{i1} \\ E_{i2} \end{bmatrix} \qquad (10.5)$$

where

$$\begin{bmatrix} S_{11}\ S_{12} \\ S_{21}\ S_{22} \end{bmatrix}$$

is referred to as the scattering matrix.

If the device is perfectly matched and there are no reflections, then $S_{11} = S_{22} = 0$; and if the device is linear and reciprocal, then $S_{12} = S_{21}$. These basic characteristics can be extended to multiple port devices.

Figure 10.2. A Two-Port Device

The scattering matrix of a perfect directional coupler or hybrid coupler (Figure 10.1) can be written as

$$\begin{bmatrix} 0 & S_{12} & S_{13} & 0 \\ S_{21} & 0 & 0 & S_{24} \\ S_{31} & 0 & 0 & S_{34} \\ 0 & S_{42} & S_{43} & 0 \end{bmatrix}$$

The values of S are determined by the coupling factor and output phase relation. In the above scattering matrix, the input impedance of all ports are perfectly matched; there is infinite isolation between the input and the isolated ports and infinite isolation between the transmitted and coupled ports.

The most popular directional couplers at microwave frequency are the parallel line, branch line, and hybrid ring types. However, at lower frequency it is impractical to fabricate a directional coupler with size comparable to wavelength. Under such conditions, lumped circuits can be used to build a directional coupler.

10.2 PARALLEL LINE DIRECTIONAL COUPLERS [4-19]

A parallel line directional coupler is shown in Figure 10.3. Two parallel conductors, 1-2 and 3-4, are placed close to each other (Figure 10.4). The length of the conductor is about $\lambda_0/4$ where λ_0 is the wavelength at the center of the operating frequency band. The electromagnetic energy can be coupled from one conductor to the other through two different ways: inductive and capacitive couplings as shown in Figure 10.4. The inductive couping induces current I_L flowing from 4 to 3, while the capacitive coupling induces current I_C flowing from the conductor center to both 3 and 4. If the device is properly designed so that the capacitively induced current is equal in magnitude to the inductively induced current, then the voltages across port 4 will cancel each other, while the voltages at port 3 are additive. If port 1 is referred to as the input port, then port 2 is the transmitted port, 3 is the coupled output, and 4 is the isolated port. The output at 3 is 90 degrees leading the output from port 2. Since the coupled output power flows in the opposite direction of the input signal, this coupler is sometimes called a backward directional coupler. It is also referred to as a quarter-wavelength directional coupler because the coupling conductors are $\lambda_0/4$ long. This directional coupler has an inherited 90 degree phase shift.

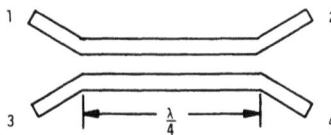

Figure 10.3. Parallel Line Directional Coupler

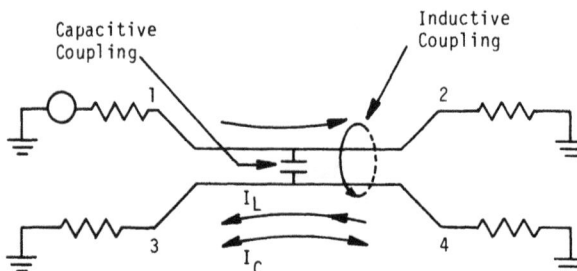

Figure 10.4. Capacitive and Inductive Coupling Between Two Parallel Conductors

A general expression for a 90 degree backward directional coupler is

$$V_3 = V_{coupled} = j\sin\theta\exp\left[-j(\beta\ell + \varepsilon)\right] \tag{10.6}$$

$$V_2 = V_{dc} = \cos\theta\exp\left[-j(\beta\ell + \varepsilon)\right] \tag{10.7}$$

where port 1 is the input with an amplitude of 1, β is the propagation constant $(2\pi/\lambda)$, ℓ is the coupled length, ε is a small phase dispersion error term, and θ is the coupling angle. For an ideal case, $\varepsilon = 0$ and $\ell = \lambda/4$, then

$$V_3 = \sin\theta \tag{10.8}$$

$$V_2 = -j\cos\theta \tag{10.9}$$

If the coupler is a 3 dB coupler, then $\theta = 45$ degrees and

$$V_3 = \frac{1}{\sqrt{2}} \tag{10.10}$$

$$V_2 = -\frac{j}{\sqrt{2}} \tag{10.11}$$

To change this coupler to a 180 degree hybrid, a 90 degree phase shift (discussed in Chapter 11) should be added to port 2. Under this condition, when the input is port 1, the output from port 2 will lag the output from 3 by 180 degrees; and port 1 is referred to as the difference (or Δ) port. When the input is from port 4, then the outputs from ports 2 and 3 are in phase and port 1 is the isolated port. Thus, port 4 is referred to as the sum (or Σ) port.

A wider bandwidth with more uniform coupling can be achieved when several basic couplers are connected in cascade. With the center quarter-wavelength coupler more tightly coupled than the outer quarter-wavelength coupler, a maximally flat or equal-ripple coupling response may be obtained. This idea can extend to many sections. Figure 10.5a shows a single section directional coupler with its typical frequency response, while Figure 10.5b shows a three-section directional coupler with its improved performance. Another approach (Reference 8) to improve the bandwidth is to combine several loose couplers to form a single, tightly coupled broadband unit. A coupler with bandwidths up to 17:1 was achieved by this approach. A typical tandem directional coupler with two 5 section couplers is shown in Figure 10.6.

Although there are many different physical configurations of coupling two parallel lines (References 5 and 6) the most commonly used ones for directional couplers are shown in Figure 10.7. Figure 10.7a is for weak coupling while Figure 10.7b is for strong coupling (Reference 9). These two configurations can be fabricated through printed circuit techniques. Tight coupling will couple more power out from the main transmission line. The configuration shown in Figure 10.7b has a disadvantage; the individual lines are unsymmetrically located with respect to the ground plates. This tends to excite the ground-plane mode. This problem can be prevented if the structure is closed in at the sides so only the desired transverse electromagnetic (TEM) mode can propagate. A 3 dB 90 degree hybrid of parallel line structure can be symbolically represented in Figure 10.8a, while a 180 degree hybrid is shown in Figure 10.8b. Sometimes a 180 degree hybrid can also be represented by Figure 10.8a with the amount of phase shift marked. Figure 10.8c shows a directional coupler with unequal power output. These symbols are not accepted universally, and there are other symbols for the directional couplers.

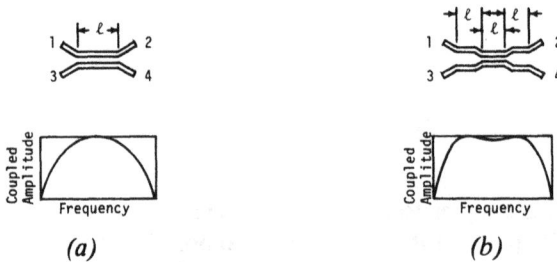

Figure 10.5. (a) Quarter-Wavelength and (b) Three Quarter-Wavelength Coupled Transmission Line Directional Couplers

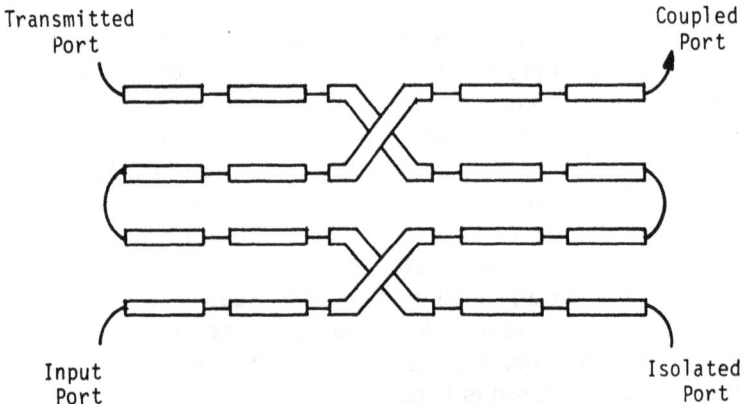

10.6   A Tandem Directional Coupler

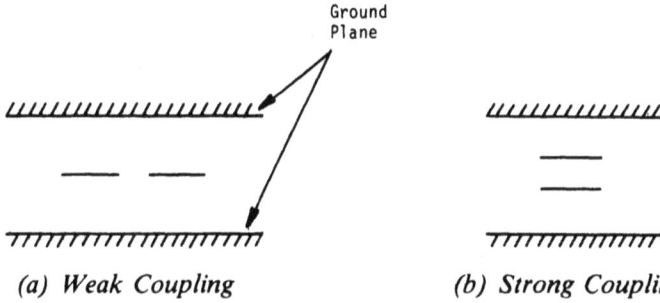

(a) Weak Coupling   (b) Strong Coupling

Figure 10.7. Cross Sections of Parallel-Line Coupler Coupling
Configuration

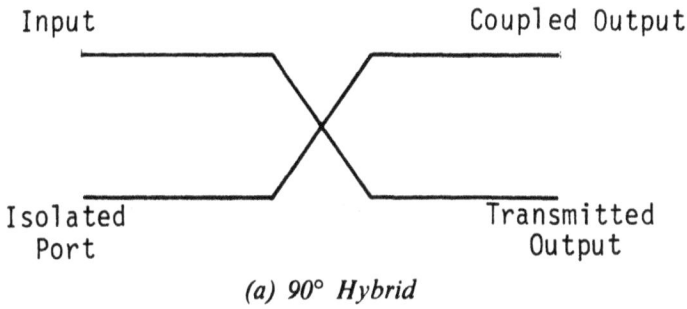

(a) 90° Hybrid

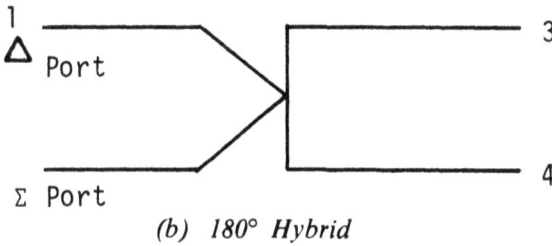

(b) 180° Hybrid

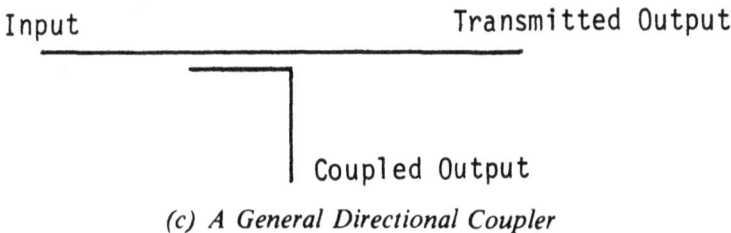

(c) A General Directional Coupler

Figure 10.8. Symbolic Representations of a 3 dB Hybrid Coupler

10.3 INTERDIGITATED DIRECTIONAL COUPLERS [20-29]

Tight coupling in a Microwave Integrated Circuit (MIC) or microstrip line form is a real challenge because of spacing limits. The simple parallel line coupler in microstrip form usually cannot produce a 3 dB coupler because of loose coupling. To overcome this difficulty, increase the total number of parallel lines as shown in Figure 10.9. This coupling configuration is called the interdigitated directional coupler.

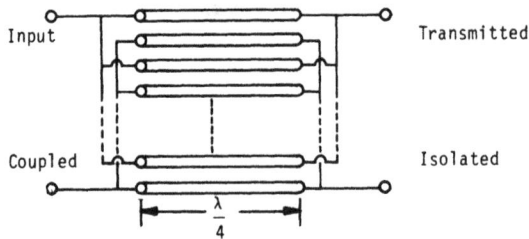

Figure 10.9. *Interdigitated Directional Coupler*

The actual directional coupler fabricated by printed circuit techniques is shown in Figure 10.10. Since the strip lines and microstrip lines are two-dimensional arrangements, binding wires are used to connect the parallel conductors into two groups. This kind of directional coupler can be made broadband with relatively low loss. The basic operational principle of this coupler is similar to the parallel line coupled type, since the only difference is the coupling scheme.

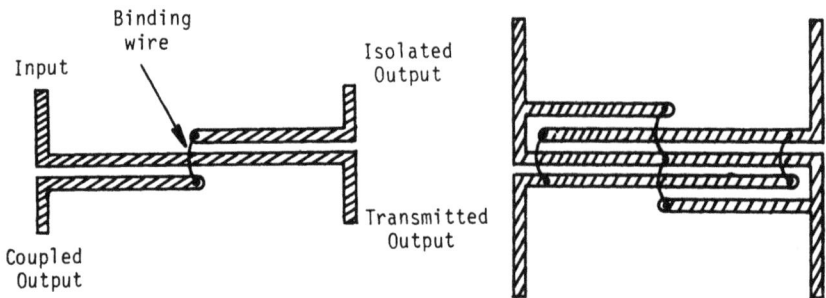

Figure 10.10. *Interdigitated Directional Coupler Configuration in MIC Form*

Ou (Reference 22) derived a set of equations for designing the interdigitated couplers. The results are summarized as

$$Y_o^2 = \frac{[(k-1)Y_{oo}^2 + Y_{oo}Y_{oe}][(k-1)Y_{oe}^2 + Y_{oo}Y_{oe}]}{(Y_{oo} + Y_{oe})^2} \tag{10.12}$$

$$\frac{P_c}{P_i} = \frac{(k-1)Y_{oo}^2 - (k-1)Y_{oe}^2}{(k-1)Y_{oo}^2 + 2Y_{oo}Y_{oe} + (k-1)Y_{oe}^2} \tag{10.13}$$

where Y_o is the characteristic admittance of the directional coupler equaling 0.02 mhos for a 50 ohm input impedance; k is the total number of coupled lines (see Figure 10.10); Y_{oo}, Y_{oe} are the odd and even mode admittances for a pair of the k coupled lines; P_c is the coupled output power; and P_i is the input power.

In actual design Y_o, P_c, P_i and k are chosen. Then Y_{oo} and Y_{oe} are calculated from Equations 10.12 and 10.13. From Y_{oo} and Y_{oe}, the coupled line configuration can be determined as discussed in Chapter 8.

10.4 BRANCH-LINE DIRECTIONAL COUPLERS[6, 30-33]

A branch-line directional coupler is shown in Figure 10.11. This is considered as a two (N = 2) branch directional coupler. The lengths of the branches and the distance between them are all quarter-wavelength at the center of the operating frequency band. The principle of operation can be explained as follows. If port 1 is the input, then port 2 is the transmitted port. Power can flow to port 3 by two paths: abc and adc. The signal at point c from these two paths are in phase; therefore, port 3 is the coupled output. Since the signal path lengths from 1 to 2 and 1 to 3 are different by $\lambda/4$, the outputs from 2 leads that of port 3 by 90 degrees. The power flow from 1 to 4 also has two paths: abcd and ad. These two paths are 180 degrees out-of-phase, and the signals from these two paths cancel each other

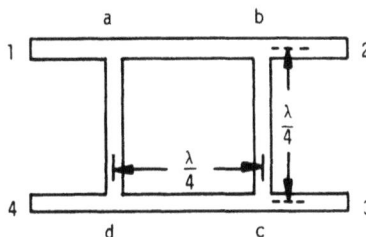

Figure 10.11. A Simple Branch-Line Directional Coupler

at port 4. Therefore, port 4 is the isolated port. A branch-line coupler is usually narrow band (less than 10 percent). The number of branches can be increased to improve the bandwidth.

The basic theoretical analysis of branch-line couplers was discussed by Reed and Wheeler (Reference 30). They treat the problem as a two-port network as shown in Figure 10.12, and the results of the study will be expressed as follows. In Figure 10.13, an ABCD matrix is defined as

$$\begin{bmatrix} E_1 \\ I_1 \end{bmatrix} = \begin{bmatrix} A & B \\ C & D \end{bmatrix} \begin{bmatrix} E_2 \\ I_2 \end{bmatrix} \tag{10.14}$$

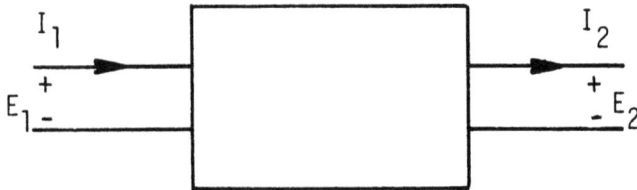

Figure 10.12. A General Two-Port Network

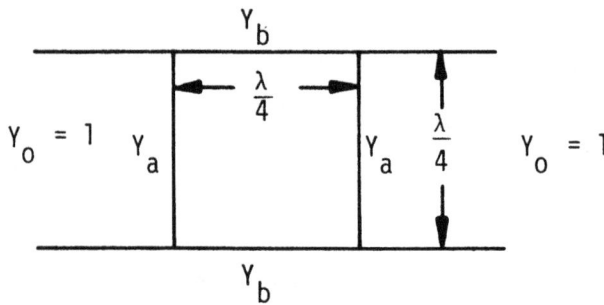

Figure 10.13. A Symmetrical Branch-Line Coupler with Normalized Input and Output Admittances

For a symmetrical branch-line coupler with normalized input and output admittances, the matrix for the even and odd mode transmission can be written as

$$M\pm = \begin{bmatrix} \mp \dfrac{Y_a}{Y_b} & \dfrac{j}{Y_b} \\[2em] j\left(Y_b - \dfrac{Y_a^2}{Y_b}\right) & \mp \dfrac{Y_a}{Y_b} \end{bmatrix} \qquad (10.15)$$

where $M+$ represents the even mode matrix and $M-$ is the odd mode. Y_a is the normalized parallel branch admittance, and Y_b is the normalized series branch admittance. In order to have a perfect isolation at port 4, B = C in both the even and odd mode matrix. This condition leads to

$$Y_b^2 - Y_a^2 = 1 \qquad (10.16)$$

The power coupled to port 3 is expressed in dB below the input power at port 1 as

$$P_3\,(\text{dB}) = 20\log \frac{\sqrt{Y_b^2 - 1}}{Y_b} \qquad (10.17)$$

If $Y_a = 1$ and $Y_b = \sqrt{2}$, then Equation 10.17 is fulfilled and P_3 (dB) = 3 dB which represents a 3 dB directional coupler. The values of Y_o, Y_a, and Y_b can be obtained by a properly designed width of strip line. Levy and Lind (Reference 31) provided design equations and computation tables for symmetrical branch-line directional couplers of multiple sections. Lind (Reference 32) extended the design to include asymmetrical branch-line couplers which can be considered as an impedance matching network as well.

10. 5 HYBRID-RING DIRECTIONAL COUPLERS[2, 3, 34]

A hybrid-ring directional coupler uses the signal path difference from the input to the output to generate either constructive interference or destructive interference at the output ports. The port where destructive interference occurs is considered the isolation port. The hybrid-ring coupler is also referred to as the rat-race.

A hybrid-ring directional coupler is shown in Figure 10.14. Any port can be used as the input port. Two adjacent ports are the outputs and the third is the isolation port. This phenomena can be easily explained. If port 1 is the input terminal, the power reaches port 4 through two paths (upper and lower half circle) of equal length and the signal is delayed $3\lambda/4$ from its input; therefore, the signal will come out at port 4. The signal from port 1 reaches port 3 through two paths (1-2-3 of $\lambda/2$ and 1-4-3 of λ).

The u.fference between the two paths is $\lambda/2$; therefore, there is no output from port 3. The two paths from port 1 to 2 are $\lambda/4$ and $5\lambda/4$; therefore, the signal is additive at port 2 and delayed by $\lambda/4$ with respect to the input signal at port 1. Thus, when port 1 is the input, the outputs are port 2 and 4; and they are 180 degrees out-of-phase.

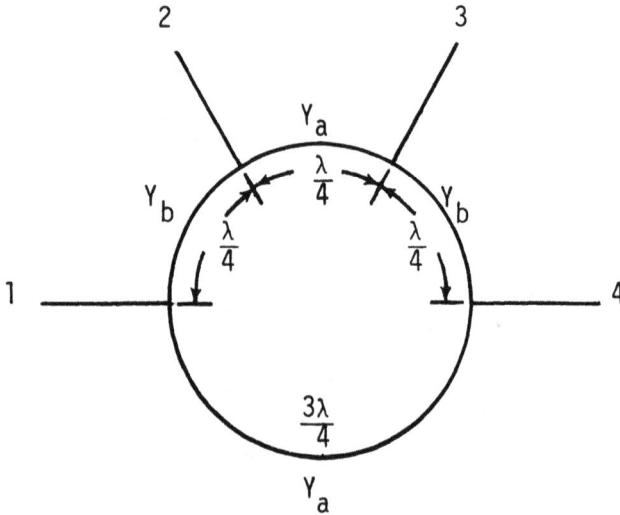

Figure 10.14. A Hybrid-Ring Directional Coupler With Normalized Input Admittance

If port 2 is the input, then port 1 and 3 are the outputs, and are in phase. Port 4 is the isolated port. Trace the signal paths to reach this conclusion.

The import features of this coupler are:

1. The input impedance is properly matched when its two adjacent arms are terminated by matched loads.
2. The two outputs are isolated from each other.
3. The two outputs are either in phase or 180 degrees out-of-phase depending on the input arm chosen.
4. The power split ratio is adjusted by varying the impedances between the arms. They are represented by Y_a and Y_b as normalized admittances.

The conditions to generate perfect isolation are derived by Pon (Reference 34), and the results are

$$Y_a{}^2 + Y_b{}^2 = 1 \qquad (10.18)$$

where Y_a and Y_b are the admittances shown in Figure 10.14.

If port 1 is the input, then the output voltage ratio between ports 2 and 4 is

$$\frac{V_2}{V_4} = -\frac{Y_b}{Y_a} \qquad (10.19)$$

If port 2 is the input, then the voltage ratio between ports 1 and 3 is

$$\frac{V_1}{V_3} = \frac{Y_b}{Y_a} \qquad (10.20)$$

Equations 10.19 and 10.20 also state the phase relation between the output signals. The minus sign represents 180 degrees out-of-phase while the plus sign represents the two outputs in phase.

10.6 LUMPED ELEMENT DIRECTIONAL COUPLERS[30, 35, 36, 37]

This type of directional coupler is very useful for the lower microwave frequency range, where the conventional strip line or microstrip couplers become too large and impractical. A lumped element directional coupler is shown in Figure 10.15. It is equivalent to a parallel line directional coupler at higher microwave frequency.

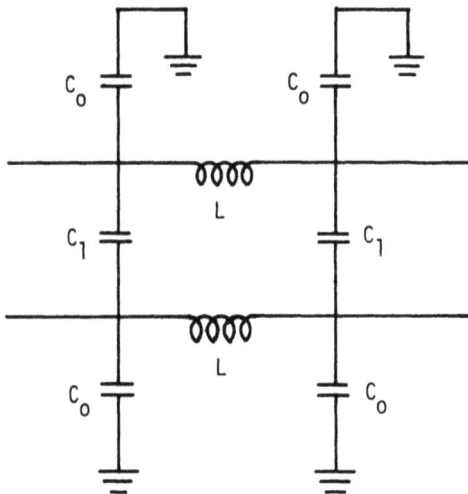

Figure 10.15. Lumped Element Directional Coupler

The analysis of this circuit is based on References 30 and 35. The circuit (Figure 10.15) can be considered for the even and odd mode excitations separately, and their equivalent circuits are shown in Figure 10.16a and b. The ABCD matrix defined in Equation 10.14 can be written for the even and odd modes with the help of Equation 10.15 and Figure 10.16. The results are

$$M+ \ = \ \begin{bmatrix} 1 - \omega_0^2 LC_0 & j\omega_0 L \\ j\omega C_0 (2 - \omega_0^2 LC_0) & 1 - \omega_0^2 LC_0 \end{bmatrix} \tag{10.21}$$

and

$$M- = \begin{bmatrix} 1 - \omega_0^2 L (C_0 + 2C_1) & j\omega_0 L \\ j\omega_0 (C_0 + 2C_1)[2 - \omega_0^2 L (C_0 + 2C_1)] & 1 - \omega_0^2 L(C_0 + 2C_1) \end{bmatrix} \tag{10.22}$$

where $M+$ and $M-$ are the even and odd matrix, and ω_0 is the angular frequency at the center of the band.

The capacitance and inductance are shown in Figure 10.16.

Considering the input and output impedances as Z_0 and the isolation condition of $B = C$, the following conditions can be obtained

$$C_1 = \frac{1}{Z_0 \omega_0 \sqrt{k}} \tag{10.23}$$

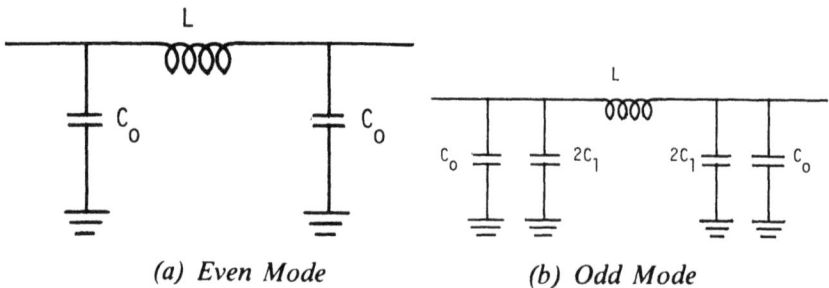

(a) Even Mode   (b) Odd Mode

Figure 10.16. Excitation Modes of the Lump Element Directional Coupler

$$\omega_0^2 L (C_0 + C_1) = 1 \qquad (10.24)$$

and

$$L = \frac{2 C_0^2 Z_0^2}{1 + \omega_0^2 C_0^2 Z_0^2} \qquad (10.25)$$

where $k = P_2/P_3$ is the power ratio at port 2 to that of port 3.

From Equations 10.23, 10.24, and 10.25, the values of L, C_0, and C_1 can be calculated for given values of Z_0 and ω_0.

10.7 MAGIC-T[30, 35, 37-40]

A magic-T usually refers to a section of waveguide with E-plane and H-plane couplings as shown in Figure 10.17. It is a 3 dB hybrid coupler with the following properties.

1. Input from port 1 (or 2) will divide equally between ports 3 and 4 and no output at 2 (or 1).
2. Input from 3 will divide equally between ports 1 and 2 with 180 degrees phase difference. Port 4 is the isolated port.
3. Input from 4 will divide equally between ports 1 and 2 and the outputs are in phase. Port 3 is the isolated port.

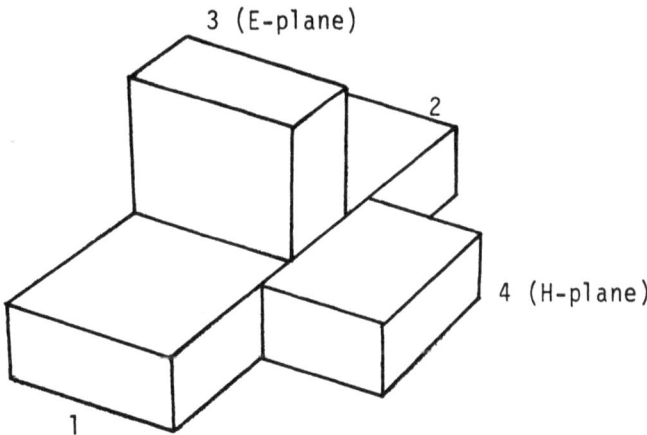

Figure 10.17. Waveguide Magic-T

This section will discuss a magic-T in microstrip line and slotline forms. Several different configurations have been proposed and fabricated (Reference 37). Only one configuration will be discussed. Figure 10.18 shows the circuit configuration of the magic-T. In this magic-T, both strip line and slotline are used. Solid lines in Figure 10.18 represent slotlines,

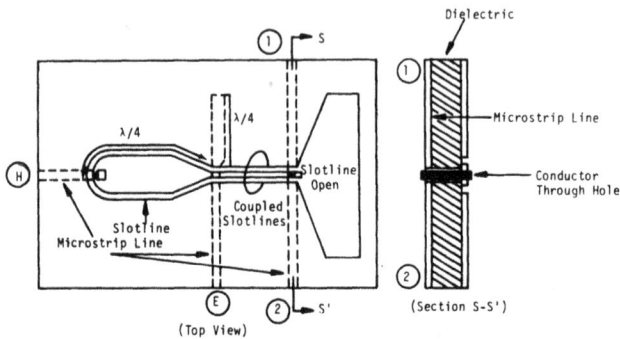

Figure 10.18. MIC Magic-T's (Based on Aikawa and Ogawa, Reference 37)

coupled slotlines, coplanar waveguides on the surface of a substrate. Dotted lines represent microstrip lines on the reverse side of the substrate. Ports 1, 2, E, and H correspond to the four ports of a conventional magic-T in Figure 10.17. Figure 10.18 can be used to explain the phase relations among the different ports. An input from port H is converted to the slot-line mode and divided into two quarter-waveguide slotlines in parallel. After propagating in the odd mode of the coupled slotlines, it is further divided into two microstrip line ports, 1 and 2 in phase (Figure 10.19a). An input from port E is converted to the even mode of the coupled slotlines, and thereafter is divided into the two microstrip line 1 and 2 which are 180 degrees out-of-phase (Figure 10.19b). A magic-T operating at 6 GHz with 50Ω input impedance at all four ports was fabricated. Isolation of 30 dB between E and H ports and 25 dB between ports 1 and 2 was realized.

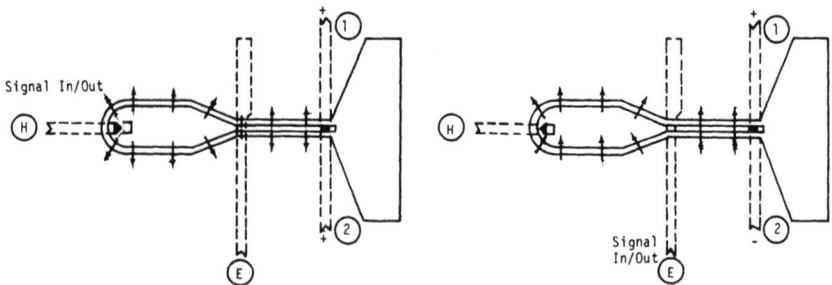

(a) *In phase coupling (Odd-Mode in the Coupled Slotlines)*

(b) *180 Degree Out-of-Phase Coupling (Even-Mode in the Coupled Slotlines)*

10.19. Schematic Behaviors for In Phase and Out-of-Phase Coupling (Based on Aikawa and Ogawa, Reference 37)

10.8 APPLICATIONS OF HYBRID COUPLERS[7]

The most popular use of hybrid and directional couplers is to split the input power into two different paths. This is a very popular application in receiver design. Described in the following paragraphs are other very important applications which use the 90 degrees and 180 degrees phase shift of the couplers.

1. Improve the input impedance matching of mismatched components by 90 degree hybrid couplers. As shown in Figure 10.20a, the input signal is divided equally in amplitude to ports 2 and 3. When ports 2 and 3 are shorted, the signals reflected from these ports are as shown in Figure 10.20b and c, respectively. Applying superposition theory in Figures 10.20b and c, shows that even with ports 2 and 3 shorted, the input signal from port 1 will come out from port 4 and will not reflect back at port 1 as shown in Figure 10.20. Therefore, the impedance at port 1 is well matched since there is no reflection at port 1. In order to simplify the discussion above, the signals at the input and the coupled output are assumed in phase. One can use this feature of the 90 degree hybrid to build balanced amplifiers, phase shifters, balanced detectors, attenuators, and balanced mixers with improved input impedance matching. Figure 10.21 shows an attenuator with improved input impedance matching in the reflective mode while Figure 10.22 shows a balanced amplifier in the transmission mode. In either case, the reflected signals at the inputs and outputs will be absorbed by the resistive loads, R.

2. Combine and separate dc and radio frequency (RF) through 90 degree hybrid couplers. If two 90 degree hybrid couplers are connected as shown in Figure 10.23, the RF input fror port 1 and the dc input from port 4 will be combined at port 3, because the RF input signal will have a 90 degrees difference in phase at points 5 and 6, and cancel each other at port 2. The dc input cannot be coupled to point 6, and it will reach port 3 through port 5. Applying the reciprocal theory, if RF signals and dc voltage including reasonably low frequency are applied at port 3, it will be separated at ports 1 and 4.

3. A very important component in an instantaneous frequency measurement (IFM) receiver is a phase correlator which can be fabricated through hybrid couplers. Detailed discussion can be found in Chapter 4.

4. Another popular application of hybrid couplers is for mixers. Different combinations of hybrid couplers and diodes can produce mixers with special characteristics. This topic will be discussed in Chapter 14.

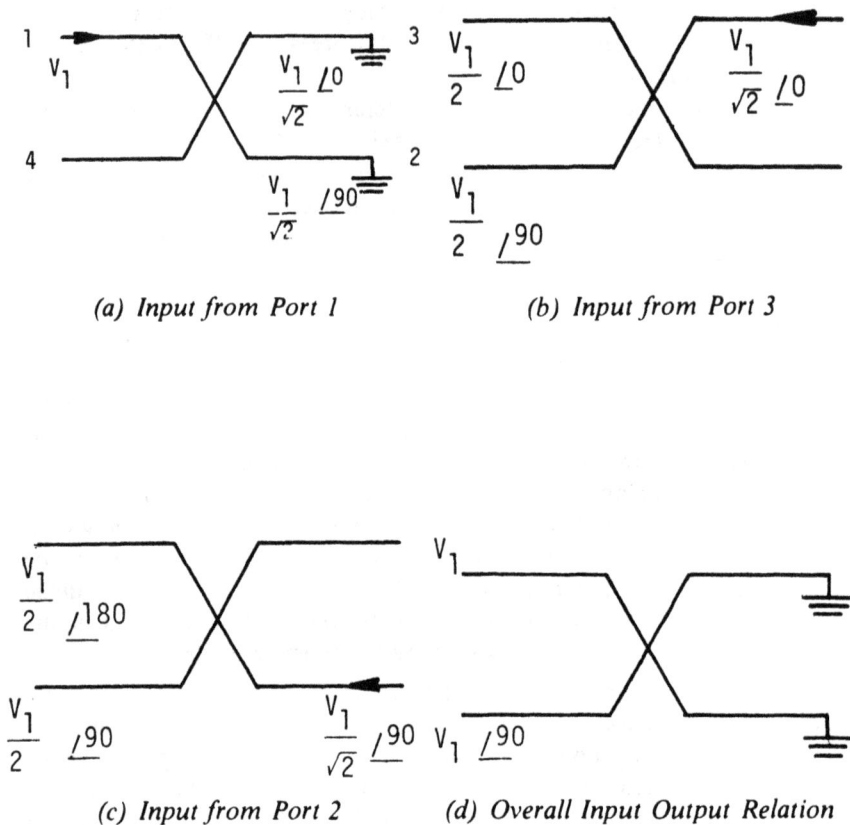

(a) Input from Port 1   (b) Input from Port 3

(c) Input from Port 2   (d) Overall Input Output Relation

Figure 10.20. Demonstrations of Total Reflection from the Outputs of a Hybrid Coupler

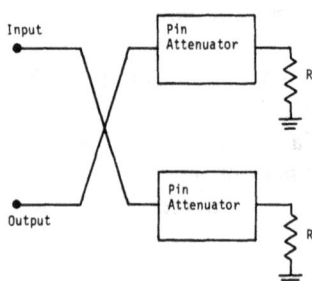

Figure 10.21. Attenuator with Improved Input Impedance

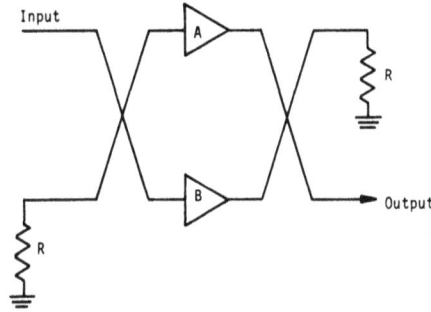

Figure 10.22. *A Balanced Amplifier using 90 Degree Hybrid Couplers*

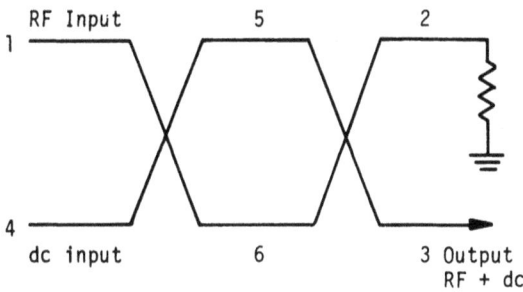

Figure 10.23. *RF and dc Combination Through Cascade of Two 90 Degree Hybrid Couplers*

10.9 POWER DIVIDERS AND THEIR APPLICATIONS[41-43]

A microwave power divider is a passive device which will divide the input signal into a number of equiphase outputs. In general, all of the outputs have equal amplitudes; however, sometimes the outputs are of different amplitudes. The general requirements of a power divider are wide frequency bandwidth, low input and output voltage standing wave ratios (VSWRs), and good isolations between different outputs. The two commonly used power dividers are the T-junctions and Wilkinson's power dividers.

In a microwave receiver, the power dividers can be used to separate an input signal into many parallel outputs for further processing. It can be used to combine several signals together; and in this application, it is often referred to as a power combiner. If the phase relation is not of primary interest, a hybrid coupler can be used to replace the in phase power divider, since the structure of a hybrid coupler is usually simpler than that of an in phase power divider. However, in many applications, the local oscillator (LO) signal is applied to a dual diode image rejection mixer (Chapter 14), the phase relation is of primary importance. Then the hybrid coupler and power divider are not interchangeable. One of the most common applications of the in phase divider is in solid-state microwave power amplifiers. Since the power handled by a single microwave device is limited, parallel combinations of a number of single amplifiers as shown in Figure 10.24 can improve the output power. In Figure 10.24 the power amplifier requires two n-way power dividers. Special power dividers/combiners are designed for this purpose (References 41, 42, and 43). However, the basic design approaches of these power combiners are the same as the Wilkinson's dividers which will be discussed in the following sections.

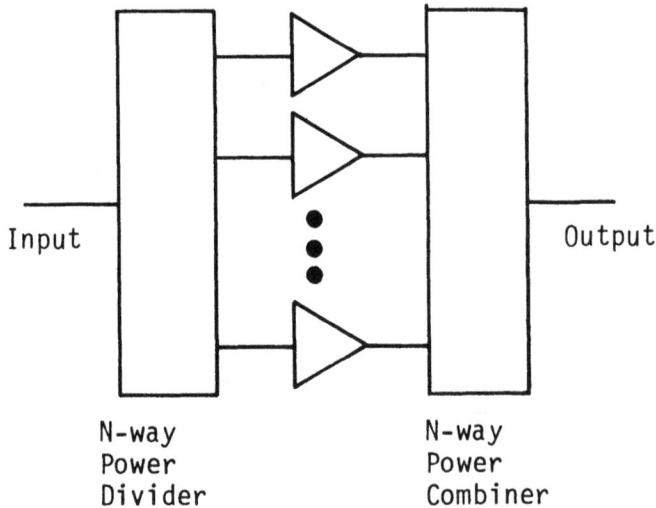

Figure 10.24. Power Amplifiers uses n-Parallel Individual Amplifiers to Improve Output Power

10.10 T-JUNCTION POWER DIVIDERS [44]

A simple T-junction is shown in Figure 10.25. Since the structure is symmetrical, the input power is divided equally to the two outputs. This structure has rather poor isolation between the output ports and poor impedance matching at the input as well as the output ports. To improve the performance of the T-junction, resistors are inserted in the circuits. There are two ways to put the resistors in the T-junctions: a three-resistor construction and a two-resistor construction as shown in Figure 10.26. The resistors used are related to the impedance of the power divider. For the three-resistor power divider the resistance $R = Z_0/3$; and for the two-resistor type, the resistance $R = Z_0$ where Z_0 is the characteristic impedance of the system.

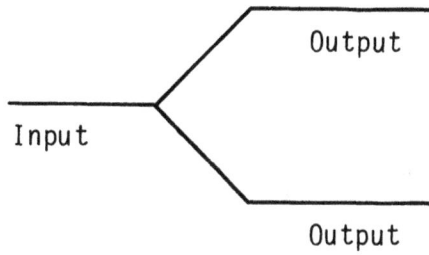

Figure 10.25. A Simple T-Junction Power Divider

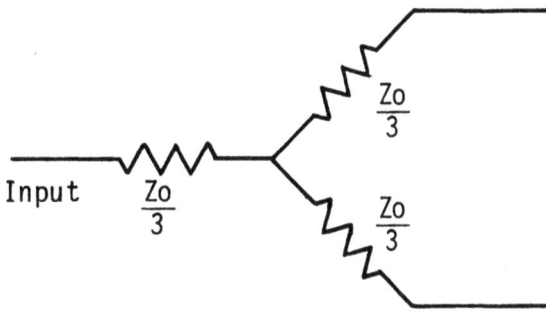

(a) Three-Resistor Construction

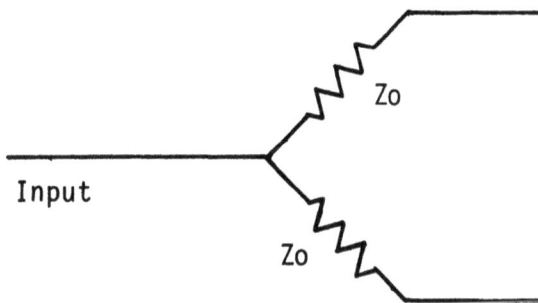

(b) Two-Resistor Construction

Figure 10.26. T-Junction Resistive Power Dividers

The three-resistor power divider provides good impedance match at both output arms when the input terminal is properly matched. Once a good source match has been achieved, the three-resistor power divider may be used to divide the output into equal signals. This kind of power divider has relatively poor isolation between the two output ports. The resistors used in the T-junction will increase the insertion loss. The three-resistor power divider is fundamentally frequency independent; therefore, the operation bandwidth can be extremely broad.

The two-resistor power divider can be used in leveling or ratio measurement. The major advantages of using a two-resistor power divider versus the directional coupler are the good source match characteristics, flat tracking between output ports, and wide operation frequency bandwidth from dc to microwave frequency. A leveling loop or power ratio measurement in a microwave circuit is used to keep the power level constant at the load through a reference port as shown in Figure 10.27. The portion of the load reflection signal, re-reflected by the power divider, is equal to the portion transmitted to the reference arm. This equality means that any change in output power caused by the load mismatch also appears in the reference arm, permitting the leveling loop or ratio meter to compensate for this variation. The three-resistor power divider is not suitable for leveling loop operation, because the portion of the load reflection signal re-reflected by the power divider is much smaller than the portion transmitted to the reference port.

The T-junction power dividers can be cascaded to provide more outputs as shown in Figure 10.28. However, the total number of outputs is limited to an integer power of 2. If a three-way power divider is needed, one of the four outputs will be terminated by a matching load and the power will be dissipated in the load as waste.

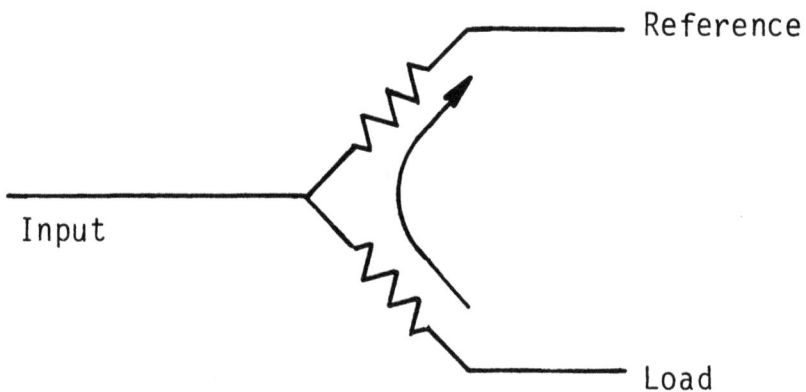

Figure 10.27. Two-Resistor Power Divider Used in a Leveling Loop

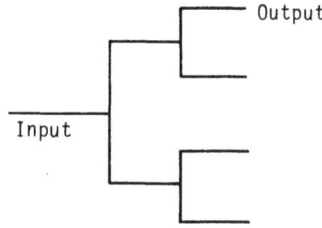

Figure 10.28. Cascade of T-Junction Power Dividers

10.11 N-WAY HYBRID POWER DIVIDERS (WILKINSON'S POWER DIVIDERS) [45-48]

This configuration of a special power divider was first discussed by Wilkinson in 1960 (Reference 45). Therefore, it is often referred to as Wilkinson's power divider. The power divider schematic shown in Figure 10.29 is broadband hybrid which has low VSWR at all ports and high isolations between output ports. The input and output impedances of the power divider are Z_0. The input signal is applied to n-parallel quarter-wavelength transmission lines of characteristic impedance Z_{01} each. At the output ports, resistors of value R are connected to a common point b. In the actual configuration the output ports are physically close together. The results of Reference 45 are presented here. The resistor R is determined as

$$R = Z_0 \qquad (10.26)$$

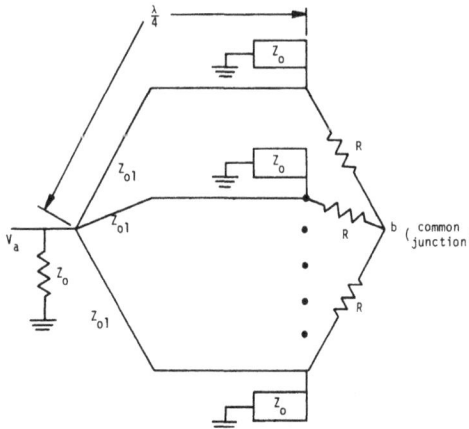

Figure 10.29. A Wilkinson's Power Divider

The characteristic impedance of the transmission line is

$$Z_{01} = \sqrt{n}\, Z_0 \qquad (10.27)$$

where Z_0 is the input and output impedance at all the ports, Z_{01} is the characteristic impedance of the transmission line and n is the total number of parallel paths.

If the internal loads R and the characteristic impedance Z_{01} of the power divider are adjusted according to Equations 10.26 and 10.27, the outputs will be completely isolated and matched. The input impedance under these conditions will also be matched to the resistance R.

The principal operation of the Wilkinson's power divider is explained as follows. When a signal is applied at the input of the power divider, it divides by virtue of symmetry into n equiphase equiamplitude ports. No power is dissipated by the resistors when matched loads are connected to the outputs, since all the transmission lines will be at the same potential. However, if a reflection occurs at one of the output terminals, the reflected signal will split; part of it will travel directly to the remaining output terminals via the resistors; and the rest of it will travel back to the input, splitting again at the junction of the transmission lines and then returning to the remaining output terminals. Thus, the reflected wave arrives at the remaining output terminals in two parts, and the length difference between the two paths will be 180 degrees when the transmission lines are $\lambda/4$ in length. If the resistor R, characteristic impedance Z_{01}, and input impedance Z_0 fulfill the relations given by Equations 10.26 and 10.27, the reflected signals from the two different paths will completely cancel each other at all the loads. Therefore, there is good isolation between the output ports.

The Wilkinson's power divider is suitable for the three-dimensional structure. In many applications it is desirable to have planar configuration which can be fabricated through strip line or microstrip line techniques. The two-way Wilkinson's power divider is suitable for planar construction. A general two-way power divider with different input and output impedances is shown in Figure 10.30. The resistance and the characteristic impedance are given by

$$R = 2Z_2 \qquad (10.28)$$

$$Z_{01} = \sqrt{2Z_1 Z_2} \qquad (10.29)$$

where R is the internal resistor shown in Figure 10.30, Z_1 is the input impedance, Z_2 is the output impedance, and Z_{01} is the characteristic impedance of the line.

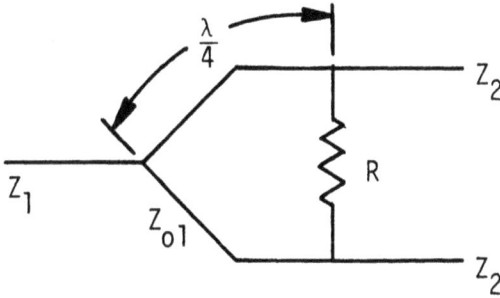

Figure 10.30. *Two-Way Wilkinson's Power Divider with Different Input and Output Impedances*

The Wilkinson's power divider design approach is extended to uneven power output (Reference 47). The configurations are shown in Figure 10.31a and b. The input and output impedances are assumed to be Z_0. Let k be the voltage ratio of port 3 to port 2, then

$$k^2 = \frac{P_3}{P_2} > 1$$

where P_3 and P_2 are the power outputs at port 3 and 2, respectively; and it is assumed that $P_3 > P_2$.

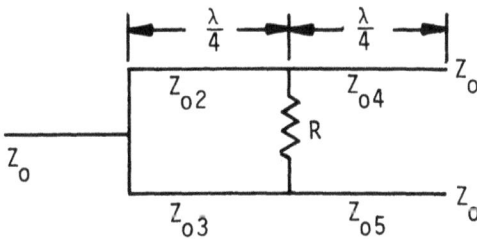

(a) *Fundamental Design*

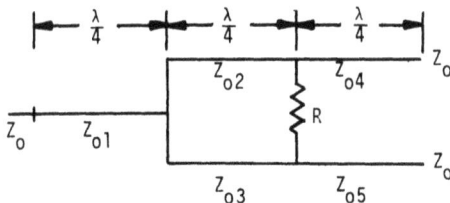

(b) *Improved Design*

Figure 10.31. *Power Divider with Uneven Power Outputs*

For Figure 10.31a the characteristic impedances of the transmission lines are

$$Z_{02} = Z_0 \sqrt{k(1 + k^2)} \qquad (10.30)$$

$$Z_{03} = Z_0 \sqrt{\frac{1 + k^2}{k^3}} \qquad (10.31)$$

$$Z_{04} = Z_0 \sqrt{k} \qquad (10.32)$$

$$Z_{05} = \frac{Z_0}{\sqrt{k}} \qquad (10.33)$$

and

$$R = Z_0 \frac{1 + k^2}{k} \qquad (10.34)$$

The transmission line lengths are all $\lambda/4$ at the center of the operation band. Sometimes the impedances calculated from the above equations are rather high in the hundreds of ohms. It requires narrow conductors to generate these high characteristic impedances. The small conductor could increase the insertion loss and cause a power handling problem. In addition, the input VSWR of this design is rather high. To improve the input VSWR characteristics, an input $\lambda/4$ transformer is added at the input as shown in Figure 10.31b. The characteristic impedances and the isolation resistor are

$$Z_{01} = \left(\frac{k}{1 + k^2}\right)^{1/4} Z_0 \qquad (10.35)$$

$$Z_{02} = (k^3 + k^5)^{1/4} Z_0 \qquad (10.36)$$

$$Z_{03} = \left(\frac{1 + k^2}{k^5}\right)^{1/4} Z_0 \qquad (10.37)$$

$$Z_{04} = Z_0 \sqrt{k} \qquad (10.38)$$

$$Z_{05} = \frac{Z_0}{\sqrt{k}} \qquad (10.39)$$

$$R = Z_o \frac{1 + k^2}{k} \qquad (10.40)$$

From the above equations, the impedance of each line can be calculated. The characteristic impedances of the transmission lines will determine the configuration of the power divider.

10.12 MODIFIED WILKINSON POWER DIVIDERS[49-56]

The Wilkinson's power divider has been further extended to n-way planar construction as shown in Figure 10.32. The input power is divided into n-parallel paths and each path contains m-sections. The synthesis for n = 3, m = 2, and n = 4, m = 3 have been formulated in Reference 49. The transmission lines are all quarter-wavelength at the center of the operating frequency range.

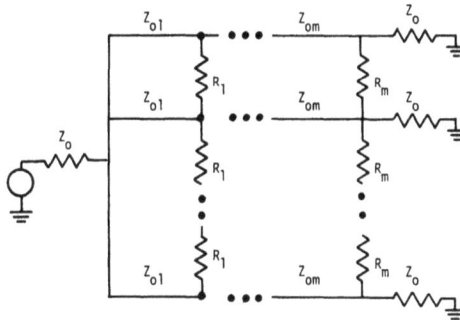

Figure 10.32. N-Way Hybrid Power Divider of Planar Structure

To improve the power handling capability of the power divider, a configuration as shown in Figure 10.33 is proposed (Reference 50). In the conventional Wilkinson's power dividers, the isolation resistors must be physically small in size and are difficult to heat sink for high power applications. In this approach, instead of a star of resistors connected between the output ports, an additional network of transmission lines (all quarter-wavelength at midband) and shunt resistors which can be made as external loads are used. This approach creates two effects: (1) the circuit becomes narrow in bandwidth compared to the simple Wilkinson's structure, and (2) the resistor is no longer a critical portion of the circuit and can be placed in shunt-to-ground at the end of a matched transmission line of arbitrary length. Thus, the use of standard high power external loads in this configuration is made possible.

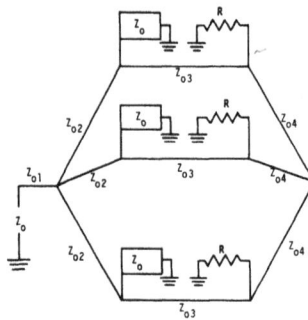

Figure 10.33. High Power N-Way Power Divider

To broaden the bandwidth of the two-way Wilkinson's power divider, multisection design or continuous tapered design can be used (References 51, 52, and 53). The multisection design shown in Figure 10.34 can result in 3:1 bandwidth. Examples of 2 and 3 sections are calculated in Reference 51 and general design formulas and design data are also given. The bandwidth of the power divider can be further broadened by continuous tapered design as shown in Figure 10.35. A bandwidth of 10:1 can be accomplished through this approach.

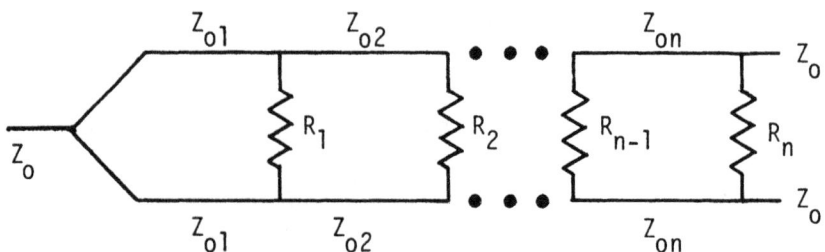

Figure 10.34. Multisection Power Divider to Broaden Bandwidth

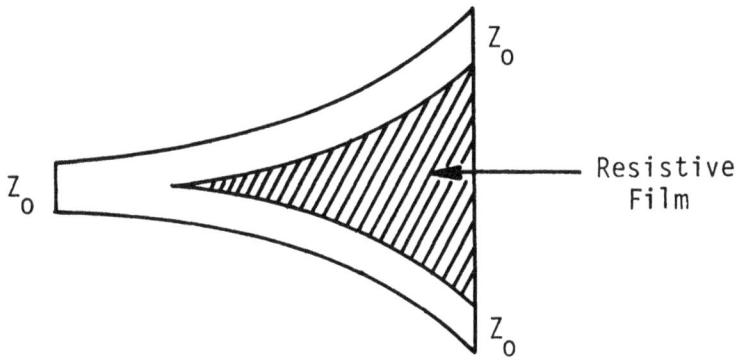

Figure 10.35. Continuous Tapered Power Divider

There are also other approaches to broaden the bandwidth of n-way power dividers (References 54 and 55). In these references, multistage power dividers are designed. Bandwidth up to 10:1 was calculated.

Low frequency to UHF power dividers can be made through lumped elements (Reference 56). The basic design is the same as that of the Wilkinson's divider. An isolation resistor is also required. The only difference is that the transmission lines are replaced by lumped circuit elements.

REFERENCES

1. Montgomery, C.G., Dicke, R.H., and Purcell, E.M., Editors, "Principles of microwave circuits," MIT Radiation Laboratory Series, Boston Technical Publishers, Inc., 1964.
2. Reich, H.J., Ordung, P.F., Krauss, H.L., and Skalnik, J.G., *Microwave theory and techniques,* D. Van Nostrand Co., Inc., 1953.
3. Ishii, T.K., *Microwave Engineering,* The Ronald Press Co., 1966.
4. Jones, E.M.T., and Bolljahn, J.T., "Coupled-strip-transmission-line filters and directional couplers," IRE Trans. on Microwave Theory and Techniques, Vol. MTT-4, pp. 75-81, April 1956.
5. Shimizu, J.K., and Jones, E.M.T., "Coupled-transmission-line directional couplers," IRE Trans. on Microwave Theory and Techniques, Vol. MIT-6, pp. 403-410, October 1968.
6. Matthaei, G.L., Young, L., and Jones, E.M.T., *Microwave filters, Impedance-matching networks and coupling structures,* Chapter 13, McGraw-Hill Book Co., New York, 1964.
7. "Anaren," published by Anaren Microwave Inc., New York (publication M9001-67), 1967.
8. Shelton, J.P., Wolfe, J., and Van Wagoner, R.C., "Tanden couplers and phase shifters for multi-octave bandwidth," Microwaves, April 1965.
9. Bahl, I.J., and Bhartia, P., "Characteristics of inhomogeneous broadside-coupled strip lines," IEEE Trans. on Microwave Theory

and Techniques, Vol. MTT-28, pp. 529-535, June 1980.

10. Rehnmark, S., "High directivity CTL-couplers and a new technique for the measurement of CTL-coupler parameters," IEEE Trans. on Microwave Theory and Techniques, Vol. MTT-25, December 1977.

11. Sheleg, B., and Spielman, B.E., "Broadband directional couplers using microstrip with dielectric overlays," IEEE Tran. on Microwave Theory and Techniques, Vol. MTT-22, pp. 1216-1219, December 1974.

12. Toulios, P.P. and Todd, A.C., "Synthesis of symmetrical TEM-mode directional couplers," IEEE Trans. on Microwave Theory and Techniques, MTT, Vol. MTT-13, pp. 536-544, September 1965.

13. Podell, A., "A high directivity coupler technique," in Proc. G-MTT-1970, Int. Microwave Symposium (May 1970), pp. 33-36.

14. Buntschuh, C., "Octave bandwidth, high directivity microstrip coupler," RADC-TR-73-396, Contract F30602-72-0282, AD777320, January 1974.

15. Carr, J.W., "Balanced-line microwave hybrids," Microwave Journal, Vol. 16, No. 5, pp. 49-52, May 1973.

16. Rehnmark, S., "Wide-band advanced line microwave hybrids," IEEE Trans. on Microwave Theory and Techniques, Vol. MTT-25, pp. 825-830, October 1977.

17. DeRonde, F.C., "Recent developments in broadband directional couplers on microstrip," in 1975 G-MTT Symposium Digest, pp. 215-217.

18. Dalley, J.E., "A strip line directional coupler utilizing a nonhomogeneous dielectric medium," IEEE Trans. on Microwave Theory and Techniques, Vol. MTT-17, pp. 706-712, September 1969.

19. Shamasundara, S.D., and Gupta, K.C., "Sensitivity analysis of coupled microstrip directional couplers," IEEE Trans. on Microwave Theory and Techniques, Vol. MTT-26, pp. 788-794, October 1978.

20. Lange, J., "Interdigitated strip line quadrature hybrid," in 1969 G-MTT Int. Microwave Symp., pp. 10-13.

21. Lange, J., "Interdigitated strip line quadrature hybrid," IEEE Trans. on Microwave Theory and Techniques, Vol. MTT-17, pp. 1150-1151, December 1969.

22. Ou, W.P., "Design equations for an interdigitated directional coupler," IEEE Trans. on Microwave Theory and Techniques, Vol. MTT-23, pp. 253-255, February 1975.

23. Waugh, R., and LaCombe, D., "Unfolding the Lange coupler," IEEE Trans. on Microwave Theory and Techniques, Vol. MTT-20, pp. 777-779, November 1972.

24. Miley, J.E., "Looking for a 3 to 8 dB microstrip coupler," Microwave p. 58, March 1974.

25. Paolino, D.D., "Design more accurate interdigitated coupler," Microwaves, pp. 34-38, May 1976.

26. Tajima, Y., and Kamihashi, S., "Multiconductor couplers," IEEE Trans. on Microwave Theory and Techniques, Vol. MTT-26, pp. 795-801, October 1978.
27. Presser, A., "Interdigitated microstrip coupler design," IEEE Trans. on Microwave Theory and Techniques, Vol. MTT-26, pp. 801-805, October 1978.
28. Kajfez, D., Paunovic, Z., and Pavlin, S., "Simplified design of Lange coupler," IEEE Trans. Microwave Theory and Techniques, Vol. MTT-26, pp. 806-808, October 1978.
29. Porter, J., "Computer coupled-microstrip line," Electronics, Vol. 25, pp. 116-118, January 1977.
30. Reed, J., and Wheeler, G.J., "A method of analysis of symmetrical four-port networks," IEEE Trans. on Microwave Theory and Techniques, Vol. MTT-4, pp. 246-252, October 1956.
31. Levy, R., and Lind, L., "Synthesis of symmetrical branch-guide directional couplers," IEEE Trans. on Microwave Theory and Techniques, Vol. MTT-16, pp. 80-89, February 1968.
32. Lind, L., and Scanlan, J.O., "Synthesis of asymmetrical branch-guide directional coupler-impedance transformers," IEEE Trans. on Microwave Theory and Techniques, Vol. MTT-17, pp. 45-48, January 1969.
33. Ho, C.Y., "Transform impedance with a branch-line coupler," Microwaves, p. 47, May 1976.
34. Pon, C.Y., "Hybrid-ring directional coupler for arbitrary power divisions," IEEE Trans. on Microwave Theory and Techniques, Vol. MTT-9, pp. 529-535, November 1961.
35. Ho, C.Y., "Design of lumped element quadrature couplers," Microwave Journal p. 67 September, 1979.
36. Peppiatt, H.J., Hall, J.A., and McDaniel, A.V., Jr., "A low-noise class C oscillator using a directional coupler," IEEE Trans. on Microwave Theory and Techniques, Vol. MTT-16, pp. 748-752, September 1968.
37. Aikawa, M. and Ogawa, H., "A new magic-T using coupled slot lines," IEEE Trans. on Microwave Theory and Techniques, Vol. MTT-28, pp. 523-528, June 1980.
38. Laughlin, G.J., "A new impedance-matched wide-band balun and magic tee," IEEE Trans. on Microwave Theory and Techniques, Vol. MTT-24, pp. 135-141, March 1976.
39. Kraker, D.I., "Asymmetric coupled-transmission-line magic-T," IEEE Trans. on Microwave Theory and Techniques, Vol. MTT-12, pp. 595-599, November 1964.
40. Jones, E.M.T., "Wide-band strip line magic-T," IEEE Trans. on Microwave Theory and Techniques, Vol. MTT-8, pp. 160-168, March 1960.

41. Niclas, K.B., "Planar power combining for medium power GaAs FET amplifiers in X/Ku-bands," Microwave Journal, p. 79, June 1979.

42. Pucel, R.A., "Power combiner performance of GaAs MESFET," Microwave Journal, p. 51, March 1980.

43. Schellenberg, J.M., "Transistor combinatorial techniques," Air Force Avionics Laboratory, Technical Report AFAL-TR-77-205, 15 May 1977.

44. Johnson, R.A., "Understanding microwave power splitters," Microwave Journal, December 1975.

45. Wilkinson, E.J., "An N-way hybrid power divider," IEEE Trans, on Microwave Theory and Techniques, Vol. MTT-52, pp. 116-118, January 1960.

46. Parad, L., and Moynihan, R., "Split-tee power divider," IEEE Trans. on Microwave Theory and Techniques, Vol. MTT-13, pp. 91-93, Janaury 1965.

47. Fike, G.F., "Designing hybrid couplers with uneven power splits," Microwaves, p. 64, November 1973.

48. Ho, C.Y., "Hybrids in planar microwave integrated circuit," Hybrid Microelectronics Seminar notes, December 1976.

49. Nagai, N., Mackawa, E., and Ono, K., "New n-way hybrid power dividers," IEEE Trans. on Microwave Theory and Techniques, Vol. MTT-25, pp. 1008-1012, December 1977.

50. Howe, H., Jr., "Simplified design of high power, n-way in-phase power divider/combiners," Microwave Journal, p. 51, December 1979.

51. Cohn, S.B., "A class of broadband three-port TEM-mode hybrids," IEEE Trans. on Microwave Theory and Techniques, Vol. MTT-16, pp. 110-116, February 1968.

52. Ekinge, R.B., "A new method of synthesizing matched broadband TEM-mode three-ports," IEEE Trans. on Microwave Theory and Techniques, Vol. MTT-19, pp. 81-88, January 1971.

53. Goodman, P.C., "A wide-band strip line matched power divider," Digest of GMTT-Symposium, pp. 16-20, 1968.

54. Yee, H.Y., Change, F.C., and Audeh, N.F., "N-way TEM-mode broadband power divider," IEEE Trans. on Microwave Theory and Techniques, Vol. MTT-18, pp. 682-688, October 1970.

55. Taub, J.J., and Kurpis, G.P., "A more general n-way hybrid power divider," IEEE Trans. on Microwave Theory and Techniques, Vol. MTT-17, pp. 406-408, July 1969.

56. Head, M.J., "Synthesize lumped element in phase power dividers," Microwave Journal, p. 111, May 1980.

CHAPTER 11
ATTENUATORS, CIRCULATORS
AND PHASE SHIFTERS

11.1 GENERAL CHARACTERISTICS OF ATTENUATORS

An attenuator is a device which will attenuate the signal passing through it. In general, a microwave engineer is only concerned with the amplifications of the signal. In a transmitter, a signal of high power is desired. In a receiver, a high signal-to-noise (S/N) is desired. Then why does one want to attenuate the signal in a microwave receiver? The answer is that the attenuators are used to adjust the signal level in a receiver or to improve the voltage standing wave ratio (VSWR) of interconnected microwave circuits. There are two kinds of attenuators: fixed and variable. The variable attenuator can be subdivided into two kinds: step attenuator and continuously variable attenuator. In a step attenuator, the attenuation is changed in steps such as 10 dB steps or 1 dB step. In a continuously variable attenuator, the attenuation is changed continuously and a dial is usually available to read the attenuation either directly or indirectly from a calibration chart.

The fixed and variable attenuators are available in both waveguide and coaxial systems. Most of the receivers under 20 GHz use coaxial cable type attenuators. The performance characteristics of a fixed attenuator are:
1. input and output impedances
2. flatness with frequency
3. average and peak power handling capability
4. temperature dependence

For a variable attenuator, additional characteristics should be considered, such as:
1. amount or range of attenuations
2. insertion loss in the minimum attenuation position
3. incremental attenuation for step attenuator
4. accuracy of attenuation versus attenuator setting
5. attenuator switching speed and switching noise.

Most of these performance characteristics are self explanatory. It is not necessary to discuss them in detail.

11.2 FIXED ATTENUATOR [1]

A fixed coaxial attenuation is made of a resistive network. Most of the fixed attenuators are symmetric types which means that both ends match the same characteristic impedance. There are T- and π-network attenuators as shown in Figures 11.1 and 11.2. For the T-network the required resistance values are

$$R_1 = Z_0 \frac{\sqrt{N} - 1}{\sqrt{N} + 1} \qquad (11.1)$$

$$R_3 = \frac{2 Z_0 \sqrt{N}}{N - 1} \qquad (11.2)$$

where Z_0 is the characteristic impedance and N is the power ratio from input to output which is related to the attenuation A by

$$A = 10 \log N \qquad \text{or} \qquad N = 10^{-\frac{A}{10}} \qquad (11.3)$$

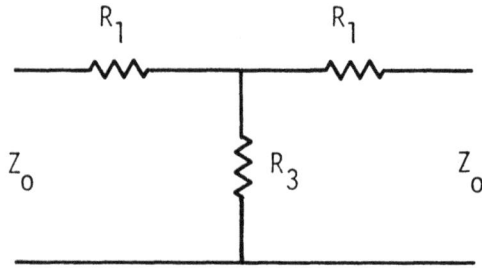

Figure 11.1. Symmetrical T-Network Attenuator

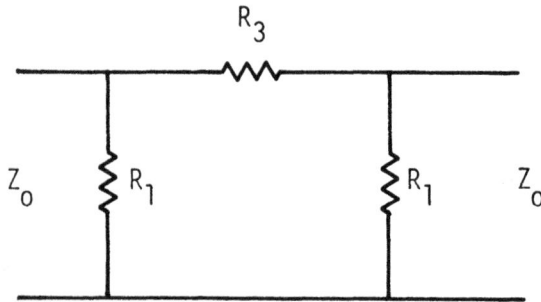

Figure 11.2. Symmetric π-Network Attenuator

For a 3 dB attenuator which matches 50Ω impedance, the values calculated from Equations 11.1, 11.2 and 11.3 are

$$N = 1.995 \qquad R_1 = 8.55\Omega \qquad \text{and} \qquad R_3 = 141.9\Omega$$

For the symmetric π-network

$$R_1 = Z_0 \frac{\sqrt{N} + 1}{\sqrt{N} - 1} \qquad (11.4)$$

$$R_3 = Z_0 \frac{N - 1}{2\sqrt{N}} \qquad (11.5)$$

The resistors used in these attenuators must have very low inductance, otherwise, the attenuator cannot operate properly at high frequency. Most of the resistors used are made of carbon. The shunt resistors used in the fixed coaxial attenuators are usually carbon rings.

11.3 CONTINUOUSLY VARIABLE ATTENUATORS(2-5)

In this kind of attenuator, the attenuation is often adjusted manually. This kind of attenuator is used in manually operated receivers to adjust the gain (or the sensitivity) of the receiver. They are seldom used in modern receivers with automatic gain control; however, they are very popular in microwave laboratory setups. Several of the continuously variable attenuators will be discussed here.

A coaxial resistive film tube attenuator is shown in Figure 11.3. The attenuation can be adjusted by sliding the two sections of coaxial cable and changing the distance L. The input power is absorbed by the resistive center tube. The attenuation is proportional to the length L.

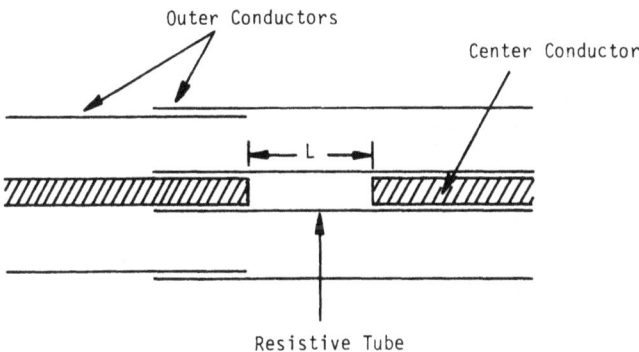

Figure 11.3. Coaxial Resistive Film Tube Attenuator

A coaxial cut-off variable attenuator is shown in Figure 11.4. In this arrangement there is no resistive element. The attenuation is caused by reflection. Therefore, this type of attenuator is also referred to as a zero loss cut-off variable attenuator. The transverse electromagnetic (TEM) mode in the coaxial section is changed into transverse magnetic (TM_{01}) mode (Reference 2) in the cylindrical waveguide section with a frequency below the cut-off frequency of the waveguide. Therefore, the signal will experience attenuation in this section. The attenuation is proportional to the length L. Since this is a reflection type attenuator, the VSWR is high. To improve the performance of the attenuator, a single crystal or amorphous semiconducting center rod can be added to the cylindrical waveguide (Reference 3).

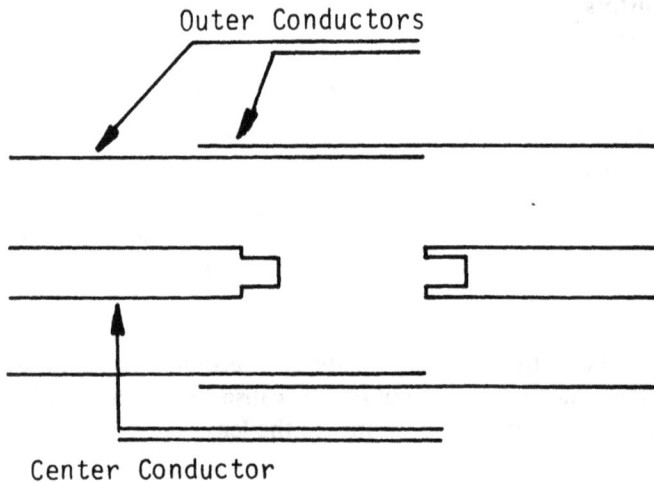

Outer Conductors

Center Conductor

Figure 11.4. Coaxial Cut-Off Variable Attenuator

A lossy wall attenuator (Reference 4) is shown in Figure 11.5. Figure 11.5a shows the longitudinal section while Figure 11.5b shows the transverse section. The attenuation is induced by the ohmic loss in the low conductive section of the wall. The attenuation can be changed by moving the center conductor in or out of the lossy wall section.

A variable coupler attenuator is characterized by its accuracy and wide dynamic range with low VSWR (Reference 4). The schematic of such a variable coupler attenuator is shown in Figure 11.6. The attenuator utilizes an adjustable coupled transmission line directional coupler. By mechanically varying the spacing between coupled lines, the coupling and attenuation of the device varies. For tight coupling from input to output, loss of less than 5 dB can be accomplished. However, the VSWR for this tight coupling becomes prohibitive for a precision attenuator. The useful attenuation range is from 5 to 90 dB. It is obvious that the major disadvantage of this kind of attenuator is the high minimum attenuation.

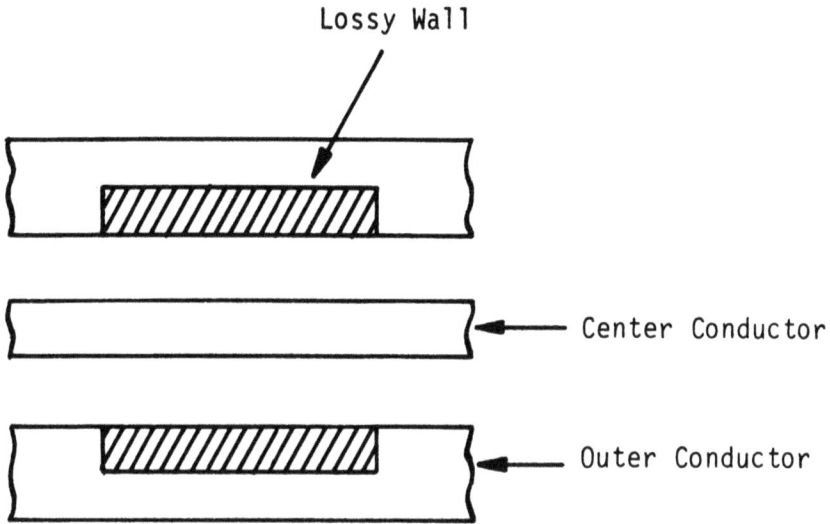

(a) Longitudinal Section

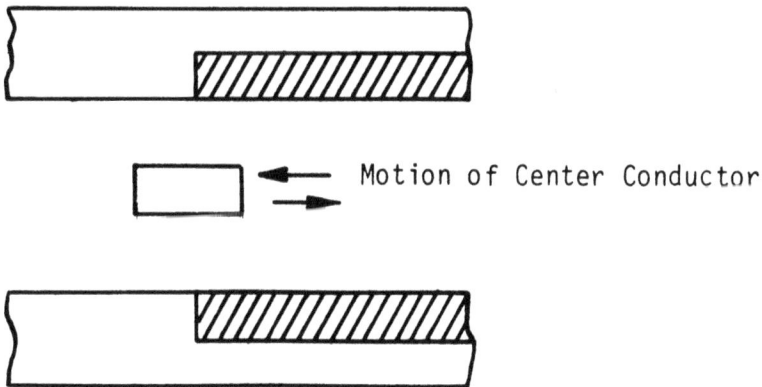

(b) Transverse Section

Figure 11.5. Lossy Wall Attenuator

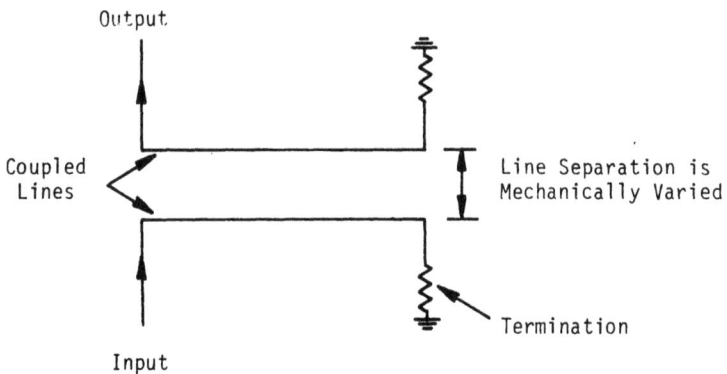

Figure 11.6. Schematic of Variable Coupled Attenuator

A zero loss cut-off attenuator (Reference 4) is made of hybrid couplers and cut-off attenuators. This attenuator is capable of both low insertion loss and very high maximum attenuation. A schematic of a zero loss cut-off attenuator is shown in Figure 11.7. The discussion of hybrid couplers which is helpful in understanding the operation of the zero loss cut-off attenuator can be found in Section 10.8.

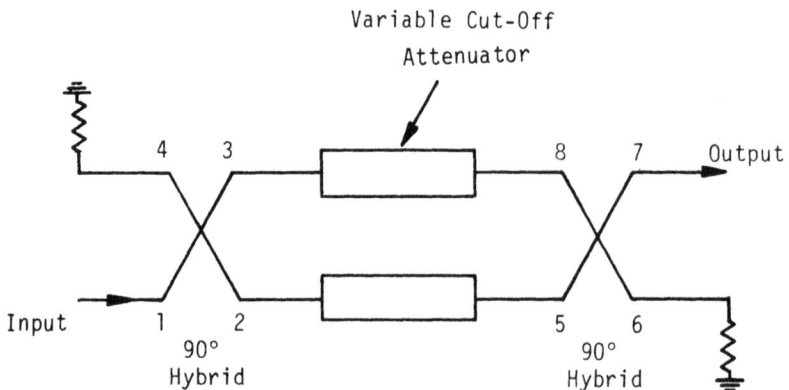

Figure 11.7. Schematic of Zero Loss Cut-Off Attenuator

An input signal from port 1 is divided equally to ports 2 and 3 through the input 3 dB hybrid. The variable cut-off attenuators reduce the input signal strength from ports 2 to 5 and 3 to 8 which will combine at the output port 7. The signal reflected by the cut-off attenuators at 2 and 3 will come out at port 4 and be absorbed by the termination. Thus, the reflected signal does not arrive at the input port 1, and a low input VSWR will be realized.

11.4 STEP ATTENUATOR WITH SWITCHABLE FIXED ATTENUATORS

A step attenuator is a variable attenuator. Its attenuation changes a fixed amount at each step. A common example is a 1 dB attenuation per step for a total of 10 dB change or at 10 dB steps for a total range of 100 dB. The attenuator can be either controlled manually or by programmable relays. This kind of step attenuator is made of many individual fixed attenuators. The desired attenuation value is obtained by switching to the proper fixed attenuators. In a programmable attenuator, the switching is accomplished by an external control signal. Since the switches are mechanical, the switching time is relatively long in the millisecond range. Both the manual and programmable attenuators are of little use in receivers. However, they are indispensible components in a microwave laboratory setup and in receiver evaluations.

11.5 PIN ATTENUATORS [6,7]

The word "pin" stands for P-material, Intrinsic, and N-material. A pin diode is a silicon semiconductor device consisting of a layer of nearly intrinsic (high resistivity) material of finite thickness (10-200 μm) between the highly doped p- and n-materials. The microwave resistance of the pin diode depends on the bias voltage. The resistance will decrease when it is forward biased, and this resistance change can be accomplished in nanoseconds. The pin diode can be used to build either microwave attenuators or switches. Since the pin attenuator can be switched at very high speed, noise is usually generated. The noise spectrum is usually below 1 GHz. When the attenuator is operating in the frequency range above the noise spectrum, the noise can be filtered out through high-pass filters. However, when the attenuator operates in the hundreds of megahertz range, the noise generated by the switching cannot be filtered out as easily. Under such conditions, the spectrum, the amplitude, and the time domain responses of the noise are very important. A delay line can be placed in front of the pin switch to accommodate the problem. The delay time in the delay line should be long enough that when the signal arrives at the attenuators, the correct attenuation is set and the noise generated by the switching has ceased. (See Section 11.6.)

In Figure 11.8a, a simple pin attenuator is shown in a series arrangement, while Figure 11.8b shows a parallel arrangement. In the series arrangement when the bias voltage is zero, the pin diode is in a high resistance state; and the input radio frequency (RF) signal will be attenuated (the highest attenuation level). When the control bias increases, the resistance of the pin diode decreases and so does the attenuation. In Figure 11.8b, when the bias is zero, the high resistance of the pin diode does not affect the RF signal, and the attenuator is at its lowest level. When the bias voltage increases, the pin diode resistance decreases and part of the RF input signal will be reflected. Therefore, attenuation is increased. If the diode is assumed pure resistive, the RF attenuation is given by

$$\alpha_{(series)} = 20 \log (1 + \frac{R_I}{2Z_0}) \tag{11.6}$$

$$\alpha_{(shunt)} = 20 \log (1 + \frac{Z_0}{2R_I}) \tag{11.7}$$

where R_I is the resistance of the pin diode and Z_0 is the characteristic impedance of the transmission line. The input impedances of these attenuators are obviously not constants; they depend on the resistance of the pin diode. Attenuation is achieved primarily through reflection and partially by dissipation in the pin diode. Therefore, these attenuators are

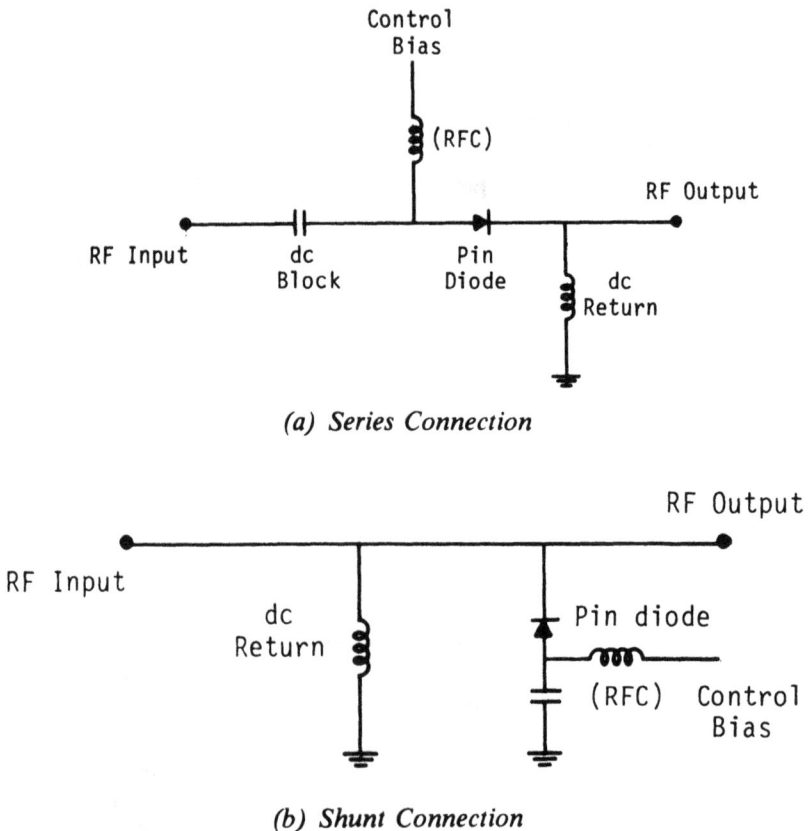

(a) Series Connection

(b) Shunt Connection

Figure 11.8. Schematic of Pin Attenuators

generally referred to as reflective attenuators. In a real diode there are always parasitic reactive elements such as lead inductance and junction capacitance. Although Equations 11.6 and 11.7 are frequency independent, in reality they are frequency dependent. If a single pin diode cannot provide the required attenuation, multiple pin diode arrangements can increase the total attenuation.

Pin attenuators of constant impedance can be built by using two basic approaches: (1) the pin diode themselves absorb the RF energy, and (2) the pin diodes reflect part of the RF energy to a different RF port and dissipate the rest of the energy in resistive loads. In Figure 11.9, the three pin diodes are arranged in a pi-network similar to the arrangement of fixed attenuation as shown in Figure 11.3. The resistances of the pin diodes are designed such that Equations 11.4 and 11.5 will be fulfilled. Thus, the RF signal is absorbed by the pin diodes in the pi-network, and the attenuator will have constant input and output impedances.

Figure 11.10 shows a single hybrid coupled pin attenuator. The input signal is split equally between B and C. Depending on the resistances of the pin diodes, part of the input signal is reflected to the output port and part of the input signal is absorbed in the pin diode and resistances Z_0. In this arrangement the input and output are isolated and constant impedances are obtained.

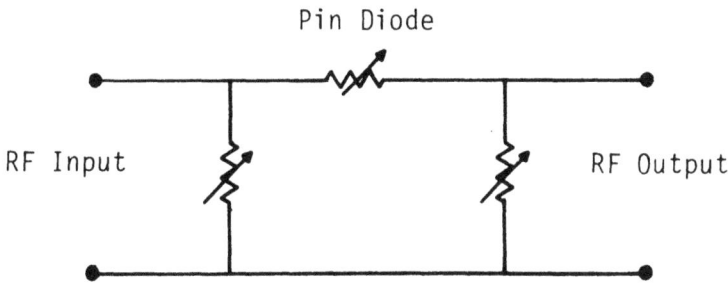

Figure 11.9. Pin Attenuator with pi-Network Arrangement

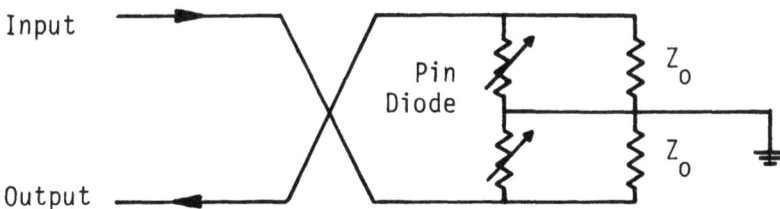

Figure 11.10. Single Hybrid Coupled Pin Attenuator

11.6 APPLICATIONS OF ATTENUATORS

As mentioned in the introduction, an attenuator is commonly used to adjust the signal level. In general, a microwave amplifier has fixed gain. If the right amount of gain is not available in a certain group of amplifiers, obtain an amplifier with a slightly higher gain than the required value, and add proper attenuation at the output of the amplifier. For example, a group of small signal amplifiers covering 2 to 4 GHz frequency range with a noise figure of 3 dB have gains of 7, 14, 25, 33, and 40 dB (Avantek ABG-4030, 4033, 4034, 4035, and 4036). If in a certain design, a 20 dB gain is required, the 25 dB gain amplifier should be used with a 5 dB fixed attenuator (sometimes referred to as a 5 dB pad) as shown in Figure 11.11. The attenuator will also improve the impedance matching of the circuit.

In some systems, the gain in the RF chain depends on the incoming signal strength. If the input signal is weak, a high gain is required to improve the signal strength. If the input signal is strong, a low gain is desirable so that the receiver will not be driven into saturation. The arrangement shown in Figure 11.12 can accomplish this. Sometimes this

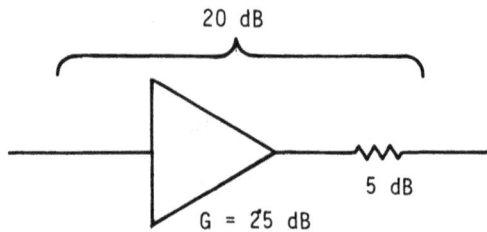

Figure 11.11. Amplifier with Fixed Attenuator to Obtain Desired Total Gain

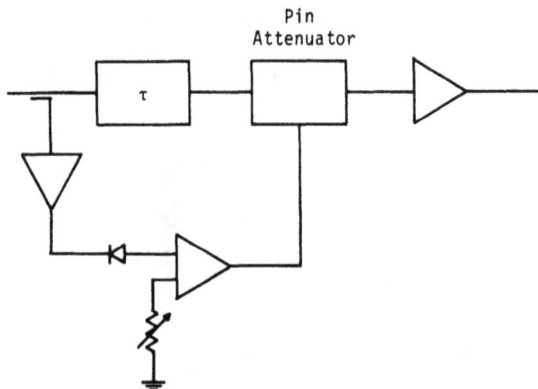

Figure 11.12. IAGC Using a Step Attenuator

scheme is referred to as instantaneous automatic gain control (IAGC). The input signal is partially coupled out to another path, amplified, detected, and compared with a reference level through a comparator. The output from the comparator controls the attenuator. If the input signal is weak and below the reference voltage, the attenuator is at a low attenuation level. If the input signal is strong and above the reference level, a fixed attenuation will be switched in. The added attenuation can be in multiple stages through several comparators. A delay line is desirable in the main signal path, which will provide time for the attenuator to settle before the signal arrives. Otherwise, if the input signal is a pulse, the leading edge of the pulse will reach or pass the attenuator before the proper attenuation is chosen. The IAGC circuit can react in ten to several hundred nanoseconds. When a fast step attenuator switches, there is transient noise generated which should be handled very carefully in the receiver design. This problem can also be avoided by the added delay line in the signal path, since the delay line can delay the signal until the transient noise generated by the switch fades out. This arrangement cannot improve the two signal instantaneous dynamic range of the receiver; however, it can move the instantaneous dynamic range to a different input power level as discussed in Chapter 2. The pin diode attenuator is very useful in microwave receiver designs to attenuate strong signals that are outside the receiver's dynamic range.

An attenuator can be used between a source and its load to improve the isolation. As shown in Figure 11.13a, an oscillator is connected directly to a load. If the impedance of the load is not properly matched, power will be reflected from the load. This reflected power may change the frequency and the power level of the source. This phenomena is generally referred to as the pulling effect on the oscillator (see Chapter 15). The ideal approach to reduce the pulling effect is to add an isolator between the source and the load. The isolator will reduce the reflected power drastically and have little

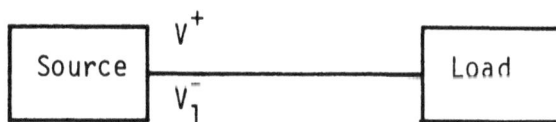

(a) Signal Source Directly Connected to Load

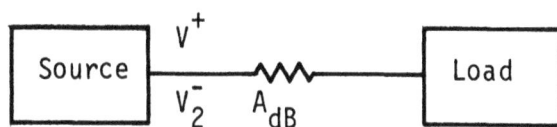

(b) An Attenuator is Added Between the Source and Load

Figure 11.13. Attenuator to Improve VSWR

effect on the incident power. The isolator will be discussed later in this chapter. However, if the circuit in Figure 11.13a covers a wide frequency range and there is no isolator readily available for this frequency range, an attenuator can be used to reduce the VSWR. The disadvantage of using an attenuator is that it reduces the incident power as much as the reflected power. The improvement of VSWR by adding the attenuator is discussed as follows.

The VSWR is defined as the ratio of the maximum voltage to the minimum voltage on a transmission line as

$$S = \frac{V_{max}}{V_{min}} = \frac{V^+ + V^-}{V^+ - V^-} = \frac{1 + \dfrac{V^-}{V^+}}{1 - \dfrac{V^-}{V^+}} = \frac{1 + \zeta}{1 - \zeta} \tag{11.8}$$

where V_{max}, V_{min} are the maximum and minimum voltages, respectively; V^+ and V^- are the voltage of the incident and reflected waves, respectively; ζ is the magnitude of the voltage reflection coefficient defined as V^-/V^+. If the line has no loss, the VSWR is independent of the measurement location. If the transmission line has loss, then the VSWR is dependent on the location of the measurement.

The value of S ranges from 1 to ∞. When $S = 1$, the impedance is perfectly matched.

For the arrangement in Figure 11.13a, the VSWR at the source is

$$S_1 = \frac{V^+ + V_1^-}{V^+ - V_1^-} \tag{11.9}$$

where S_1 is the VSWR of the circuit without the attenuator, and V^+ and V_1^- are the incident and reflected voltage.

When the attenuator is added the new VSWR becomes

$$S_2 = \frac{V^+ + V_2^-}{V^+ - V_2^-} \tag{11.10}$$

where V_2^- is the new reflected voltage and related to V_1^- by

$$V_2^- = V_1^- 10^{-\frac{A}{10}} \tag{11.11}$$

where A is the attenuation added in dB. It should be noted that V_2^- is attenuated twice by attenuator A.

Substituting Equations 11.9 and 11.11 into Equation 11.10 obtains

$$S_2 = \frac{(S_1 + 1) + (S_1 - 1)\, 10^{-\frac{A}{10}}}{(S_1 + 1) - (S_1 - 1)\, 10^{-\frac{A}{10}}} \qquad (11.12)$$

Equation 11.12 can be rewritten as

$$A = 10 \log \frac{(S_1 - 1)(S_2 + 1)}{(S_1 + 1)(S_2 - 1)} \; dB \qquad (11.13)$$

Equation 11.12 can be used to determine the new VSWR at the source if A is given, while Equation 11.13 can be used to calculate the amount of attenuation required to obtain a certain VSWR.

11.7 TERMINATIONS

A termination is a one-port device with an impedance that matches the characteristic impedance of a given transmission line. It is used in a certain terminal of a device to absorb the power to that terminal. As in Figures 11.6 and 11.7, the terminations are used to absorb the power reflected to the terminals. The VSWR of a termination is of primary concern. Termination can also be used as a dummy load in a power transmitting system. In this application, in addition to the VSWR, the power handling capacity is also very important. In a receiver, terminations are usually placed in various unconnected ports of components such as hybrid and power dividers to keep the VSWR of the signal path low. It is extremely important that the isolated port in a directional coupler and the unused port of a power divider (i.e., only three ports of a four-way power divider are used) be properly terminated. As discussed in Chapter 10, all of the design considerations of directional couplers and power dividers are based on the fact that all ports are terminated with matched loads. If an unused port is not properly terminated, then the isolations between the output ports will be reduced which may severely degrade the performance of the receiver.

11.8 CHARACTERISTICS OF CIRCULATORS[8-15]

A microwave circulator is a nonreciprocal ferrite device which contains three or four ports. The input from port n will come out at port n + 1 but not out at port n − 1. A three-port ferrite junction circulator, usually called the Y-junction circulator, is most commonly used. They are available in either rectangular waveguide or strip line forms. The signal flow in the three-port circulator is assumed as 1→2, 2→3, and 3→1.

If port 1 is the input, then the signal will come out of port 2; in an ideal situation, no signal should come out of port 3 which is called the isolated port. The insertion loss of the circulation is the loss from 1 to 2, while the insertion loss from 1 to 3 is referred to as isolation. When the input is port 2, the signal will come out of port 3 and port 1 is the isolated port. Similar discussions can be applied to port 3. The scattering matrix of a perfect circulator is

$$S = \begin{bmatrix} 0 & 0 & 1 \\ 1 & 0 & 0 \\ 0 & 1 & 0 \end{bmatrix} \tag{11.14}$$

which is symbolically represented in Figure 11.14.

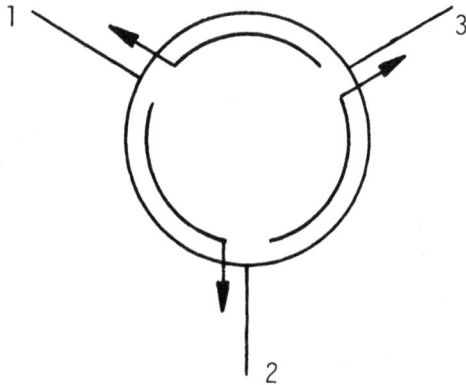

Figure 11.14. Symbolic Expression of a Y-Junction Circulator

The ferrite junction circulator was in use for a number of years before the theory of operation received much attention. Early experiments found that waveguide T-junctions that have a transversely magnetized ferrite slab suitably placed in the junction could, with proper matching and adjustment of the magnetic field, behave like circulators. The bandwidth of such devices was very narrow. The general properties of the circulators, including the tuning procedure for the symmetrical junction circulators, were studied by Auld (Reference 8) through the consideration of their scattering matrices. Milano et. al., (Reference 9) have applied these concepts in the design of a Y-junction strip line circulator. Bosma (Reference 10) made an analysis of the strip line Y-circulator. The circulator was solved as a boundary value problem, and the circulation parameters and frequency bandwidth were discussed. Fay and Comstock (Reference 11) gave a phenomenological explanation of the circulator operation in terms of the rotation of the magnetic field pattern which will be discussed later. Simon (Reference 12) reported experimental results of an octave bandwidth circulator in the frequency range of 0.6 to 8 GHz. Circulators of frequencies from 1 to 2 GHz and 2 to 4 GHz with minimum isolation of 20 dB have been fabricated. Wu and Rosenbaum (Reference 13) used Bosma's approach and reported an octave bandwidth Y-junction circulator design without external turning elements. A circulator operating from 7 to 12.4 GHz with a 15 dB isolation was reported. Ayasli (Reference 14) analyzed the wide-band strip line circulators through integral equations. Miyoshi and Miyanchi (Reference 15) investigated the optimum shape of a planar circulator for wide-band operation and predicted that a triangular shaped circulator with slightly concave sides can realize the best performance for wide-band operation. The general requirements on circulators are a wide operating frequency band, a low insertion loss in the forward direction, and a very high isolation. The principle of operation on junction circulators and their applications will be discussed in the next section.

11.9 FERRITE CIRCULATORS [11, 16]

This discussion is based on Reference 11. A strip line Y-junction ferrite circulator is shown in Figure 11.15. It consists of two ferrite cylinders filling the space between a metallic conducting center disk and two conducting ground plates. The connections to the center disk are in the form of three strip line center conductors attached to the disk at points 120 degrees apart around its circumference. A dc magnetizing field is applied parallel to the axis of the ferrite cylinders. In this simple arrangement, the device will perform as a circulator.

Experimentally, it was found that the Y-circulator has some but not all of the properties of a low-loss transmission cavity. The maximum isolation

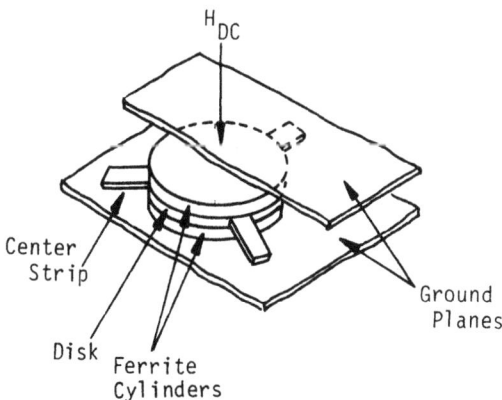

Figure 11.15. Strip Line Y-Junction Circulator (Based on Fay and Comstock, Reference 11)

occurred almost at the frequency at which the insertion loss was minimum. These observations suggest a resonance of the center disk structure as being an essential feature of the operation of the circulator. The lowest frequency resonance of the circular disk structure is the dipolar mode as shown in Figure 11.16. The electric field vectors E are perpendicular to the plane of the disk. The E field, in the upper ferrite disk, is shown as circles with a dot or a cross in the center to represent outward and inward direction of the paper, respectively. The E field vectors, in the lower ferrite disk, are mirror images of the upper one. The RF magnetic field vector H lies parallel to the plane of the disk shown as solid lines in Figure 11.16. The H field curves over the edge of the metal disk and continues back on the underside in the opposite direction as the solid lines indicate and back on the upper side again to make a closed loop. Therefore, the electric and magnetic fields in the upper and lower ferrite disks are the same, and the analysis of the device need only be concerned with one ferrite disk. Figure 11.16a shows the E and H fields of the resonance mode without the dc magnetic bias field. Port 1 is the input. Ports 2 and 3, if open-circuited, will see voltages which are 180 degrees out-of-phase with the input voltage. The input power from port 1 will be divided equally among ports 2 and 3. Under this condition, there is no circulation effect in the device.

The standing wave pattern of Figure 11.16a in which the field varies as exp(jωt) can be generated by two counter-rotating field patterns of the same configuration. Each of these patterns would involve an RF magnetic field pattern which is circularly polarized at the center of the disk, becomes more elliptical as the radius increases, and is linearly polarized at the edge of the disk. If a dc magnetic biasing field is applied in the direction of the axis of the disk, the relative permeabilities μ_r^+ in the clockwise direction and μ_r^- in the counterclockwise direction can be written as

$$\mu_r^{\pm} = 1 + \frac{\gamma_e M_0}{|\gamma_e| H_{dc} \pm \omega} \tag{11.15}$$

where γ_e is the gyromagnetic ratio of an electron, M_0 is the saturation magnetization, H_{dc} is the dc bias magnetic field, and ω is the angular frequency of the input signal.

The relative permeability is plotted in Figure 11.17. If μ_r^+ is larger than μ_r^-, the clockwise wave will rotate more than the counterclockwise wave and a net rotation will result. Therefore, the field pattern shown in Figure 11.16a will be rotated as shown in Figure 11.16b under the influence of the dc magnetic field. Under this condition, port 3 is situated at the voltage null of the disk; and the voltages at ports 1 and 2 are equal. The device is equivalent to a transmission cavity between ports 1 and 2, and port 3 is isolated.

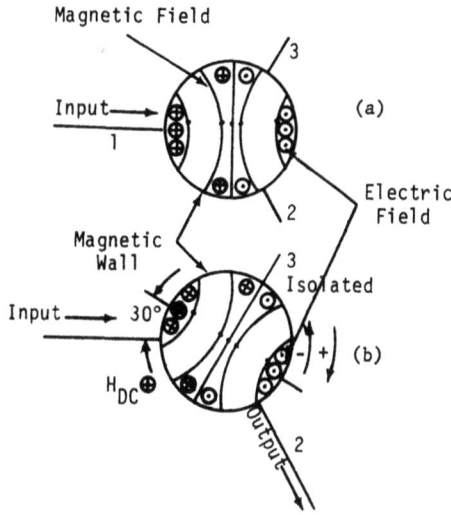

(a) Dipolar Mode of a Dielectric Disk $H_{int} = 0$

(b) Analogous Pattern of a Magnetized Disk. H_{int} for Circulation

Figure 11.16. E and H Fields in a Y-junction Circulator (Based on Fay and Comstock, Reference 11)

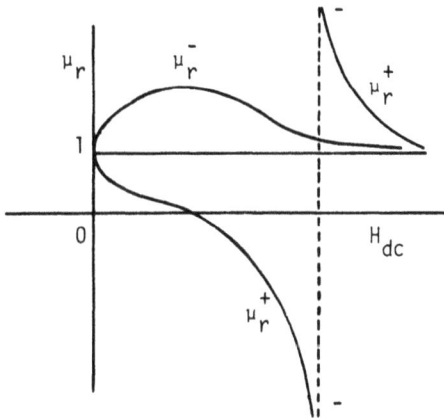

Figure 11.17. Graph of μ_r Versus H_{dc}

The simple explanation given above illustrates how a circulator works physically. However, in the conducting disk, there are also other resonant modes existing in order to fulfill the boundary condition between the strip line and the disk. Detailed discussion can be found in the references listed in Section 11.8.

11.10 APPLICATIONS OF CIRCULATORS AND ISOLATORS[17]

There are many applications of circulators in connection with receivers. They can be used either at system level or component level. Some of the common uses will be discussed.

If the same antenna is used, in a transmitting and receiving system, a circulator can separate the transmitter and the receiver as shown in Figure 11.18. The power from the transmitter will go to the antenna rather than the receiver, and the received signal will go to the receiver. In a practical system, this arrangement usually lacks sufficient isolation. The power generated from the transmitter is usually very strong. If there is not enough isolation provided by the circulator, the transmitter power will saturate the receiver or even damage it. The isolation in a circulator is generally 20 dB. Another possible problem is that if there is a mismatch between the transmitter and the antenna, the reflected power will also reach the receiver. Therefore, there are limited applications for using a circulator to isolate a transmitter and receiver sharing the same antenna.

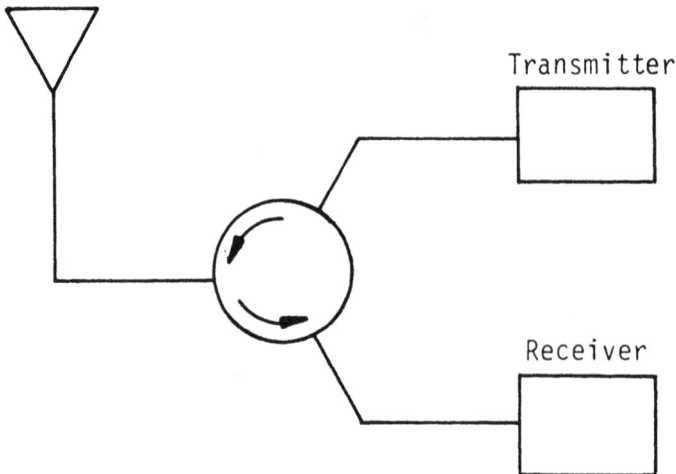

Figure 11.18. Common Antenna for Transmitter and Receiver

There are many devices with negative RF resistance (i.e., Gunn diode, tunnel diode) that can be used as amplifiers. These devices can be used to build one-port amplifiers as discussed in Chapter 13. The input and output of the amplifier share the same port which will cause some inconvenience in circuit designs. A circulator is commonly used for these kinds of amplifiers to provide input and output isolation as shown in Figure 11.19.

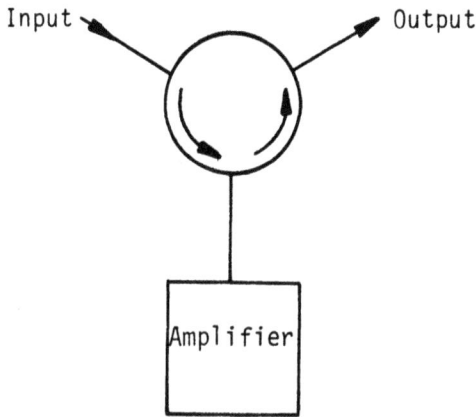

Figure 11.19. Circulators to Provide Input and Output Isolation for a
One-Port Amplifier

Circulation can be used as an isolator. An isolator is a non-reciprocal
two-port device. The power will flow in one direction and be absorbed by
the device in the opposite direction. There are two types of isolators in
waveguide configurations: the Faraday rotation ferrite isolator and the
field displacement ferrite isolator (Reference 17). In strip line configurations, a circulator can be made into an isolator by terminating one port of
the circulator with a matched load as shown in Figure 11.20. The signal
from port 1 will reach port 2, but the input from port 2 will reach port 3
and be absorbed. The scattering matrix of a perfect isolator is

$$S = \begin{bmatrix} 0 & 1 \\ 0 & 0 \end{bmatrix} \qquad (11.16)$$

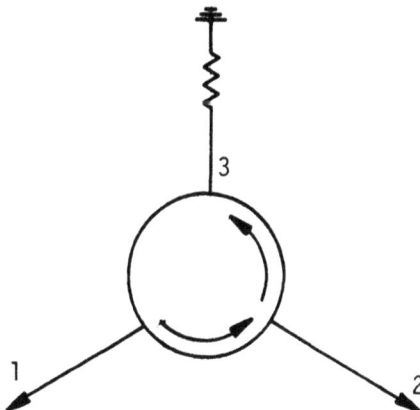

Figure 11.20. Circulator Working as an Isolator

An isolator is often used between an oscillator and its load, as discussed early in this chapter and in Chapter 15, in order to minimize the pulling effect of the oscillator. At the mixer terminals, isolators are sometimes used to reduce the power reflected back to the mixer. The reflected power may produce undesired spurs in the mixer. Only if an isolator is not available or to reduce cost, a fixed attenuator can be used as a substitute for an isolator to reduce the VSWR in a circuit. The attenuator is not very efficient if it is used as an isolator to reduce VSWR.

Frequency multiplexing: In a channelized receiver, the input signal is multiplexed into many parallel filters of contiguous frequencies. If all the filters are directly connected together as shown in Figure 11.21a, the input VSWR will be rather poor because the center frequencies of the filters are close together. One way to improve the VSWR is to separate the contiguous filters into two groups (or banks) and connect them through a circulator as shown in Figure 11.21b. In the two banks of filters, bank 1 contains all the even number filters, and bank 2 contains all the odd number filters. Since the frequency separation between filters in each bank is increased, the coupling between adjacent filters will decrease and the VSWR at the input of the filter bank will improve. In Figure 11.21b the input from port 1 will reach port 2. If the signal frequency matches the frequency of any filter in bank 1, it will pass through the proper filter. Signals whose frequency does not match the bank 1 filters will be reflected and reach port 3 where it will pass the proper filter in bank 2. This is a common practice in channelized receivers.

11.11 CHARACTERISTICS OF PHASE SHIFTERS[18, 19]

A phase shifter is a two-port microwave device which is used to control the phase of the signal. The phase of the signal can be manually controlled in a microwave circuit by a piece of dielectric film inserted in a waveguide. This approach is mostly suitable for laboratory test setups. At the present time, most of the phase shifters are built with pin diodes and ferrite devices. They are often controlled through a current source. Phase shifters are used as the major components in electronically controlled array antennas. An array antenna contains many individual radiating elements. By properly controlling the phase of each element, the resulting beam of the antenna can be steered electronically. According to the law of reciprocity, the same antenna can be used as a receiving antenna with steerable beam.

In a microwave receiver, an electronically controlled phase shifter is seldom used. Some limited applications are suggested in connection with the antenna applications. The phase shifters are discussed here because they are very important microwave components, and they will probably be used in future receiving systems.

According to performance characteristics, the diode phase shifters can be divided into two categories: (1) the constant time delay (sometimes

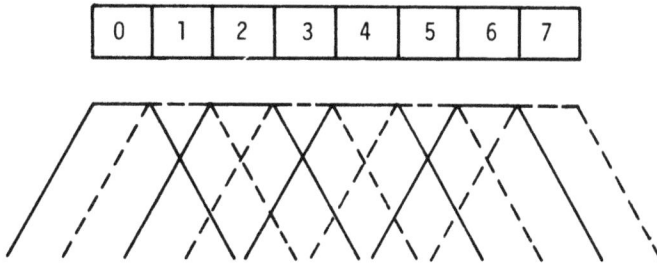

(a) One Contiguous Filter Bank

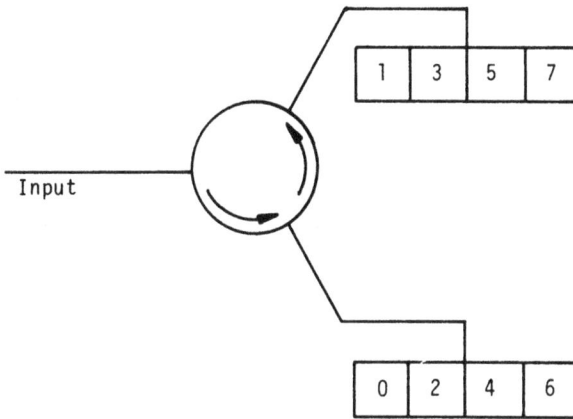

(b) Use of Circulator to Connect Two Banks of Filters

Figure 11.21. Parallel Filters of Contiguous Frequencies

referred to as nondispersive) phase shifters, and (2) constant phase shift
(or dispersive) phase shifters. For phased array antennas, a constant time
delay is generally required, while for phase comparison networks a constant phase shift is needed. According to the physical construction, the
diode phase shifters can be divided into four groups: (1) reflection type,
(2) switched line, (3) loaded line, and (4) high-low pass. The first two kinds
of diode phase shifters are basically constant time delay, although they can
be made into constant phase devices. The loaded line and high-low pass
phase shifters will provide constant phase shift.

11.12 FERRITE PHASE SHIFTERS[19, 20]

Most ferrite phase shifters are made by placing ferrite materials in a
section of waveguide. The phase shifters can be either a reciprocal or
nonreciprocal device depending on the applied dc magnetic bias and the
configuration. Phase shifting is accomplished by changing the magnetic
permeability of the ferrite through changing the external applied dc

magnetic field. The ferrite is commonly used as a bulk control medium which means the microwave travels in the ferrite itself. The ferrite is usually several wavelengths long. The change in relative permeability for the ferrite comes about because of the difference in propagation constant of right-hand and left-hand circularly polarized RF magnetic fields. A plane wave can be represented by two circularly polarized waves: left-hand and right-hand. Since the propagation constants are different for left-hand and right-hand circularly polarized wave, a plane wave traveling through the ferrite material will change its phase. The two primary advantages of a ferrite phase shifter in comparison with a diode phase shifter are: (1) The ferrite phase shifter has higher power handling capability. Since the phase control is accomplished in bulk medium, the power density in the ferrite is low. The device has a limiting effect; the insertion loss will suddenly increase if the input signal is beyond a certain power level. (2) The ferrite phase shifter has lower insertion loss because of the waveguide configuration. The major disadvantages of the ferrite phase shifter are its relatively large size and slow switching speed, since the driver must change the control current in order to change the dc magnetic field. In a receiver system, the power handling capability is usually not a major problem and the switching speed is of more concern; thus, diode phase shifters which have faster switching time are often preferred over ferrite shifters in a receiver.

11.13 REFLECTION TYPE PHASE SHIFTERS[21-27]

A reflection type phase shifter is shown in Figure 11.22. Figure 11.22a shows a circulator type while Figure 11.22b shows a 90 degree hybrid type. Sections of transmission lines are connected from the diodes to the ground. The length of the transmission lines determine the amount of phase shift. The pin diodes can be biased either in the low resistance mode or high resistance mode. If the diodes are in the low resistance mode, the input signal will travel to the ends of the transmission lines and be reflected there. If the pin diodes are in the high resistance mode, the input signal will be reflected at the diodes. The reflected signals will emerge from the output port in Figures 11.22a and b. The path difference between these two conditions is Δl which corresponds to a phase shift of

$$\Delta\phi = 2\pi\Delta l/\lambda \qquad (11.17)$$

where Δl is twice the transmission line length from the pin diode to ground and λ is the wavelength of the input signal.

The bandwidth of these phase shifters is generally limited by the circulator or 90 degree hybrid rather than the pin diodes. Since a broadband hybrid is easier to fabricate than a circulator, it is more desirable to use the hybrid although it uses two pin diodes in the phase shifter.

The reflection type phase shifter, using a transmission line to produce

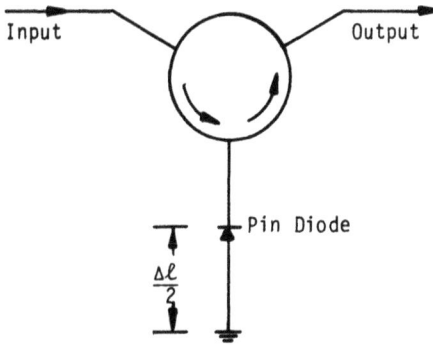

(a) Circulator Type

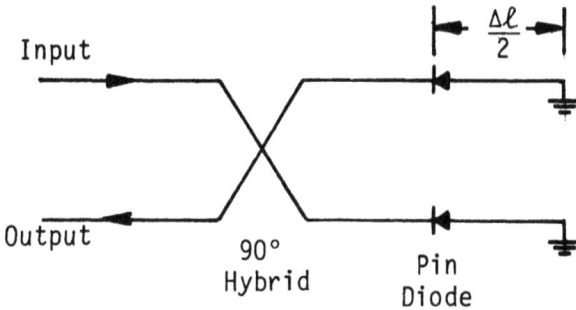

(b) Hybrid-Coupler Type

Figure 11.22. Reflection Type Phase Shifter

the necessary phase shift, is a constant delay time type. If the transmission lines are replaced by lumped circuits which can provide a constant phase shift, the phase shifter will generate a constant phase shift (i.e., a constant phase shifter proposed by Schiffman, Reference 27). The network takes the phase difference between a uniform transmission line and a coupled line to produce a constant phase. Figure 11.23 shows such a configuration with the phase difference between them. The phase difference is approximately a constant over wide frequency range. In Figure 11.23, ϕ represents the phase angle and $\theta = \beta l$; β is the phase constant and l is the length of the coupled line. ϱ is defined as $\varrho = Z_{oe}/Z_{oo}$, and Z_{oe}, Z_{oo} are the even and odd mode characteristic impedances of a coupled line, respectively. k is the ratio of the length of the uniform line to that of the coupled lines. Sometimes this network is added to one of the outputs of a 90 degree hybrid coupler to make a 180 degree hybrid coupler as discussed in Chapter 10.

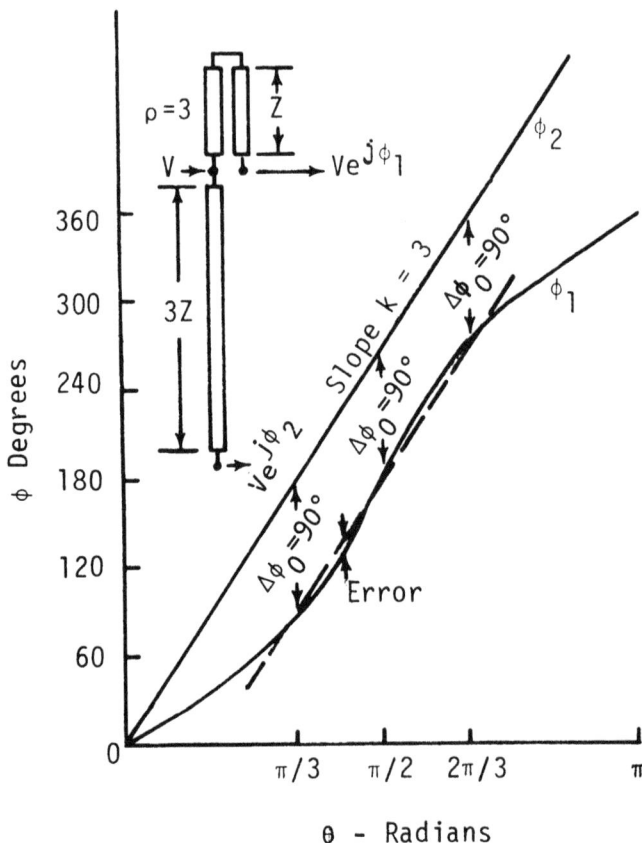

Figure 11.23. Network and Curves of Phase Response for Each of its Two Branches (Based on Schiffman, Reference 27)

11.14 SWITCHED LINE PHASE SHIFTERS

A switched line phase shifter is shown in Figure 11.24. There are two single-pole-double-throw (SPDT) pin diode switches which switch the signal path between l and $l + \Delta l$. Since the path length difference is Δl, the phase shift generated is expressed in Equation 11.17. In this device four diodes are used; the top two and bottom two must be complementarily biased to implement the two SPDT switches. The insertion loss difference between the two paths caused by diodes will be canceled, because in either path the input signal must pass two pin diodes. The loss variation is caused only by the length difference of the two paths. This phase shifter is a constant-delay-time type by nature. It can be converted to a constant phase delay by using a constant phase circuit in the delayed path.

(a) Symbolic Notation

(b) With Pin Switches

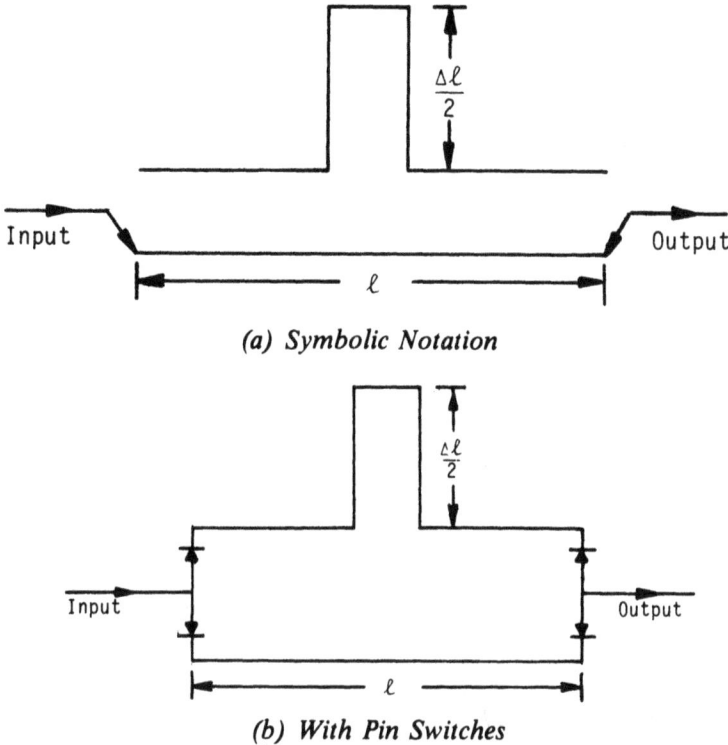

Figure 11.24. Switched Delay Line Phase Shifter

11.15 LOADED-LINE PHASE SHIFTERS

A loaded-line phase shifter is shown in Figure 11.25. A transmission line of approximately $\lambda/4$ is loaded at both ends with pure susceptances B. The principle operation of this phase shifter design arises from two factors: (1) any symmetric pair of quarter-wavelength spaced shunt susceptances will have mutually canceling reflections provided their normalized susceptances

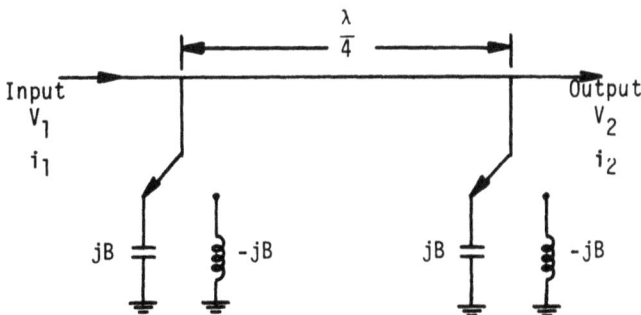

Figure 11.25. Loaded-Line Phase Shifter

are small compared with unity. This feature provides the phase shifter section with good match in both control states, regardless of the susceptance sign or value, provided the magnitude is small and (2) shunt capacitance elements will lengthen a transmission line electrically while inductive elements will shorten it. Thus, switching from inductive to capacitive elements produces an increase in electric length with a corresponding phase shift.

The loaded line can be considered as three sections: the two end sections with the shunt susceptances and the center section transmission line. The transformation matrix of the two end sections can be written as

$$
\begin{bmatrix} v_1 \\ i_1 \end{bmatrix} = \begin{bmatrix} 1 & 0 \\ jB & 1 \end{bmatrix} \begin{bmatrix} v_2 \\ i_2 \end{bmatrix}
\tag{11.18}
$$

where v_1, i_1 are the voltage and current at one end of this section and v_2, i_2 are the voltage and current of the other end. Similarly, the transformation matrix of the transmission line with normalized characteristic impedance $Z_0 = 1$ can be written as

$$
\begin{bmatrix} v_1 \\ i_1 \end{bmatrix} = \begin{bmatrix} \cos\theta & j\sin\theta \\ j\sin\theta & \cos\theta \end{bmatrix} \begin{bmatrix} v_2 \\ i_2 \end{bmatrix}
\tag{11.19}
$$

where $\theta = 2\pi l/\lambda$ and l is the length of the transmission line.

The transformation matrix of the phase shifter is the cascade of these three sections which can be written as

$$
\begin{bmatrix} v_1 \\ i_1 \end{bmatrix} = \begin{bmatrix} 1 & 0 \\ jB & 1 \end{bmatrix} \begin{bmatrix} \cos\theta & j\sin\theta \\ j\sin\theta & \cos\theta \end{bmatrix} \begin{bmatrix} 1 & 0 \\ jB & 1 \end{bmatrix} \begin{bmatrix} v_2 \\ i_2 \end{bmatrix}
$$

$$
= \begin{bmatrix} \cos\theta - B\sin\theta & j\sin\theta \\ j[2B\cos\theta + (1 - B^2)\sin\theta] & \cos\theta - B\sin\theta \end{bmatrix} \begin{bmatrix} v_2 \\ i_2 \end{bmatrix}
$$

$$
= \begin{bmatrix} A & B \\ C & D \end{bmatrix} \begin{bmatrix} v_2 \\ i_2 \end{bmatrix}
\tag{11.20}
$$

where ABCD is referred to as the normalized matrix. The transmission coefficient S_{21} of the phase shifter can be expressed in terms of the normalized matrix as

$$S_{21} = \frac{2}{A + B + C + D}$$

$$= \frac{1}{\left[\cos\theta - B\sin\theta \right] + j \left[B\cos\theta + \left(1 - \frac{B^2}{2} \right) \sin\theta \right]} \qquad (11.21)$$

The phase ϕ of the loaded transmission line section can be found through the real and imaginary parts of S_{21} and expressed as

$$\phi = \tan^{-1}\left[-\frac{B\cos\theta + \left(1 - \frac{B^2}{2} \right)\sin\theta}{\cos\theta - B\sin\theta} \right]$$

$$= \tan^{-1}\left[-\frac{B + \left(1 - \tfrac{1}{2}B^2 \right)\tan\theta}{1 - B\tan\theta} \right] \qquad (11.22)$$

where positive phase corresponds to phase advance, and negative phase to phase delay. When the transmission line is $\lambda/4$, then $\theta = \pi/2$ and Equation 11.22 reduces to

$$\phi = \tan^{-1}\left[\frac{1 - \tfrac{1}{2}B^2}{B} \right] \qquad (11.23)$$

If the value of B is switched from B_1 to B_2, the phase shift will be given by

$$\Delta\phi = \tan^{-1}\left[\frac{1 - \tfrac{1}{2}B_1^2}{B_1} \right] - \tan^{-1}\left[\frac{1 - \tfrac{1}{2}B_2^2}{B_2} \right] \qquad (11.24)$$

This phase shifter basically provides a constant phase shift.

11.16. HIGH-LOW PASS PHASE SHIFTERS

This phase shifter is quite similar to the loaded-line phase shifter. Since at low frequency (i.e., UHF) it is impractical to use transmission lines of the same length as the wavelength to build devices, the transmission line in the loaded-line phase shifter (Figure 11.16) is replaced by lumped circuits. The high-low pass phase shifter can be either a T- or a π-network. Figure 11.26 shows a T-network. The three switches (S_1, S_2, and S_3) are controlled in synchronization.

As mentioned before, while shunt capacitance lengthens the electrical path and shunt inductance shortens it, series inductance lengthens it and series capacitance shortens it. The transformation matrix of Figure 11.26 can be written as

$$\begin{bmatrix} A & B \\ C & D \end{bmatrix} = \begin{bmatrix} 1 & jX \\ 0 & 1 \end{bmatrix}\begin{bmatrix} 1 & 0 \\ jB & 1 \end{bmatrix}\begin{bmatrix} 1 & jX \\ 0 & 1 \end{bmatrix} \tag{11.25}$$

$$= \begin{bmatrix} 1 - BX & j(2X - BX^2) \\ jB & 1 - BX \end{bmatrix}$$

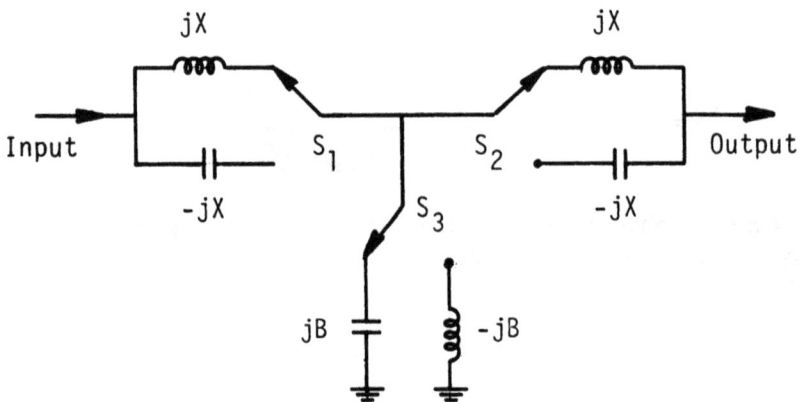

Figure 11.26. High-Low Pass Phase Shifter (T-Network)

The corresponding transmission coefficient is

$$S_{21} = \frac{2}{2(1 - BX) + j(B + 2X - BX^2)} \tag{11.26}$$

The phase delay is

$$\phi = \tan^{-1}\left[-\frac{B + 2X - BX^2}{2(1 - BX)} \right] \tag{11.27}$$

In order to match the input impedance, $|S_{21}| \equiv 1$, implies that

$$B = \frac{2X}{1 + X^2} \qquad (11.28)$$

Substituting Equation 11.28 into 11.27 then

$$\phi = \tan^{-1} \left[\frac{2X}{X^2 - 1} \right] \qquad (11.29)$$

If the value of X is changed to $-X$, the net phase shift is

$$\Delta\phi = 2 \tan^{-1} \left[\frac{2X}{X^2 - 1} \right] \qquad (11.30)$$

It should be noted that in order to have good input impedance matching, the inductance X and susceptance B must be mutually related through Equation 11.28. This phase shifter provides a constant phase shift.

REFERENCES

1. Reference data for radio engineers, pp. 254-262, 4th edition, ITT Corp., July 1957.
2. Ishii, T.K., *Microwave engineering,* p. 59, The Ronald Press Company, 1966.
3. Cristal, E.G., "A continuously variable coaxial-line attenuator," IEEE Trans. on Microwave Theory and Techniques, Vol. MTT-28, pp. 191-199, March 1980.
4. Mohr, R.J., "New coaxial variable attenuators," Microwave Journal, March 1965, Narda Probe, Vol. 2, No. 4, September 1965, published by the Narda Microwave Corp.
5. "Narda catalog no. 21," The Narda Microwave Corporation, Plainview, N.Y.
6. Watson, H.A., Editor, *Microwave semiconductor devices and their circuit applications,* McGraw-Hill Book Co., 1969.
7. *Applications of Pin Diodes,* Hewlett Packard, Application Note 922.
8. Auld, B.A., "The synthesis of symmetrical waveguide circulators," IRE Trans. on Microwave Theory and Techniques, Vol. 7, pp. 238-247, April 1959.
9. Milano, U., Saunders, J., and Davis, L., "A Y-junction strip line circulator," IRE Trans. on Microwave Theory and Techniques, Vol. 8, pp. 346-351, May 1960.

10. Bosma, H., "On strip line Y-circulator at UHF," IEEE Trans. on Microwave Theory and Techniques, Vol. MTT-12, pp. 61-72, January 1964.

11. Fay, C.E., and Comstock, R.L., "Operation of the ferrite junction circulator," IEEE Trans. on Microwave Theory and Techniques, Vol. MTT-13, pp. 15-27, January 1965.

12. Simon, J.W., "Broadband strip line transmission Y-junction circulators," IEEE Trans. on Microwave Theory and Techniques, Vol. MTT-13, pp. 335-345, May 1965.

13. Wu, Y.S., and Rosenbaum, F.J., "Wide-band operation of microstrip circulators," IEEE Trans. on Microwave Theory and Techniques, Vol. MTT-22, pp. 849-856, October 1974.

14. Ayasli, "Analysis of wide-band strip line circulators by integral equation technique," IEEE Trans. on Microwave Theory and Techniques, Vol. MTT-28, pp. 200-209, March 1980.

15. Miyoshi, T., and Miyanchi, S., "The design of planar circulars for wide-band operation," IEEE Trans. on Microwave Theory and Techniques, Vol. MTT-28, pp. 210-214, March 1980.

16. Soohoo, R.F., *Theory and application of ferrites,* Prentice-Hall, Inc., Englewood Cliffs, New Jersey, 1960.

17. Ishii, T.K., *Microwave engineering,* The Ronald Press Company, New York, 1966.

18. Watson, H.A., *Microwave semiconductor devices and their circuit applications,* Chapter 10, McGraw-Hill Book Co., 1969.

19. Skolnik, M.I., *Radar handbook,* Chapter 12, McGraw-Hill Book Co., 1970.

20. Ince, W.J., Temme, D.H., Willwenth, F.G., and Dibartolo, J., "A method of fabrication for waveguide nonreciprocal toroidal ferrite phasers." IEEE Trans. on Microwave Theory and Techniques Vol. MTT-20 pp. 705-707, Oct. 1972.

21. Garver, R.V., "Broadband diode phase shifters," IEEE Trans. on Microwave Theory and Techniques, Vol. MTT-20, pp. 314-323, May 1972.

22. White, J.F., "Diode phase shifters for array antennas," IEEE Trans. on Microwave Theory and Techniques, Vol. MTT-22, pp. 658-674, June 1974.

23. Kamihashi, S., Juroda, M., and Hirai, K., "Novel pin diode for MIC phase shifter," Microwave Journal, p. 49, December 1980.

24. Glance, B., "A fast low-loss microstrip p-i-n phase shifter," IEEE Trans. on Microwave Theory and Techniques, Vol. MTT-27, pp. 14-16, January 1979.

25. Davis, W.A., "Design equations and bandwidth of loaded-line phase shifters," IEEE Trans. on Microwave Theory and Techniques, Vol. MTT-22, pp. 561-563, May 1974.

26. Bahl, I.J., and Gupta, K.C., "Design of loaded-line p-i-n diode phase shifter circuits," IEEE Trans. on Microwave Theory and Techniques, Vol. MTT-28, pp. 219-224, March 1980.
27. Schiffman, B.M., "A new class of broad-band microwave 90-degree phase shifters," IRE Trans. on Microwave Theory and Techniques, Vol. MTT-6, pp. 232-237, April 1958.

CHAPTER 12
MICROWAVE FILTERS

12.1 INTRODUCTION AND FUNDAMENTAL CHARACTERISTICS OF MICROWAVE FILTERS[1-12]

Microwave filters are one of the most popular components in receivers. Almost all receivers have some filters at one stage or another. The main functions of the filters are: (1) to reject undesirable signals outside the filter pass band and (2) to separate or combine signals according to their frequency. A good example for the latter application is the channelized receiver in which banks of filters are used to separate input signals. Sometimes filters are also used for impedance matching. Filters are almost always used before and after a mixer to reduce spurs. There are many books (References 1, 2, 3, and 5) and articles (References 4 and 6) which are devoted to filter designs. There are many kinds of filters used in microwave receivers, so it is impossible to cover all of them. This chapter will introduce some of the popular microwave filters used in microwave receivers, their operating principles, and their performance characteristics. The ways to connect filters together to form a multiplexer will also be discussed. Since many microwave receivers are designed to receive a pulsed signal from radar, the transient phenomenon in a filter is very important; it determines the detection scheme of many receivers (see Chapter 5). The general approaches to predict the filter transient response will also be discussed.

The commonly used terminologies related to filters will be discussed first. A filter is a two-port network which will pass and reject signals according to their frequencies. There are four kinds of filters according to their frequency selectivities. Their names reflect their characteristics, and they are: (1) a low-pass filter which passes the low frequency signals below a predetermined value, (2) a high-pass filter which passes the high frequency signals above a predetermined value, (3) a band-pass filter which passes signals between two predetermined frequencies, and (4) a band reject filter (sometimes referred to as notch filter) which rejects signals between two predetermined frequencies. A band-pass filter with different skirt slopes on the two sides of the pass band is sometimes referred to as an asymmetrical filter. In this filter the sharpness of the rejection band attenuation is significantly different above and below the center frequency. Since these filters are commonly used in single side-band communication applications, they are also called single side-band (SSB) filters. A typical SSB filter frequency response is shown in Figure 12.1.

The attenuation of a filter is specified in the pass band as well as the stop band. The more out-of-band rejection needed, the more filter sections are required and the design of the filter is more complicated. A filter with high out-of-band attenuation will effectively reduce unwanted signals;

however, the transient response time of a pulsed signal through this filter is also longer. In designing receivers for pulsed signals, the transient time of a filter must be evaluated against its out-of-band rejection.

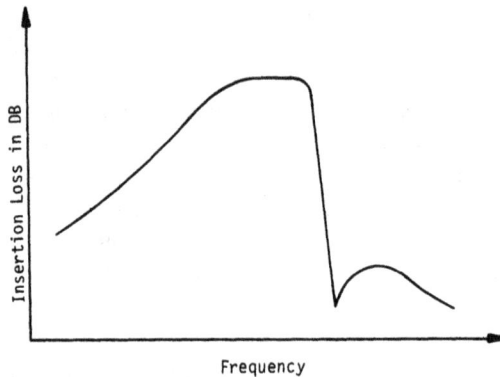

Figure 12.1 Typical Performance of a Single Side-Band Filter

According to the inband and out-of-band attenuation characteristics, the filters are commonly divided into two groups: Butterworth and Chebyshev filters. A Butterworth filter is also called a maximally flat filter; in its pass band the attenuation increases montonically toward the stop band. The output voltage of a Butterworth filter can be expressed as

$$\left(\frac{V}{V_p}\right)^2 = \frac{1}{1 + \left(\frac{X}{X_{3dB}}\right)^{2n}} \tag{12.1}$$

A Chebyshev filter is also referred to as an equal ripple filter. In the pass band of the filter, the attenuation varies between two predetermined values. The output voltage of a Chebyshev filter can be expressed as

$$\left(\frac{V}{V_p}\right)^2 = \frac{1}{1 + \left[\left(\frac{V_p}{V_v}\right)^2 - 1\right] \cosh^2\left(n \cosh^{-1}\frac{X}{X_v}\right)} \tag{12.2}$$

where V is the output voltage at point X, V_p is the peak output voltage in pass band, V_v is the valley output in the pass band of a Chebyshev filter. The ripple factor is defined as

$$R_{dB} = \text{ripple factor in dB} = 20 \log \left(\frac{V_p}{V_v}\right) \tag{12.3}$$

n is the number of poles for a low-pass and high-pass filter and n = number of reactances in the filter, for band-pass and band rejection 2n equals the number of poles and n = number of resonators in the filter, X is a variable defined such that X_v is the value of X at the point on the skirt where the attenuation equals the valley attenuation; X_{3dB} is the value of X at the point on the skirt where the attenuation is 3 dB below V_p.

For a low-pass filter

$$X = 2\pi f \tag{12.4}$$

For a high-pass filter

$$X = -\frac{1}{2\pi f} \tag{12.5}$$

For a band-pass filter

$$X = \frac{f}{f_0} - \frac{f_0}{f} \tag{12.6}$$

For a band reject filter

$$X = -\frac{1}{\left(\dfrac{f}{f_0} - \dfrac{f_0}{f}\right)} \tag{12.7}$$

where f is the operational frequency and f_0 is the midfrequency of the pass band or the reject band. If f_1 and f_2 are two frequencies where the characteristic exhibits the same attenuation, the relation between f_0, f_1, and f_2 is

$$f_0^2 = f_1 f_2 \tag{12.8}$$

Equations 12.1 and 12.2 can be rewritten in terms of decibels

$$\left(\frac{V}{V_p}\right)_{dB} = 20 \log\left(\frac{V}{V_p}\right) = 10 \log \frac{1}{1 + \left(\dfrac{X}{X_{3dB}}\right)^{2n}} \tag{12.9}$$

and

$$\left(\frac{V}{V_p}\right)_{dB} = 20 \log\left(\frac{V}{V_p}\right)$$

$$\tag{12.10}$$

$$= 10 \log \frac{1}{1 + \left[\left(\dfrac{V_p}{V_v}\right)^2 - 1\right]\cosh^2\left(n \cosh^{-1}\dfrac{X}{X_v}\right)}$$

Figures 12.2 and 12.3 show the attenuation versus frequency of Butterworth and Chebyshev filters, respectively. For the same number of poles, a Chebyshev filter has more out-of-band rejection than a Butterworth filter near the band edges.

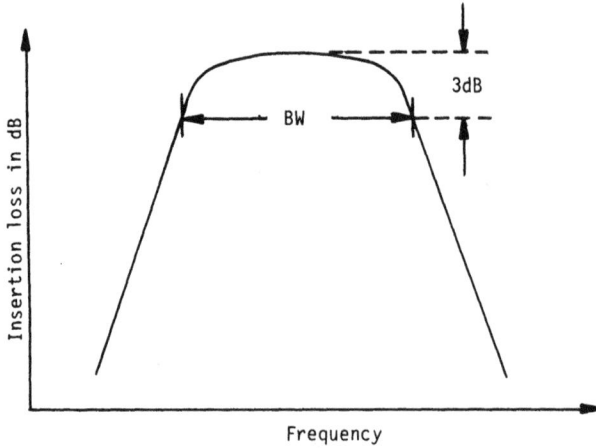

Figure 12.2. Frequency versus Attenuation of a Butterworth Filter

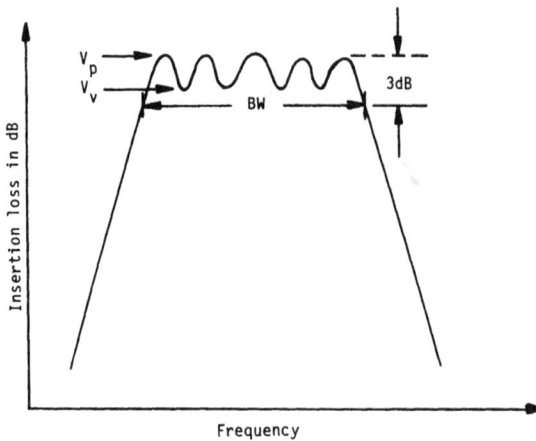

Figure 12.3. Frequency versus Attenuation of a Chebyshev Filter

On the asymptote of the skirt of the filter, the attenuation of the filter changes by a multiple number (n) of 6 dB per octave of frequency change. This n is the number of poles used in Equations 12.1 and 12.2. The number of poles of a filter can be obtained by measuring the performance of the filter.

Besides the Butterworth and Chebyshev filters, there is another kind of filter called an elliptic filter (References 9-12) because the filter has an elliptic-function response. The filter has its poles near the cut-off frequencies to provide sharp roll-off on its skirts in comparison with that of Butterworth and Chebyshev filters with a similar number of elements. The attenuation in both the pass and stop bands are of equal ripple as shown in Figure 12.4. An elliptic filter operating in the microwave frequency range is generally compact in size. The elliptic filter has inherently a wide bandwidth (>30 percent) and is not suitable for narrow band designs due to size limitations. This filter is usually designed for operating below 10 GHz.

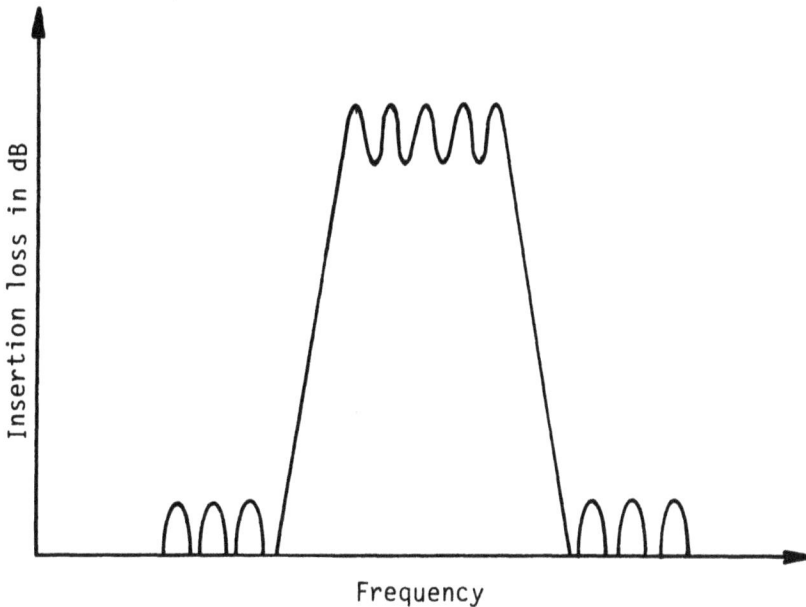

Figure 12.4. *Frequency Response of an Elliptic Band-Pass Filter*

The bandwidth of a band-pass (or band reject) filter is generally referred to as the difference between two frequencies at which the attenuation is 3 dB below (or above) that of the center of the band. These 3 dB points are called the cut-off frequencies. The Q of a band-pass or band reject filter is defined

$$Q = \frac{f_0}{BW} \tag{12.11}$$

where f_0 is the center frequency of the filter and BW is the bandwidth. It should be noted that the insertion loss of the filter does not affect its Q. For example, a surface acoustic wave (SAW) filter with 15 dB insertion loss and 2 MHz bandwidth at 400 MHz has a Q of 200.

If the input voltage to the filter is E_i and the output voltage is E_0, the transmission phase of the filter is defined as

$$\phi = \arg \frac{E_0}{E_i} \quad \text{radians} \tag{12.12}$$

The phase delay of the filter at any given angular frequency ω is

$$t_p = \frac{\phi}{\omega} \quad \text{seconds} \tag{12.13}$$

while the group delay is

$$t_g = \frac{d\phi}{d\omega} \quad \text{seconds} \tag{12.4}$$

The group delay represents the delay time of the modulation signal passing through the filter. If the relation between ϕ and ω is linear, the delay time is constant at different frequencies. If the relation between ϕ and ω is not linear, then each frequency has a slightly different delay time through the filter, and dispersion will occur. The group delay of the Butterworth and Chebyshev filters was studied by Kudsia and Chitre (Reference 9).

The input and output impedances of a filter are frequency dependent. Most filters are built with pure reactive elements only. Therefore, the filtering effect is accomplished through reflection of the input signal rather than absorption. The impedances at the input and output are designed to match the characteristic impedance of the system in the pass band and the reflection is minimized. At the stop band, the impedances are poorly matched; therefore, loss will occur through reflection.

The shape factor of a filter is used to determine the sharpness of the skirt. It is usually referred to as the shape factor x at y dB below the minimum insertion loss for a band-pass filter which can be expressed as

$$x = \frac{\text{bandwidth at y dB}}{\text{bandwidth at 3 dB}} \tag{12.15}$$

For example, a filter of 10 MHz bandwidth has a shape factor of 1.5 at 40 dB down, meaning the bandwidth of the filter is 15 MHz wide at 40 dB attenuation below that of the center frequency.

12.2 CONSTANT k, m-DERIVED, BUTTERWORTH AND CHEBYSHEV FILTERS(1, 2, 8, 9, 13)

Traditionally, electric filters are built by lumped circuits (LC). Although an LC filter is seldom used in microwave circuits because of the frequency limitation, it is the basis of all filter designs. The LC filter can be discussed through a ladder network (References 7 and 8). An L-section of the ladder is shown in Figure 12.5. Two L-sections can be connected in cascade to form either a T-section or π-section as shown in Figures 12.6 and 12.7. The filtering condition is determined by the value of $Z_1/4Z_2$. When

$$-1 \leqslant \frac{Z_1}{4Z_2} \leqslant 0 \tag{12.16}$$

the attenuation (α) through the section is 0, which defines the pass band.

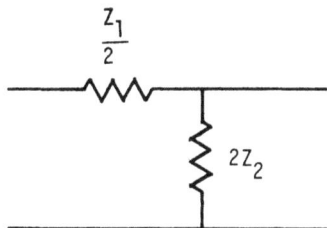

Figure 12.5.　An L-Section of a Ladder Network

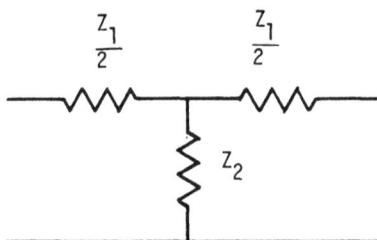

Figure 12.6.　A T-Section of a Ladder Network

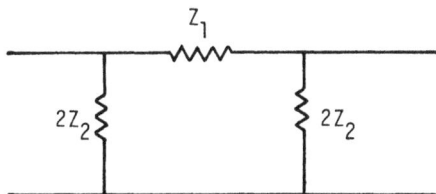

Figure 12.7.　An π-Section of a Ladder Network

When

$$\frac{Z_1}{4Z_2} < -1 \tag{12.17}$$

$$\alpha = 2\cosh^{-1}\sqrt{-Z_1/4Z_2} \tag{12.18}$$

$$0 < \frac{Z_1}{4Z_2} \tag{12.19}$$

$$\alpha = 2\sinh^{-1}\sqrt{Z_1/4Z_2} \tag{12.20}$$

which are referred to as the stop band.

When the impedance Z_1 and Z_2 fulfill the condition

$$Z_1 Z_2 = R^2 \tag{12.21}$$

where R is a positive real constant, the filter is called a constant k filter. Under this condition, Z_1 and Z_2 will be purely reactive and of opposite sign. This is quite a restriction on the impedances Z_1 and Z_2. In order to improve attenuation in the stop band near the cut-off frequency, m-derived filters are introduced. Let the impedances in the m-derived filter be Z_1' and Z_2'. For a T-section they are related to impedances Z_1 and Z_2 by the following equations:

$$Z_1' = mZ_1 \tag{12.22}$$

and

$$Z_2' = \left(\frac{1 - m^2}{4m}\right)Z_1 + \frac{1}{m}Z_2 \tag{12.23}$$

and for a π-section

$$Z_2' = \frac{Z_2}{m} \tag{12.24}$$

and

$$Z_1' = \cfrac{1}{\cfrac{1}{mZ_1} + \cfrac{1}{\cfrac{4m}{1-m^2}Z_2}} \qquad (12.25)$$

The m-derived T- and π-section are shown in Figures 12.8 and 12.9. Comparing Figures 12.6 and 12.7, notice that the m-derived filter uses more components. An example will show the difference between a constant k filter and an m-derived filter. Figure 12.10 shows a constant k low-pass filter, while Figure 12.11 shows an m-derived low-pass filter. In the m-derived low-pass filter, there is one additional indicator. When the series LC is in resonance, the impedance is zero, and the corresponding attenuation is infinite. The resonance frequency can be placed near the cut-off frequency and increases the attenuation near the cut-off frequency. However, the attenuation far from the cut-off frequency in the stop band is slightly higher for the constant k filter than that of the m-derived filter.

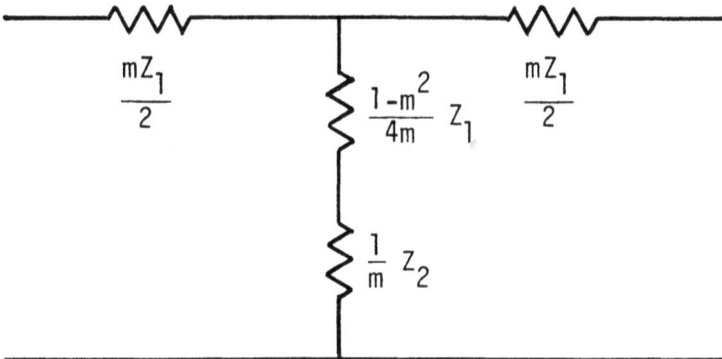

Figure 12.8. *m-Derived T-Section*

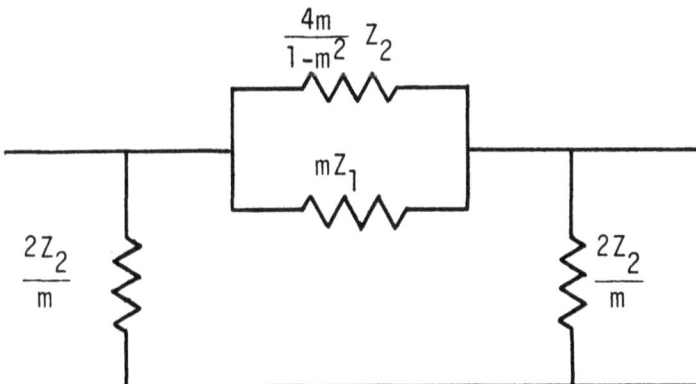

Figure 12.9. *m-Derived π Section*

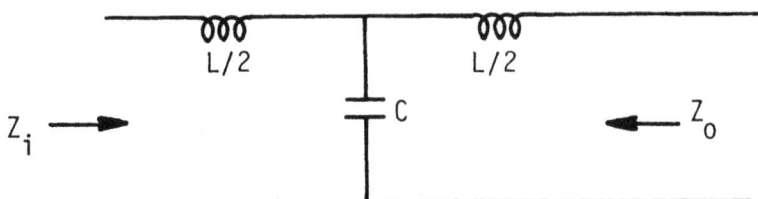

Figure 12.10. Constant k Low-Pass Filter

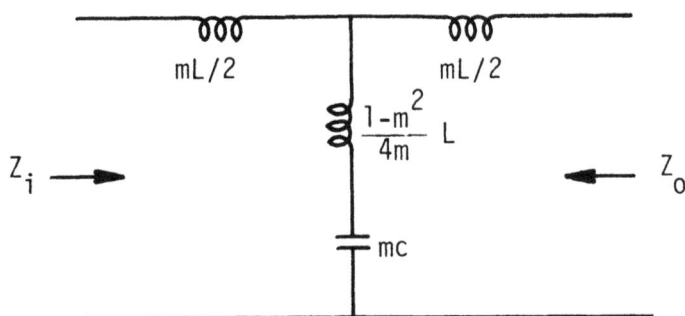

Figure 12.11. m-Derived Low-Pass Filter

The basic assumption of constant k and m-derived filters is that the input and output of the filter must be terminated with matched impedances Z_i and Z_0 as shown in Figures 12.10 and 12.11. This requirement is impractical to fulfill over the operation frequency range. In general, the best that can be done is to terminate the filter with a constant resistive load which will degrade the performance of the filter. Other approaches to filter designs are through the Butterworth and Chebyshev methods. The approaches to these designs are (Reference 13):

1. From the desired out-of-band attenuation, determine the number of poles required. For a Chebyshev filter, the desired ripple factor must also be assigned.
2. The locations of the poles on a S plane are calculated. The S plane consists of real axis σ and imaginary axis $j\omega$.
3. Generate the transfer function of the filter and synthesize the filter.

The number of poles can be determined by Equations 12.1 and 12.2. The locations of the poles for low-pass filters can be located as follows:
For a Butterworth filter let

$$S_k = \Delta\omega \left[\cos\left(\frac{2k-1+n}{2n} \pi \right) + j \sin\left(\frac{2k-1+n}{2n} \pi \right) \right] \qquad (12.26)$$

where $k = 1, 2, \ldots 2n$ (an integer), $\Delta\omega = 2\pi$ times the bandwidth of the filter, and n is the number of poles of the filter. If the real part of S_k is positive, we discard it. Only S_k with negative real parts are kept.

For a Chebyshev filter

$$S_k = \Delta\omega \, (\tanh V \sin U_k + j \cos U_k) \tag{12.27}$$

where

$$U_k = \frac{\pi}{2n} \, (2k - 1) \tag{12.28}$$

$$V = \frac{1}{n} \sinh^{-1}\left(\frac{1}{\varepsilon}\right)$$

and

$$\varepsilon = \sqrt{10^{R_{dB}/10} - 1} \tag{12.29}$$

R_{dB} is the ripple factor in dB.

Similarly, only S_k with negative real parts are kept. The poles of the Butterworth and Chebyshev filters can be shown in Figures 12.12 and 12.13, respectively. In Figure 12.12 a circle with $\Delta\omega$ radius is divided into 2n parts starting from the real axis σ. Only the points on the left plane are considered as the poles. The poles of the Chebyshev filter shown in Figure 12.13 can be obtained by first locating the poles of the Butterworth filter. Draw a smaller circle of radius $\Delta\omega \tanh V$, where V is given by Equation 12.28. Draw radial lines from the Butterworth poles to the origin. From the intersection of the radial lines with the smaller circle, draw lines parallel to the $j\omega$ axis. The intersection of these lines with the lines drawn parallel to the σ axis from the Butterworth poles determine the poles for the Chebyshev filter. After the poles are determined, the transfer function of the filter is given as

$$H(s) = \frac{(\Delta\omega)^n S^n}{(S - S_1)(S - S_2)\ldots(S - S_n)} \tag{12.30}$$

where $S_1, S_2, \ldots$ are the poles

For a band-pass filter, the poles can be determined as

$$S_b = \frac{S_k}{2} \pm j \sqrt{\omega_0^2 - \left(\frac{S_k}{2}\right)^2} \tag{12.31}$$

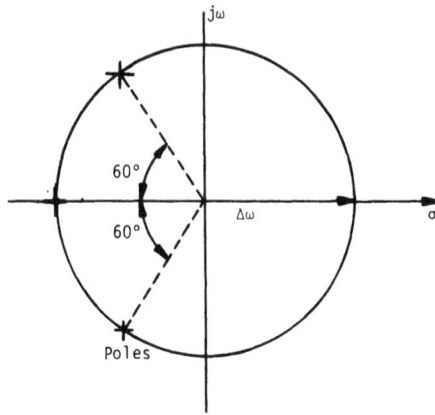

Figure 12.12. *Pole Locations of a Three-Pole Butterworth Filter*

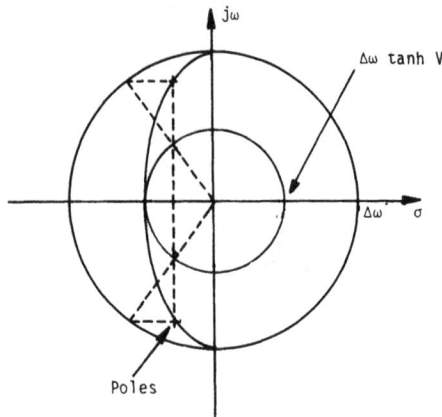

Figure 12.13. *Pole Locations of a Three-Pole Chebyshev Filter*

where S_b are the poles of the band-pass filter, S_k is the corresponding pole of a low-pass filter, and ω_0 is the center frequency of the pass band filter.

It is obvious from Equation 12.30 that the number of poles in the band-pass filter doubles that of the low-pass filter. If the band-pass filter has a narrow bandwidth, Equation 12.31 can be approximated by

$$S_b \cong \frac{S}{2} \pm j\omega_0 \qquad (12.32)$$

The transfer function can be obtained from Equation 12.30. From the transfer function, the filter can be synthesized. Tables and graphs with design date are available for these filters (References 1 and 2).

12.3 HELICAL FILTERS[7, 14-16]

High quality lumped elements, inductance, and capacitance are hard to obtain at high frequency because of the resistive loss. Thus, a lumped element filter is impractical at high frequency range. At VHF and UHF frequencies a helical filter (Reference 14) of reasonable size and high-Q (~1000) can be constructed. Basically, a helical resonator resembles a quarter-wave coaxial resonator, except that the inner conductor is in the form of a single layer solenoid, or helix. The outer conductor can be either a cylindrical or a cubic shaped cavity. One end of the helix is connected to the outer conductor; the other end is open. The equivalent inductance L and capacitance C for a cylindrical helical resonator as shown in Figure 12.14 can be expressed empirically as (Reference 16)

$$L = 0.025 \, n^2 d^2 \left[1 - \left(\frac{d}{D} \right) \right]^2 \; \mu H \text{ per axial inch} \qquad (12.33)$$

$$C = \frac{0.75}{\log (D/d)} \; \mu\mu F \text{ per axial inch} \qquad (12.34)$$

where d is the diameter of the helix in inches, D is the diameter of the cylindrical cavity in inches, $n = 1/\tau$ is the number of turns of the helix per inch, where τ is the pitch of the winding in inches.

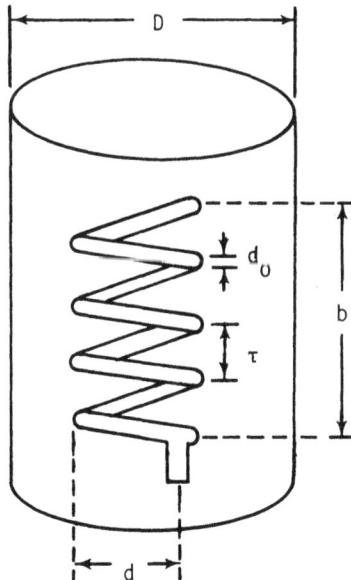

Figure 12.14. A Helical Resonator with Cylindrical Cavity

Equation 12.34 is valid only for the following condition

$$\frac{b}{d} = 1.5 \tag{12.35}$$

where b is the axial length of the coil in inches.

These equations and all those below are accurate for the resonator when it is realized between the following limits:

$$1.0 < \frac{b}{d} < 4.0$$

$$0.45 < \frac{d}{D} < 0.6$$

$$0.4 < \frac{d_0}{\tau} < 0.6 \text{ at } \frac{b}{d} = 1.5$$

$$0.5 < \frac{d_0}{\tau} < 0.7 \text{ at } \frac{b}{d} = 4.0$$

$$\tau < \frac{d}{2}$$

where d_0 is the diameter of the center conductor in inches.

The length of the coil in inches can be expressed as

$$b = \frac{235}{f_0 \sqrt{LC}} \tag{12.36}$$

where f_0 is the resonant frequency in MHz and the characteristic impedance of the resonator is given by

$$Z = 1000 \sqrt{\frac{L}{C}}$$

$$= 183\, nd \left\{ \left[1 - \left(\frac{d}{D}\right)^2 \right] \log \frac{D}{d} \right\}^{1/2} \text{ ohms} \tag{12.37}$$

The unloaded Q can be expressed as

$$Q_u \cong 220 \frac{\left(\frac{d}{D}\right) - \left(\frac{d}{D}\right)^3}{1.5 + \left(\frac{d}{D}\right)^3} D\sqrt{f_0} \tag{12.38}$$

$$\cong 50 \, D\sqrt{f_0}$$

The number of turns per inch n is given as

$$n \cong \frac{1720}{f_0 \, b \, d} \left[\frac{\log (D/d)}{1 - \left(\frac{d}{D}\right)^2} \right]^{1/2} \text{ turns/inch} \tag{12.39}$$

In order to make a filter out of the helical resonator, proper couplings at both the input and output are required. One helical resonator is equivalent to one section in a filter. If more sections are required, multiple numbers of resonators must be connected in cascade, and proper couplings must be provided between adjacent ones. Figure 12.15 shows a two section helical filter. The input and output of the filter are coupled out through tapping on the helix. The coupling between the two cavities is through a coupling window. The center frequency of the filter can be adjusted slightly around the designed frequency through the turning screws. This kind of filter is very popular for a channelized receiver because of its relatively small size and the operation frequency in the UHF range. The filters shown in Figures 5.16, 5.17 and 5.18 are helical filters with two and six sections. The frequency response curve of a five resonator filter is shown in Figure 12.16.

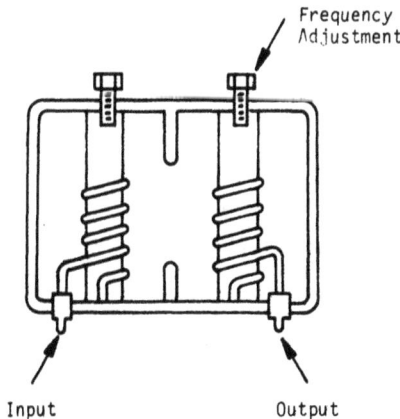

Figure 12.15. A Two-Section Helical Filter

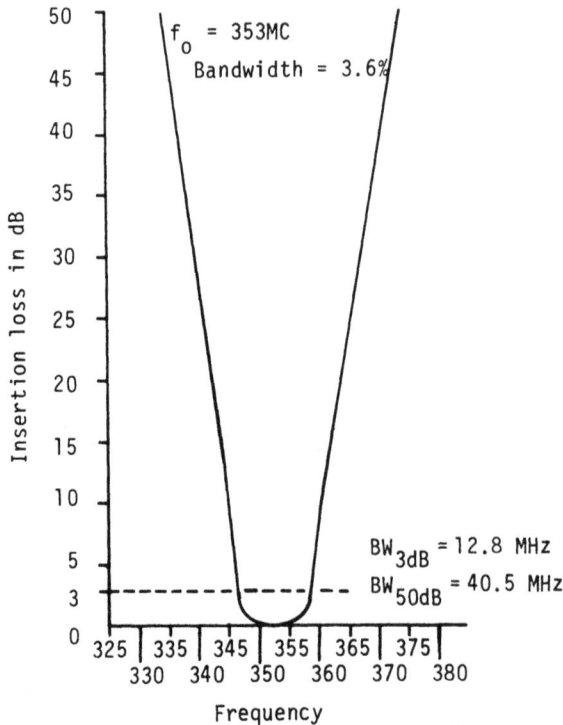

Figure 12.16. *Response Curve of a Five Resonator Filter (Based on Zverev and Blinchikoff, Reference 15)*

12.4 STRIP LINE FILTERS[1, 17-31]

Since strip lines and microstrip lines have become the most popular transmission lines in microwave circuits, especially for receivers because of the low power handling requirements, filters built of strip lines and microstrip lines are the most commonly used. The advantages of strip line filters are: (1) the filter structure is simple, and (2) it can become part of a microwave integrated circuit (MIC). However, if low insertion loss in the pass band is desired, strip line filters with relative high dielectric loss may not be a suitable approach. There are many different ways strip lines can be connected to make a two-terminal network, because the ends of a strip line can be either open circuit or grounded. Ozaki and Ishii (Reference 17), and others (Reference 18), suggested 10 canonical sections. There are many papers (References 19-22) discussing different kinds of filters using strip line designs. This chapter will mainly discuss band-pass filters since they are most commonly used in microwave receivers. The most popular strip line design is the parallel-coupled resonator filter. Sometimes the interdigital and combline filters can also be made in strip line form.

A parallel-coupled resonator filter is shown in Figure 12.17. The filter consists of half-wavelength strip line resonators, positioned so that adjacent resonators parallel each other along half of their length. This arrangement provides relatively large coupling for a given spacing between resonator strips and the bandwidth of the filter can reach 30 percent. Increasing the thickness of the strip will increase the coupling and frequency bandwidth. The number of sections of the filter is equal to the number of resonators. Since these strips also resonant at $3\lambda_0/2$, there is a second pass band which is at the third harmonic of the fundamental frequency. The example in Figure 12.17 has three resonators; therefore, the filter is a three section filter. To design this kind of filter, there are several different approaches (References 19, 23, and 24). In all of the approaches, tables, equations, and graphs are generally used. The general approach is to find the value of the impedance of each line section as shown in Figure 12.17. Once the impedance value is determined, the line can be designed as discussed in Chapter 8. Ho (Reference 30) discussed several approaches. The following discussion is based on References 23 and 30. An example will also be presented.

1. The specifications of the filter must be determined, i.e., the type of filter, center frequency, bandwidth, number of sections, and ripple factor (if a Chebyshev filter).

2. Find the prototype element g_1, g_2,... from tables (i.e., in Reference 1 (Chapter 4) and Reference 30) then determine g_0 by

$$g_0 = \frac{\pi(f_H - f_L)}{(f_H + f_L)} \qquad (12.40)$$

where f_H and f_L are the high and low cut-off frequencies, respectively, and the center frequency is

$$f_0 = \frac{f_H + f_L}{2} \qquad (12.41)$$

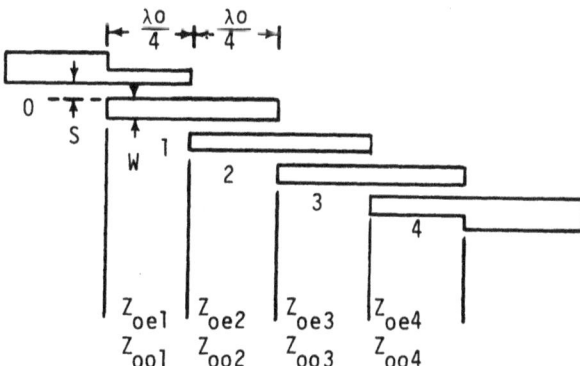

Figure 12.17. Configuration of Parallel-Coupled Resonator Filter

3. Calculate the following ratios to determine Z_{oe} and Z_{oo}

$$\frac{Z_0}{K_{i-1,i}} = \frac{\pi(f_H - f_L)}{(f_H + f_L)} \frac{1}{\sqrt{g_{i-1}, g_i}} \quad \text{for } i = 1, 2, \ldots n + 1, \quad (12.42)$$

and the even Z_{oe} and odd Z_{oo} impedances of the ith section can be found through

$$Z_{oei} = Z_0 \left[1 + \left(\frac{Z_0}{K_{i-1,i}} \right) + \left(\frac{Z_0}{K_{i-1,i}} \right)^2 \right] \quad (12.43)$$

$$Z_{ooi} = Z_0 \left[1 - \left(\frac{Z_0}{K_{i-1,i}} \right) + \left(\frac{Z_0}{K_{i-1,i}} \right)^2 \right] \quad (12.44)$$

4. From Z_{oe} and Z_{oo}, the physical dimensions of each of the $\lambda/4$ sections can be found from data available in strip line designs.

The following is an example to demonstrate the use of these equations. A Chebyshev filter has 10 percent bandwidth, three sections, and a ripple factor of 0.01 dB. Let $f_0 = 1$, $f_H = 1.05$, and $f_L = .95$; the values of g_1, g_2, g_3 are found as $g_1 = g_3 = 0.6291$, $g_2 = 0.9702$ and

$$g_0 = g_4 = \frac{\pi(f_H + f_L)}{(f_H + f_L)} = 0.1571$$

Since $g_0 = g_4$ and $g_1 = g_3$, then

$$\frac{Z_0}{K_{0,1}} = \frac{Z_0}{K_{3,4}} = \frac{\pi(f_H - f_L)}{(f_H + f_L)} \cdot \frac{1}{\sqrt{g_0 g_1}} = .4997$$

$$\frac{Z_0}{K_{1,2}} = \frac{Z_0}{K_{2,3}} = \frac{\pi(f_H - f_L)}{(f_H + f_L)} \frac{1}{\sqrt{g_1 g_2}} = 0.2011$$

From Equations 12.43 and 12.44

$$Z_{oe1} = Z_{oe4} = 50 \left[1 + \frac{Z_0}{K_{0,1}} + \left(\frac{Z_0}{K_{0,1}} \right)^2 \right] = 87.47 \, \Omega$$

$$Z_{oo1} = Z_{oo4} = 50 \left[1 - \frac{Z_0}{K_{0,1}} + \left(\frac{Z_0}{K_{0,1}} \right)^2 \right] = 37.50 \, \Omega$$

$$Z_{oe2} = Z_{oe3} = 50 \left[1 + \frac{Z_0}{K_{1,2}} + \left(\frac{Z_0}{K_{1,2}} \right)^2 \right] = 62.08 \ \Omega$$

$$Z_{oo2} = Z_{oo3} = 50 \left[1 - \frac{Z_0}{K_{1,2}} + \left(\frac{Z_0}{K_{1,2}} \right)^2 \right] = 41.97 \ \Omega$$

From these values, the physical dimensions of the strip lines (see Figure 12.17) can be determined, if the material of the substrate is known. The width of the strip line W and space S between lines can be determined graphically if Z_{oe} and Z_{oo} are known (References 25-27). The length of the strip line is $\lambda/4$ at the midband. However, the resonators are usually slightly shortened at the ends to components for fringing capacitance.

Other approaches to design the parallel-coupled band-pass filters are quite similar to the mentioned method (References 28 and 29). Starting from the specifications of the filter to generate the impedances Z_{oe} and Z_{oo}, only the equations to obtain Z_{oe} and Z_{oo} are different in each approach. Figure 12.18 shows the frequency response of the filter comparing with theoretical results.

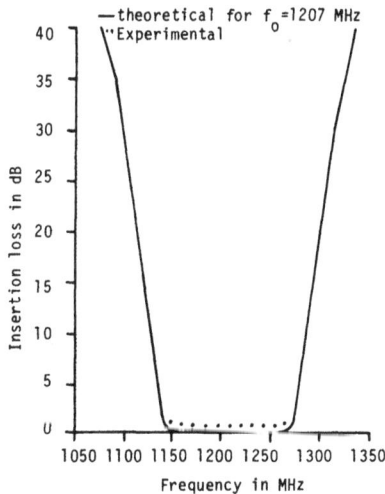

Figure 12.18. *Insertion Loss of Parallel-Coupled Resonator Filter (Based on Cohn, Reference 23)*

12.5 HAIRPIN-LINE FILTERS (31, 32)

There is another kind of filter which is very much like the parallel-coupled line filter and is called the hairpin-line filter. In the hairpin-line filter, the half-wave coupling resonators are bent into a U shape (hairpin) rather than the straight line resonators in the parallel-coupled resonators. The configurations of hairpin filters are shown in Figure 12.19. There are two kinds of couplings at the input and output of the filter. In Figure 12.19a, the input and output lines are open circuited at the opposite ends of the feeding points. In Figure 12.19b, the input and output lines are grounded at the opposite ends. The hairpin filter with the open-circuited feed lines will yield practical impedance levels for narrow to approximately 25 percent bandwidths (Reference 31). In Reference 31, hairpin filters with 5 percent bandwidths and 20 percent bandwidths and center frequencies from 1 to 1.5 GHz are fabricated in strip line and microstrip line forms. The number of resonators varies from 4 to 7. The insertion loss at the center of the strip line hairpin filters (for center frequency = 1.5 GHz, 5 percent bandwidth, 4 resonators) was measured 2.5 dB. Spurious response of typical 40 dB below the main response was measured. Noticeable spurious responses are present at 2.8 and 4.5 GHz which are approximately the second and third harmonics as shown in Figure 12.20. Filters with hybrid circuits between parallel-coupled and hairpin-lines can also be built. One of the possible examples is shown in Figure 12.21. Reference 32 presented the theoretical study results of hairpin filters. The theoretical calculation is useful up to 20 percent bandwidth.

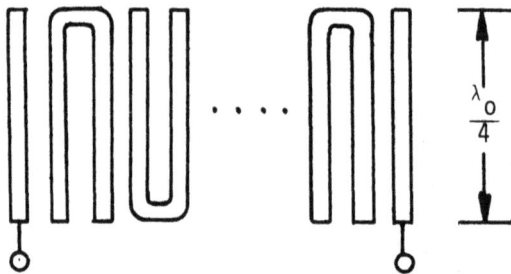

(a) With Open-Ended Transformers

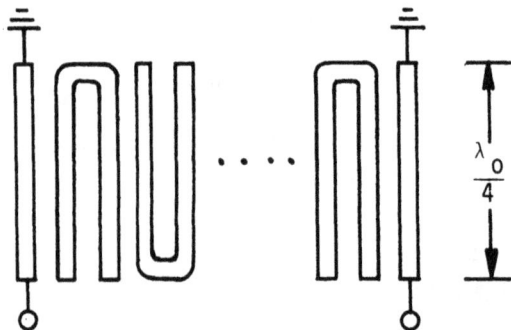

(b) With Short-Ended Transformers
Figure 12.19. Hairpin Filters

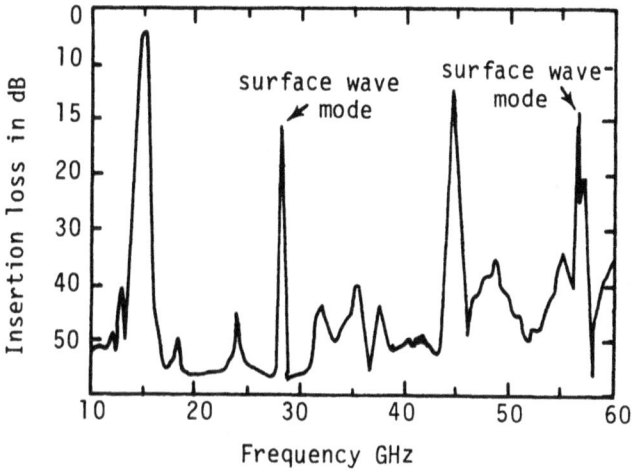

Figure 12.20. *Frequency Response of a Hairpin Filter (Based on Cristol and Frankel, Reference 31)*

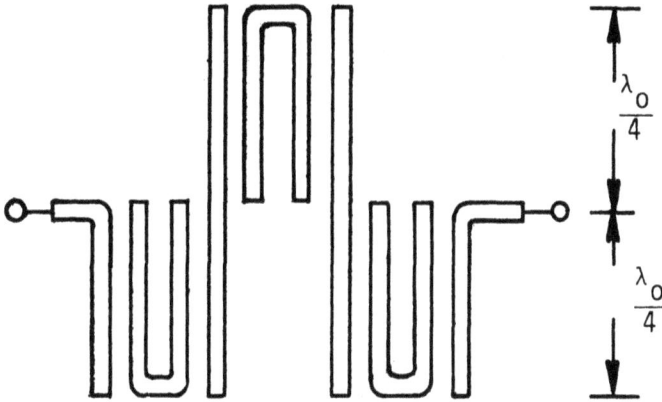

Figure 12.21. *Example of Hybrid Parallel-Coupled/Hairpin Filter*

12.6 INTERDIGITAL AND COMBLINE FILTERS (1, 4, 33-37)

There are two other popular kinds of filters which can be made in strip line forms. They are the interdigital filters and combline filters.

The configuration of an interdigital filter is shown in Figure 12.22. It consists of parallel strip lines resonators of $\lambda_0/4$ at the midband. One end of each strip line is grounded and the other end is open circuit. Two adjacent strip lines always ground at opposite ends. The ungrounded end of the strip line can be either open circuit or capacitively loaded to ground. Since each strip line has one end grounded, the strip line can be replaced by self-supporting solid conductors. In the interdigital filter, the coupling is achieved by way of the fields fringing between adjacent resonator elements. Lines 0 and n + 1 operate as impedance transforming sections and not as resonators. Thus, an n element interdigital field will have n + 2 line elements. This kind of filter has narrow and moderate bandwidth. Thirty percent or more bandwidth can be approached. The main drawback in designing a wide bandwidth interdigital filter is that the gaps between the input and output elements and their adjacent elements tend to be inconveniently small and the widths of elements 1 to n also tend to be small. There is a second pass band in these filters at the third harmonic of the fundamental frequency.

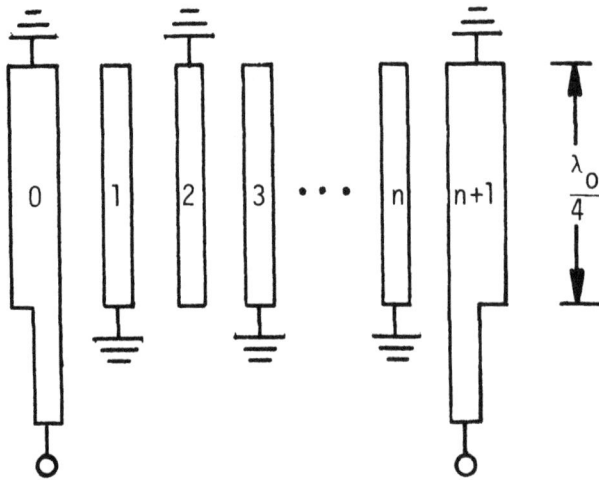

Figure 12.22. Configuration of an Interdigital Filter

Capacitive loading can be added to the open end of the interdigital filter as shown in Figure 12.23. The capacitance added at the ends of the elements will reduce the length of the element required. The resonator length can be reduced to less than $\lambda_0/16$ with capacitive loading (Reference 4). Therefore, capacitive loading can reduce the size of the filter. Design formulas for interdigital filters can be found in References 1, 4, and 33. Interdigital filters with round-rod shaped resonators were discussed in Reference 34 and in microstrip lines in Reference 35.

Another filter design which is similar to the interdigital filter is the combline filter (Reference 37). The configuration of a combline filter is shown in Figure 12.24. The resonators consist of line elements which are short-circuited to ground at one end and the other ends are capacitively loaded to ground. The difference between a combline filter and an interdigital filter, is that in a combline filter resonator all of the elements are grounded on the same ends. The filter can be built either in strip line form or

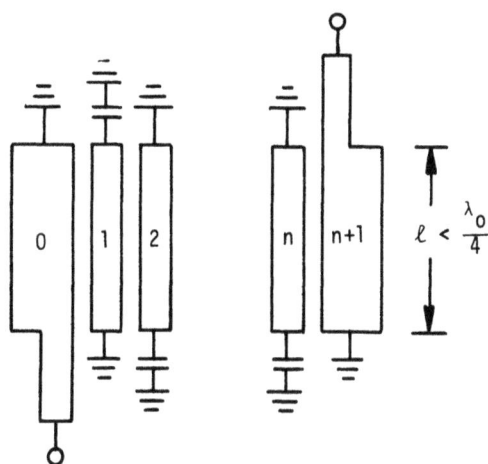

Figure 12.23. Capacitively Loaded Interdigital Filter

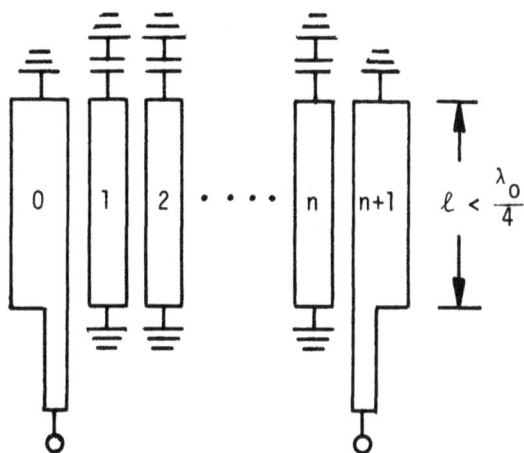

Figure 12.24. A Combline Band-Pass Filter

with conducting rods. The resonators of the filter stand-up like a comb. Line 0 and n + 1 are not resonators but parts of impedance-transforming sections at the ends of the filter. Coupling between resonators is achieved through the fringing fields between resonator lines. With the lumped capacitor loading, the resonator lines are less than $\lambda_0/4$ long at resonance. Without the capacitive loading at the end of each resonators, the magnetic and electric coupling effects cancel each other out and the combline structure no longer works as a filter, because the input cannot couple to the output at all.

The second pass band of the filter occurs when

$$\omega \gtrsim \omega_0 \left(\frac{\lambda_0}{2l} \right) \tag{12.45}$$

where ω_0 is the center angular frequency of the pass band, λ_0 is the corresponding wavelength, and l is the resonator line length. Usually, it is desirable to make the loading capacitance sufficiently large such that the resonator line will be $\lambda_0/8$ or less. Under this condition, the second pass band frequency from Equation 12.45 is

$$\omega \geqslant \omega_0 \ \frac{\lambda_0}{2\frac{\lambda_0}{8}} = 4\omega_0 \tag{12.46}$$

which is far away from the primary pass band. In addition the filter will be smaller for the shorter resonator line length. Detailed design information on combline filters is available from References 1, 4, and 37.

In general, fine alignment on the combline filter is required through adjusting the loading capacitors. The adjustment will bring the pass band to the desired value.

12.7 SURFACE ACOUSTIC WAVE (SAW) FILTERS[38-43]

A SAW filter is a band-pass filter fabricated on a piezoelectric substrate. Like a SAW delay line mentioned in Chapter 9, the input electric signal in a SAW filter is changed into an acoustic wave, filtered in the acoustic domain, and changed back into an electric signal at the output of the filter. The advantage of using acoustic waves to build a filter is the extremely small size and the possibility of mass production through integrated circuit techniques. A SAW filter can operate from a few megahertz up to 1 GHz. The characteristic of the filter in the frequency domain is determined by the geometric shape of the transducers.

The operating principle of the filter will be explained here (References 38 and 39). A typical interdigital transducer (as shown in Figure 12.25) is fabricated on the surface of a piezoelectric surface. When an electric voltage is applied to this transducer, the electric field between the individual fingers causes the surface of the piezoelectric material to distort. Therefore, an acoustic wave is induced on the surface of the substrate. The acoustic wave will travel in both directions perpendicular to the fingers. The overall acoustic wave is the summation of all the individual waves generated by the electrodes. If the overlap of the electrodes (or the length of electrodes) varies in the wave propagation direction (referred to as apodization) as shown in Figure 12.26, the contribution to the overall acoustic wave is proportional to the electrode overlap. The frequency response of the filter can be expressed as (Reference 40)

$$H(f) = \sum_{n=1}^{N} a_n \exp (j2\pi X_n f/v) \qquad (12.47)$$

where a_n is the overlap weighting at each electrode, X_n is the location of the electrode, v is the surface acoustic wave velocity, and f is the frequency.

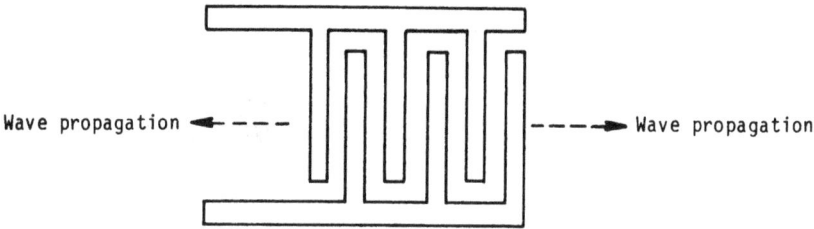

Figure 12.25. A Typical Interdigital Transducer

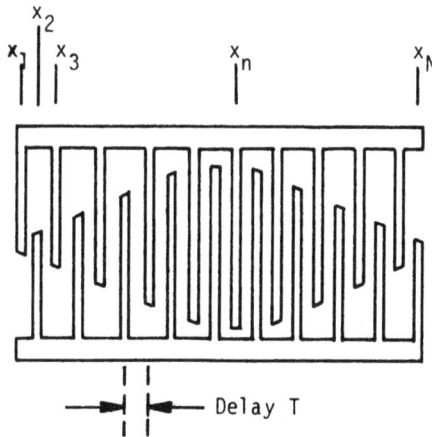

Figure 12.26. Apodization of Transducer

The distance between two adjacent electrodes is one-half wavelength and the corresponding delay time is T. If the transducer is symmetrical with 2M + 1 uniformly spaced electrodes, the transfer function is of the form

$$H(f) = 1 + 2 \sum_{n=1}^{M} a_n \cos (2\pi f T n) \qquad (12.48)$$

where T is the delay time between electrodes. From Equation 12.48, it can be seen that H(f) is a real function and is dispersionless for any transducer which is symmetric.

The frequency response of the SAW filter (represented by Equations 12.47 and 12.48) has a repetition pattern as shown in Figure 12.27. The repetition is due to the discrete nature of the electrodes, because the wave is sampled only at distinct points rather than sampling the entire acoustic wave. Since the electrodes are placed at half wavelength intervals, the response repeats in a frequency interval equal to twice the fundamental frequency. In other words, there is a peak at the third harmonic, fifth harmonic, etc.

The operating principle of the SAW filter can also be explained using Fourier transforms. It is well known that a rectangular pulse in the time domain can be represented by a sinc function (sinx/x) in the frequency

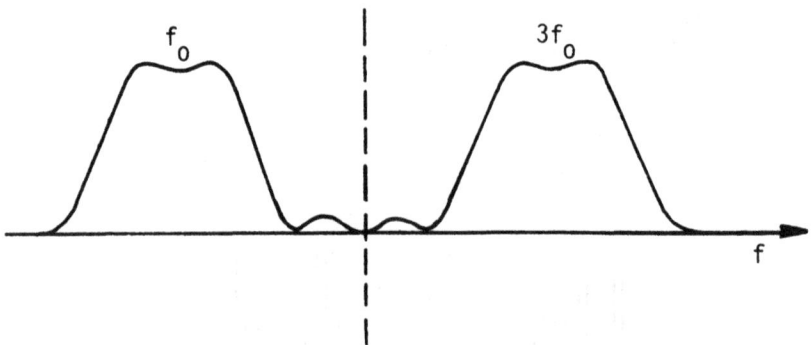

Figure 12.27. Frequency Response of an SAW Filter

domain. A sinc function in the time domain will generate a rectangular response in the frequency domain. Adapting this approach, one can generate a sinc function in the time domain on the transducer; and a rectangular response in the frequency domain will be obtained which represents an ideal filter with infinite slope on the skirt of the filter. Theoretically, a sinc function will extend to infinity in the time domain which is impractical. The sinc function must be truncated at both ends, and the corresponding frequency response will deviate from an ideal filter. Unwanted ripples in the pass band and side lobes in the stop band will be generated because of the truncation. The frequency responses of a simple sinc function and truncated sinc function are shown in Figure 12.28. The frequency responses and their corresponding time domain responses are shown side-by-side. The weighting function is a rectangular response in the time domain which truncates the sinc function abruptly. Various weighting functions are available. The filter response can be tailored

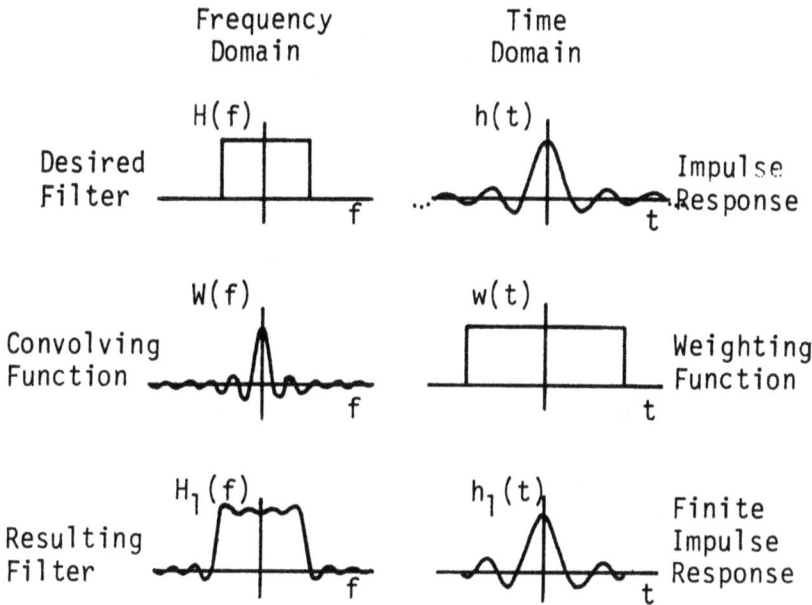

Figure 12.28. Filter Design Using Truncation of Infinite Impulse Response (Based on Reference 40)

through different weighting functions. Other approaches (References 41 and 42) are available for optimizing the frequency response of a SAW filter. Since the SAW filter design uses a different principle than that of the conventional filter, it is possible to design a filter with special responses (Reference 43).

12.8 SAW FILTER MULTIPLEXING (44-50)

One of the major applications of SAW filters is to build channelized receivers which can include several hundred filters. The size, weight, and cost of the filters are very important factors; SAW filters are very attractive based on these requirements. Basically, a channelized receiver is a filter bank which is wide open to receive signals of unknown frequency and automatically routes the signals to a number of output channels according to frequency (see Chapter 5). The ways of multiplexing the input into many parallel channels are important and will be discussed next.

1. The input transducers of SAW filters of different frequencies are connected together directly (References 44 and 45). A wide-band amplifier is used to drive these filters. In Reference 45, one amplifier is used to drive four continuous SAW filters in parallel. The insertion loss of the SAW filters can be compensated by the amplifier. The output impedance of the amplifier is usually designed

to match the input impedance of the combination of SAW filters for maximum power transfer. Therefore, specially designed amplifiers are required.

2. A wide bandwidth power divider can be used in front of the filters. The input is divided into many parallel outputs by the power divider, and each output is followed by a band-pass filter of proper center frequency. This approach is straight-forward, but with relatively high insertion loss and lots of hardware; however, the input impedance is very well matched (Reference 46).

3. Another approach (Reference 47) uses the series and parallel combination of transducers as shown in Figure 12.29. In this figure, two of the input transducers are connected in series and then they are connected in parallel. The input impedance of this approach depends on the individual impedance of each input transducer and how they are connected.

4. The multiplexing can also be accomplished through a constant-k ladder network as shown in Figure 12.30. In this connection, there is loss along the feeding line; and at the individual input transducer, there is usually no special circuit to match the impedance to the feeding line.

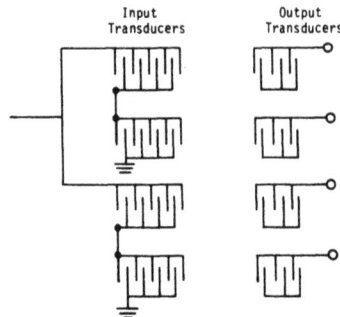

Figure 12.29. Series/Parallel Connections of the Input Transducers of SAW Filter

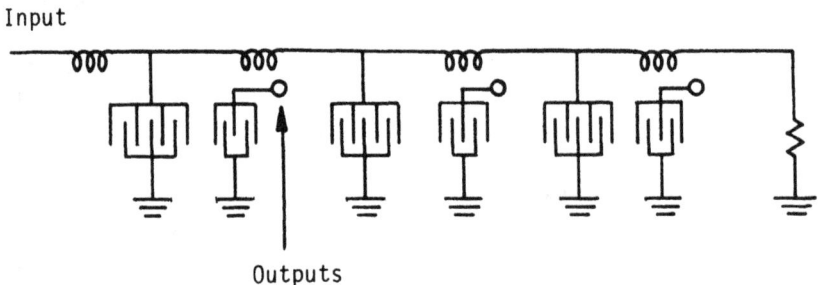

Figure 12.30. Constant-k Network as the Input Matching Circuit for SAW Filter

5. There is another approach (References 44, 48, and 49) to multiplex the input signal into different channels through a multistrip coupler. The multistrip lines in front of the transducer form a 3 dB coupler as shown in Figure 12.31. A surface wave incidental upon a multistrip coupler in track B will emerge from tracks A and B with equal amplitudes. The surface wave emerging from track A is advanced 90 degrees in phase relative to the wave emerging from track B. A frequency diplexer is shown in Figure 12.32. With the delay time τ built in the multistrip coupler, the frequency outputs from A and B are shown in Figure 12.32. This approach can be applied repeatedly as shown in Figure 12.33. A four-channel SAW multiplexer can be fabricated by using the multistrip coupler.

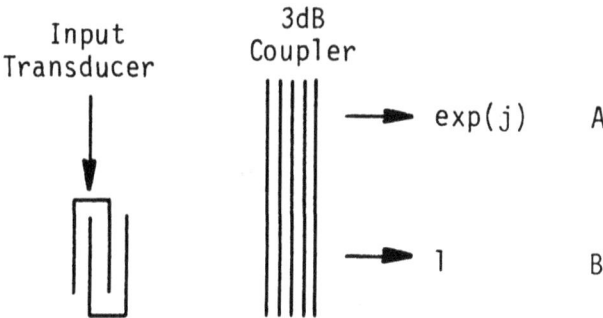

Figure 12.31. 3 dB Multistrip Acoustic Coupler

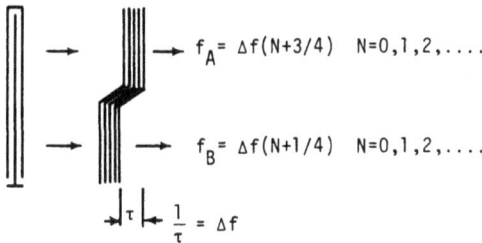

(a) Filter Configuration

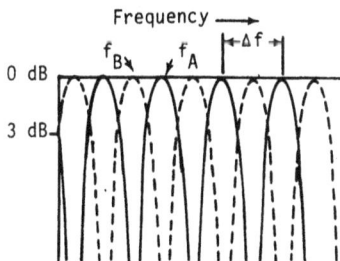

(b) Frequency Response

Figure 12.32. Frequency Response of the Offset Multistrip Coupler. The Solid Line is for Track A and the Broken Line for Track B. (Based on Van de Vaart and Solie, Reference 49)

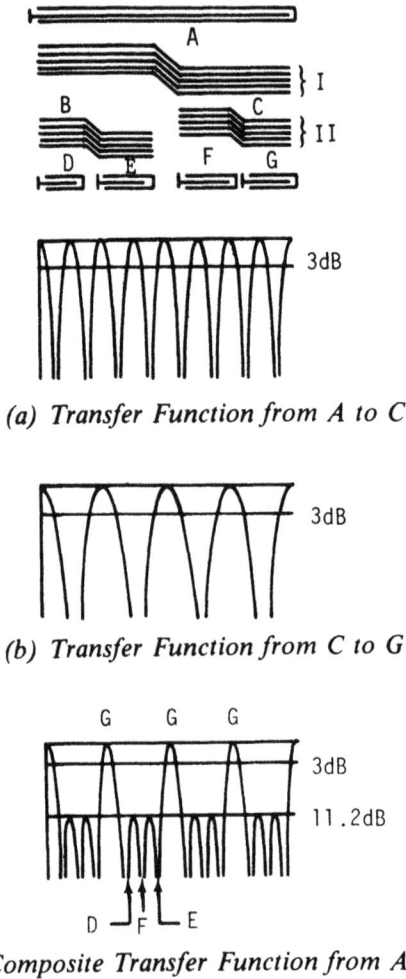

(a) Transfer Function from A to C

(b) Transfer Function from C to G

(c) Composite Transfer Function from A to G

Figure 12.33. Schematic of a Four-Channel Multistrip Coupler Multiplexer Together with the Predicted Transfer Function (Based on Van de Vaart and Solie, Reference 49)

6. A different approach (Reference 50) uses reflectors as shown in Figure 12.34. Frequency dependent reflectors are used to reflect signals of different frequencies to different output ports. The reflectors can be either metal strips or grooves etched in the substrate. A similar technique is used to generate a SAW dispersive delay line as in Chapter 8.

The spaces between the reflecting strips are not uniform. As in Figure 12.34, the high frequency is deflected to output 1 while the low frequency

Input
Transducer   Reflectors

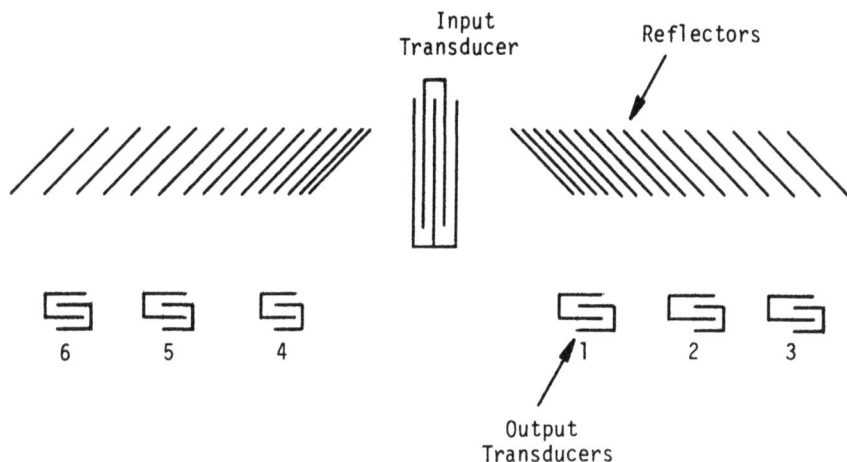

6   5   4   1   2   3

Output
Transducers

*Figure 12.34. SAW Filter Multiplexing Through Frequency Dependent
Reflectors*

is deflected to output 3. In this arrangement, the outputs from 1 to 3 all
have different delay times. Some receiver designs can even take advantage of this phenomena. If equal delay time is desired from all the outputs,
the distances between outputs 1, 2, and the deflector array can be properly
designed to make the delay time equal to that of port 3. Since the input
transducer is bidirectional, deflectors on the opposite side can be added to
obtain more channelized outputs such as 4, 5, and 6.

12.9 CHARACTERISTICS OF SAW FILTERS (51-55)

The characteristics of a SAW filter are different in many ways from a
conventional filter. These differences can be discussed in the frequency
domain and time domain (Reference 51). They have serious impacts on
receiver designs. The characteristics of SAW filters are discussed in
References 38-43.

A. Frequency Domain Responses of SAW Filters

The insertion loss of a SAW filter is relatively high ($\sim$10 dB) compared
to a conventional filter ($\sim$1 dB). A typical result is shown in Figure 12.35a.
Because of the limited transducer size and different time delay paths, there
are sidelobes in the frequency domain as predicted in Section 12.7. The
out-of-band attenuation of the filter does not increase monotonically as in
conventional filters. These sidelobes in frequency domain can severely
degrade the receiver performance (i.e., dynamic range) if the SAW filters
are used in the receiver. There are usually pass bands at the third or second
harmonics of the fundamental frequency. The SAW filter has a long delay
time from input to output ranging from hundreds of nanoseconds to a few
microseconds because of the relative slow acoustic wave velocity. This

delay time can be measured as phase delay as shown in Figure 12.35b. The phase variation is quite linear versus frequency in the pass band.

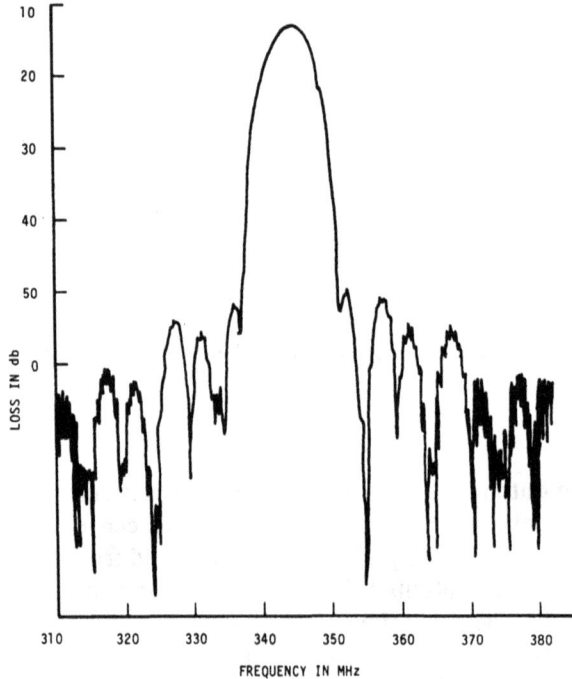

(a) Insertion Loss versus Frequency

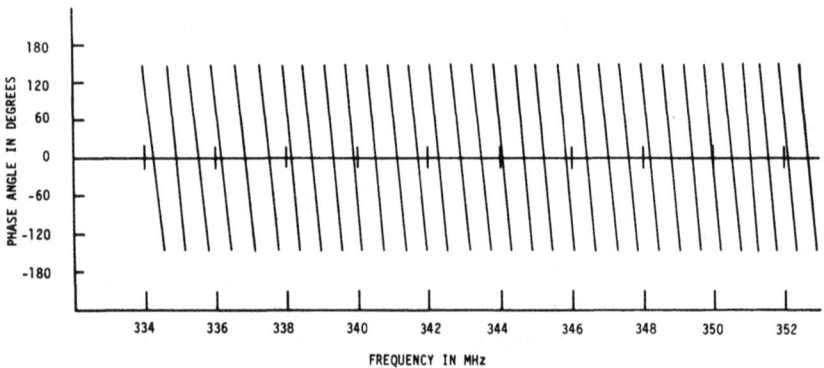

(b) Phase versus Frequency

Figure 12.35. SAW Filter Response in Frequency Domain

B. Time Domain Responses of SAW Filters

The response in the time domain is similar to that of a SAW delay line (Chapter 8). The signals from the output of a SAW filter as shown in

Figure 12.36 are in the following order: the direct feed through, the main signal, triple transit, and fifth transit. There are also other spurious outputs following the main signal. The direct feed through is caused by the capacitive coupling between the input and output transducers. The triple and fifth transits are caused by the signal reflected back and forth from the transducers, and the spurious responses are caused by edge reflections from the substrate. In a receiver, the detection circuits following the SAW filters must wait until all the responses following the main signal are below a certain level in order to process another signal. The detection circuits must also not respond to the direct feed through signal. These responses in the time domain will slow the process rate of the receiver.

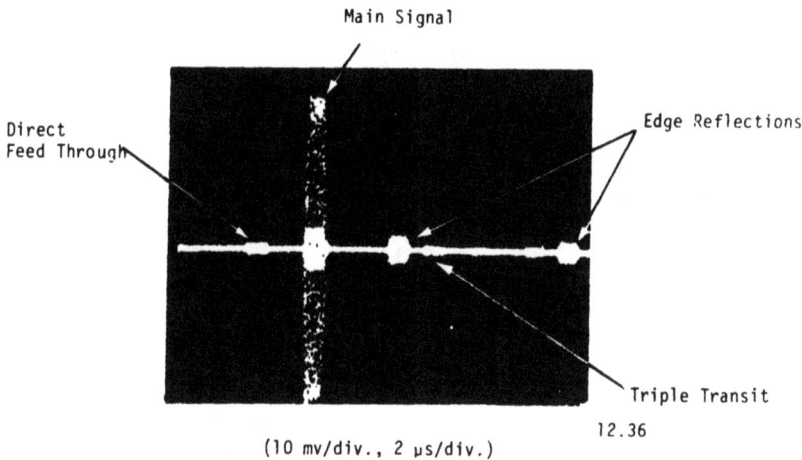

(10 mv/div., 2 µs/div.)

Figure 12.36. Signals Output from a SAW Filter in Time Domain

The input and output impedances of a SAW filter are most capacitive. Matching circuits are required to use the filters in a system. Usually the size of the matching networks are larger than the filters. This problem needs further investigation. For additional discussions on the SAW filter characteristics see References 52 through 55.

12.10 YTTRIUM IRON GARNET (YIG) FILTERS[56-63]

A YIG filter is a band-pass filter. Its center frequency can be electronically tuned over a wide frequency range, sometimes over an octave bandwidth (for example, 2-18 GHz). YIG filters are available in the 0.5 to 40 GHz frequency range. The bandwidth of the filter ranges from 10 MHz to a few hundred megahertz. The insertion loss of the filter is from a few to 10 dB. Band rejection YIG filters are also available. The band-pass YIG filter is commonly used in superheterdyne receivers in front of the mixer. This filter will limit the spurs generated by the mixer. The band rejection YIG filter is usually used in front of a wide-band microwave receiver to reject certain known signals.

The principle operation of YIG filters is as follows. YIG is a ferrite material. Its single crystal form is cut into cubes. The cubes are ground into spheres the size of 20 to 50 mils in diameter. It is the quality of the surface finish which determines the Q-factor of the sphere and hence its microwave properties, a typical finish being in the order of 0.1 μm. The unloaded Q of a generalized ellipsoidal ferrimagnetic sample excited in a uniform precision magnetostatic mode is (References 56 and 58)

$$Q_u = \frac{\omega_0}{2\eta_0} \qquad (12.49)$$

where η_0 is a loss parameter which in the case of a sphere is given by $\mu_0\gamma_e\Delta H/2$ where ΔH is the resonance line width, and ω_0 is the uniform precession resonance frequency

$$\omega_0 = \mu_0\gamma_e H_0 \qquad (12.50)$$

for spheres, where μ_0 is the permeability of free space, γ_e is the gyromagnetic ratio and is close to 1.759×10^{11} for electrons in most ferrites, and H_0 is the external dc magnetic field. Equation 12.50 states that the operational frequency of a YIG filter is linearly proportional to the applied dc magnetic field.

To build a filter, the YIG sphere is put in the center of two coils as shown in Figure 12.31. One coil is in the y-z plane (the input coil); the other

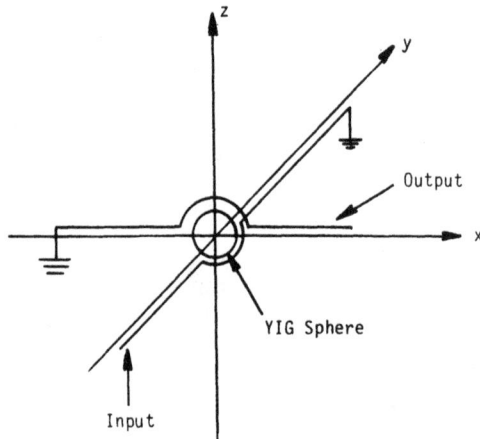

Figure 12.37. Single Stage YIG Pass Band Filter

one is in the x-y plane (the output coil). These two coils are coupled
through the YIG sphere only. Without the sphere the two coils are not
coupled, because their axes are perpendicular to each other. As a dc
magnetic field, H_0, is applied along the z-axis, and with proper field
strength, resonance will occur in the sphere which coincides with the applied microwave frequency. As this occurs, the applied rf power is coupled
from the input coil to the output coil through the YIG sphere. The
characteristic of the filter depends on the coupling between the coil and the
sphere, and the sphere itself. The filter in Figure 12.37 is a one-stage filter.
The center frequency of the filter is related to the magnetic field through
Equation 12.50. A single sphere filter is a one-section filter and the attenuation on the asymptote of the skirt changes 6 dB per octave. If more than
one stage is desired, additional YIG spheres can be incorporated in the
filter. A three-stage YIG filter is shown symbolically in Figure 12.38. The
dc magnetic field containing the YIG spheres must be uniform and the frequency responses of each of the spheres must be tracked in a multistaged
YIG filter.

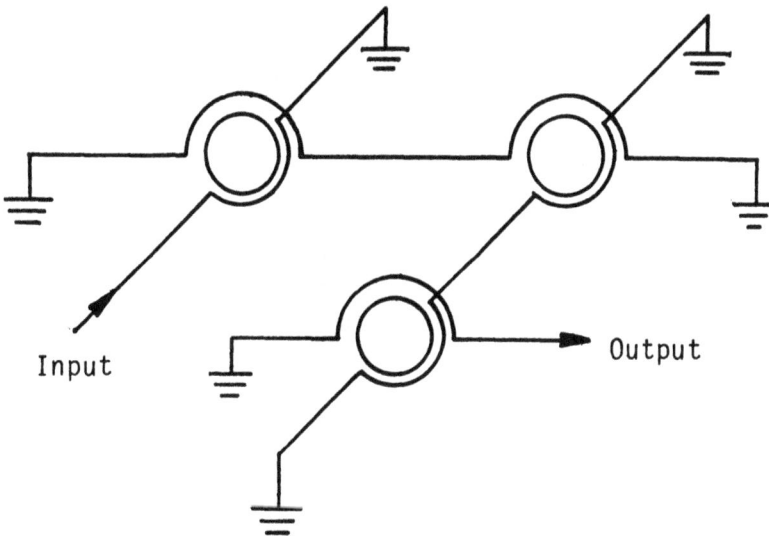

Figure 12.38. A Three-Stage YIG Filter

YIG material is temperature sensitive. Therefore, if the ultimate in
frequency stability is required, the temperature dependence must be reduced
to a minimum. The temperature stability can be achieved by including a
heater within the YIG filter. The YIG sphere is glued to a ceramic rod
which passes through the heater, the whole assembly is thermally insulated
from the rest of the circuit.

12.11 CHARACTERISTICS OF YIG FILTERS[64]

A YIG filter has some special characteristics which do not exist in conventional filters. These special characteristics and terminologies (limiting level saturation of YIG sphere, inband and out-of-band spurious responses, sensitivity, time constant, linearity, and hystersis effect) will be discussed here.

A. Limiting Level

The limiting level refers to the input power level at which the output power stops increasing linearly while the input power increases. Under this condition, the precession angle of the electrons in the YIG ferrite reaches its maximum and no further increase is possible. Therefore, there is no further increasing in coupling between the input and output coils and the output power does not increase linearly anymore beyond a certain input power.

B. Inband and Out-of-Band Spurious Response

There is more than one resonant mode occurring with the YIG crystal. These other modes lead to spurious resonances, which result in a typical pass band resonant shape as shown in Figure 12.39. Some of these modes stay at a fixed frequency from the main resonance while others travel through the pass band causing pass band ripple. For similar reasons there are spurs existing in the rejection bands which are referred to as out-of-band spurs.

C. Sensitivity

The sensitivity of a YIG filter is the ratio of the operating frequency range to the differential current required to tune across it. It is often expressed in MHz/mA. The larger the sensitivity value, the less the required current to tune the filter.

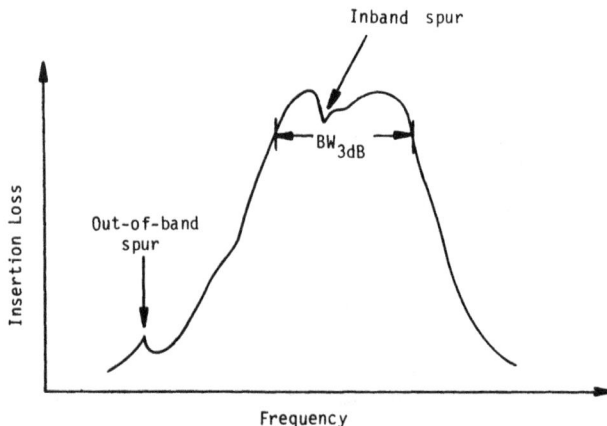

Figure 12.39. A Typical Frequency Response of a YIG Filter

D. Tuning Response Time

The center frequency of a YIG filter is tuned by changing the dc magnetic field. Because of the inductance of the tuning coil, it takes time for the current to reach its final desired value. This delay in current change reflects the tuning speed of the YIG filter. The time required for reaching the final frequency (being less than 0.1 percent of step size) is often referred to as full band tuning response time. This time ranges from ten microseconds to a few milliseconds.

E. Linearity

Equation 12.50 shows that the frequency is linearly proportional to the applied dc magnetic field. If the magnetic field is linearly proportional to the applied current, the frequency of the filter should be linearly proportional to the tuning current. The deviation from this linear relationship is measured in megahertz.

F. Hysteresis

In order to generate the required dc magnetic field, pole pieces of soft magnetic materials are needed. These magnetic materials always have hysteresis effects which reflect to the performance of the YIG filter as shown in Figure 12.40. The same current can tune the filter to two different frequencies depending on the direction of tuning. The hysteresis of a YIG filter is referred to as the maximum differential frequency (at a fixed coil current) due to the hysteresis of the magnetic circuit when tuned in both directions through the operating frequency range. As shown in Figure 12.40, the hysteresis of a YIG filter is expressed in megahertz.

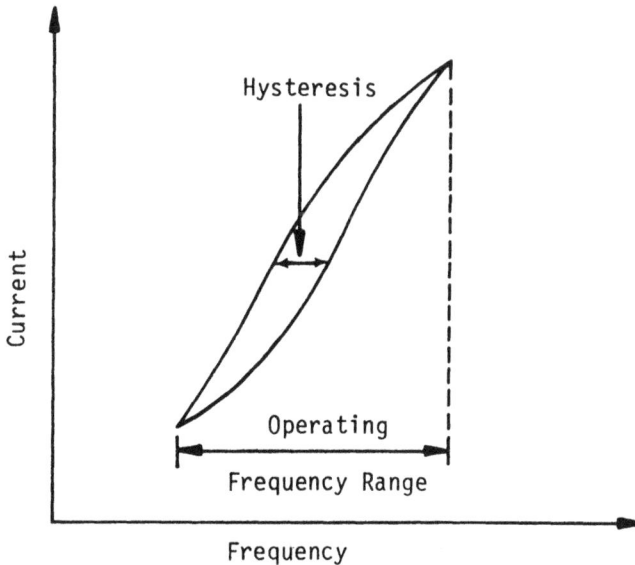

Figure 12.40. A Typical Hysteresis Effect of a YIG Filter

12.12 DIELECTRIC FILTERS[65-83]

It is known that the electromagnetic field stays inside of dielectric or ferrimagnetic materials when they have free space boundaries. Materials of high dielectric constant or high permeability can be used to fabricate resonators. The electromagnetic field of a given resonant mode will be confined in and near the resonator. The field outside the resonator will attenuate to negligible values at a distance less than the free space wavelength. Therefore, the main loss in the resonator will be caused by the loss of the dielectric material rather than through radiation. The unloaded Q of such a resonator is given by

$$Q_u = \frac{1}{\tan \delta} \tag{12.51}$$

where $\tan \delta$ is the loss tangent of the material. Materials of high dielectric constant are of primary interest. Polycrystalline TiO_2 ceramic, also referred to as rutile, has a ε_r of about 100 and the unloaded Q_u of 5000 to 10,000. The Q_u of high-purity TiO_2 can be as high as 12,000. However, the main problem with TiO_2 is that the dielectric constant varies with temperature and is about 1000 parts per million per degree Celsius (1000 ppm/°C). The resonant frequency is related to the dielectric constant (ε_r) by

$$f_0 \simeq \sqrt{\varepsilon_r} = \sqrt{\varepsilon_{r0} \pm \Delta\varepsilon_r} \simeq \varepsilon_{r0} \pm \tfrac{1}{2}\Delta\varepsilon_r + \dots \tag{12.52}$$

where ε_{r0} is the dielectric constant at temperature T_0 and $\Delta\varepsilon_r$ is the change of ε_r due to temperature change. From the second term in Equation 12.52, it is seen that a 1000 ppm/°C change in ε_r will cause approximately a 500 ppm/°C change in frequency. For example, if the center frequency of a dielectric filter is at 3000 MHz, a temperature change of 20 °C, will cause the center frequency to shift by

$$500 \times 10^{-6} \times 3000 \times 10^6 \times 20 = 30 \times 10^6$$

or 30 MHz which is unacceptable in most receiver applications. Materials with higher ε_r such as strontium titanate ($\varepsilon_r \simeq 250$) and barium titanate ($\varepsilon_r > 1000$) have even worse temperature dependence which means that they are not suitable for fabricating dielectric filters. However, recently developed material (Reference 74) ($Ba_2Ti_9O_{20}$) with $\varepsilon_r \simeq 40$ and temperature stability of 0-2 ppm/°C can be used to build dielectric filters with sufficient stability. Band rejection filters can also be built by using high dielectric constant materials (Reference 69).

A common dielectric resonator is a cylindrical rod of diameter D and height L as shown in Figure 12.41. When $L \ll D$, $TE_{01\delta}$ is the dominant mode. The E and H fields are also shown. In $TE_{01\delta}$ the first two subscript integers denote the waveguide mode and δ is the noninteger ratio $= 2L/\lambda_g$ < 1 where λ_g is the waveguide wavelength.

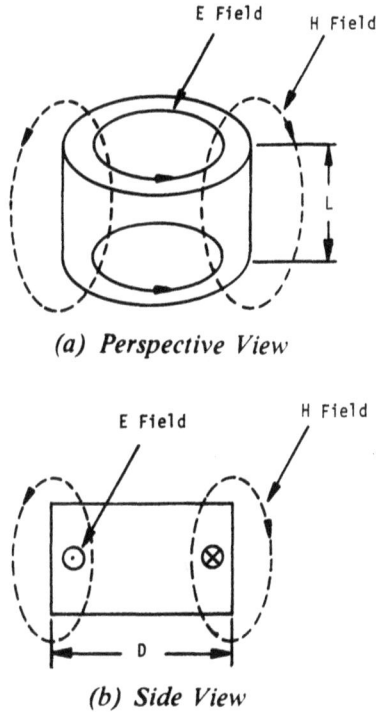

(a) Perspective View

(b) Side View

Figure 12.41. $TE_{01\delta}$ Mode of Dielectric Resonator

The resonant frequency of dielectric resonator was calculated by many authors (References 65, and 71-74) and compared with experimental results. There is no closed form solution for the resonant frequency of the dielectric resonator. Cohn (Reference 65) used the magnetic wall approach and the result is about 10 percent lower than the measured values. Konishi, et. al., (Reference 71) used variation methods to calculate the resonant frequency of the $TE_{01\delta}$ mode, and the results were within 1 percent of experimental data. Itoh and Rudokas, (Reference 72) derived a simple numerical approach to solve the problem, and the calculated values agreed well with experimental results. Pospieszalski (Reference 75) included conducting plates which were placed near the dielectric resonator in their calculation, and the relative discrepancy between frequencies measured and computed is within 3.5 percent. Here the approach of Reference 72 is presented, because it is relatively simple compared with other approaches. In this approach the field problem is solved in four regions (see Figure 12.42b). The dielectric is set on a dielectric sheet of ε_{r2} and thickness t above a ground plane. The H_z field in each region may be written as

$$H_z = \begin{cases} A_1 \sin \beta (z - z_0) \, J_0 (hr) & \text{region (1)} & (12.53a) \\[2mm] A_2 \sin \beta (z - z_0) \, K_0 (pr) & \text{region (2)} & (12.53b) \\[2mm] A_3 \exp [- \gamma(z - L)] \, J_0(hr) & \text{region (3)} & (12.53c) \\[2mm] A_4 \sinh [\xi(z + t)] \, J_0(hr) & \text{region (4)} & (12.53d) \end{cases}$$

where

$$\beta^2 = \varepsilon_{r1} k_0^2 - h^2 = k_0^2 + p^2$$

$$\gamma^2 = h^2 - k_0^2$$

$$\xi^2 = h^2 - \varepsilon_{r2} k_0^2$$

$$k_0 = \omega_0 \sqrt{\mu_0 \varepsilon_0} \qquad\qquad (12.54)$$

where L is the thickness of the dielectric disk, and t is the distance from the conducting plate (Figure 12.45), and z_0, A_1, A_2, A_3, and A_4 are constants to be determined. J_0 and K_0 are the Bessel and the modified Hankel functions of order zero, respectively. ω_0 is the angular resonant frequency. All of the other field components can be derived from Equation 12.53. Note that $E_z = E_r = H_\theta = 0$ for TE mode with no circumferential field variation. The next task is to apply the continuity conditions on H_z and E_θ at $r = D/2$, $0 < z < L$; and E_θ and H_r at $z = 0$ and L, $0 < r < D/2$. When these conditions are applied, A_1, A_2, A_3, A_4, and z_0 can be eliminated to yield two coupled eigenvalue equations

$$\frac{J_0'\left(h \dfrac{D}{2}\right)}{h J_0\left(h \dfrac{D}{2}\right)} + \frac{K_0'\left(p \dfrac{D}{2}\right)}{p K_0\left(p \dfrac{D}{2}\right)} = 0 \qquad\qquad (12.55a)$$

$$\beta L = q\pi + \tan^{-1}\left(\frac{\gamma}{\beta}\right) + \tan^{-1}\left(\frac{\xi}{\beta}\coth \xi t\right), \qquad (12.55b)$$

$$q = 0, 1, 2, \ldots$$

where the principle branches are to be taken for the arctangent functions. The primes in Equation 12.55a indicate differentiation with respect to the arguments. Equations 12.55a and 12.55b together with 11.54 are then solved for the resonant frequency ω_0.

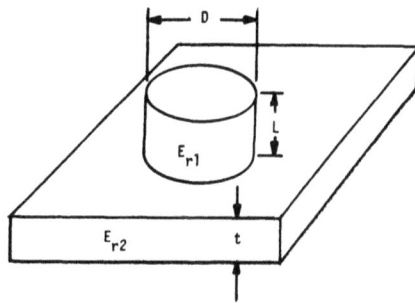

(a) Perspective View

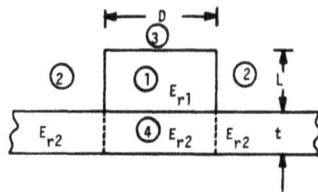

(b) Regions Assigned (Side View)

Figure 12.42. Field Problem is Solved in Four Regions for a Dielectric
Filter

For a given k_0, Equation 12.55a has a finite number of roots which correspond to different radial field distributions. The order of these roots are
numbered with the index m starting from 1. On the other hand, the field
variation in the axial (z) direction is governed by the index q. In the circumferential direction, the field does not vary; and hence, the index takes
only the value zero. The resonant modes are now designed as TE_{0mq}, and
the dominant mode TE_{010} is often referred to as $TE_{01\delta}$.

The resonant frequency of the dominant mode in the dielectric resonator
placed in free space has been computed. For this case $\varepsilon_{r2} = 1$ and $t \to \infty$ so
that Equation 12.55b becomes

$$\beta \tan \left(\frac{\beta L}{2} \right) = \gamma \qquad\qquad (12.56)$$

for q = 0. The results are plotted in Figure 12.43. The resonant frequencies
calculated agree well with experimental data.

Sethares and Stiglitz (Reference 77) developed a scheme to visually
observe the resonant mode in dielectric resonator. A liquid crystal layer is
used to detect the temperature which is related to the energy distribution in
the resonator. The liquid crystal used in their experiment is an ELC (en-
capsulated liquid crystal) solution with an active temperature range of 33°

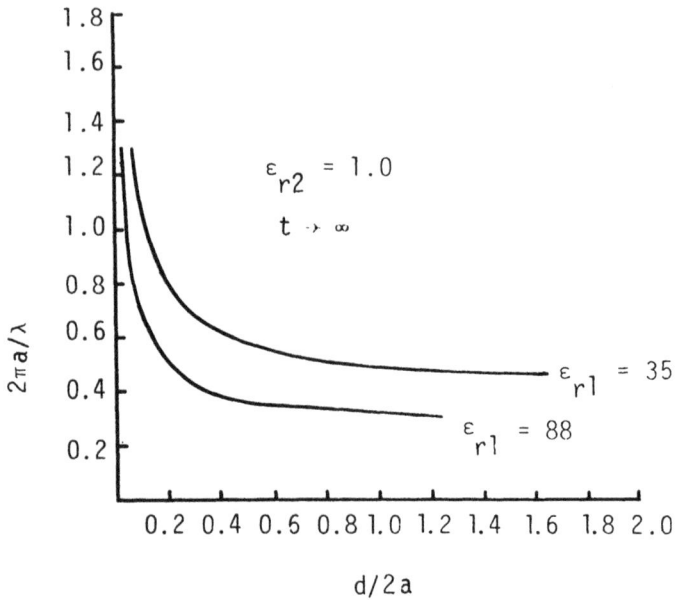

Figure 12.43. Numerical Results of the Resonant Frequency (Based on Itoh and Rudokas, Reference 72)

to 37°C. The thickness of the liquid crystal is between 25μm and 75μm. The results of the $TE_{11\delta}$ mode were displayed. This technique is suited for the investigation of the resonances of complicated or special geometry. Karp et. al., (Reference 78) presented the data for measured resonant frequencies and unloaded Qs of rutile and strontium titanate ($SrTiO_3$) in various geometrical shapes. Frequency tuning was discussed in References 78 and 79. Masse and Pucel (Reference 80) used a dielectric material (K38) to build a band-pass filter. The typical properties of K38 material are: ε_r = 38, tanδ = 4 $\times$ 10^{-4}, and temperature coefficient of ε is -50 ppm/°C which is about ten times better than that of rutile. The filter consists of four resonators with the following properties: center frequency, 9.65 GHz; bandwidth, 50 MHz; rejection, greater than 20 dB at 9.6 GHz. Plourde and Linn (Reference 74) used the $Ba_2Ti_9O_{20}$ ceramics to build a three-section band-pass filter with center frequency at 4.11 GHz and 26 MHz bandwidth. The temperature coefficient is approximately 1.4 ppm/°C with a reference temperature of 25 °C. A similar study on $Ba_2Ti_9O_{20}$ was also reported by Ren (Reference 82) and a filter built by Wakino et. al., (Reference 83).

    Lammers and Stiglitz (Reference 81) used the dielectric (strontium titanate) resonators to build a 96 multiplexer that covered approximately 1 GHz centered at 5 GHz. Since the dielectric thermal coefficient of strontium titanate is 2000 ppm, the center frequency of the filter drifts approximately

10 MHz per centigrade, which is the main shortcoming of the multiplexer.

It can be concluded that although the dielectric resonator has high potential to build band-pass filters or channelized receivers, at higher frequency (GHz range) the temperature stability problem of the dielectric constant must be solved. It seems that new materials with high dielectric constant, low-loss, and good temperature stability are the key factors for better dielectric filters.

12.13 CRYSTAL AND CERAMIC BAND-PASS FILTERS[84-90]

A crystal filter is an electric band-pass filter in which at least one of the components is piezoelectrically resonant. Because the resonator is a mechanical device, the operational frequency of the crystal filter is generally below 100 MHz which should not be considered microwave frequency. However, this kind of filter has high selectivity and stability; it is commonly used in microwave receivers to sort continuous wave (CW) signals. It is a very important component in the field of microwave receivers. Thus, a general discussion on the structure of the filter and its application to microwave receivers will be included here.

The common material used in a crystal filter is quartz which has a very high-Q. The equivalent circuit of a crystal resonator is shown in Figure 12.44. The operational frequency of the quartz crystal resonator depends on the crystal cut (Reference 84). The operational frequency of a quartz crystal ranges from 1.2 KHz to 100 MHz. The highest frequency response from 0.5 to 100 MHz can be obtained through AT-cut. A quartz crystal is shown in Figure 12.45a. An AT-cut is shown in Figure 12.45b which is parallel to the x-axis and makes a 35°15' angle to the z-axis. An x-cut is perpendicular to the x-axis which will produce a resonator with operational frequency of 60 to 300 KHz. Lukaszek (Reference 86) reported that by using

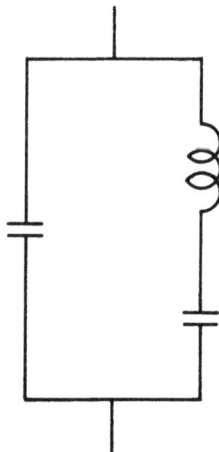

Figure 12.44. Equivalent Circuit of a Crystal Resonator

ion-etching techniques to reduce the resistance of the thin film electrodes, fabrication of a quartz crystal resonator at frequencies up to 200
MHz can be accomplished. He also published the relation between electrode diameter, crystal resistance, and frequency for fifth and seventh harmonic units which is shown in Figure 12.46. As seen from this figure for a
fixed frequency, the crystal resistance depends primarily on the electrode
diameter. As an example, to obtain a crystal resistance below 100 ohms at
210 MHz with a seventh harmonic unit, an electrode diameter of approximately
2.0 mm must be utilized.

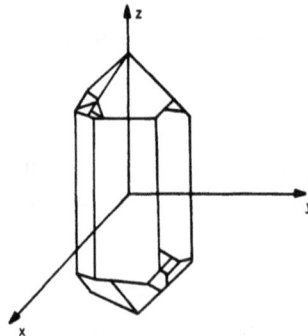

(a) Quartz Crystal

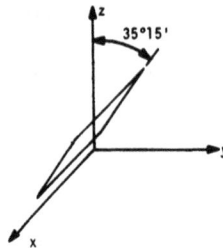

(b) AT Cut

Figure 12.45. Quartz Crystal and the AT Cut

Other materials such as lead-zirconate-titanate (PZT) (Reference 87), a
ceramic material, is also used to build a band-pass filter. The operation
frequency of PZT is in the hundred KHz. The physical configuration of
PZT filter is shown in Figure 12.47. The circular ceramic disc carries two
concentratic electrodes (dot and ring) on one face and a full electrode c
the opposite face to operate at the first overtone of its radial vibration

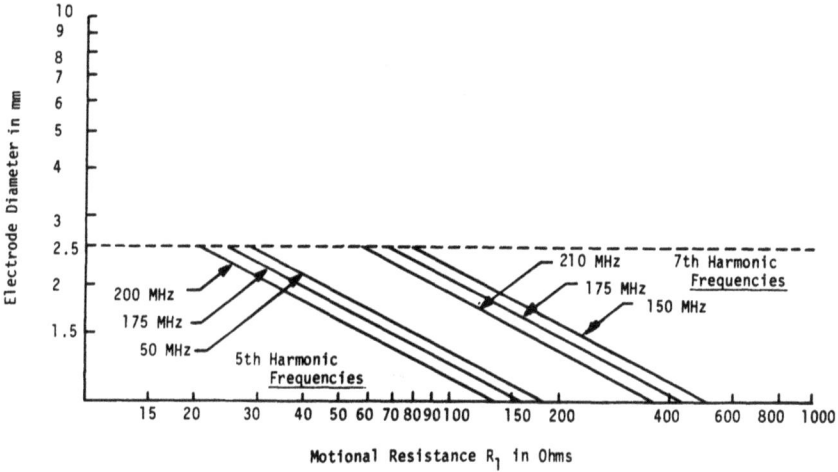

Figure 12.46. Relation Between Electrode Diameter, Crystal Resistance, and Frequency for Fifth and Seventh Harmonic Units (Based on Lukaszek, Reference 86)

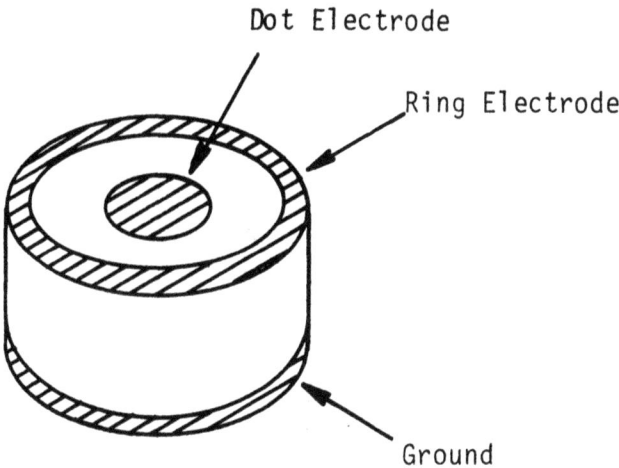

Figure 12.47. Physical Configuration of PZT Ceramic Disc

mode. The electrodes provide both impedance transformation and a band-pass characteristic.

The crystals can be incorporated in the filter design as resonant elements. One way of using the crystal in a filter is shown in Figure 12.48. This kind of circuit is also referred to as the lattice structure (Reference 88). It can be analyzed as a generalized lattice network. Figure 12.48 shows the schematic of a single section, and these sections can be connected in cascade.

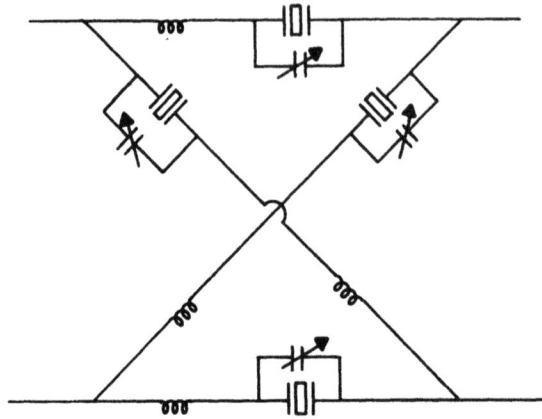

Figure 12.48. Lattice Circuit for Filter with Crystals and Series Inductors in Each Arm

There is another approach (References 89 and 90) to build crystal filters which is referred to as stacked-crystal filters. One of the filters in Reference 89 will be discussed here. The filter is composed of two square AT-cut quartz plates bonded into a sandwich arrangement by a beeswax composition approximately 2 μm thick. The plates are 25 mm on a side and possess individual fundamental resonances at 2.16 MHz. The x-axes

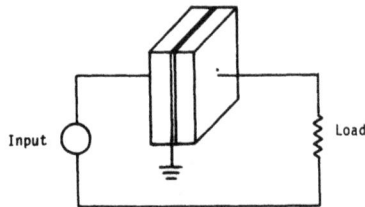

(a) Filter Configuration

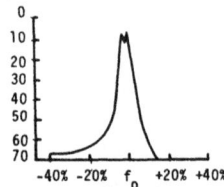

(b) Attenuation Versus Frequency

Figure 12.49. Symmetrical Band-Pass Response of 2.16 MHz Stacked-Crystal Filter; Bandwidth is 3.4 Percent (Based on Ballato and Lukaszek, Reference 89)

of the plates are parallel. Each plate is individually electroded with 1000 Å of chronium in the standard keyhole pattern; in each case, the circular portions are concentric and are 10 and 19 mm in diameter. In forming the stack, the 19 mm electrodes are mated back-to-back to provide shielding and are connected to a common ground. The outer electrodes of the complete sandwich are used as input and output. The filter and its performance are shown in Figure 12.49a and b.

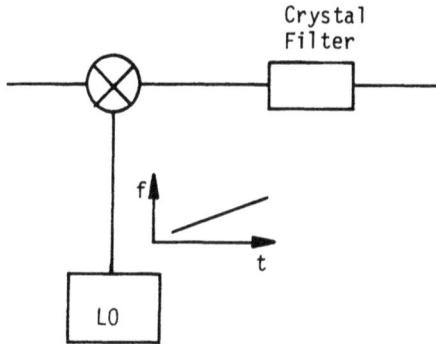

Figure 12.50. Crystal Filter for Measuring CW Signals

The crystal filter is usually used in the last stage of a microwave receiver. The main purpose of the filter is to receive CW signals. Because of its relative narrow bandwidth, pulsed signals may not be properly processed through the crystal filter. One design to search CW signals by using crystal filters is shown in Figure 12.50 as in a superheterodyne receiver discussed in Chapter 3. A mixer is used in front of the crystal filter. If the local oscillator sweeps across a certain bandwidth, the rf signals in the band-width will be converted to the center frequency of the filter and be detected sequencially.

12.14 FILTER MULTIPLEXING(91-98)

In microwave receiver designs, it is often desirable to have multiplexers which separate frequencies into certain ranges from a spectrum of signals covering a large range of frequencies. If a receiving antenna covers 8 to 18 GHz, sometimes it is desirable to divide the input from the antenna into 2 GHz bandwidths for further processing. In a channelized receiver, the function of the receiver is to use banks of filters to separate signals accord-ing to their frequencies. In both of the above applications, band-pass multiplexers are needed.

It usually appears that the design of multiplexers could be easily accom-plished by simply designing the required number of filters individually and then connecting them in series or in parallel. However, in such an approach the input voltage standing wave ratio (VSWR) is very poor. For example, if two band-pass filters of adjacent center frequencies that cross each other at 3 dB point are connected in parallel, the input admittance of the filter, in the stop band, will be a susceptance that varies with frequency. One filter will seriously affect the performance of the other filter.

The basic idea behind the problem is that if there are n filters connected in parallel as shown in Figure 12.51 or in series as shown in Figure 12.52, the input admittance and impedance should be for a normalized system as follows:

$$Y_{in} = Y_{in1} + Y_{in2} + \ldots + Y_{inn} = 1 \qquad (12.57)$$

$$Z_{in} = Z_{in1} + Z_{in2} + \ldots + Z_{inn} = 1 \qquad (12.58)$$

where Y_{ini} and Z_{ini} are the input admittance and impedance of the ith filter. Consequently, the sum of the real parts of the input admittances (or input impedances) must be constant and the sum of reactive parts must be zero. In actual design, it is difficult to fulfill Equations 12.57 and 12.58. Therefore, the frequency multiplexing of filters requires special design procedures. A few of the basic approaches will be discussed next.

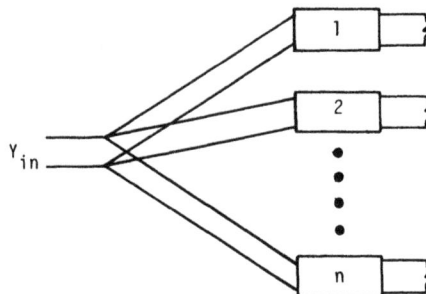

Figure 12.51. n Filters Connected in Parallel

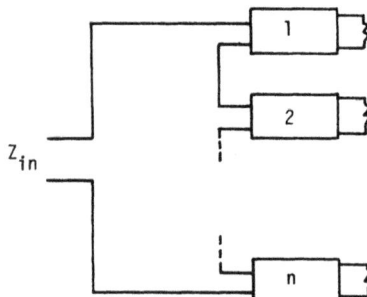

Figure 12.52. n Filters Connected in Series

12.15 MULTIPLEXING THROUGH DIRECTIONAL FILTERS(93)

A directional filter is a four-port device which is particularly designed for multiplexers. Figures 12.53a and b show a directional filter with the coupling between ports 1-4 and 1-2. Power incident at port 1 emerges from port 4 with the frequency response of a band-pass filter, while the remaining power emerges from port 2 with the response of a band reject filter. No power reaches port 3, and there is no reflection from port 1.

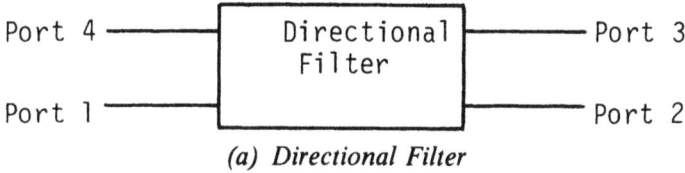

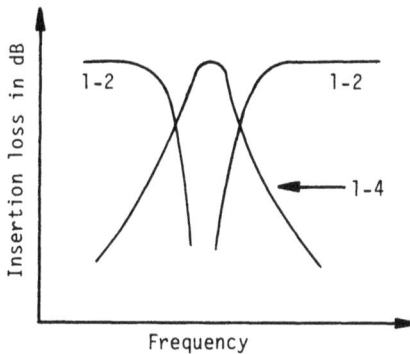

Port 4 ──────┌──────────────┐────────── Port 3
             │ Directional │
             │   Filter   │
Port 1 ──────└──────────────┘────────── Port 2

(a) Directional Filter

Insertion loss in dB

1-2   1-2

1-4

Frequency

(b) Coupling Between Ports 1-4 and 1-2

Figure 12.53.   A Directional Filter and its Coupling Between Ports 1-4 and 1-2

One commonly used directional filter in strip line form is shown in Figure 12.54. The input signal can be considered under two conditions: resonant frequency and off-resonant frequency. First consider the resonant condition in Figure 12.54. When the input signal wavelength equals twice the resonators AD and BC, energy will be coupled to port 4 rather than port 3. This phenomenon can be explained by the superposition principle of the even and odd modes as shown in Figure 12.55. Even mode excitation of ports 1 and 4 at the center frequency causes only the BC strip to resonant which means the reflected wave has amplitude $V/2$ at port 1 and port 4. On the other hand, odd mode excitation causes only the AD strip to resonant which means that reflecting wave have amplitudes $-V/2$ at port 1 and $V/2$ at port 4. Therefore, at the resonant frequency of the strips, the amplitude of the reflected wave at port 4 is V while that at port 1 is zero. In other words, if a signal is incident from port 1, with frequency at the center of the filter, ideally all the energy will reach port 4. When the input frequency is off-resonant with respect to the strips AD, BC, there is no energy coupled from strip line 12 to strip line 43; and the input signal will pass through unattenuated to port 2.

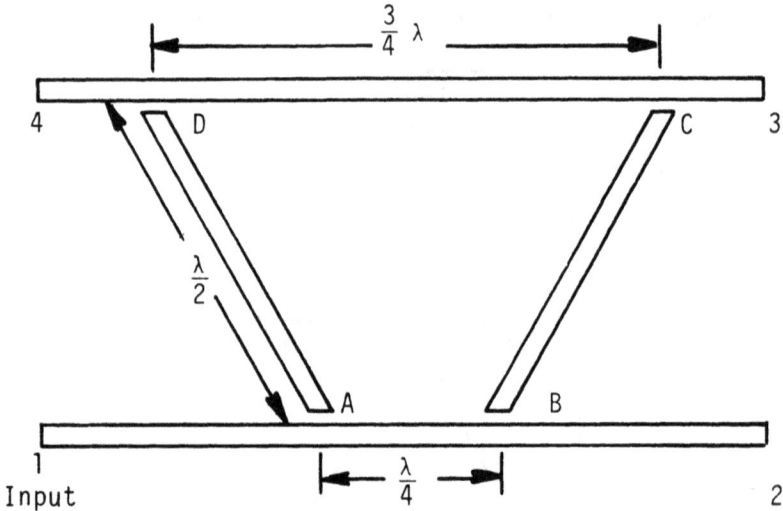

Figure 12.54. A Directional Filter in Strip Line Form

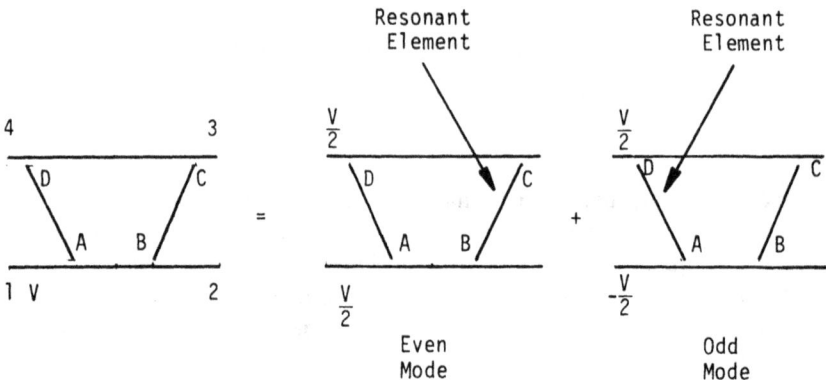

Figure 12.55. Using Superposition of Explain the Characteristic of Directional Filter

Another form of strip line directional filters is known as the traveling wave loop directional filter as shown in Figure 12.56. In this figure the directional filter has two resonators. The number of resonators can be increased to improve the off-frequency rejection. The mean circumference of the loops is a multiple of 360 degree at the midband frequency. Coupling between the loops is obtained by using the principle of quarter-wavelength directional couplers. At the midband frequency a signal incident on port 1 excites clockwise traveling waves in each of the loops. The traveling wave in the last loop excites a signal at port 4. At frequencies well removed from the resonator frequency of the loops, a signal incident on port 1 is transferred to port 2.

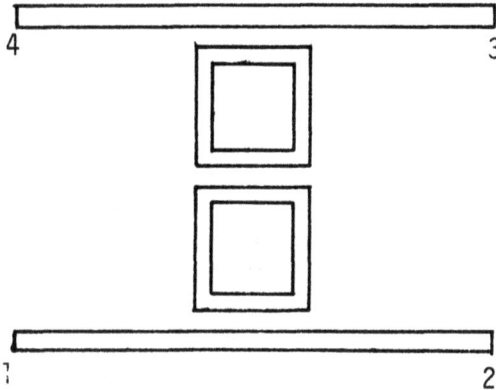

Figure 12.56. A Traveling Wave Loop Directional Loop

The directional filters can be connected in cascade to form a frequency multiplexer as shown in Figure 12.57. If the input signal contains frequencies f_a, f_b, f_c, f_d, and f_e, the f_a will emerge from filter a and the remaining frequencies will pass filter b and f_b will emerge from filter b and so on. The main practical drawback of directional filters is that each resonator of each filter has two different orthogonal modes, and if more than two resonators are required per filter, the tuning of the filters may be difficult.

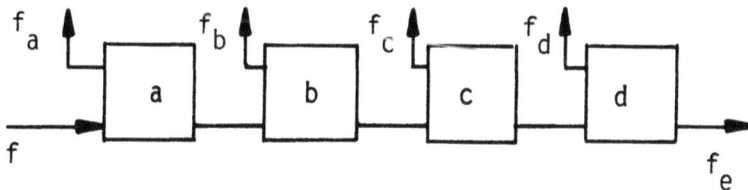

Figure 12.57. A Multiplexer Using Directional Filters

12.16 MULTIPLEXING THROUGH MINIMUM REACTANCE OR MINIMUM SUSCEPTANCE NETWORKS[94]

The input impedance of a network can be expressed in a general form as

$$Z_{in}(S) = H \frac{\prod_{i=1}^{n}(S - z_i)}{\prod_{i=1}^{m}(S - p_i)} \qquad (12.59)$$

where H is a constant, and z_i and p_i are the zeros and poles of the function, respectively. In Equation 12.59 if none of the p_i are pure imaginary and n ⩽ m, the impedance function is referred to as a minimum reactance network. If none of the z_i are pure imaginary and m ⩽ n, the impedance function is referred to as minimum susceptance network. Stated in another way, the input impedance of a minimum reactance network is finite for all real frequencies ω, including infinity. Similarly, the input admittance of a minimum susceptance network is finite for all real frequencies ω, including infinity. For practical purposes, the minimum reactance network starts out with a shunt susceptance, while the minimum susceptance network starts out with a series reactance as shown in Figure 12.58a and b.

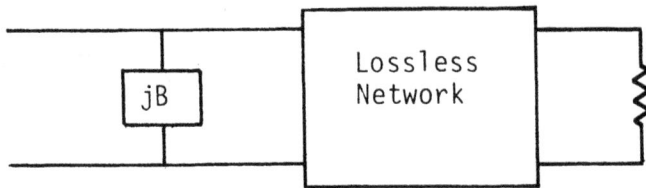

(a) A Minimum Reactance Network

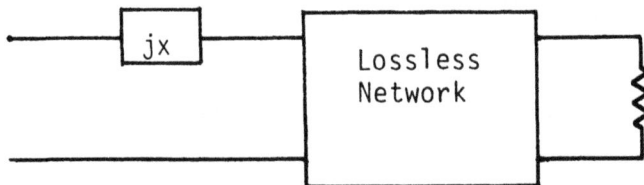

(b) A Minimum Susceptance Network

Figure 12.58. Networks

If two minimum reactance networks are connected in series, the resulting network is also a minimum reactance network. If two minimum susceptance networks are connected in parallel, the resulting network is a minimum susceptance network. An important property of the minimum reactance network is that the real and imaginary parts are related by

$$X(\omega) = \omega\frac{2}{\pi} \int_0^\infty \frac{R(u) - R(\omega)}{u^2 - \omega^2} \, du \qquad (12.60)$$

where $R(\omega)$ is the real part of the input impedance of a minimum reactance network, $X(\omega)$ is the imaginary part of the input impedance of a minimum reactance network, and u is a dummy variable.

If $R(\omega)$ is constant over a band of frequencies (referred to as the pass band) and then falls to zero outside the pass band, then the imaginary part $X(\omega)$ has

1. On the average, a negative slope in the pass band.
2. An extremum in the vicinity of the transition from the pass band to stop band.
3. On the average, a positive slope in the stop band.

This result is illustrated qualitatively in Figure 12.59. It is obvious that if the proper minimum reactance filters are designed and connected in series, the real part of the input impedance of each individual filter will approximately add up to a constant value, while the imaginary part of each individual input impedance will cancel one another. The resulting network will have approximately a real constant input impedance.

12.17 MULTIPLEXING OF NARROW BAND FILTERS[93]

If the channels of the multiplexer are very narrow (one percent bandwidth or less) and separated by guard bands which are several times the pass band width of the individual filters, then some relatively simple approaches can be used to multiplex them. One way of coupling them together is shown in Figure 12.60. Although the filters are shown in lumped circuit form, they can be various kinds of microwave filters. In this approach decoupling between filters is achieved by a decoupling resonator adjacent to each filter. The decoupling network is a band-stop filter with the same center frequency of the band-pass filter adjacent to it. Since the filters and the decoupling resonators have narrow bandwidth, they are loosely coupled to the main transmission line. Therefore, when the filters and decoupling networks are off-resonance, they have little effect on the transmission line.

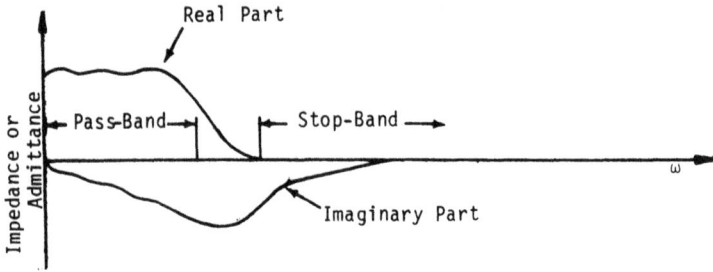

(a) Low-Pass Network Characteristics

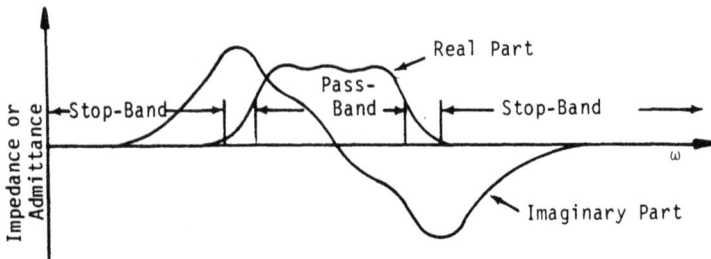

(b) Bandpass Network Characteristics

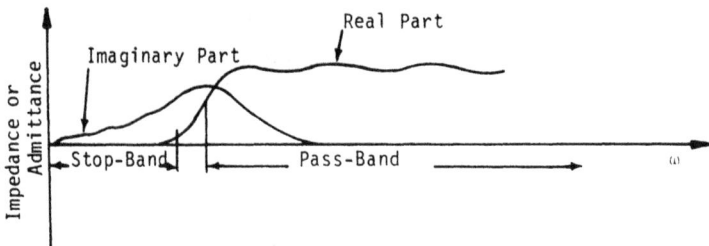

(c) High-Pass Network Characteristics

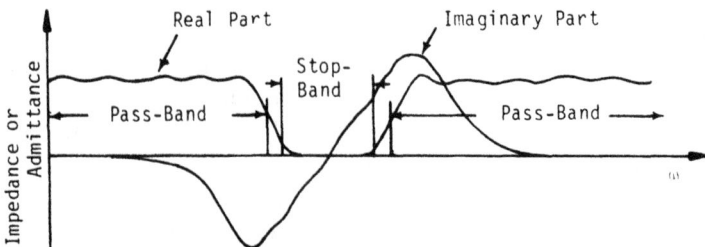

(d) Bandstop Network Characteristics

Figure 12.59. Illustrations of Input Impedance (Admittance) of Minimum-Reactance (Susceptance) Networks of Common Filter Types (Not to Scale)

If a signal of f_2 enters the multiplexer, it will bypass filter f_1 and its decoupling resonator. The signal emerges from filter f_2 and the decoupling resonator f_2 will short (will open in Figure 12.60b) all the circuits beyond it. In this manner, the input signal will emerge from the desired channels. The separation $\Delta\lambda$ between the channel filter and its decoupling resonator should be kept as small as possible to avoid the undesirable transmission line effect.

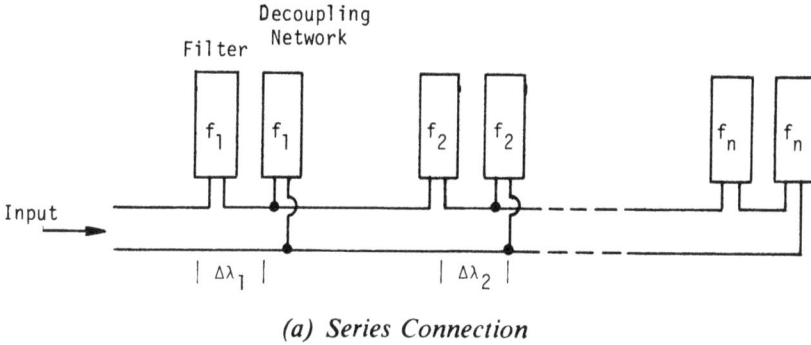

(a) Series Connection

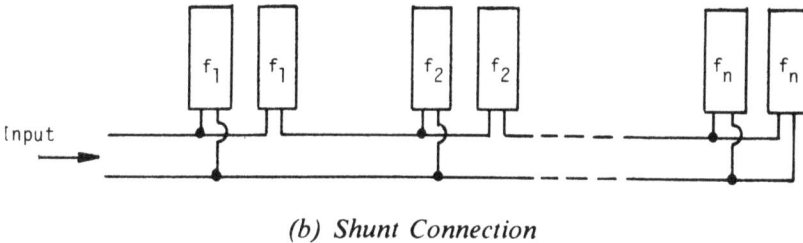

(b) Shunt Connection

Figure 12.60. Multiplexers Using Decoupling Resonators

12.18 DIRECT ANALYTICAL DESIGN OF MULTIPLEXER[95-98]

This approach is to connect a number of band-pass filters to a common point. In order to compensate the interactions among the filters, each filter is internally modified and annulling networks are introduced. The design procedure forces the reflection at the common input port to be zero at a finite set of frequencies in each channel. The detailed mathematics will not be presented. However, the limitations of this approach will be discussed as follows:

1. The channels may not be spaced too close in frequency. The relative spacing of channel B from its adjacent channel A is defined as the ratio of the center frequency separation to the bandwidth of channel A. Relative spacings of 2:1 or more give excellent results, and quite acceptable results are obtained down to 1.5:1.
2. The channel return loss specification cannot be too high. In general, a return loss of approximately 20 dB (corresponding to VSWR of 1.22) gives acceptable results, but 26 dB (VSWR of 1.1) or more is difficult to achieve. The return loss is related to the VSWR as

$$ \text{VSWR} = \frac{1 + \sqrt{R}}{1 - \sqrt{R}} \qquad (12.61) $$

where R is the return loss in power ratio. Twenty dB return loss corresponds to $R = 1/100$.

3. There is an upper limit to the number of channels in the multiplexer, and the outer (i.e., lowest and highest) frequency channels deteriorate the most.

Criterion 1 gives the same conclusion as Section 12.17 that filters with guard bands between them can be multiplexed relatively easy, since there is less interaction between the channels compared to that of contiguous channels. The approach mentioned in Section 11.10 to separate the odd and even channels of contiguous filters by a circulator should be applicable here. Although the maximum number of channels was not mentioned in criterion 3, the maximum number discussed is 8 (Reference 98).

12.19 MICROWAVE LENS[99, 100]

The microwave lens mentioned here is an analogy to an optic lens. It is a two-dimensional device which can be built in strip line or microstrip forms. The microwave lens was originally designed to measure angle-of-arrival (AOA) of the incoming signal. The basic operation principle can be illustrated in Figure 12.61. The principle part of the microwave lens is a beam forming network. The beam forming network is a metal sheet on a dielectric plate which is sandwiched between two ground plates. The inputs to the beam forming network are connected from the left side with antennas and coaxial cables. All the input antennas are aligned in a straight line and all the feeding coaxial cables are of the same length. In such an arrangement, a wavefront (#5 corresponding to output port 5) parallel to the plane of the antenna array will reach the lens arc ABC in equal phase. If the lens is designed with path length A5 = B5 = C5, then the input wavefront #5 will emerge from output port 5. If a wavefront #1 is incident on the lens, the path length is designed so that DA1 = EB1 = FC1 and the output emerges from port 1. It should be noted that only three input ports (A, B, and C) are considered here. It should also be noted

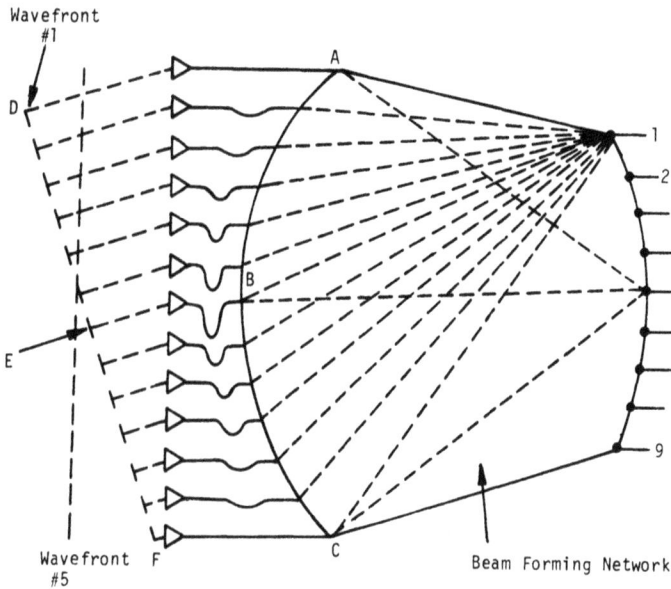

Figure 12.61. Basic Configuration of Microwave Lens

that in this design the signal coming out of a certain port is independent of the input frequency, but depends only on the spatial direction of the input signal. Therefore, the lens is a device which measures the AOA of the input signals over a wide frequency range.

If the feeding mechanism of the microwave lens is properly modified, it can be used to separate signals to different output ports according to their frequencies. The feeding network can be a nondispersive delay line as shown in Figure 12.62. In Figure 12.62a, the input signal is separated into many parallel channels through power dividers which keep all the outputs in phase. The outputs from the power dividers are delayed consecutively with nondispersive delay lines and the difference between adjacent lines is ΔL. If $\Delta L = \lambda_0$ (in general $\Delta L = n\lambda_0$ where n is an integer), where λ_0 is the wavelength corresponding to the center frequency of the microwave lens, a signal with frequency f_0 will arrive at line AB in phase and emerges from the center port. If the input frequency $f_1 > f_0$, the wavefront is tilted with respect to AB and emerges from output ports in the upper half of the lens. Similarly, if the input frequency $f_2 < f_0$, it will come out from ports in the lower half of the lens.

The input network can be changed to a series delay line as shown in Figure 12.62b. The input is coupled through the delay line to each individual input port. If the differential delay between adjacent input ports is ΔL, the same results in Figure 12.62a will be accomplished.

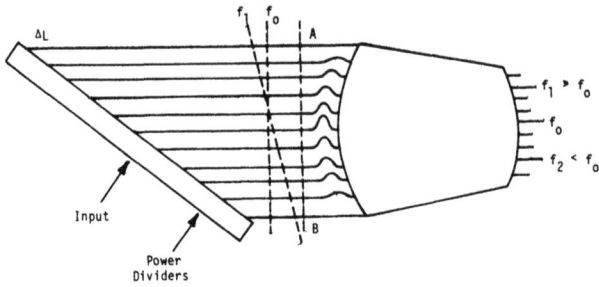

(a) Parallel Feed

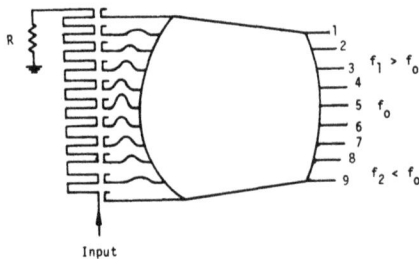

(b) Series Feed

Figure 12.62. Delay Lines to Feed Microwave Lens

One of the basic approaches to lens design can be expressed as follows (see Figure 12.63):

1. Choose the center frequency of the lens and determine the delay time required.
2. Choose arc ABC on a circle (ABC in general can be any arc) then choose the total number of input ports which is related to side lobes. The side lobe energy will emerge from output ports adjacent to the desired one (which is undesirable). In general the larger the number of inputs ports, the lower the side lobes.
3. The locations of the output ports are determined from individual input frequency. If the input frequency is f_N, find the phase delay caused by the delay lines from A to C and B to C and label them as 2n and n.
4. Find a point N such that the line length

$$CN = BN + n = AN + 2n \qquad (12.62)$$

The output at point N is corresponding to the output port of frequency f_N.

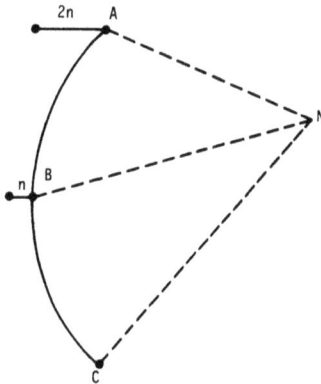

Figure 12.63. Basic Design Approach to a Microwave Lens

It should be noted that only three inputs are used; therefore, the lens will have an abrasive effect, which means that for any input signal only three ports ABC will focus on the desired output; other input ports will only focus near the desired output port. If the feeding to the input ports are properly weighed rather than all uniformly fed with the same amplitude, the side lobe levels from the output ports can be reduced. The detail of the weighting effect is similar to that discussed in Chapter 6. A microwave lens with tapered input and output ports is shown in Figure 12.64. In order to improve the impedance matching at the input and output of the lens, the edge of the lens is star-shaped. The actual size of the microwave lens depends on the operating frequency and the dielectric material on which the lens is fabricated. In general, the lens is limited in an octave bandwidth, otherwise signals of frequency f_1 and f_2 will emerge from the same output port when $f_1 \neq 2f_2$, since the two signals have the same equiphase plane at the input of the beam forming network.

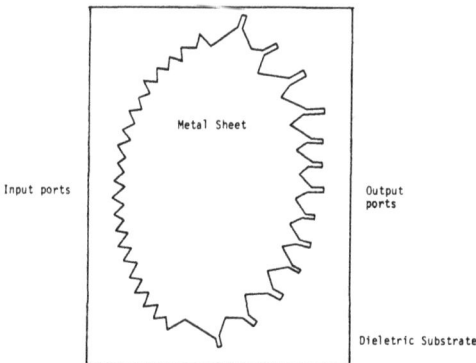

Figure 12.64. A Microwave Lens Configuration

12.20 TRANSIENT RESPONSE OF FILTERS[101-105]

When a CW signal passes through a filter, the output is relatively easy to predict. That is, the output signal is related to the input signal with a known attenuation and phase shift provided by the filter. However, if the input signals are pulse modulated, the outputs of the pulse modulated signals are complex functions of time. In a channelized receiver, the transient phenomenon through filters is one of the most challenging tasks in the design, especially if the receiver is designed to measure the most accurate frequency information on short pulses.

Henderson and Kautz (Reference 102) plotted the impulse responses and step responses of several kinds of filters up to ten poles. As expected, the more poles a filter has, the more out-of-band rejection it provides; however, the more severe the transient response. One popular approach to solve the transient is to use Fast Fourier Transform (FFT) (Reference 105). The output from a filter not only depends on the filter itself, it is also a function of the input signal frequency and pulse shape. Reference 104 solves the transient response of a filter with three kinds of pulses: a square, a composite with finite leading and trailing edges, and a sine square signal. The result of the square signal will be discussed here.

In Figure 12.65, the input to the filter is R(t), and the filter response is H(t). The output of the filter is given by the convolution integral

$$Y(t) = \int_0^t R(\tau) H(t - \tau) \, d\tau \qquad (12.63)$$

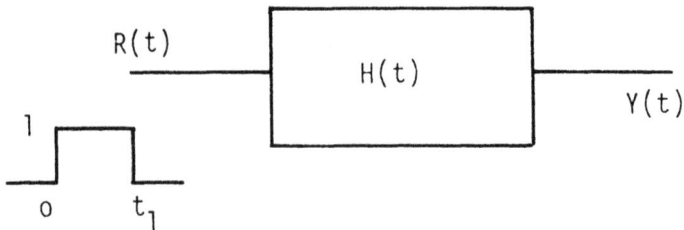

Figure 12.65. *Filter Response to Input Signal R(t)*

For a square pulse of unit amplitude and width t_1, the pulse can be expressed as

$$R(t) = U(t) - U(t - t_1) \qquad (12.64)$$

where U(t) is the step function.

The filter response H(t) in the S plane can be written as Equation 12.30

$$H(S) = \frac{A_1}{S - S_1} + \frac{A_2}{S - S_2} + \cdots + \frac{A_k}{S - S_k} + \cdots \frac{A_{2n}}{S - S_{2n}} \quad (12.65)$$

where $S_1, S_2 \ldots S_k$ are the roots of the filter and $A_1, A_2 \ldots A_k$ are constants. Substituting Equations 12.64 and 12.65 into Equation 12.63 obtains

$$Y(t) = U(t) \int_0^t R(\tau) H(t - \tau) d\tau$$

$$(12.66)$$

$$- U(t - t_1) \int_{t_1}^t R(\tau) H(t - \tau) d\tau$$

For $t < t_1$

$$Y(t) = \sum_{k=1}^{2n} [Y_K(t) - Y_K(0)] \quad (12.67)$$

For $t \geqslant t_1$

$$Y(t) = \sum_{k=1}^{2n} [Y_k(t_1) - Y_k(0)] \quad (12.68)$$

where

$$Y_k(\tau) = \frac{A_k \exp (S_k t)}{2} \left[\frac{\exp (B_k \tau)}{B_k} + \frac{\exp (C_k \tau)}{C_k} \right] \quad (12.69)$$

where A_k is the coefficient in Equation 12.65

$$B_k = -S_k + j\omega \quad (12.70)$$

$$C_k = -S_k - j\omega \quad (12.71)$$

Substituting Equation 12.69 into Equations 12.67 and 12.68, the response of a square pulse passing through a band-pass filter with known poles can be plotted with the help of a computer. The calculated and measured results of a 3-pole Chebyshev filter with center frequency $f_o = 300.8$ MHz, BW = 17.8 MHz, ripple factor of 0.25 dB with a square input signal of 200 ns at frequency of 309.8 MHz are shown in Figure 12.66. The experimental and calculated results agree well.

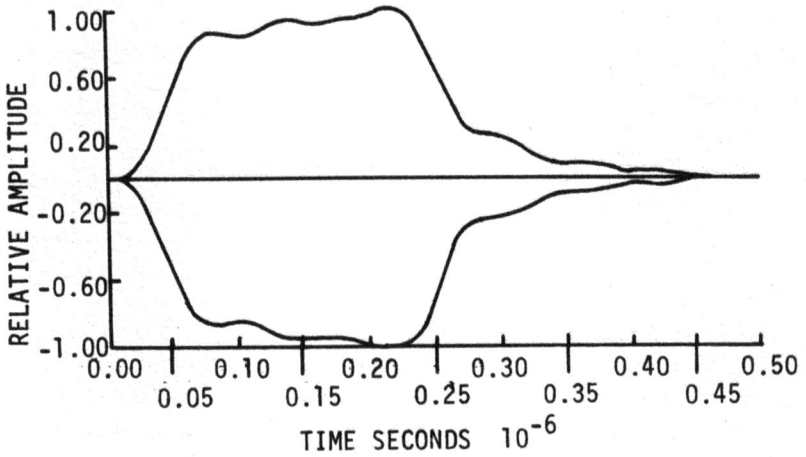

(a) Calculated Results

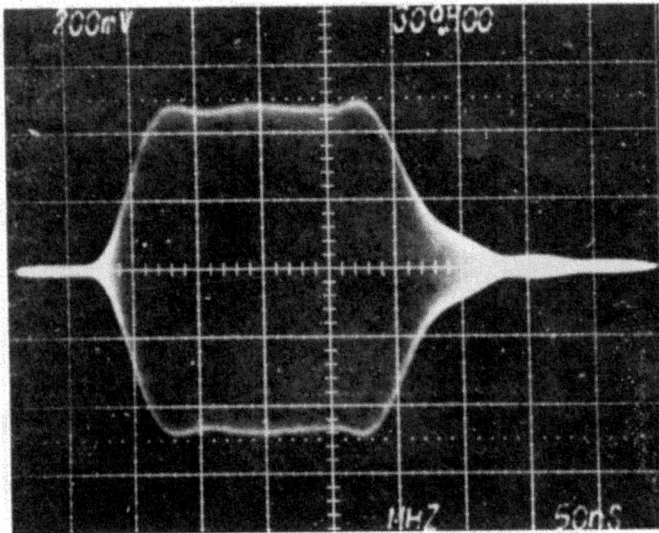

(b) Measured Results

Figure 12.66. Filter Output of a Pulsed Signal

REFERENCES

1. Matthaei, G.L., Young, L., and Jones, E.T.M., *Microwave filters, imped-ance-matching networks and coupled structures,* McGraw-Hill Book Co., 1964.
2. Skwirzynski, J.K., *Design theory and data for electrical filters,* D. Van Nostrand Co., 1965.
3. Rhodes, J.D., *Theory of electrical filters,* Wiley & Sons, New York, 1976.
4. Young, L., Matthaei, G.L., Cristal, E.G., Weller, D.B., Robinson, L.A., Schiffman, B.M., and Adams, D.K., "Novel microwave filter design techniques," Defense Documentation Center No. 462155, SRI Project 4344, Contract DA-36-039-AMC-00084(E), February 1965.
5. Geffe, P.R., *Simplified modern filter design,* John F. Rider (publisher), 1964.
6. Graver, R.V., "Basic microwave filter theory," Microwave Journal, p. 87, September 1979, reprint from *Microwave diode control devices,* Appendix C, Artech House Books, 1976.
7. Westman, H.P., editor, *Reference data for radio engineers,* Chapters 7, 8, and 9, Howard W. Sams & Co., Inc., Fifth edition, 1972.
8. Van Valkenburg, M.E. *Network Analysis,* Chapters 13 and 14, Prentice-Hall, Inc., 1962.
9. Kudsia, C.M., and Chitre, N.K.M., "Transmission and reflection group delay of Butterworth, Chebyshev, and elliptic filters," RCA review, pp. 248-267, June 1969.
10. Levy, R., and Whiteley, I., "Synthesis of distributed elliptic-function filters from dumped-constant prototypes," IEEE Trans. Microwave Theory and Techniques, Vol. MTT-14, pp. 506-517, November 1966.
11. Horton, M.C., and Wenzel, R.J., "The digital elliptic filter-a compact sharp-cutoff design for wide band-stop or band-pass requirements," IEEE Trans. Microwave Theory and Techniques, Vol. MTT-15, pp. 307-314, May 1967.
12. Levy, R., and Rhodes, J.D., "A comb-line elliptic filter," IEEE Trans. Microwave Theory and Techniques, Vol. MTT-19, pp. 26-29, January 1971.
13. Chausi, M.S., *Principles and design of linear active circuits,* Chapters 4 and 15, McGraw-Hill, Inc., 1965.
14. Macalpine, W.W., and Schildknecht, R.O., "Coaxial resonators with helical inner conductor," Proc. IRE, Vol. 47, pp. 2099-2105, 1959.
15. Zverev, A.I., and Blinchikoff, H.J., "Realization of a filter with helical components, IEEE Trans. Components Parts, Vol. CP-8, pp. 99-110, September 1961.

16. Zverev, A.I., "Handbook of Filter synthesis" Chapter 9, Wiley, 1967.
17. Ozaki, H., and Ishii, J., "Synthesis of a class of strip line filters," IRE Trans. Circuit Theory, Vol. CT-5, pp. 104-109, June 1958.
18. Wenzel, R.J., "Exact design of TEM microwave networks using quarter-wave lines," IEEE Trans. Microwave Theory and Techniques, Vol. MTT-12, pp. 94-111, January 1964.
19. Matthaei, G.L., "Design of wide-band (and narrow-band) band-pass microwave filters on the insertion loss basis," IEEE Trans. Microwave Theory and Techniques, Vol. MTT-8, pp. 580-593, November 1960.
20. Schiffman, B.M., and Matthaei, G.L., "Exact design of band-stop microwave filters," IEEE Trans. Microwave Theory and Techniques, Vol. MTT-12, pp. 6-15, January 1964.
21. Kirton, P.A., and Pang, K.K., "Extending the realizable bandwidth of edge-coupled strip line filters," IEEE Trans. Microwave Theory and Techniques, Vol. MTT-5, pp. 672-676, August 1977.
22. Horton, M.C., and Wenzel, R.J., "General theory and design of optimum quarter-wave TEM filters," IEEE Trans. Microwave Theory and Techniques, Vol. MTT-13, pp. 316-327, May 1965.
23. Cohn, S.B., "Parallel-coupled transmission line resonator filters," IEEE Trans. Microwave Theory and Techniques, Vol. MTT-6, pp. 223-231, April 1958.
24. Cristal, E.G., "New design equations for a class of microwave filters," IEEE Trans. Microwave Theory and Techniques, Vol. MTT-19, pp. 486-490, May 1971.
25. Judd, S.V., Whiteley, I., Clowes, R.J., and Richard, D.C., "An analytical method for calculating microstrip transmission line parameters," IEEE Trans. Microwave Theory and Techniques, Vol. MTT-18, pp. 78-87, February 1970.
26. Bryant, T.G., and Weiss, J.A., "Parameters of microstrip transmission lines and of coupled pairs of microstrip lines," IEEE Trans. Microwave Theory and Techniques, Vol. MTT-16, pp. 1021-1027, December 1968.
27. Cohn, S.B., "Shielded coupled strip transmission lines," IRE Trans. Microwave Theory and Techniques, Vol. MTT-3, pp. 29-38, October 1955.
28. Mara, J.H., "Broadband microstrip parallel-coupled filters using multiline sections," Microwave Journal, p. 97, April 1979.
29. Gupta, C., "Design of parallel coupled line filters," Microwave Journal, p. 39, December 1979.
30. Ho, C., "Band-pass filter design using MIC techniques," Hybrid Microelectronics Seminar No. 3, June 10, 1977.

31. Cristal, E.G., and Frankel, S., "Hairpin-line and hybrid hairpin-line/half-wave parallel-coupled line filters," IEEE Trans. Microwave Theory and Techniques, Vol. MTT-20, pp. 719-728, November 1972.

32. Gysel, U.H., "New theory and design for hairpin-line filters," IEEE Trans. Microwave Theory and Techniques, Vol. MTT-22, pp. 523-531, May 1974.

33. Wenzel, R.J., "Exact theory of interdigital band-pass filters and related coupled structures," IEEE Trans. Microwave Theory and Techniques, Vol. MTT-13, pp. 559-575, September 1965.

34. Dishal, M., "A simple design procedure for small percentage bandwidth round-rod interdigital filters," IEEE Trans. Microwave Theory and Techniques, Vol. MTT-13, September 1965.

35. Milligan, T.A., "Dimensions of microstrip coupled lines and interdigital structures," IEEE Trans. Microwave Theory and Techniques, Vol. MTT-25, pp. 405-410, May 1977.

36. Matthaei, G.L., "Interdigital band-pass filters," IEEE Trans. Microwave Theory and Techniques, Vol. MTT-10, pp. 479-491, November 1962.

37. Wenzel, R.J., "Synthesis of combline and capacitively loaded interdigital band-pass filters of arbitrary bandwidth," IEEE Tran. Microwave Theory and Techniques, Vol. MTT-19, pp. 678-686, August 1971.

38. Tancrell, R.H., and Holland, M.G., "Acoustic surface wave filters," Proceedings of the IEEE, Vol. 59, pp. 393-409, March 1971.

39. Hartmann, C.S., Bell, D.T., and Rosenfeld, R.C., "Impulse model design of acoustic surface-wave filters," IEEE Trans. Microwave Theory and Techniques, Vol. MTT-21, pp. 162-175, April 1973.

40. Tancrell, R.H., "Analytic design of surface wave band-pass filters," IEEE Trans. Sonics and Ultrasonics, Vol. SU-21, pp. 12-22, January 1974.

41. Engan, H., "Surface acoustic wave multielectrode transducers," IEEE Trans. Sonics and Ultrasonics, Vol. SU-22, pp. 395-401, November 1975.

42. El-Diwany, M.H., and Campbell, C.K., "Modification of optimum impulse response techniques for application to SAW filter design," IEEE Trans. Sonics and Ultrasonics, Vol. SU-24, pp. 277-279, July 1977.

43. Reilly, J.P., Campbell, C.K., and Suthers, M.S., "The design of SAW band-pass filters exhibiting arbitrary phase and amplitude response characteristics," IEEE Trans. Sonics and Ultrasonics, Vol. SU-24, pp. 301-305, September 1977.

44. Jones, R.R., Schellenberg, J., Tanski, W.J., and Moore, R.A., "Transplexing SAW filters for ECM, Part I," Microwaves, p. 43, December 1974.

45. Jones, R.R., Schellenberg, J., Tanski, W.J., and Moore, R.A., "Transplexing SAW filters for ECM, Part II," Microwaves, p. 68, January 1975.

46. Huber, C., Lane, J., Newman, B.A., Godfrey, J.T., Grauling, C.H., and Moore, R.A., "A low side lobe SAW continuous filter bank using MDC LiTaO$_3$," 1977 Ultrasonics Symposium Proceedings, IEEE Cat. #77CH1264-1SU, pp. 568-572.

47. Private communication with Dr. A. Slobodnik and P. Carr, Air Force Cambridge Research Laboratory.

48. Solie, L.P., "A surface acoustic wave multiplexer using offset multistrip couplers," 1974 Ultrasonics Symposium Proceedings, pp. 153-156.

49. Van de Vaart, H., and Solie, L.P., "Surface acoustic wave multiplexing techniques," 1975 Ultrasonic Symposium Proceedings, IEEE Cat. #75CH0994-4SU.

50. Private communication with Dr. E. Stern, Lincoln Laboratory.

51. Krueger, C.H., and Tsui, J.B.Y., "Application of surface acoustic wave filter to channelized receivers," NAECON, 1975.

52. Matthaei, G.L., Wong, D.Y., and O'Shaughnessy, B.P., "Simplifications for the analysis of interdigital surface-wave devices," IEEE Trans. Sonics and Ultrasonics, Vol. SU-22, pp. 105-114, March 1975.

53. Laker, K.R., and Slobodnik, A.J., "Impact of SAW filters on RF pulse frequency measurement by double detection," AFCRL-TR-75-0011, January 1975.

54. Wang, W.C., Schachter, H., and Cassara, F.A., "Application of surface acoustic wave devices to communication receivers," POLY-MRI-1403-79, AFOSR-77-3353, November 1979.

55. Sandy, F., "Extensions of the capabilities of the transducer circuit model and SAW device analysis program," RADC-TR-79-145, May 1979.

56. Carter, P.S., "Magnetically-tunable microwave filter using single-crystal yttrium-iron-garnet resonators," IEEE Tran. Microwave Theory and Techniques, Vol. MTT-9, pp. 252-260, May 1961.

57. Skeie, H., "Nonreciprocal coupling with single-crystal ferrites," IEEE Trans. Microwave Theory and Techniques, Vol. MTT-12, pp. 587-594, November 1964.

58. Comstock, R.L., "Synthesis of filter-limiters using ferrimagnetic resonators," IEEE Trans. Microwave Theory and Techniques, Vol. MTT-12, pp. 599-607, November 1974.

59. Carter, P.S., "Equivalent circuit of orthogonal-loop-coupled magnetic resonance filters and bandwidth narrowing due to coupling inductance," IEEE Trans. Microwave Theory and Techniques, Vol. MTT-18, pp. 100-105, February 1970.

60. Igarashi, M., and Naito, Y., "Properties of a four-port nonreciprocal circuit utilizing YIG on strip line filters and circulator," IEEE Trans. Microwave Theory and Techniques, Vol. MTT-20, pp. 828-833, December 1972.

61. Igarashi, M., and Naito, Y., "Theoretical analysis of magnetic resonance nonreciprocal circuits-limitations of 3 dB bandwidth and available range," IEEE Trans. Microwave Theory and Techniques, Vol. MTT-22, pp. 821-829, September 1974.

62. Helszain, J., "Scattering parameters of looped coupled YIG resonators," Microwave Journal, p. 53, December 1978.

63. Hurst, G.J., "YIG and what it means to microwave instruments," Marconi Instrumentation, p. 76, Vol. 15, No. 4.

64. YIG devices, Watkins-Johnson Company, 1980.

65. Cohn, S.B., "Microwave band-pass filters containing high-Q dielectric resonators," IEEE Trans. Microwave Theory and Techniques, Vol. MTT-16, pp. 218-227, April 1968.

66. Harrison, W.B., "A miniature high-Q band-pass filter employing dielectric resonators," IEEE Trans. Microwave Theory and Techniques, Vol. MTT-16, pp. 210-218, April 1968.

67. Iveland, T.D., "Dielectric resonator filters for application in microwave integrated circuits," IEEE Trans. Microwave Theory and Techniques, Vol. MTT-19, pp. 643-652, July 1971.

68. Naumann, S.J., and Sethares, J.C., "Microwave dielectric resonators," Air Force Cambridge Research Laboratories, AFCRL-65-867, November 1965.

69. Gerdine, M.A., "A frequency stabilized microwave band-rejection filter using high dielectric constant resonators," IEEE Trans. Microwave Theory and Techniques, Vol. MTT-17, pp. 354-359, July 1969.

70. Yee, H.Y., "Natural resonance frequency of microwave dielectric resonator," IEEE Trans. Microwave Theory and Techniques, Vol. MTT-13, p. 256, March 1965.

71. Konishi, Y., Hoshino, N., and Utsumi, Y., "Resonant frequency of a $T_{01\delta}$ dielectric resonator," IEEE Trans. Microwave Theory and Techniques, Vol. MTT-24, pp. 112-114, February 1976.

72. Itoh, T., and Rudokas, R.S., "New method for computing the resonant frequencies of dielectric resonators," IEEE Trans. Microwave Theory and Techniques, Vol. MTT-25, pp. 52-55, January 1977.

73. Pospieszalski, M.W., "On the theory and application of the dielectric post resonator," IEEE Trans. Microwave Theory and Techniques, Vol. MTT-25, pp. 228-231, March 1977.

74. Plourde, J.K., and Linn, D.F., "Microwave dielectric resonator filters utilizing $Ba_2Ti_9O_{20}$ ceramics," IEEE MTT-S, International Microwave Symposium, pp. 290-293, 1977.

75. Pospieszalski, M.W., "Cylindrical dielectric resonators and their applications in TEM line microwave circuits," IEEE Trans. Microwave Theory and Techniques, Vol. MTT-27, pp. 233-238, March 1979.

76. Stiglitz, M.R., and Sethares, J.C., "A hybrid ferrimagnetic dielectric microwave filter," Proceedings IEEE, Vol. 55, pp. 1734-1735, October 1967.

77. Sethares, J.C., and Stiglitz, M.R., "Visual observation of high dielectric resonator modes," Applied Optics, Vol. 8, pp. 2560-2562, December 1969.

78. Karp, A., Shaw, H.J., and Winslow, D.K., "Circuit properties of microwave dielectric resonators," IEEE Trans. Microwave Theory and Techniques, Vol. MTT-16, pp. 818-828, October 1968.

79. Stiglitz, M.R., "Frequency tuning of rutile resonators," Proceedings IEEE, pp. 413-414, March 1966.

80. Masse, D.J., and Pucel, R.A., "A temperature stable band-pass filter using dielectric resonators," Proceedings IEEE, pp. 730-731, June 1972.

81. Lammers, U.H., and Stiglitz, M.R., "A 96-channel multiplexer with dielectric resonators," Microwave Journal, p. 59, October 1979.

82. Ren, C.L., "Waveguide band-stop filter utilizing $Ba_2Ti_9O_{20}$ resonators," IEEE, Vol. MTT-5, International Microwave Symposium, pp. 227-229, 1978.

83. Wakino, K., Nishikawa, T., Matsumoto, H., and Ishikawa, Y., "Miniaturized band-pass filters using half-wave dielectric resonators with improved spurious response," IEEE, Vol. MTT-5, International Microwave Symposium, pp. 230-232, 1978.

84. Mason, W.P., *Physical acoustics: principles and methods*, Vol. 1, Chapter 5, Academic Press, New York, 1964.

85. Reference 16, Chapter 8.

86. Lukaszek, T.J., "Mode control and related studies of VHF quartz filter crystal," IEEE Trans. Sonics and Ultrasonics, Vol. SU-18, pp. 238-246, October 1971.

87. Lungo, A., and Sauerland, F., "A ceramic band-pass transformer and filter element," Clevite Electronic Components, Clevite Corporation, Technical Paper TP-46, Cleveland, Ohio.

88. Van Valkenburg, M.E., *Introduction to modern network synthesis*, Chapter 12, John Wiley and Sons, Inc., 1960.

89. Ballato, A., and Lukaszek, T., "Stacked-crystal filters," Proceedings IEEE, pp. 1495-1496, October 1973.

90. Ballato, A., Bertoni, H.L., and Tamir, T., "Systematic design of stacked-crystal filters by microwave network methods," IEEE

Trans. Microwave Theory and Techniques, Vol. MTT-22, pp. 14-25, January 1974.

91. Cristal, E.G., and Matthaei, G.L., "A technique for the design of multiplexers having contiguous channels," IEEE Trans. Microwave Theory and Techniques, Vol. MTT-12, pp. 88-93, January 1964.

92. Wenzel, R.J., "Application of exact synthesis method to multichannel filter design," IEEE Trans. Microwave Theory and Techniques, Vol. MTT-13, pp. 5-15, January 1965.

93. Reference 1, Chapters 14 and 16.

94. Buntschuh, C., "Octave bandwidth, high directivity microstrip coupler," RADC-TR-73-396, Contract F 30602-72-C-0282, AD777320, January 1974.

95. Haine, J.L., and Rhodes, J.D., "Direct design formulas for asymmetric band-pass channel diplexers," IEEE Trans. Microwave Theory and Techniques, Vol. MTT-25, pp. 807-812, October 1977.

96. Rhodes, J.D., and Levy, R., "A generalized multiplexer theory," IEEE Trans. Microwave Theory and Techniques, Vol. MTT-27, pp. 99-111, February 1979.

97. Rhodes, J.D., and Levy, R., "Design of general manifold multiplexers," IEEE Trans. Microwave Theory and Techniques, Vol. MTT-27, pp. 111-123, February 1979.

98. Rhodes, J.D., and Alseyab, S.A., "A design procedure for band-pass channel multiplexer connected at a common junction," IEEE Trans. Microwave Theory and Techniques, Vol. MTT-28, pp. 246-253, March 1980.

99. Rotman, W., and Turner, R.F., "Wide-angle microwave lens for line source application," IEEE Trans. Antennas and Propagation, Vol. pp. 623-632, November 1963.

100. Archer, D.H., and Prickett, J., "Signal spectrum analyzer," U.S. Patent 3,735,256, May 22, 1973.

101. Murakami, T., and Sonnenfeldt, R.W., "Transient response of detectors in symmetric and asymmetric side band systems," RCA Revew, pp. 581-611, December 1955.

102. Henderson, K.W., and Kautz, W.H., "Transient response of conventional filters," IRE Trans. Circuit Theory, Vol. pp. 333-347, December 1958.

103. Lecklider, T.H., "Program gives filter time response electronic design," Electronic Design, Vol. 9, p. 192, April 1974.

104. Tsui, J.B.Y., Adair, J.E., Hawkins, J.E., and LaFleur, S.J., "Transient response of filters," Air Force Avionics Laboratory Technical Report AFAL-TR-77-249, December 1977.

105. Rabiner, L.R., and Radar, C.M., "Digital signal processing," IEEE Press, 1972.

CHAPTER 13
LINEAR MICROWAVE AMPLIFIERS, LOGARITHMIC AMPLIFIERS AND LIMITERS

13.1 INTRODUCTION

Amplifiers are very important components in receivers. They have been used in almost all the modern receivers. The main purpose of an amplifier is to amplify the input signal strength and improve the sensitivity of the receiver (see Chapter 2). Amplifiers can be used at different stages of a receiver. If it is used at the front end of the receiver, it is usually referred to as the radio frequency (RF) amplifier. If it is used after the mixer which usually works as a frequency down converter to convert the RF to an intermediate frequency (IF), the amplifier is referred to as an IF amplifier. If the amplifier is used after the crystal detector, it is referred to as the video amplifier. In this chapter, the discussion will be limited to RF and IF amplifiers. The approach is from a system point of view and concentrates on the performance and effects to receivers.

In this chapter, the general performance of a linear amplifier will be discussed. The amplifier operates within a small signal (linear) region of its transfer characteristic. In a linear amplifier, a 1 dB increase/decrease in input power results in a 1 dB increase/decrease in output power. Many of the basic concepts of receivers such as sensitivity and noise figure that have been discussed in Chapter 2 will be discussed again here in detail. Since amplifiers are one of the fundamental building blocks of a receiver, they share many of the same general characteristics as a receiver.

13.2 FREQUENCY BANDWIDTH, GAIN, GAIN FLATNESS[1-7]

The specified bandwidth of an amplifier is often referred to as the 3 dB bandwidth. The gain of the amplifier usually decreases at both the low and high frequency ends. The 3 dB bandwidth is a bandwidth between two frequencies where the amplification of the amplifier is 3 dB below its normal (or specified) value. However, it must be emphasized here that the bandwidth of an amplifier is also often referred to as a frequency band for which the amplifier will meet all electrical specifications including gain, gain flatness, noise figure, power output, voltage standing wave ratio (VSWR), and intercept point. In most microwave amplifiers, the bandwidth is limited to an octave bandwidth or less, where the high frequency cut-off equals two times the low frequency cut-off. For example, 2-4 GHz and 4-8 GHz amplifiers are octave bandwidth. However, there are also many amplifiers with the bandwidth wider than octave (i.e., 1-1500 MHz and 2-8 GHz).

The gain of an amplifier is defined as the power output to the power input which can be written as

$$P_0 = GP_i \qquad (13.1)$$

or expressed in decibels as

$$G = 10 \log \frac{P_0}{P_i} = 10 \log \frac{V_0^2/R_0}{V_i^2/R_i} \qquad (13.2)$$

where P_0 and V_0 are the output power and voltage of the amplifier, and P_i and V_i are the input power and voltage of the amplifier. The power P is often replaced by symbol S for discussion of noise figure. R_0 and R_i are the output and input resistances of the amplifier, respectively. Usually $R_0 = R_i$ and Equation 13.2 can be written as

$$G = 20 \log \frac{V_0}{V_i} \, dB \qquad (13.3)$$

The gain flatness is defined as the maximum excursion of the gain over the specified frequency range at a constant temperature, expressed as plus/minus decibels from the nominal gain level. An amplifier with ± 1 dB gain flatness will have less than 2 dB total gain variation over the specified frequency range.

If the input signal strength keeps increasing, the gain of the amplifier will start to decrease, which results from the onset of saturation at the output. A 1 dB compression point is used to specify the power output capability of an amplifier. The input power is increased until the gain decreases deviate from linearity by 1 dB. The output power of the amplifier is then measured and expressed in dBm which is called the 1 dB compression point. The dBm is an absolute power level and expressed as

$$dBm = 10 \log P \qquad (13.4)$$

where P is power in milliwatts (mw).

The input power and the output power relation can be written as

$$10 \log P_0 = 10 \log G + 10 \log P_i \qquad (13.5)$$

or

$$P_0(dBm) = G(dB) + P_i(dBm)$$

The plot of P_0 versus P_i (sometimes referred to as the transfer characteristic) is shown in Figure 13.1 with the 1 dB compression point indicated. In the linear region, the slope is 1. The gain of the amplifier can be read from the difference between the output and input. In Figure

13.1 the gain is 30 dB and the 1 dB compression point is about 12 dBm. As a rule of thumb (Reference 4), the saturated gain of the amplifier G_{sat} is typically 6 dB less than the small signal gain of the amplifier. The saturated power output P_{sat} is usually approximately 6 dB above the power output at 1 dB gain compression. This approximation will provide some general ideas of how an amplifier will perform when it is driven into saturation.

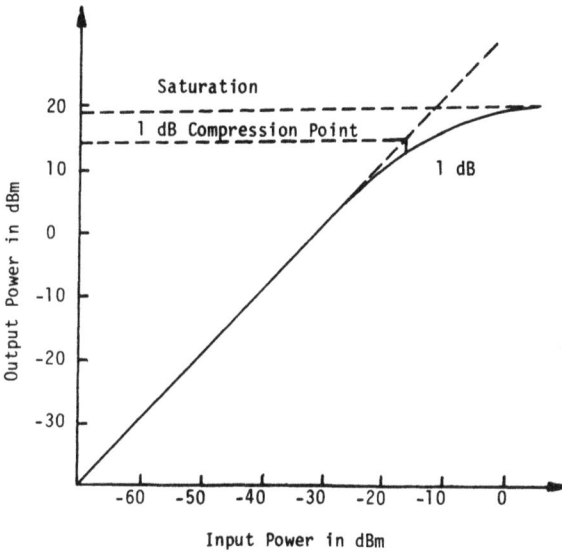

Figure 13.1. Input and Output Signal Relation of an Amplifier

13.3 NOISE FIGURE[8, 9]

An ideal amplifier adds no noise of its own to the signal being amplified. However, all practical amplifiers generate noise to some extent. A measure of the noise produced by a practical amplifier compared with the noise of an ideal amplifier is called the noise figure. The available thermal noise power input to the amplifier can be expresed as

$$N_i = kTB \tag{13.6}$$

where k is Boltzmann's constant 1.37×10^{-23} watts/deg K, T is equivalent noise temperature in kelvin (290 °K = room temperature), and B is frequency bandwidth of the amplifier.

The noise figure is defined as

$$F = \frac{S_i/N_i}{S_0/N_0} = \frac{\text{signal-to-noise ratio at the input}}{\text{signal-to-noise ratio at the output}} \qquad (13.7)$$

Since the output signal is related to the input signal through

$$S_0 = GS_i \qquad (13.8)$$

by substituting Equation 13.8 into 13.7, the noise figure can be expressed as

$$F = \frac{N_0}{kTBG} \qquad (13.9)$$

where S_i is the available input signal power, S_0 is the available output signal power, and N_0 is the available output noise power.

The available power refers to the power which would be delivered to a matched load. If there is no noise generated by the amplifier, the signal-to-noise ratio at the input and the output will be the same and the noise figure in Equation 13.7 will be 1 (or noise figure = 0 dB). The output noise power can be expressed in terms of the input noise power as

$$N_0 = GN_i + \Delta N \qquad (13.10)$$

where ΔN is the noise generated by the amplifier. Substituting Equation 13.10 into Equation 13.9 obtains

$$F = \frac{GN_i + \Delta N}{kTBG} = 1 + \frac{\Delta N}{kTBG} \qquad (13.11)$$

The noise figure of the amplifier in most cases determines the sensitivity of the receiver. The noise figure varies with frequency and temperature. In general, the noise figure at 25 °C is given by the manufacture's data sheet.

13.4 INTERCEPT POINTS[2,3]

If an amplifier is a perfectly linear device, and when two signals are amplified simultaneously, there will be two output signals and no harmonically related spurious signals generated. As discussed in the preceding paragraphs, the gain of the amplifier becomes nonlinear when the output power reaches saturation. Thus, in any RF amplifier, when two or more signals are present simultaneously, harmonically related spurious signals can be generated by intermodulation due to the nonlinear characteristic of the amplifier. These spurious signals are called intermodulation (intermod) products, and their amplitude is a function of the amplitude of the input signals and is also related to the power output capability of the amplifier. Intermod products become a concern at higher input signal levels, when the signal power levels fall within the nonlinear gain region. In the nonlinear region, the relationship between the input and output of the amplifier can be written as

$$e_0 = a_1 e_i + a_2 e_i^2 + a_3 e_i^3 + \ldots \tag{13.12}$$

where e_0 and e_i are the output and input voltage of the amplifier, respectively; and a_1, a_2, $\ldots$ are coefficients.

The second harmonic which is generated by the second term of Equation 13.12 is a signal with frequency double that of the fundamental signal. Usually, the second harmonic is not a major concern in amplifiers with less than an octave bandwidth since the second harmonic will fall outside the frequency band of interest. If the amplifier bandwidth is over an octave, then the second order harmonic is of great concern. The second order intercept point which is a point at which the first and the second order asymptotes intersect will be used to determine the amplitude of the second order harmonic.

Third order intermod products are produced by two simultaneous input signals of different frequencies. In order to simplify the discussion, the amplitudes of the two signals are assumed equal. The input signal can be expressed as

$$e_i = \cos\omega_1 t + \cos(\omega_2 t + \theta) \tag{13.13}$$

The third term $(a_3 e_i^3)$ of Equation 13.12 generates terms containing

$$\cos(2\omega_1 t - \omega_2 t + \theta) \text{ and } \cos(2\omega_2 t + 2\theta - \omega_1 t) \tag{13.14}$$

where ω_1 and ω_2 are the angular frequencies of the input signals, and θ is the relative phase angle. The third order intermod harmonics are shown in Figure 13.2.

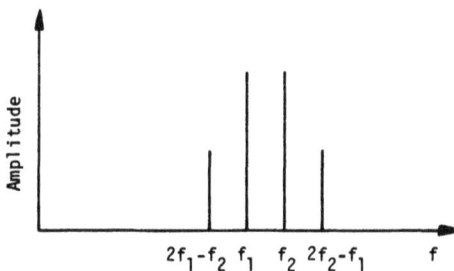

Figure 13.2. Third Order Intermodulation Harmonics

The third order intermod products are the greatest concern since they are the lowest order intermodulation products which can fall within the pass band of the amplifier. Thus, the third order intermod products are often used to impose an upper bound on the spurious free dynamic range of an amplifier. For this reason, the third order intercept point is usually indicated in an amplifier. Figure 13.3 shows the second and third intercept points of an amplifier. The second order products have a slope of 2:1 while the third order products have a slope of 3:1. Their asymptotes intercept the asymptote of the fundamental and the points are referred to as the second and third order intercept points. By knowing the second and third order intercept points and the input level (or output level) of the desired signals, the amplitude of the second and third order intermodulation products can be calculated from Figure 13.3. Given the input signal level, the third order intercept point, and the third order intermod products (reference at the output) can be obtained through the intersection of two straight lines.

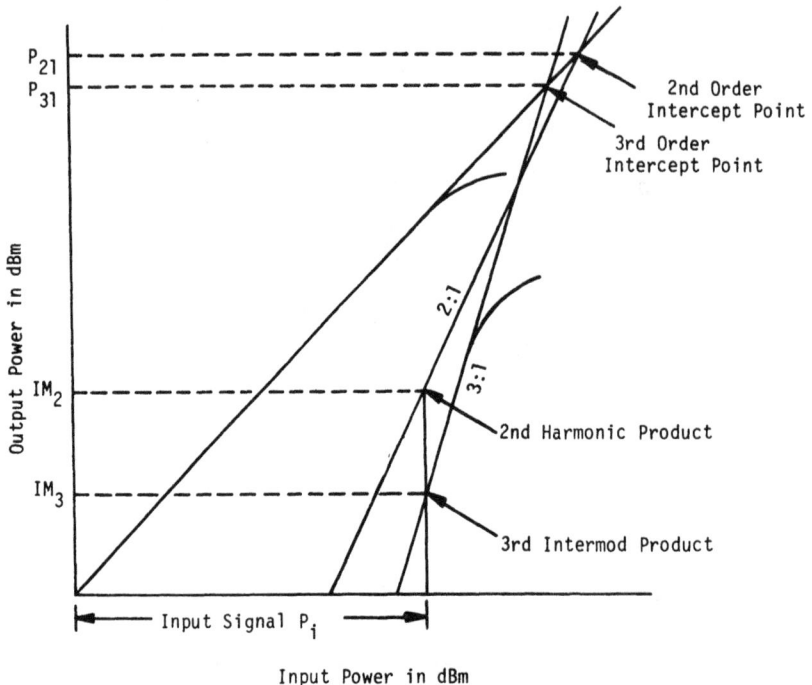

Figure 13.3. Second and Third Order Intercept Points of Amplifier

One of the straight lines is

$$\frac{y - P_{31}}{x - (P_{31} - G)} = 3 \qquad (13.15)$$

The other one is

$$x = P_i \qquad (13.16)$$

Combining Equations 13.15 and 13.16 obtains

$$y = IM_3 = 3(P_i + G) - 2P_{31} \text{ dBm} \qquad (13.17)$$

where P_i is the input power level of either signal since both signals have the same amplitude. G is the small signal gain of the amplifier, and P_{31} is the third order intercept point.

If the output power level P_0 is given, then $P_i = P_0 - G_s$ can be substituted in Equation 13.17 to find third order intermod IM_3 where G_s is the gain of the amplifier at that specific output power level.

Similarly, the second harmonic product (referenced to output) is

$$IM_2 = 2(P_i + G) - P_{21} \text{ dBm} \qquad (13.18)$$

where P_{21} is the second order intercept point.

For the more general cases where the two input signals under consideration are of unequal magnitude, third order levels may be computed through approximation from the following formulas (Reference 2):

$$IM_3 = 2P_1 + P_2 - 2P_{31} + 3G \text{ dBm} \quad \text{for } (2f_1 - f_2) \qquad (13.19)$$

and

$$IM_3 = 2P_2 + P_1 - 2P_{31} + 3G \text{ dBm} \quad \text{for } (2f_2 - f_1) \qquad (13.20)$$

where f_1 and f_2 are the signal frequencies which correspond to power level P_1 and P_2.

As an example, consider an amplifier with the following parameters:

$$G = 30 \text{ dB}, P_{31} = 20 \text{ dBm}, NF = 4 \text{ dB}, B = 100 \text{ MHz}$$

(a) If $P_1 = P_2 = -40 \text{ dBm} = P_i$

From Equation 13.17 the third intermodulation product is

$$IM_3 = 3(-40 + 30) - 2 \times 20 = -70 \text{ dBm}$$

(b) If $P_1 = -40$ dBm, $P_2 = -43$ dBm, then the two intermodulations products are from Equations 13.19 and 13.20

$$IM_3 = 2 \times (-40) + (-43) - 2 \times 20 + 3 \times 30 = -73 \text{ dBm}$$
for $(2f_1 - f_2)$

$$IM_3 = 2(-43) + (-40) - 2 \times 20 + 3 \times 30 = -76 \text{ dBm}$$
for $(2f_2 - f_1)$

13.5 DYNAMIC RANGE

From the above discussion, it is realized that the input RF power to an amplifier is limited to a certain region. If the signal is too weak, it will be buried in the noise. If it is too strong, spurious response will generate. The range of RF input power to the amplifier over which two signals can be unambiguously separated from noise, harmonics or intermodulation products is referred to as two signal spurious free dynamic range. The lower limit for the input power could be the equivalent noise level (NE) which can be expressed as

$$NE = NF - 114 + 10 \log B \text{ dBm} \qquad (13.21)$$

where the -114 is the noise floor for 1 MHz bandwidth system. It is derived from Equation 13.6

$$10 \log kTB = 10 \log 1.37 \times 10^{-23} \times 290 \times 10^6 = -144 \text{ dBw} = -114 \text{ dBm}$$

NF is the noise figure of the amplifier and B is the bandwidth of the amplifier in MHz.

The upper limit is chosen as the condition at which two equal signals produce third order intermodulation products equal to the NE. The condition for the NE which equals the third order intermodulation product is

$$IM_3 = NE + G = 3 P_i + 3G - 2 P_{31} \text{ dBm} \qquad (13.22)$$

or the input power level to have this condition is

$$P_I = P_i = \frac{NE}{3} - \frac{2G}{3} + \frac{2P_{31}}{3} \text{ dBm} \qquad (13.23)$$

The spurious free dynamic range is often defined from this input power level to the noise floor; therefore,

$$DR = P_I - NE = \frac{2}{3} (P_{31} - G - NE) \qquad (13.24)$$

where DR represents the dynamic range.

Use the same example in the last section with G = 30 dB, P_{31} = 20 dBm, NF = 4 dB, and B = 100 MHz. Then

$$NE = 4 + (-114) + 10 \log 100 = -90 \text{ dBm}$$

The input power that will generate a third order intermodulation products equal to the noise floor is from Equation 13.23.

$$P_I = \frac{-90}{3} - \frac{2 \times 30}{3} + \frac{2 \times 20}{3} = -36.7 \text{ dBm}$$

The corresponding dynamic range is

$$DR = P_I - NE = -36.9 - (-90) = 53.3 \text{ dB}$$

13.6 INPUT, OUTPUT IMPEDANCES AND VOLTAGE STANDING WAVE RATIO (VSWR)[9-13]

The input and output impedances of an RF amplifier are designed to be a pure resistive value and match the characteristic impedance of a given transmission line. The most common value of the input and output impedances is 50 ohms. However, it is very seldom that we have a microwave amplifier which has a 50 ohm impedance across the entire working frequency range. Instead of listing the ranges of the impedance value, the VSWR is often used to represent the impedance values.

The VSWR can be written as (Reference Equation 11.8)

$$S = \frac{|V^+| + |V^-|}{|V^+| - |V^-|} = \frac{V^+(1 + |\varrho|)}{V^+(1 - |\varrho|)} = \frac{1 + |\varrho|}{1 - |\varrho|} \tag{13.25}$$

where V^+ and V^- represent the incidental and reflected voltage respectively.

The reflection coefficient can be expressed in terms of impedance as

$$\varrho = \frac{Z - Z_0}{Z + Z_0} \tag{13.26}$$

where Z can be considered as the input impedance or the output impedance of the amplifier. Z_0 is the characteristic impedance of the line and is 50 ohms in most microwave circuits. Combining Equations 13.25 and 13.26, the VSWR can be written as

$$S = \frac{|Z + Z_0| + |Z - Z_0|}{|Z + Z_0| - |Z - Z_0|} \tag{13.27}$$

or Equation 13.27 can be written as

$$S = \frac{Z}{Z_0} \quad \text{for } Z \leqslant Z_0$$

$$\text{and} \quad S = \frac{Z_0}{Z} \quad \text{for } Z_0 \leqslant Z$$

(13.28)

If the VSWR at the input of an amplifier is 2, for a 50 ohm system the input impedance can be either 100 ohm or 25 ohm for a pure resistive case from Equation 13.28. If over the working frequency range and the input of the amplifier has a maximum VSWR of 2, then the input impedance can vary from 25 to 100 ohms. The VSWR of an amplifier will affect the gain flatness of a receiver over the operating frequency range.

13.7 PHASE LINEARITY AND GROUP DELAY[3, 11-13]

If a single frequency is applied to the input of any device, active or passive, its relative phase at the output port will be shifted by an amount related to the time it takes to propagate through the device.

The delay time may be expressed in terms of the phase shift as

$$t_f = \frac{\Delta\phi}{\omega}$$

(13.29)

where t_f is single frequency time delay (in seconds), $\Delta\phi$ is the phase shift between the input and output ports, and ω is the angular frequency.

In an ideal amplifier, the amount of phase shift between input and output would be directly proportional to frequency and the delay time t_f remains constant, regardless of the frequency of the input signal. A plot of the phase shift versus frequency of an ideal amplifier will be a straight line with a slope dependent on the delay time between the input and output. In practical situations, the measured curve will not be a perfectly straight line, but will exhibit perturbations of a few degrees to either side of the linear curve. Phase linearity of device is usually expressed as the maximum deviation from the linear curve within the band of interest or simply by comparing the measured phase angle to the theoretically predicted values at a number of discrete frequencies.

If two frequencies are applied to the input of an amplifier, each will propagate at a different velocity and the modulation envelope (caused by the beating of the two frequencies) will be propagated at a different velocity—the envelope or group velocity. The group velocity can be expressed as

$$v_g = \frac{d\omega}{d\phi}$$

(13.30)

The group delay of a device is directly related to the rate of change of phase shift versus frequency and can be expressed as

$$t_d = \frac{d\phi}{d\omega} \tag{13.31}$$

From Equations 13.29 and 13.31, it is seen that for an ideal amplifier with linear phase shift versus frequency the single frequency time delay t_f and the group delay t_d are equal.

For a system to accurately measure the phase of the input signal (i.e., in a phase interferometry system) the phase linearity is very important. The phase among all the channels must be properly matched.

13.8 CASCADE OF AMPLIFIERS[5-7]

Usually many amplifiers are required in a receiver to provide enough gain for the detector to function properly and at the same time to keep the noise figure at a minimum. In general, a low-noise amplifier also has relatively low output power. When amplifiers are connected in cascade, the low-power, low-noise amplifier is placed at the beginning of the chain. The amplifier with the highest output power will be at the last stage of the chain. This section will discuss the performance of cascade amplifier (i.e., noise figure, gain, bandwidth, and output power level).

Some of the discussions are based on the theoretical analysis and others are based on experimental observations. The gain of the cascade is the sum of the unit gains so long as none of the individual stages are driven into saturation. The total gain of the cascade chain G_T is

$$G_T = G_1 G_2 G_3\, G_n \ldots$$

or

$$G_T(dB) = G_1(dB) + G_2(dB) + \ldots + G_n(dB) \tag{13.32}$$

The exact gain flatness of a cascade is difficult to calculate since in the cascade amplifier chain none of the stages except for the first are driven by a 50 ohm source and the gain versus frequency may not follow the performance specified on the data sheet. The gain flatness is basically accumulative of each individual amplifier. However, some compensation effect should be taken into consideration. As an example (Reference 5), if every stage has a specified gain flatness of ± 1 dB, the band-pass flatness of 2, 3, or 4 stages in cascade is approximately ± 1.5 dB, ± 2.0 dB, and ± 2.5 dB, respectively. Of course the worst case is still ± 2 dB, ± 3 dB, and ± 4 dB, respectively.

The overall bandwidth of a cascade amplifier chain will be determined by the narrowest bandwidth amplifier used in the cascade. If all amplifiers have the same bandwidth, the overall bandwidth will stay approximately the same. However, there may be some roll-off of the edges of the band-width depending on the specific units cascaded and the number of stages.

13.9 NOISE FIGURE OF CASCADE AMPLIFIERS[8, 9, 14]

The noise figure of a cascade amplifier chain can be demonstrated by considering only two stages. Let G_1 and F_1 be the gain and noise figure of the first stage while G_2 and F_2 represent the same parameters of the second stage. The problem is to find the overall noise figure (F_T) of the two stages in cascade. From Equation 13.10 the noise at the output of stage 2 can be expressed as

$$N_0 = \text{noise from stage 1 at output of stage 2} \\ + \text{noise } \Delta N_2 \text{ introduced by stage 2} \tag{13.33}$$

The noise from amplifier 1 is $kTBF_1$ which is amplified by G_1 and G_2 while the noise generated by amplifier 2 is $kTBF_2$. However, the term $kTBF_2$ contains the noise kTB from generator 1; therefore, the noise generated from generator 2 alone is $F_2kTB - kTB$. Equation 13.33 can then be written as

$$N_0 = kTBF_1G_1G_2 + \Delta N_2 = kTBF_1G_1G_2 + (F_2 - 1)\,kTBG_2 \tag{13.34}$$

The noise figure can be obtained from Equatins 13.10 and 13.11 as

$$F_T = \frac{kTBF_1G_1G_2 + (F_2 - 1)\,kTBG_2}{kTB\,G_1G_2} \tag{13.35}$$

$$= F_1 + \frac{F_2 - 1}{G_1}$$

The contribution of the second stage to the overall noise figure may be negligible if the gain of the first stage is very high. This is a very important concept in receiver design. The noise figure of N stages of amplifiers in cascade can be shown as

$$F = F_1 + \frac{F_2 - 1}{G_1} + \frac{F_3 - 1}{G_1G_2} + \ldots + \frac{F_N - 1}{G_1G_2\ldots G_{N-1}} \tag{13.36}$$

In Equation 13.36 the gains G_1, G_2, ..., and noise figures F_1, F_2, ..., are in power ratio rather than in dB. Examples of using this equation can be found in Chapter 2.

13.10 DYNAMIC RANGE OF CASCADE AMPLIFIERS(14, 15)

The intermod product generated from a cascade amplifier chain can be found by letting the input voltage equal to the sum of V_1 and V_2 with angular frequencies ω_1 and ω_2, respectively.

$$V_i = V_1 \sin\omega_1 t + V_2 \sin(\omega_2 t + \theta) \tag{13.37}$$

The output will be obtained from Equation 13.12 as (for simplicity let $\theta = 0$)

$$V_{out} = a_1 V_i + a_2 V_i^2 + a_3 V_i^3 + \ldots$$

$$= a_1 [V_1 \sin\omega_1 t + V_2 \sin\omega_2 t]$$

$$+ a_2 [V_1^2 \sin^2\omega_1 t + V_2^2 \sin^2\omega_2 t$$

$$+ V_1 V_2 \sin(\omega_1 - \omega_2)t - V_1 V_2 \sin(\omega_1 + \omega_2)t]$$

$$+ a_3 \{V_1^3 \sin^3 \omega_1 t$$

$$+ 3V_1^2 V_2 [\tfrac{1}{2}\sin\omega_2 t + \tfrac{1}{4}\sin(2\omega_1 - \omega_2)t - \tfrac{1}{4}\sin(2\omega_1 + \omega_2)t]$$

$$+ 3V_1 V_2^2 [\tfrac{1}{2}\sin\omega_1 t + \tfrac{1}{4}\sin(2\omega_2 - \omega_1)t - \tfrac{1}{4}\sin(2\omega_1 + \omega_2)t]$$

$$+ V_2^3 \sin^3\omega_2 t\} \tag{13.38}$$

For $V_1 = V_2 = V$ and rearrangement of the terms, then

$$V_{out} = a_1 V [\sin\omega_1 t + \sin\omega_2 t]$$

$$+ a_2 V^2 [-\tfrac{1}{2}\cos2\omega_1 t - \tfrac{1}{2}\cos2\omega_2 t + \sin(\omega_1 - \omega_2)t$$

$$- \sin(\omega_1 + \omega_2)t + 1]$$

$$+ a_3 V^3 [\sin^3\omega_1 t + \sin^3\omega_2 t + \tfrac{3}{2}\sin\omega_1 t + \tfrac{3}{2}\sin\omega_2 t$$

$$- \tfrac{3}{4}\sin(2\omega_1 + \omega_2)t - \tfrac{3}{4}\sin(2\omega_2 + \omega_1)t$$

$$+ \tfrac{3}{4}\sin(2\omega_1 - \omega_2)t + \tfrac{3}{4}\sin(2\omega_2 - \omega_1)t] \tag{13.39}$$

The input power can be written as

$$P_{in} = \left(\frac{V}{2\sqrt{2}}\right)^2 \Big/ R_s = \frac{V^2}{8R_s} \tag{13.40}$$

where R_s is the input impedance. In the above equation, it is assumed that the voltage is divided equally between the input and the source impedance.
    The useful output is

$$P_{01} = \frac{a_1^2 V^2}{8R_0} = G\,P_{in} \tag{13.41}$$

where R_0 is the load impedance and G is the gain of the amplifier. If one assumes that $R_0 = R_s$, then from Equations 13.40 and 13.41 the gain can be written as

$$G = a_1^2 \tag{13.42}$$

The power output corresponding to frequency $2\omega_1$ or $2\omega_2$ is from Equation 13.39

$$P_{02} = \frac{a_2^2 V^4}{8\cdot 4R_s} = \frac{a_1^2}{64R_s^{\,2}} \frac{V^4}{a_1^2} \frac{2a_2^2 R_s}{a_1^2} = \frac{2a_2^2 R_s}{a_1^2} G\,P_{in}^{\;2} \tag{13.43}$$

The power output corresponding to frequency $(2\omega_1 - \omega_2)$ and $(2\omega_2 - \omega_1)$ is

$$P_{03} = \frac{a_3^2 V^6\, 9}{8\cdot 16\,R_s} = \frac{9a_3^2 V^6}{128\,R_s} = a_1^2 \frac{V^6}{512R_s^{\,3}} \frac{9\cdot 4R_s^{\,2}\,a_3^2}{a_1^2}$$

$$= \frac{36\,a_3^2 R_s^{\,2}}{a_1^2} G\,P_{in}^{\;3} \tag{13.44}$$

Let P_{21} be a certain value of P_{in} such that $P_{01} = P_{02}$. Thus, P_{21} is corresponding to the second order intermod intercept point. By equating Equations 13.41 and 13.43

$$G\,P_{21} = \frac{2a_2^2 R_s}{a_1^2} G\,P_{21}^{\;2} \tag{13.45}$$

$$\frac{1}{P_{21}} = \frac{2a_2^2 R_s}{a_1^2}$$ (13.46)

Let P_{31} be a certain value of P_{in} such that $P_{01} = P_{03}$. Thus, P_{31} is the third order intermod intercept point.

$$G P_{31} = \frac{36\,a_3^2 R_s^{\,2}}{a_1^2}\,G P_{31}^3$$ (13.47)

$$\frac{1}{P_{31}} = \frac{6a_3 R_s}{a_1}$$ (13.48)

For cascade amplifiers as shown in Figure 13.4 that

$$V_a = b_1 V_i + b_2 V_i^2 + b_3 V_i^3 + \dots$$ (13.49)

and

$$V_{out} = c_1 V_a + c_2 V_a^2 + c_3 V_a^3 + \dots$$ (13.50)

where b_i and C_i are constants, V_i is the input of amplifier X and V_a is the input of amplifier Y.

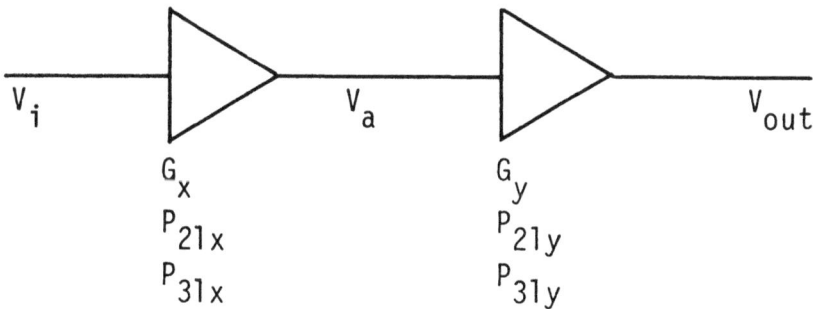

$$\begin{array}{llll} V_i & & V_a & V_{out} \\ & G_x & & G_y \\ & P_{21x} & & P_{21y} \\ & P_{31x} & & P_{31y} \end{array}$$

Figure 13.4. Cascade of Two Amplifiers

Substituting 13.49 into 13.50 then

$$V_{out} = c_1 (b_1 V_i + b_2 V_i^2 + b_3 V_i^3)$$
$$+ c_2 (b_1 V_i + b_2 V_i^2 + b_3 V_i^3)^2$$
$$+ c_3 (b_1 V_i + b_2 V_i^2 + b_3 V_i^3)^3$$

(13.51)

which can be rewritten as

$$V_{out} \simeq b_1 c_1 V_i + (b_2 c_1 + b_1^2 c_2) V_i^2 + (b_3 c_1 + b_1^3 c_1) V_i^3 + \ldots$$

(13.52)

where the term $2b_1 b_2 c_2 V_i^3$ belonging to the third term is neglected since it results from two nonlinear terms b_2 and c_2. Comparing Equation 13.52 with 13.38 it is seen that

$$a_1 = b_1 c_1$$

(13.53)

$$a_2 = b_2 c_1 + b_1^2 c_2$$

(13.54)

and

$$a_3 = b_3 c_1 + b_1^3 c_1$$

(13.55)

Then

$$a_2^2 = (b_2 c_1)^2 + (b_1^2 c_2)^2 + 2b_2 c_1 b_1^2 c_2$$

(13.56)

The overall second order intermod is given by Equations 13.46 and 13.53 as

$$\frac{1}{P_{21T}} = \frac{2a_2^2 R_s}{a_1^2}$$

$$= \frac{2b_2^2 c_1^2 R_s}{b_1^2 c_1^2} + \frac{2b_1^4 c_2^2 R_s}{b_1^2 c_1^2} + \frac{4b_2 c_1 b_1^2 c_2 R_s}{b_1^2 c_1^2}$$

$$= \frac{2b_2^2 R_s}{b_1^2} + \frac{2b_1^2 c_2^2 R_s}{c_1^2} + 2\sqrt{\frac{2b_2^2 R_s}{b_1^2} \times \frac{2c_2^2 R_s b_1^2}{c_1^2}}$$

(13.57)

With the help of Equation 13.46, the above equation can be written as

$$\frac{1}{P_{21T}} = \frac{1}{P_{21x}} + \frac{G_x}{P_{21y}} + 2\sqrt{\frac{G_x}{P_{21x}P_{21y}}} \qquad (13.58)$$

Since $G_2 = b_1^2$ (see Equation 13.42) it can be written as

$$\sqrt{\frac{1}{P_{21T}}} = \sqrt{\frac{1}{P_{21x}}} + \sqrt{\frac{G_x}{P_{21y}}} \qquad [(13.58)]$$

where P_{21x} and P_{21y} are the second order intermodulation intercept point of amplifiers x and y, respectively; and G_x is the gain of amplifier X as shown in Figure 13.5.

The third order intermodulation intercept point is from Equations 13.48, 13.53, and 13.55.

$$\frac{1}{P_{31T}} = \frac{6a_3R_s}{a_1} = \frac{6b_3c_1R_s}{b_1c_1} + \frac{6b_1^3c_1R_s}{b_1c_1} = \frac{1}{P_{31x}} + \frac{G_x}{P_{31y}} \qquad (13.59)$$

where P_{31x} and P_{31y} are the third order intermod intercept point of amplifiers x and y, respectively. Equations 13.58 and 13.59 can be generalized as

$$\sqrt{\frac{1}{P_{21T}}} = \sqrt{\frac{1}{P_{21x}}} + \sqrt{\frac{G_x}{P_{21y}}} + \sqrt{\frac{G_xG_y}{P_{21z}}} + \ldots \qquad (13.60)$$

and

$$\frac{1}{P_{31T}} = \frac{1}{P_{31x}} + \frac{G_x}{P_{31y}} + \frac{G_xG_y}{P_{31z}} + \ldots \qquad (13.61)$$

Equations 13.60 and 13.61 are used to calculate the intermod intercept points for cascade chain of amplifiers. In the above equations, the basic assumption is that there is in phase coherency between signals 1 and 2. Under this assumption, the second order and third order intermods are for conservative design. In other words, the results obtained will be the worst case.

13.11 POWER OUTPUT, STABILITY, AND GROUNDING OF CASCADE AMPLIFIERS[3]

In a normal cascade, where amplifier modules are arranged in ascending order according to output power, the minimum cascade output power will be very close to the specific value of the output stage. Sometimes when an amplifier is to be driven to its full rated output power, it is recommended that excess drive be available from the preceding driver stage (Reference 3). This will assure that full output power will be available with no danger of saturating the driver stage or adversely affecting the dynamic range of the cascade.

Stability and grounding are also very important in practical design. In general, RF amplifiers are designed to be unconditionally stable with load VSWRs of any phase angle. However, if they are used in a cascade, oscillation may break out and cause instability. Instability may fall typically into two types: (1) subband oscillations which are below the amplifier frequency band; (2) out-of-band oscillations which are above the frequency band. Occasionally in-band oscillations can occur which may be caused by radiative feedback of a high gain amplifier chain where the RF output voltage is in phase with the input voltage. These oscillations can deprive the power from the amplifiers, and there will be no amplification at the desired frequency. Most of the instability can be stopped through proper grounding of the amplifiers. The necessity for good grounding can never be overemphasized in cascade of amplifiers. Decoupling of the dc power supply through capacitors will also help to stop oscillations.

13.12 TYPES OF AMPLIFIERS[16, 17]

There are a great deal of articles on the subject of amplifiers. Amplifiers can be classified according to different classification methods. As discussed in the introduction, amplifiers can be classified as RF and IF amplifiers according to where they are used in a receiver and their operating frequency. According to output power levels, they can be divided into small signal and power amplifiers but there is no definite power level to separate them. Sometimes they are classified as class A, B, and C depending on their operation points on the characteristic curve of the device. According to how the amplifiers are designed and fabricated, they can be divided as single stage (one transistor), multistage (transistors in cascade), parallel stage (using a power divider to separate the input into many branches amplified then power combined to obtain high output power), and push-pull arrangement (two transistors are 180 degrees out-of-phase to improve efficiency and output power). Classified by what devices are used in the amplifiers, they can be divided into bipolar transistor, field effective transistor (FET), tunnel diode, parametric amplifiers, Avanlanch transit-time diode, and traveling wave tube (TWT) amplifier. The most commonly used amplifiers in receiver applications are the bipolar and FET transistors. The amplifiers of different kinds of devices will be discussed briefly in the following paragraphs.

13.13 MICROWAVE FIELD EFFECTIVE TRANSISTOR (FET) AMPLIFIERS(17-24)

There are many kinds of FETs such as a junction field effective transistor (JFET) (References 17-19), metal oxide semiconductor FET (MOSFET) which is used mostly in logic circuits, and metal Schottky-barrier FET (MESFET) (References 21-24). Since the FETs are very important microwave devices, their physical operational principle will be discussed briefly. In the following sections, similar discussions will be given to bipolar junction transistors and tunnel diodes. All FETs have the same basic operating principle that the conductivity of a layer of semiconductor is modulated by a transverse electric field. In an FET, the current flow is carried by only one type of carrier (usually the n-type); thus, it is also referred to as an unipolar transistor. The most popular FET for microwave applications is the GaAs MESFET. The basic structure of a GaAs MESFET is shown in Figure 13.5a. An n-type epitaxial GaAs layer is grown on a semi-insulating

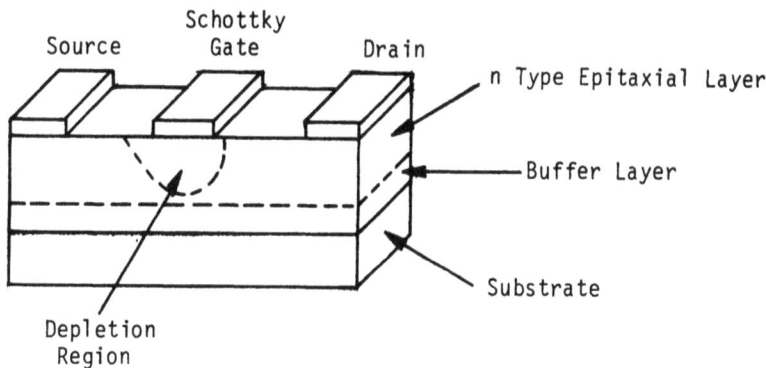

(a) Basic Structure of MESFET

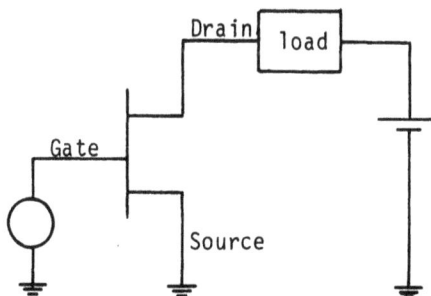

(b) FET in a Circuit

Figure 13.5. Metal Schottky Field Effective Transistor

GaAs substrate. The source, gate, and drain are evaporated on the n-type GaAs. The source and gate are reverse-biased while the source and drain are forward-biased. The majority of carriers (electrons) flow from the source to the drain. A depletion region is created under the gate by the voltage on the gate. When the reverse bias on the source and gate increases, the height of the depletion layer also increases, and the channel depth below the depletion layer decreases which will increase the resistance between the source and drain. Therefore, the gate source voltage Vgs controls the current flow from the drain to the source which goes through the load resistor R (Figure 13.5b). The amplification is thus accomplished.

Compared with other microwave amplifier devices, the FET transistor is a two-port device with relative high-gain bandwidth product, low-noise figure, high input impedance, and the input and output impedances are less sensitive to temperature variation. At first, the MESFETs were used to build amplifiers in the 2-4 GHz range, then the working frequency of the amplifier kept moving up to 20 GHz and down to 500 MHz. The MESFET will operate at 40 GHz with 3.3 dB noise figure with an approximate gain of 9 dB (Reference 24). It was predicted that GaAs FET will work up to 50 GHz. The power level of the GaAs FET also advanced very promptly through recent development. A single chip FET at 10 GHz yields an output power of 4 watts with 6 dB gain and 40 percent power-added efficiency (η) which is defined as

$$\eta = \frac{\text{RF power outport} - \text{RF power input}}{\text{dc power input}}$$

It is anticipated that most of the small signal microwave amplifiers for receiver applications will be GaAs MESFET, especially for the operation frequency over a few gigahertz.

13.14 MICROWAVE BIPOLAR JUNCTION TRANSISTOR (BJT) AMPLIFIERS (25-33)

Since the invention of junction resistor (References 24 and 25), there was a revolutionary impact on electronics. Almost all of the small signal amplifiers are now made of transistors including both FETs and BJTs. In a transistor, both the majority and minority carriers are involved; therefore, it is usually referred to as a bipolar junction device. A BJT can be considered as two diodes connected back-to-back. It can be either a n-p-n or a p-n-p type. A n-p-n BJT is shown in Figure 13.6a. One of the n materials is called the emitter, the other n material is called the collector, and the p material is called the base. The emitter-base junction is forward biased while the base-collector junction is reversed biased. In a n-p-n BJT, the motion of the electrons is of primary interest. The potential distribution of a n-p-n BJT is shown in Figure 13.6b. For electrons, the base has the highest potential level and the collector has the lowest. Electron injected from the emitter into the base will drift toward the collector. Since the base-collector junction is reversed biased, the electrons close to this junction will be pulled into the collector. It takes a small signal to inject the electrons from the emitter to the base. Once the electrons reach the base, most of them will arrive at the collector. If a load is placed in the base collector circuit, a stronger signal will be generated by the electrons. Thus, the input signal is amplified. The operation principle of a microwave transistor is the same as a low frequency transistor. However, the physical structure configuration is designed to accommodate the microwave operation (References 26 and 27). Almost all of the microwave BJTs are n-p-n type silicon transistors. Although the GaAs MESFETs are replacing the BJTs as microwave amplifiers, there is ongoing work to improve the power rating of BJT below 10 GHz (Reference 28). At 2 GHz, a continuous wave (CW) transistor amplifier with 11 dB gain and 60 watt output power was accomplished.

The characteristics of both the MESFET and BJT in a microwave region are most commonly presented in the S parameters. The S parameter is usually referred to as the scattering matrix in a microwave field which is used to express the property of a microwave component or device. This discussion of scattering matrix can be found in Chapter 10.

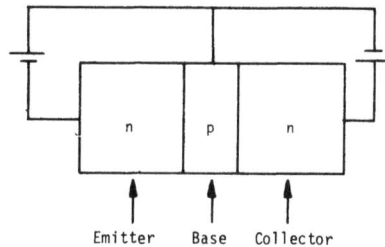

(a) A n-p-n Bipolar Junction Transistor

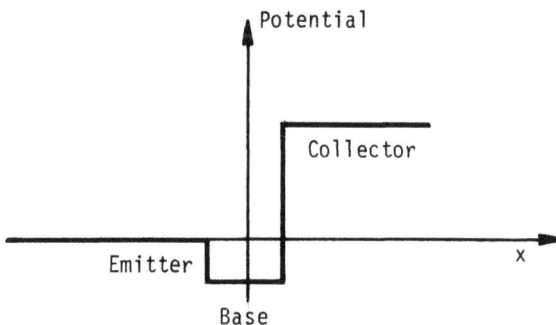

(b) Potential Distribution
Figure 13.6. Bipolar Junction Transistor

The S parameters of any two-port device (active or passive) can be measured by a network analyzer. In general, the S parameters are complex numbers. In a transistor, S_{11} and S_{22} represent the input and output impedances, while S_{21} represents the gain of the transistors. S_{12} is the isolation of the transistor. A typical S parameter is shown as

Frequency	S_{11}	S_{21}	S_{12}	S_{22}
1000 MHz	$.775 \underline{/167.0°}$,	$3.214 \underline{/52.5°}$,	$.039 \underline{/15.1°}$,	$.456 \underline{/-120.0°}$

The values of S are written in magnitudes and angles at specific frequency. The test conditions (i.e., common emitter (voltage between collector and emitter) V_{CE} and (collector current) I_C values) are also listed.

The amplifier design procedures are readily available if the S parameters are given (References 29-31). Many computer programs are available to design transistor amplifiers. The general design approach will be discussed here. A transducer power gain is defined as

$$G_T = \frac{\text{power delivered to load}}{\text{power available from source}} \qquad (13.62)$$

which can be expressed as

$$G_T = \frac{|S_{21}|^2 (1 - |\varrho_s|^2)(1 - |\varrho_L|^2)}{|(1 - S_{11}\varrho_s)(1 - S_{22}\varrho_L) - S_{12}S_{21}\varrho_L\varrho_s|^2} \qquad (13.63)$$

where ϱ_s and ϱ_L are the source and load reflection coefficient, respectively (see Equation 13.26).

$$\varrho_s = \frac{Z_s - Z_0}{Z_s + Z_0} \qquad (13.64)$$

$$\varrho_L = \frac{Z_L - Z_0}{Z_L + Z_0} \qquad (13.65)$$

where Z_s and Z_L are the source and load impedance, while Z_0 is the characteristic impedance.

In general, the amplitude of S_{12} is very small in a transistor and Equation 13.63 can be reduced to

$$G_T = G_U = \frac{|S_{21}|^2 (1 - |\varrho_s|^2)(1 - |\varrho_L|^2)}{|1 - S_{11}\varrho_s|^2 |1 - S_{22}\varrho_L|^2} \qquad (13.66)$$

G_U is the unilateral transduce power gain which is defined as G_T with S_{12} = 0. If $|S_{11}| < 1$ and $|S_{22}| < 1$ the maximum unilateral transduce power gain is

$$G_{U\,max} = \frac{|S_{21}|^2}{(1 - |S_{11}|^2)(1 - |S_{22}|^2)} \qquad (13.67)$$

This maximum unilateral transduce power gain is obtained when $\varrho_s = S_{11}^*$ and $\varrho_l = S_{22}^*$, where S_{11}^* and S_{22}^* are the complex conjugate of S_{11} and S_{22}. For narrow band transistor amplifier design, the primary task is to design the network to match Z_0 (usually 50Ω) to S_{11} at the input and Z_0 to S_{22} at the output of the transistor.

13.15 TUNNEL DIODE AMPLIFIERS(34-37)

A tunnel diode is a highly doped p-n junction diode which has a negative resistance at microwave frequency. The tunnel diode was first reported by Esaki in 1958 (Reference 34). The operating principle of a tunnel diode can be explained in Figure 13.7. In a highly doped ($10^9 \sim 10^{10}$ cm^{-3}) p type material, the Fermi level is in the valence band. In a highly doped n material, the Fermi level is in the conduction band. The Fermi level is an energy level with a 50 percent probability of being occupied by electrons. When the diode is not biased, the Fermi levels in the p and n material are lined up as shown in Figure 13.7b. When the forward bias voltage increased, the Fermi level in the n material will raise up with respect to that of the p material.

Elections from the conduction band of n material will tunnel through the barrier to the valence band of the p material as shown in Figure 13.7c. Further increasing the bias voltage will raise the Fermi level in the n material again. The election in the conduction band of the n material will face the forbidden band of the p material and tunneling effect will be stopped as shown in Figure 13.7d. The corresponding I-V curve of a tunnel diode is shown in Figure 13.8. When the applied voltage increases, the current through the diode starts to increase and reaches a peak value I_p at voltage V_p which is corresponding to Figure 13.7c. When the applied voltage is increased beyond V_p, the current starts to decrease. After further increasing the applied voltage, the current starts to increase again which resembles the conventional I-V curve of a p-n diode at forward biased condition. The I-V curve in Figure 13.8 shows that the tunnel diode has a negative differential resistance if it is properly biased. The negative resistance can be used to build microwave amplifiers and oscillators.

An equivalent circuit of a tunnel diode is shown in Figure 13.9. The lead wire resistance and inductance are represented by R_s and L_s. The junction capacitance of the diode is labeled C and the negative resistance by $-R$. Since the tunnel diode is a one-port device, a circulator will be used at the input to provide the input/output isolation for the amplifier (Figure 13.9b).

(a) Tunnel Diode is Forward Biased

(b) Zero Biased

(c) Biased at Maximum Current Condition

(d) Biased Beyond Maximum Current Condition

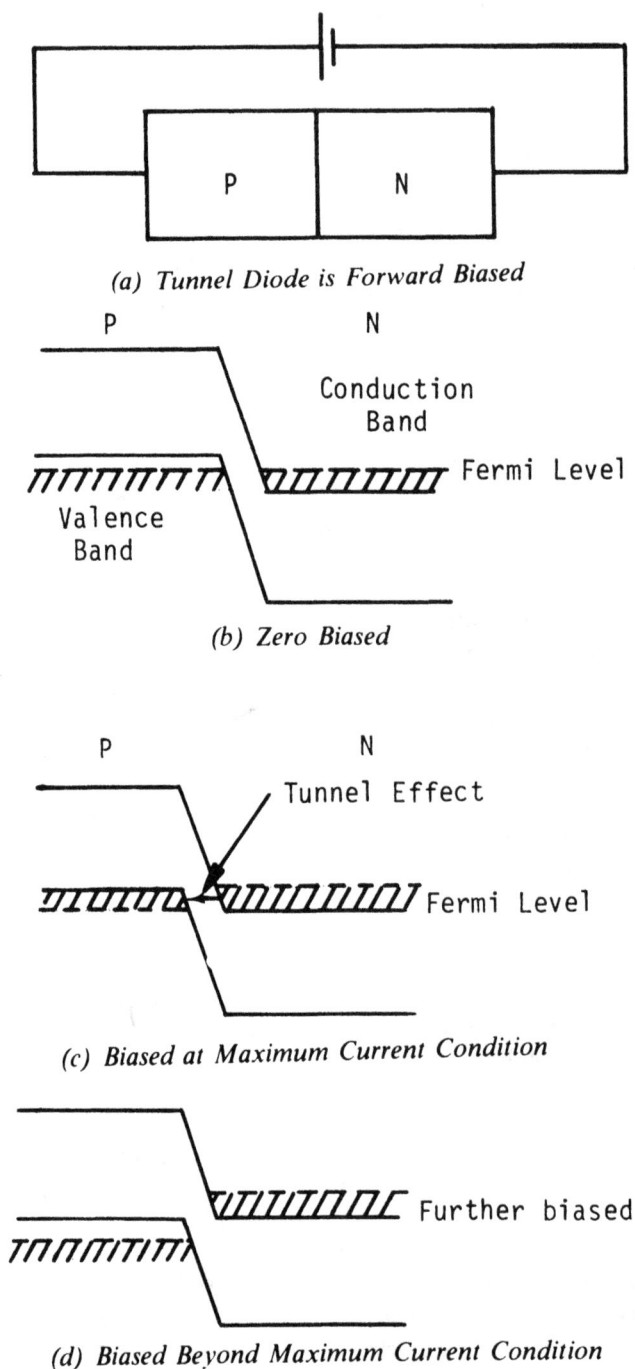

Figure 13.7. Tunnel Diode

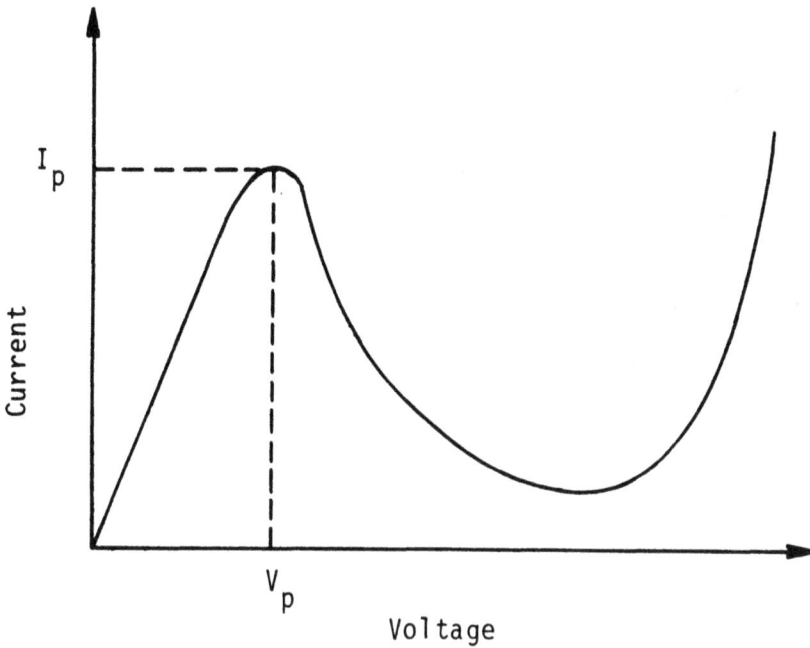

Figure 13.8. I-V Curve of a Forward Biased Tunnel Diode

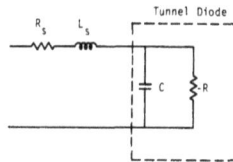

(a) Equivalent Circuit of a Tunnel Diode

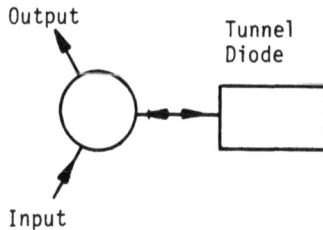

(b) Input/Output Isolation for the Amplifier
Figure 13.9. Tunnel Diode Amplifier

Since the tunnel diode has a negative resistance, it must be properly designed to have good stability. Its most popular use in the microwave field is its application in building oscillators.

13.16 PARAMETERIC AMPLIFIERS(38-44)

A parametric amplifier is a device which uses a nonlinear reactance (capacitance or inductance) to accomplish the amplification. Usually the input frequency is at one frequency, the amplified output signal is at a different frequency and the output frequency is higher than the input frequency. Although the signal frequency changes, the information in the signal remains the same. The principle of operation can be explained in Figure 13.10. In Figure 13.10a there is an ideal lossless tank circuit. The energy stored in this circuit is $\frac{1}{2}CV^2$ [$\frac{1}{2}(Q/C)^2$] or $\frac{1}{2}LI^2$ which oscillates

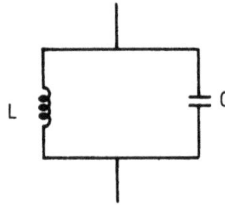

(a) A Lossless Tank Circuit

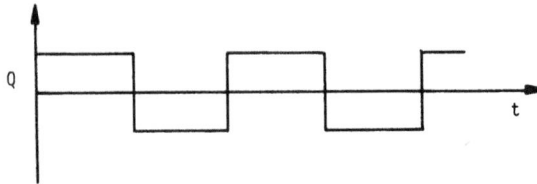

(b) Charge Q on the Capacitor Changes Periodically

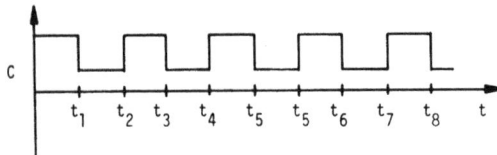

(c) Capacitance of the Capacitor Changes Periodically

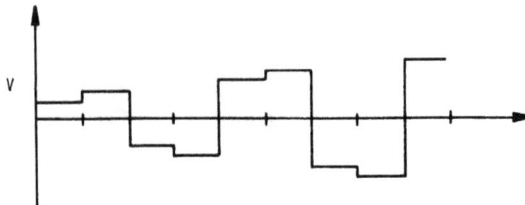

(d) Voltage Increase on the Capacitor

Figure 13.10. Parametric Amplifier

between the capacitor and the inductor, where V, Q, and I are the voltage across the capacitor, charge on the capacitor C, and current in the inductor L, respectively. If there is no outside circuit connected to this circuit, the energy should stay the same; however, one can increase the energy by changing the capacitance in a periodic manner as shown in Figure 13.10b and c. In Figure 13.10b the charge on the capacitor changes periodically. In order to simplify the discussion, the charge is assumed to be changed abruptly. The capacitance of the capacitor also changes abruptly at twice the input frequency. The change of capacitance is accomplished by mechanically changing the distance between the two plates of the capacitor. The capacitance is decreased when the charge has a value of Q and under this condition work has to be input against the Coulcomb attraction force at t_1, t_3, t_5,.... The capacitance is increased at t_0, t_2, t_4,...; under these conditions no work is required because the charge on the capacitor is zero. The work contributed through the mechanical motion of the capacitor results in a total increase in energy of the tank circuit. Figure 13.10d shows the voltage increase on the capacitor. Through this periodical change of capacitance, the input signal is amplified. A similar argument can be held for the periodical change of the inductance of the inductor to amplify the input signal.

In a practical parametric amplifier, the capacitor can be a p-n junction diode. Its junction capacitance can be controlled by an applied voltage. Thus, in a parametric amplifier, there are two input signals; one is the input signal to be amplified, the other one is a pumping signal which will change the capacitance of the junction diode. Since the parametric amplification is accomplished by nonlinear reactance, theoretically there is no resistance in this circuit and the noise of this amplifier is low.

A general power relationship among different frequencies in a parametric amplifier was derived by Manley and Rowe (the Manley-Rowe relation). The equivalent circuit used to derive the Manley-Rowe relation is shown in Figure 13.11. The relation is shown as

$$\sum_{m=0}^{\infty} \frac{nP_{mn}}{m\omega_p + n\omega_s} = 0 \qquad (13.68)$$

where ω_p and ω_s are the pumping and signal angular frequencies, respectively, and m and n are integers. P_{mn} is the power delivered or absorbed at $\omega = m\omega_p + n\omega_s$. When P_{mn} is positive, the power is flowing into the capacitor. A negative P_{mn} represents the power flowing out of the capacitor. When n = 1, there are only two values allowed for m: 0 and 1. Equation 13.68 will be reduced to

$$\frac{P_{01}}{\omega_s} + \frac{P_{11}}{\omega_s + \omega_p} = 0 \qquad (13.69)$$

$-P_{11}$ represents the power flowing into the output load. Then

$$\frac{-P_{11}}{P_{01}} = \frac{P_0}{P_s} = \frac{\omega_s + \omega_p}{\omega_s} \equiv \frac{\omega_0}{\omega_s} \qquad (13.70)$$

representing the gain of the amplifier, where ω_0 is the output signal frequency, P_0 is the output power, and P_s is the input power. Under this condition, the parametric amplifier has a positive gain. When the output signal frequency is higher than the input signal frequency, it is referred to as a parametric up-converter.

If the pumping frequency $\omega_p = 2\omega_s$ and the output frequency equals the input frequency, this parametric amplifier is referred to as a degenerate parametric amplifier. Under this condition, the amplifier has a negative input resistance and oscillation may occur.

The parametric amplifier is relatively complicated to build, because it requires a high frequency pumping source.

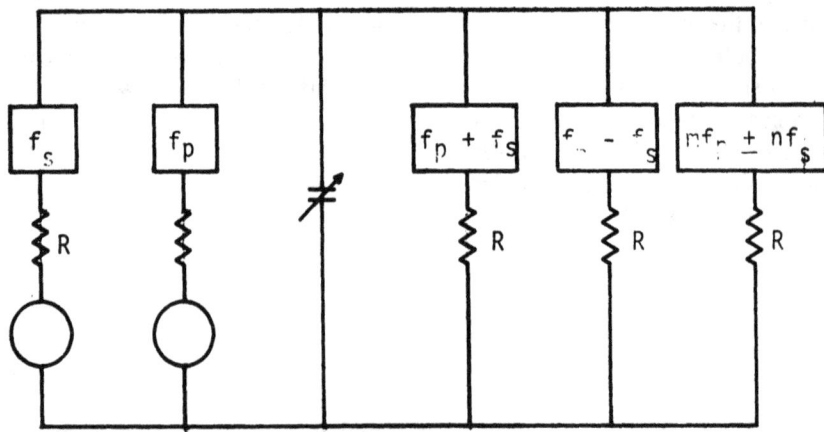

Figure 13.11. Equivalent Circuit for Manley-Rowe Relation

13.17 TRANSFER ELECTRON AMPLIFIERS [45-47]

A transferred electron device which was first discovered by J.B. Gunn in 1963 is often referred to as the Gunn device. When a dc voltage is applied to a bar of n-type GaAs or InP, the current first increases linearly with voltage. When the average electric field increases beyond a threshold field of several kilovolts per centimeter, oscillation will start. The time of this oscillation is approximately equal to the transient time of the carriers from the cathode to the anode. Subsequent research showed that these oscillations are a manifestation of the transferred electron effect. The effect occurs in GaAs and several other III-V (III and V represent chemical elements in the third and fifth column of the period table) compounds and is a consequence of the conduction-band structure of these semiconductors. This phenomena is explained next.

The simplified band structure of GaAs is shown in Figure 13.12. The conduction band is a multivalley structure and is represented here by two valleys. The energy difference between the two valleys is approximately 0.36 ev. At room temperature the electrons are at the lower valley. The electron mobility in the lower valley is approximately 7500 cm^2/μs, while in the upper valley it is about 180cm^2/μs. When the voltage across a bar of GaAs is equal to zero, most of the electrons are at the lower valley. As the voltage is increased from zero, the electrons are accelerated by the electric field and the average electron velocity increases. Therefore, the current across the GaAs will increase with the applied voltage. When the energy gained by the electron from the applied electric field is comparable to the energy gap between the two valleys, the rate of electron transfer from the lower valley to the upper valley will increase. Further increasing the applied electric will result in more electrons in the upper valley. Since the upper valley has much lower electron mobility, the current through the GaAs bar will decrease while increasing the applied voltage. Therefore, a negative differential resistance will appear. This negative resistance happens in the bulk of the material, while the negative resistance in the tunnel diode happens in the p-n junction.

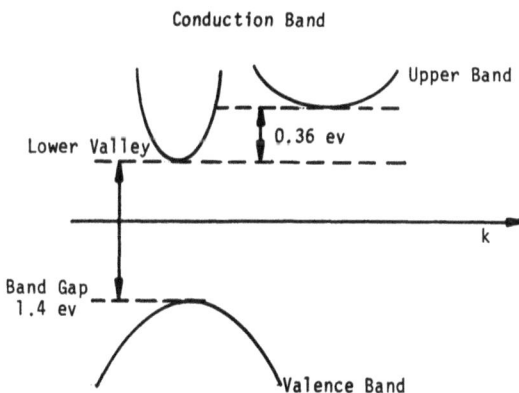

Figure 13.12. Simplified Diagram of the Band Structure of GaAs

The negative resistance of the Gunn device makes it suitable for microwave amplifiers and oscillators. A simple negative resistance transfer electron amplifier is a simple one-port device. A circulator is often used at the input to isolate the input and output of the amplifier such as in a tunnel diode amplifier. The Gunn amplifier is not very popular because it is a one-port device, but the Gunn oscillator is very popular especially at higher frequency (i.e., in millimeter (mm) wave region).

13.18 OTHER POTENTIAL MICROWAVE AMPLIFIERS[48-53]

Many solid-state microwave devices have been available since the 1960s. Most of these devices exhibit negative resistance at microwave frequency range. Theoretically, they can be used to make microwave amplifiers. Some of these devices will be discussed very briefly. In practical applications most of them are used for oscillators rather than amplifiers.

A. LSA Diode

This is a Gunn diode in a limited space-charge accumulation (LSA) mode (Reference 48). It has a negative resistance and is most commonly used for oscillators.

B. IMPATT Diode[49-51]

IMPATT stands for impact avalanche transit-time effect of a Read diode. The diode can have a negative resistance and is commonly used as an oscillator at the mm wave region.

C. TRAPATT Diode[52]

TRAPATT stands for trapped plasma avalanche triggered transit mode. This diode has a negative resistance and is used for oscillators with very high (75 percent) efficiency but also high noise.

D. BARITT Diodes[53]

BARITT stands for barrier injected transit-time. This diode is also used for oscillators, and it has relative low noise.

All of the diode amplifiers including tunnel diode and parametric amplifiers are being replaced by either the MESFET or the junction transistors. At high frequency where the FET technology is under development, diode amplifiers are used at the present time. Special receivers may still use some devices for amplifiers other than FETs.

13.19 LOGARITHMIC (LOG) AMPLIFIERS [54, 55]

A log amplifier is a microwave amplifier. The input to the amplifier is at microwave frequency or IF while the output is a video voltage. The output voltage is proportional to the logarithm of the input power as

$$V = k \log P_i \qquad (13.71)$$

where k is constant and P_i is the input power.

The output voltage versus the input RF power of an ideal log amplifier is shown in Figure 13.13. The scales are arbitrarily chosen. The log amplifier is used in a receiver to measure the amplitude of the input signal. Since a receiver usually covers a wide dynamic range (perhaps 60 dB) and a single detector does not cover such a dynamic range, one way to measure the input signal strength accurately is to use a log amplifier.

If the amplitude ratio of two inputs is desired, the log amplifiers are very useful. One of the applications to obtain the amplitude ratio of two signals is to measure the angle of arrival (AOA) of the signal through amplitude comparison (see Chapter 2). The simplest way to measure the ratio is to take the logarithm of the input signals and put them through a differential amplifier. Since log A − log B = log A/B, the ratio of the signals can be obtained.

The first factor concerned in a log amplifier is accuracy of the logarithmic response. Although the logarithmic response is obtained from approximation, the accuracy is usually rather good. The lower limit of the log amplifier output is at noise level, and the upper limit of the log amplifier is caused by saturation.

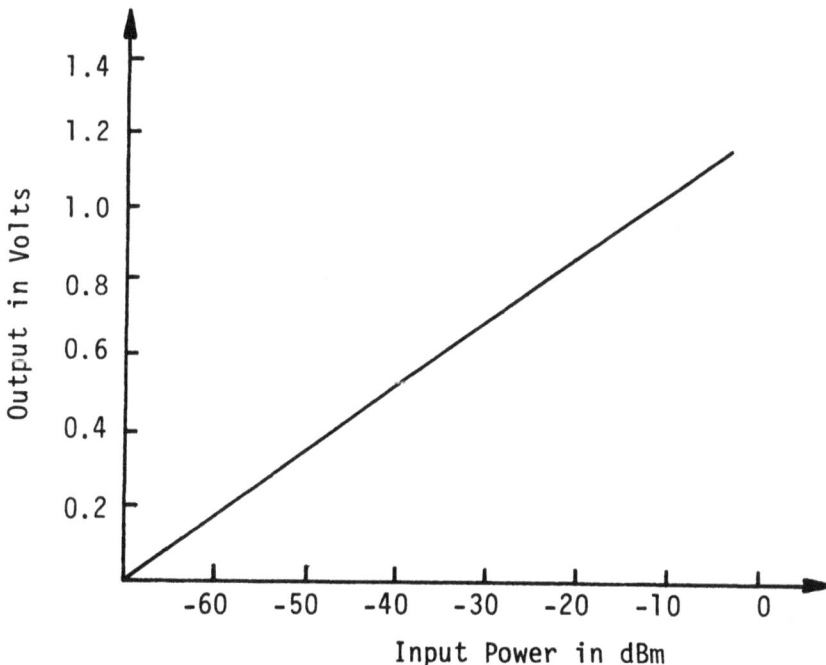

Figure 13.13. Input and Output Relations of an Ideal Log Amplifier

The frequency response of the log amplifier is determined by the RF amplifiers and detectors used to build the log amplifier. An ideal log amplifier response should be insensitive to frequency variation.

The response times of the log amplifier are usually of great concern. The rise time of the log amplifier depends on the RF bandwidth of the amplifiers and the video bandwidth of the detection circuits. The response times of the detectors or transistors used in the log amplifier are also very critical. The rise time of a few nanoseconds can be accomplished in some log amplifiers. While the rise time can be made very fast, the falling time of a log amplifier is relatively long. The falling time depends on the input pulse amplitude and width. The long falling time can prevent the log amplifier from detecting a weak signal following a strong one as shown in Figure 13.14. The falling time can be reduced through decreasing the capacitance in the circuit, because the fall time is determined by the charge draining out of the capacitors (mainly parasitic capacitance) in the detection circuits.

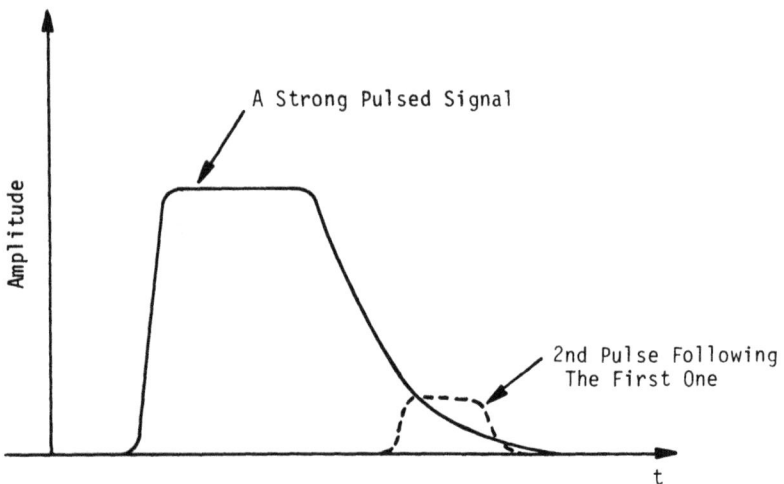

Figure 13.14. Typical Output From a Log Amplifier

The log amplifier can be either ac or dc coupled. The ac coupled amplifiers can only measure the amplitude of pulsed signals, while the dc coupled amplifiers can measure both pulsed signals as well as cw signals. The disadvantage of a dc coupled amplifier is that sometimes the baseline will drift. Occasional readjustment of the output level may be required.

When two RF signals arrive at a log amplifier simultaneously, the output amplitude will represent the composite one. Due to the logarithmic operation, the maximum error that can be induced will be 3 dB when the two signals are of the same amplitudes. In other words, if the two signals have different amplitudes, the log amplifier will measure the stronger one; and the error caused by the weak signal is always less than 3 dB.

13.20. OPERATING PRINCIPLE OF LOG AMPLIFIERS

The logarithmic relationship required for a log amplifier comes naturally from a diode, because the current that flows through an ideal diode is given by (see also Chapter 14)

$$I = I_S (\exp \frac{qV}{kT} - 1) \tag{13.72}$$

where I is the current that flows through the diode, I_S is the theoretical reverse saturation current, q is the electron charge $= 1.602 \times 10^{-19}$ C, V is the voltage applied across the diode, k is Boltzmann's constant $= 1.380 \times 10^{-23}$ joule/degree, and T is the absolute temperature.

At room temperature and forward bias conditions, the "1" in the parentheses can be neglected. Equation 13.72 can then be written as

$$V = \frac{kT}{q} \ln \frac{I}{I_S} = \frac{2.3 \, kT}{q} \log \frac{I}{I_S} \tag{13.73}$$

Therefore, the voltage across the diode is proportional to the logarithmic of the current flow through it.

If at V_1 the current is I_1 and V_2 at current I_2, then the change of voltage

$$\Delta V = V_2 - V_1 = \frac{2.3 \, kT}{q} \log \frac{I_2}{I_1} \tag{13.74}$$

For the current changes by a ratio of 10, at room temperature (T = 300° K) ΔV changes by approximately 60 mV.

A basic log circuit can be built as shown in Figure 13.15 where amplifier A is an operational amplifier. An operational amplifier has a very high input impedance and high gain. The input voltage of the amplifier is near zero. Therefore, the input current I_i must equal the feedback current I_f.

$$I_i = -I_f \tag{13.75}$$

But

$$I_i = \frac{V_i}{R_i} \tag{13.76}$$

$$I_f = I_S \left(\exp \frac{qV}{kT} - 1\right) \cong I_S \exp \frac{qV}{kT} \qquad (13.77)$$

where V_i is the input voltage, R_i is the input resistance, and V is the voltage at the output of the operational amplitude.

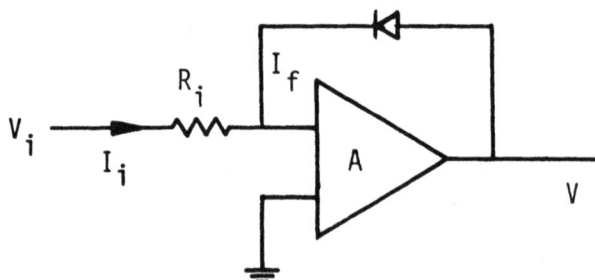

Figure 13.15. A Simple Diode Log Circuit

Substituting Equations 13.76 and 13.77 into 13.75 obtains

$$V = \frac{2.3\,kT}{q} \log\left(\frac{V_i}{R_i I_S}\right) \qquad (13.78)$$

Therefore, the output of the amplifier is proportional to the logarithm of the input voltage. The direction of the diode determines the operating polarity of the circuit. The above discussion shows that the output voltage of a diode can be either proportional to the logarithm of the input current or input voltage.

Since the input signal to a log amplifier is at microwave frequency, a detector must be used in a log amplifier as shown in Figure 13.16. Amplifier A is a linear microwave amplifier with a wide dynamic range. The detector following A can operate in either a linear or square law region. If the detector is operating in the linear region, the diode output is proportional to the voltage of the input signal. If the detector is operating in the square law region, the diode output is proportional to the power of the input signal. In either case, the output of the operational amplifier can be considered proportional to the logarithm of the input power. The difference caused by the detectors is in the slope of the output to input power

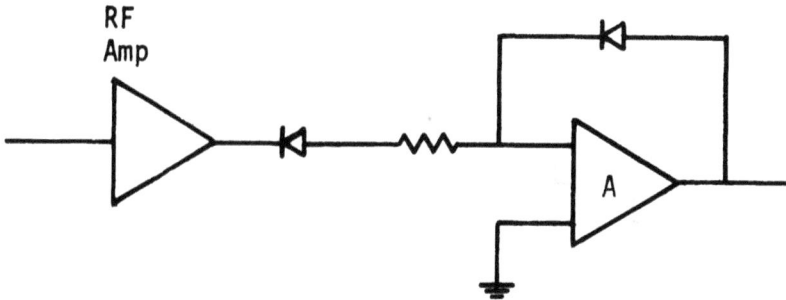

Figure 13.16. Simple Log Amplifier

relation. The output of the log amplifier can be written as

$$V = k \log P_i \qquad (13.79)$$

Different diodes will change the value of the constant k.

A shortcoming of this approach is that the microwave detector has only limited dynamic range. In this simple arrangement, the logarithmic response is obtained; however, the dynamic range of the log amplifier is not beyond that of a single detector diode.

To improve the dynamic range of the log amplifier, multiple stages of amplifiers and detectors are used as shown in Figure 13.17. The input signal is amplified in cascade and at each output a detector is provided. The outputs from the detectors are amplified and then summed up. The RF amplifier A_{n-1} will be saturated first then A_{n-2}, A_{n-3}, and so on. Detector D_1 will measure the input signal directly, D_2 will measure the output from A_1, and D_n will measure the output from A_{n-1}. The output of the summing circuit is approximately equal to the logarithm of the input power. The dynamic range of the log amplifier is improved, because each detector will only operate in a relatively small range. There are some delays in each of the RF amplifiers. In order to compensate for these delays, delay lines T_1, T_2, etc. are required.

In addition to the log video output, it is possible to have the RF output from the log amplifier in this arrangement. The RF signal is usually at a saturated level. It should be noted that when simultaneous signals are present, there will be intermodulation products at the RF output. The utilization of this RF output must be carefully handled as discussed in the following sections.

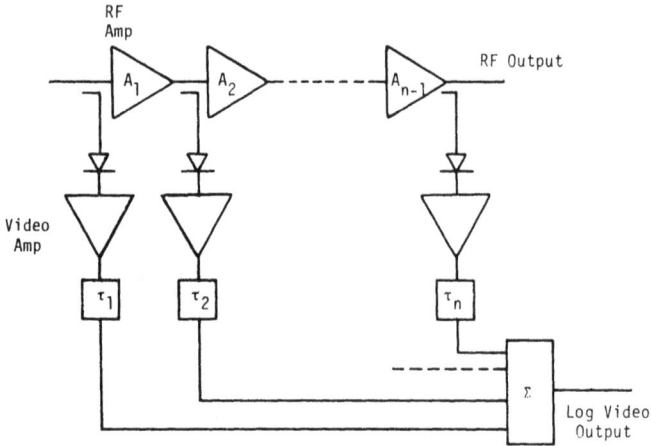

Figure 13.17. n-Stage Log Amplifier

13.21 LIMITERS AND THEIR APPLICATIONS

A microwave limiter is a two-port device. When the input signal is below a certain level, it will emerge from the output with very small attenuation. When the input signal is above a certain level, the input signal will be attenuated and the output will stay at a constant level independent of the input signal level. An ideal limiter is shown in Figure 13.18. In this ideal limiter, when the input signal is below the threshold, there is no attenuation. Sometimes the limiter is combined with amplifiers and form a limiting amplifier (amplifiers and limiters combined together). The limiting amplifier will provide power gain when the input signal is below a certain threshold. The output is limited to a constant level when the signal is above the threshold.

From the operation characteristics of the ideal limiter, it is obvious that when the input signal is below the threshold, the limiter is operating in the linear range. When the input signal is above the threshold, the limiter works in a nonlinear region. In a well-designed receiver the total gain in the RF chain will be just enough to amplify the minimum detectable signal to the saturation level of the limiter. It should be noted that the amplitude information on the signal will be lost after the limiter, and the pulse width information will be distorted also as will be shown later. Therefore, the only information remaining after the limiter is the frequency information. However, it should also be noted that at the saturation level the limiter is in the nonlinear region. If one signal is applied at the input of the limiter, higher harmonics will be generated at the output of the limiter. If simultaneous signals are applied at the input of the limiter, there are also

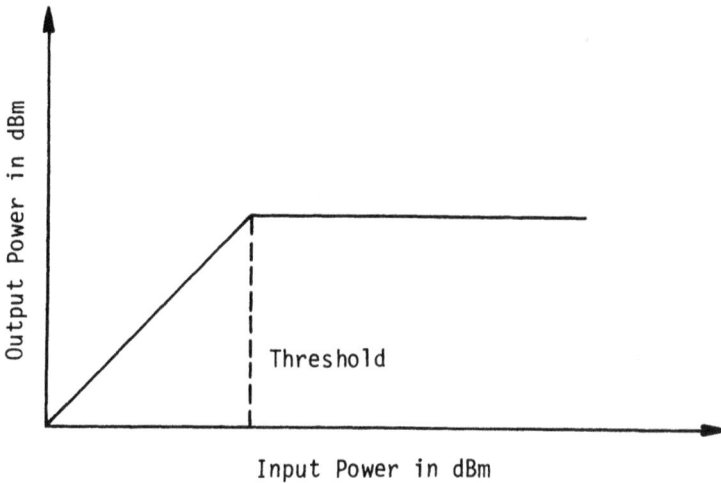

Figure 13.18. Input-Output Relation of an Ideal Limiter

intermodulation products which are difficult to filter out. In spite of all these disadvantages, a limiter is often used in a receiver anyway, because it is very difficult to measure the signal frequencies over a wide dynamic range. For example, if a receiver is designed to receive signals ranging from -60 dBm to 0 dBm it is difficult to obtain detectors to handle the entire 60 dB and encode the frequency correctly. If a limiter amplifier has a gain of 60 dB and a limited output of 0 dB, it can theoretically amplify the signals from -60 dBm to 0 dBm to a constant power level of 0 dBm at the output. It should be relatively simple to design a frequency encoder to measure frequency accurately at this constant power level.

One common use of a limiter in a receiver is as a protection device. The first element from the input of the receiver is often a limiter. When the input is above a certain level, the limiter will limit the signal amplitude from damaging the RF amplifier or mixer of the receiver. In this application the limiter works mainly in the linear region. For this application the maximum power handling capacity of the limiter is an important factor.

Another common use of a limiter is in narrow band FM (frequency modulated) receiver. In an FM signal the information is carried in the in- stantaneous frequency rather than in the amplitude of the signal; therefore; a limiter in the receiver will not lose information. The limiter will reduce the amplitude variation of the signal which behaves like noise. The bandwidth of the receiver should be narrow enough that only the signal of interest will enter the limiter and the chance of other signals in the same bandwidth with comparable power will be slim. Thus, the chance of generating intermod products is small. In addition, the higher harmonics will be outside the band of interest if the receiver is limited within octave bandwidth.

A limiter is usually used in an instantaneous frequency measurement (IFM) receiver as discussed in Chapter 4. The IFM receiver can encode the frequency of one signal at a time. If simultaneous signals are applied to the IFM receiver, the strong one will be encoded. However, when the simultaneous signals are of comparable amplitudes presenting at the receiver, erroneous information may be generated. The capture effect of the limiter that will be discussed in the next section will improve the performance of an IFM receiver on simultaneous signals. If the IFM receiver also provides amplitude and pulse width information, the limiter must be placed behind these circuits in the RF chain. In general, the limiter is installed just before the frequency discriminator (the frequency measurement circuit) of an IFM receiver.

Limiters are also commonly used in other wide-band receivers. They are not, however, used at the front end of the receiver. They are used after the fine frequency separation just before the video detectors; thus, the intermod generated in the pass band of the receiver will be greatly reduced. One example of such an application is in a channelized receiver; limiters are used after the fine frequency channelization (see Chapter 5). If there are 100 fine channels in the receiver, then at least 100 limiters will be used which definitely make the receiver bulky and expensive.

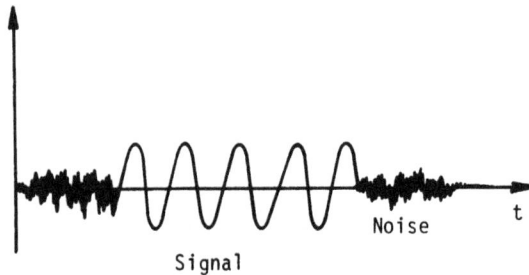

(a) Noise Does Not Reach Limiting Level

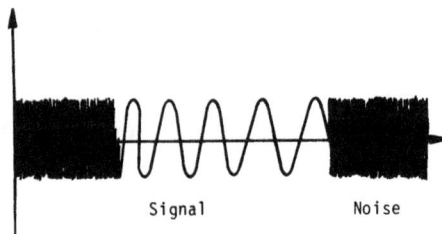

(b) Noise Reaches Limiting Level
Figure 13.19. Limiter Outputs

The outputs of a limiter usually have two situations. When there is not enough gain in front of the limiter, the limiter does not limit noise and the output is as shown in Figure 13.19a. The amplitude of the noise is less than the signal. The other case is that there is high gain in front of the limiter and the noise actually reaches the limiting level. The amplitude from the limiter does not change; only the signal will suppress the noise when it is present as shown in Figure 13.19b. In receiver applications the frequency of the output of Figure 13.19a is slightly easier to measure than that of Figure 13.19b.

13.22 DIODE LIMITERS (56-58)

The limiters can be divided into two groups according to the devices used for the limiting effect. One kind of limiter uses a diode as the limiting device and is called a diode limiter; the other kind uses a yttrium iron garnet (YIG) sphere as the limiting device and is called a YIG limiter.

The simplest way to make a limiter is to adapt the switching property of the diodes. In Figure 13.20, two diodes are connected in parallel in reverse directions across a transmission line. This arrangement can be referred to as a two diode shunting limiter. When the input signals are above a certain threshold, one of the diodes will clip one side of the signal (for example, the positive); the other diode will clip the negative side of the signal. Thus, the input signal is limited by the diodes, and the output of the limiter is

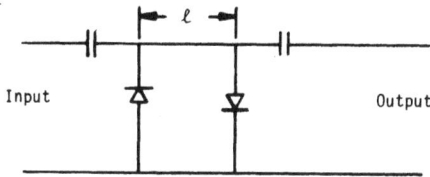

(a) Schematic Diagram

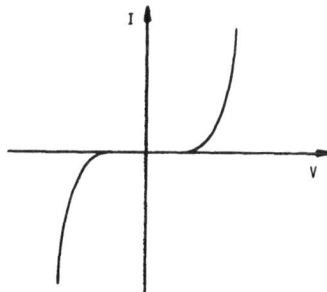

(b) I-V Characteristic Curve

Figure 13.20. Two Diode Limiters

approximately a constant. This approach can be also discussed from the impedance matching point of view. The input impedance of the limiter is a function of the input power. The higher the input power, the worse the input impedance mismatch and power will be reflected back. Therefore, when the input signal is above the limiting threshold, the input impedance match is rather poor. In this arrangement, the insertion loss of signals below the threshold may be relatively high. If the distance between the two diodes is properly adjusted and the impedance of the transmission line is properly chosen, the performance of the diode limiters can be improved.

One way to improve the input impedance matching of a limiter is to use 90 degree hybrid couplers and pin diodes (see Section 10.8). If the 90 degree hybrid and pin diodes are very well matched, the power reflected from the diodes to the input will be minimum. The reflected power from the diodes will emerge from port 4 as shown in Figure 13.21. For a perfectly matched arrangement, the power output at port 4 is

$$P_4 = \tfrac{1}{2}|\varrho|^2 |a_1|^2 \tag{13.80}$$

where ϱ is magnitude of reflection coefficient of identical loads at ports 2 and 3, and a_1 is the amplitude of the input signal at port 1.

If the ϱ decreases as the power incident on port increases, the power delivered to the load at port 4 will be approximately constant. Part of the power will go through the diodes and be consumed in the matched loads R_0. The impedance of the pin diode must be input power dependent and the output power will be determined by Equation 13.80.

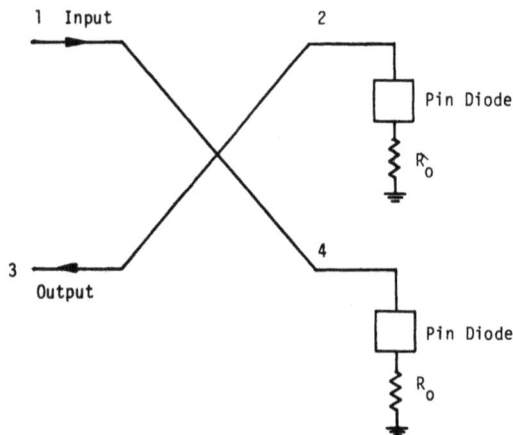

Figure 13.21. Limiter with 90 Degree Hybrid and Pin Diodes

13.23 YIG LIMITER (59-70)

A YIG limiter uses a YIG sphere (or spheres) to couple between the input and output ports. Its structure is like that of a YIG filter (see Chapter 12). A multistage YIG limiter is shown in Figure 13.22. Often the YIG limiter is referred to as the frequency selective limiter. When simultaneous signals separated far in frequency enter the limiter, they are limited independently. The performance of an ideal frequency selective limiter is shown in Figure 13.23. Each signal is limited individually as though there are no other signals. Such a device is ideal for wide-band receiver applications since there is no intermodulation generated and all the output signals are at a constant level. However, as discussed above, the limiter is a nonlinear device and an intermodulation product will be generated.

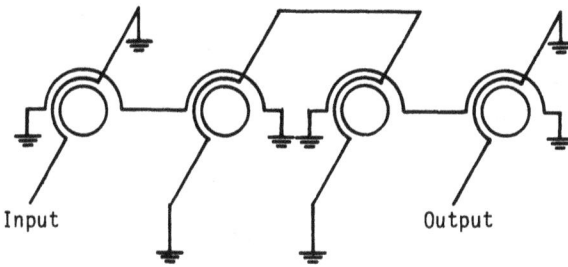

Figure 13.22. Multistage YIG Filter Selective Filter

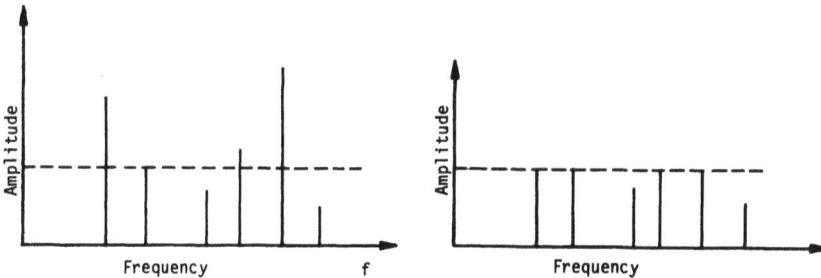

(a) Input Signals   (b) Output Signals

Figure 13.23. Inputs and Outputs of an Ideal Frequency Selection Limiter

The principle operation of the YIG limiter is as follows. In a YIG limiter the subsidiary resonance mode of operation is chosen, because it provides a reasonably low limiting threshold over broad instantaneous bandwidth. The theory of subsidiary resonance limiting was discussed in Reference 59. The external static magnetic field is adjusted so that spin waves exist at one-half the signal frequency ω. A symbolic representation of the YIG limiter is shown in Figure 13.24. There are two resonators of frequencies ω and $\omega/2$ which are coupled through a nonlinear element. When the input signal at frequency ω is below a certain threshold, it will be coupled to the output as if the subharmonic resonator is nonexistent. When the input power is above the threshold, energy will be coupled to the subharmonic resonator $\omega/2$. Under this condition, part of the RF energy is coupled through the subharmonic resonator to the crystal lattice vibration and dissipated as heat; and part is reflected to the input source because of the impedance mismatch caused by the added equivalent conductance to the ω tank by oscillation in the $\omega/2$ tank. These combined effects result as the limiting characteristic.

In a wide-band YIG limiter, many subharmonic resonators with frequencies $\omega_1/2$, $\omega_2/2$, ..., $\omega_n/2$ can be imaged to couple in the main resonator. The limiting action is frequency selective because input signals separated in frequency more than a few spin wave linewidths apart do not couple to the same subharmonic resonator. Therefore, each input signal is limited independently. However, when the input signals of close frequencies are applied to the YIG limiter simultaneously, capture effect—the strong signal suppresses the weak one—will still occur.

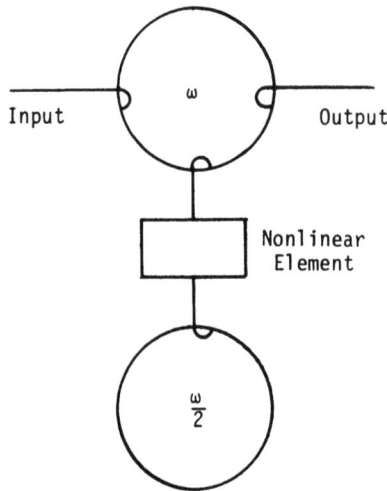

Figure 13.24. *Symbolic Representation of YIG Frequency Selective Limiter*

The critical RF magnetic field intensity for a sphere biased to subsidiary resonance is given by (Reference 61)

$$H_{crit} = \frac{2\omega \, \Delta H_k \, (\omega - \omega_0)}{\omega_m \left(\frac{\omega}{2} + \omega_0 - \frac{\omega_m}{3} \right)}$$

(13.81)

where ω is the signal angular frequency, ΔH_k is the spin wave linewidth, $\omega_0 = \gamma_e H_0$, γ_e is the gyromagnetic ratio of an electron = 2.8 MHz/Oe, H_0 is the external bias magnetic field intensity, $\omega_m = \gamma_e (4\pi M_s)$, and $4\pi M_s$ is the saturation magnetization of ferrite.

For a single YIG crystal, the lowest value of ΔH_k is 0.2 Oe and $4\pi M_s$ = 1750 gauss. For a YIG limiter with an operation frequency range of 9.275 to 9.375 GHz, the optimum field is approximately 2100 Oe.

The threshold of the YIG limiter, with a one section half wavelength resonator, shown in Figure 13.25, can be approximated as

$$P_{th} = \frac{H_{crit}^2 \, W^2 \, Z_0}{\frac{4}{\pi} \, Q_u}$$

(13.82)

where W is the width of transmission line, Z_0 is the characteristic impedance of the transmission line, and Q_u is the unloaded Q of the resonator.

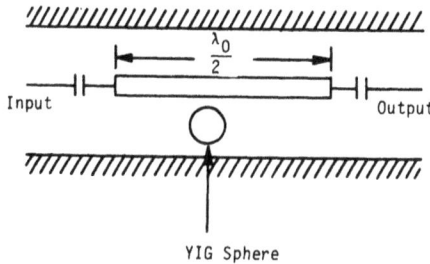

Figure 13.25. YIG Limiter with Half Wavelength Resonator

It is desirable to have the lowest possible threshold in a YIG limiter. However, the requirements for small W and Z_0 conflict, since small width implies high impedance. The optimum transmission width is approximately equal to the sphere diameter.

The YIG sphere can be replaced by varactor diodes; through parametric oscillation similar limiting effects can be achieved (Reference 73).

13.24. OPERATIONAL CHARACTERISTICS OF LIMITERS [58, 71-75]

In this section the performance of an actual limiter including the transient effect and capture effect will be discussed. The input threshold of a limiter is not a sharp break up point but a curved knee. The output will increase slightly with input power as shown in Figure 13.26. By further increasing the input power, the limiter will either burn out or become a linear device again. The difference in overload performance is dependent on whether the limiter is made of diodes or YIG devices. The former usually burns out while the latter turns into a linear device again. The normal operating range of a limiter is defined between the two knees. The threshold and saturation point of a certain limiter varies slightly over the operating frequency range. Usually the threshold of a limiter ranges from −10 dBm to 10 dBm.

A. Effects on Pulse

When a pulsed signal passes a limiter, both the pulse shape and pulse width will change at the output. Figure 13.27 shows a series of outputs from an S-band limiting amplifier. The input signal is a pulse of approximately 120 ns (Figure 13.27a) with the RF frequency at 2 GHz. When the input power level is approximately 3 dB below the 1 dB compression point of the limiting amplifier, the pulse shape is accurately maintained (Figure 13.27b).

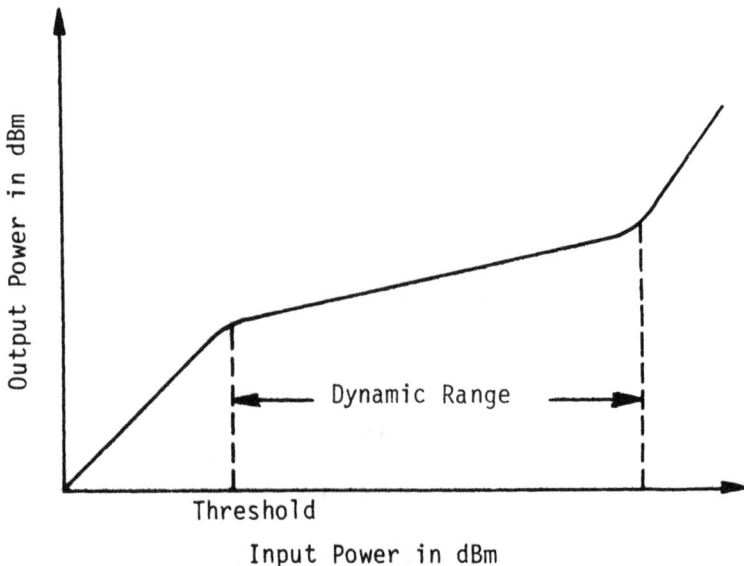

Figure 13.26. Output from a Typical Limiter

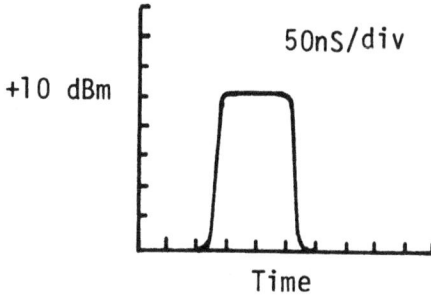

(a) Input Pulse Shape

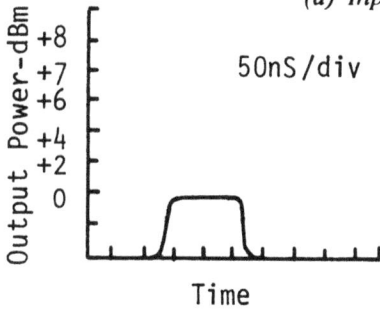

(b) Input Power of -40 dBm

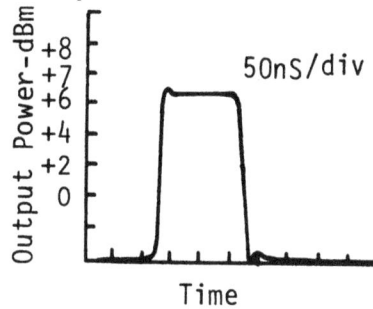

(c) Input Power of -30 dBm

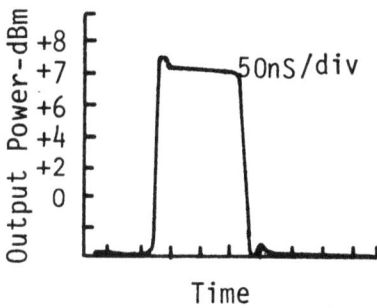

(d) Input Power of -20 dBm

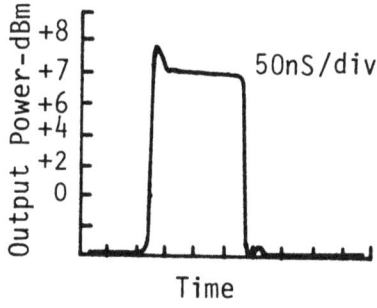

(e) Input Power of 0 dBm

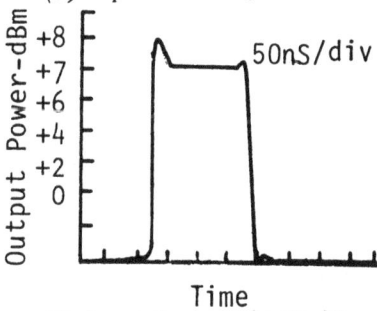

(f) Input Power of +10 dBm

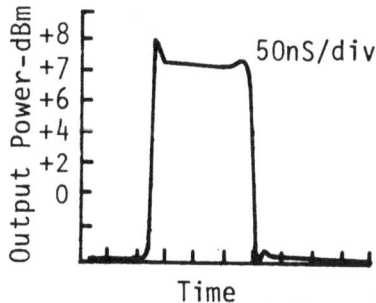

(g) Input Power of +20 dBm

Figure 13.27. Pulse Response of a Limiting Amplifier at 2 GHz. Output Power is Separately Calibrated. Time Scale, 50 ns per Division. (Based on Emergy, Reference 58)

The envelope photographs appearing in Figure 13.27c-g show the pulse response of the limiting amplifier as the input pulse power increased from light to heavy saturation. An overshoot phenomenon of approximately 1 dB existed during the initial 25 ns of the pulse output. The heavily saturated pulse is "stretched" approximately 25 ns mainly as a result of the finite fall time of the input RF pulse. The trailing edge is amplified to reach the saturation of the limiter; thus, the pulse width appears extended. Chien (Reference 68) reported a similar effect on a YIG limiter. The results are shown in Figures 13.28a and b. In Figure 13.28a the input power levels are 0, 1, 3, 6, and 9 dB above the limiting knee, while in Figure 13.28b the input power levels are 10, 15, 20, 25, and 30 dB above the limiting knee. These results show that the higher the input power level, the higher the initial peak, but the shorter the duration. The duration of the peak changes from approximately 3 μs at 6 dB above the limiting knee to 0.4 μs at 25 dB above the knee.

These initial leakages were also discussed by Ho and Siegman (Reference 73). They stated that the height of the leading edge leakage spike is equal to the nonlimited amplitude of the input signal, while the duration of the spike decreases as the input signal amplitude is increased. Thus, as the input level is increased, the leakage spike gets higher but shorter. The cause of the leakage was explained as follows. When an input signal with a very steep leading edge is applied to the YIG limiter, the output at frequency ω will at first be built up very rapidly. The rate of buildup is determined by the Q of the circuit. When the input signal is above the threshold, the subharmonic oscillations at the same time will build from thermal noise level toward their steady-state value. Therefore, there is no limiting effect on the leading edge of the pulse and the signal will leak through. The higher the input energy, the faster the subharmonic oscillation buildup. The subharmonic oscillation will take energy from the main signal and start the limiting action. That is why the higher the input signal, the narrower the leakage pike.

In either the limiting diode or YIG limiter it indicates that the limiter cannot respond instantaneously to the leading edge of the input signal. From the discussion above, it shows that the "leakage" effect is more significant in a YIG limiter than that of the diode limiter. This "leakage" effect severely limits the application of the limiter to pulsed signals, because a short pulse will not be amplitude limited properly.

B. Capture Effect (71-75)

When two simultaneous signals of different frequencies with the same amplitudes (both above the limiting threshold) reach a limiter, the output will contain the two signals and many additional frequencies. The two output signals have the same amplitudes which are approximately 3 dB below the limiting level of the limiter. The additional frequencies are the

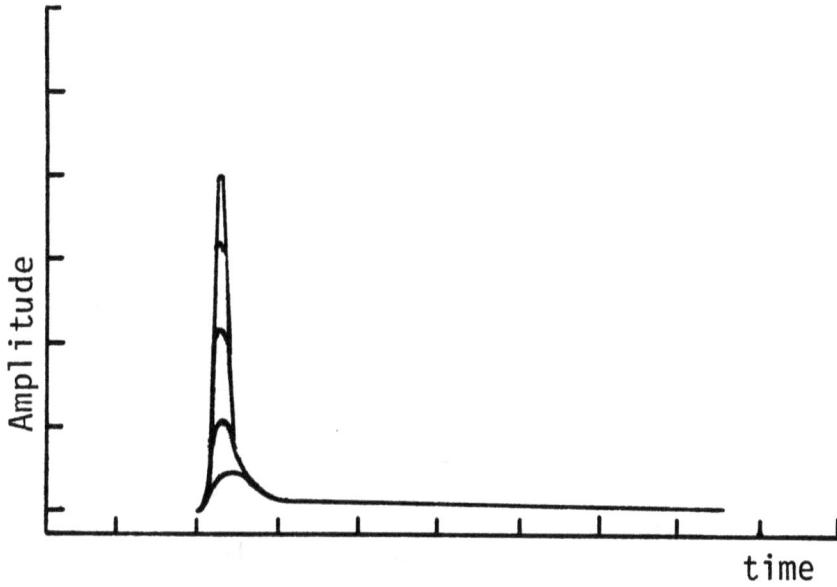

(a) *Leakage Waveforms Corresponding to Input Power 0 dB, 1 dB, 3 dB, 6 dB, and 9 dB Above the Limiting Knee*

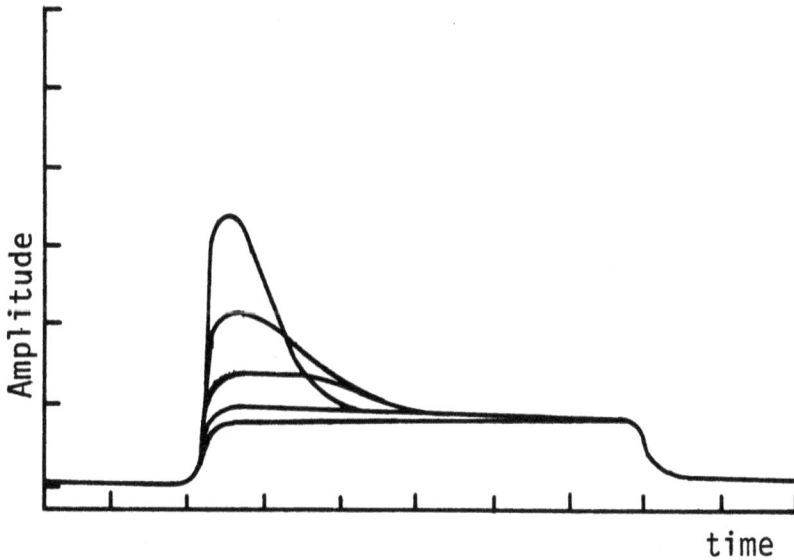

(b) *Leakage Waveforms Corresponding to Input Power 10 dB, 15 dB, 20 dB, 25 dB, and 30 dB Above the Limiting Knee*

Figure 13.28. Pulsed Outputs from a YIG Limiter (Based on Chien, Reference 68)

intermodulation products. However, if the two input signals are of different amplitudes (for example, one is 2 dB stronger than the other one) at the output of the limiter, the stronger signal will be more than 2 dB stronger than the weaker one. In other words the strong signal "captures" the limiter and suppresses the weak one. As expected, all of the frequencies caused by intermodulation are still present.

Capture effect in the YIG limiter is not as prominent as in the diode limiter, because the YIG limiter is supposed to be frequency selective. In an actual YIG limiter, when the two input frequencies are close together, capture effect becomes clear. In a diode limiter, the questions of how far the amplitude separation of the two signals should be before the capture effect can be noted and how much the strong signal suppresses the weak one can only be answered qualitatively as follows. From a theoretical calculation of the diode limiter (Reference 75), it is found that the capture effect happens gradually. When the input separation is 2.5 dB apart, the output will be approximately 5 dB apart for a back-to-back diode limiter.

Experimental results showed that in some limiters, when the input signal amplitudes are separated by 1 dB, the output signal will separate by 2 dB. However, in other limiters, the amplitude separation must be greater than 3 dB to show some capture effect. Capture effect is very useful in an IFM receiver, because the strong signal will suppress the weak one and a correct frequency reading will be obtained on the strong signal. Otherwise, when the weak signal is close to the strong one in amplitude, it will interfere with the frequency encoding process and cause the generation of erroneous frequency data.

REFERENCES

1. *Solid-state amplifiers,* Watkins-Johnson Company, June 1979.
2. *Solid-state microwave amplifiers,* Aertech Industries Catalog No. 5978.
3. "Microwave component amplications—designing with modular amplifiers," Avantek, 1976.
4. March, F., Wheeler, C., and Disman, R., "Microwave transistor amplifiers," Watkins-Johnson, Application Note 100378A, June 1973.
5. "Designing with GPD amplifiers," ATP-1002, Avantek, June 1972.
6. *Thin-film cascadable amplifiers,* Watkins-Johnson Company, 1977.
7. Cheadle, D.L., "Cascadable amplifiers," Watkins-Johnson Company, Tech Notes, Vol. 6, No. 1, January/February 1979.
8. Skolnik, M.I., *Introduction to radar systems,* McGraw-Hill Book Company, 1962.
9. Ishii, T.K., *Microwave engineering,* The Ronald Press Company, 1966.
10. Reich, H.I., Ordung, P.F., Krauss, H.L., and Skalnik, J.G., *Microwave theory and techniques,* D. Van Nostrand Company, 1953.

11. Jordan, E.C., *Electromagnetic waves and radiation systems,* Prentice-Hall, Inc., 1964.
12. Collin, R.E., *Foundations for microwave engineering,* McGraw-Hill Book Company, 1966.
13. Thomassen, K.I., *Introduction to microwave fields and circuits,* Prentice-Hall, Inc., 1971.
14. Tsui, J.B.Y., Johnson, S.J., and Brumfield, W., "Receiver analysis program," AFAL-TR-76-199.
15. Fast, S. "Advanced receiver modeling methods," AFAL-TR-76-265.
16. Geller, B., Cohn, M., "An MIC push-pull FET amplifier," IEEE Int. MTT Symposium Digest of Technical Papers, pp. 187-190, June 1977.
17. Liao, S.Y., *Microwave devices and circuits,* Chapter 6, Prentice-Hall, Inc., 1980.
18. Dacey, G.C., and Ross, I.M., "Unipolar field-effect transistor," Proc. IRE, Vol. 41, pp. 970-979, August 1953.
19. Dacey, G.C., and Ross, I.M., "The field-effect transistor," Bell System Tech. J, Vol. 34, pp. 1149-1189, November 1955.
20. Bockemuehl, R.R., "Analysis of field-effect transistor with arbitrary charge distribution," IEEE Trans. Election Devices, Vol. ED-10, pp. 31-34, January 1963.
21. Liechti, C.A., "Microwave field-effect transistors—1976," IEEE Trans. Microwave Theory and Techniques, Vol. MTT-24, pp. 279-300, June 1976.
22. Tserng, H.Q., Sokolov, V., Macksey, H.M., and Wisseman, W.R., "Microwave power GaAs FET amplifiers," IEEE Trans. Microwave Theory and Techniques, Vol. MTT-24, pp. 936-943, December 1976.
23. Liechti, C.A., "GaAs FET technology: a look into the future," Microwaves, pp. 44-49, October 1978.
24. DiLorenzo, J.V., and Wisseman, W.R., "GaAs power MESFETs: design, fabrication, and performance," IEEE Trans. Microwave Theory and Techniques, Vol. MTT-27, pp. 367-378, May 1979.
25. Bardeen, J., and Brattain, W.H., "The transistor, a semiconductor triode," Physics Review, Vol. 74, pp. 230-231, July 1948.
26. Shockley, W., "The theory of p-n junctions in semiconductors and p-n junction transistors," Bell System, Tech. J, Vol. 28, pp. 435-489, July 1949.
27. Sobal, H., and Sterzer, F., "Solid-state microwave power sources," IEEE Spectrum, 9, p. 32, April 1972.
28. Chen, J.T.C., and Snapp, C.P., "Bipolar microwave linear power transistor design," IEEE Trans. Microwave Theory and Techniques, Vol. MTT-27, pp. 423-430, May 1980.
29. Allison, R., "Silicon bipolar microwave power transistors;" IEEE Trans. Microwave Theory and Techniques, Vol. MTT-27, pp. 415-422, May 1980.

30. Weinert, F., "Scattering parameters speed design of high frequency transistor circuits," Electronics, September 1955.

31. Anderson, D., "S parameter techniques for faster, more accurate network design," Hewlett-Packard Journal, February 1967, or Hewlett-Packard Application Note 95-1.

32. Richter, K., "Predicting linear power amplifier performance," Microwaves p. 56, February 1974.

33. Kuhn, N., "CAD with graphics make circuit design a science," Microwaves. p. 42, June 1974.

34. Esaki, L., "New phenomenon in narrow germanium p-n tunnel junction," Physics Review, Vol. 109, pp. 603-604, 1958.

35. Miller, B.A., Miles, T.P., and Cox, D.C., "A design technique for realizing a microwave tunnel diode amplifier in strip line," IEEE Trans. Microwave Theory and Techniques, Vol. MTT-15, pp. 554-561, October 1967.

36. Okean, H.C., "Microwave amplifiers employing integrated tunnel-diode devices," IEEE Trans. Microwave Theory and Techniques, Vol. MTT-15, pp. 613-622, November 1967.

37. Welch, J.D., "Beam lead tunnel diode amplifiers on microstrip," IEEE Trans. Microwave Theory and Techniques, Vol. MTT-18, pp. 1077-1083, December 1970.

38. Tserng, H.Q., Coleman, D.J., Jr., Doerbeck, F.H., Shaw, D.W., and Wisseman, W.R., "A 12-w GaAs read-diode amplifier at X-band," IEEE Trans. Microwave Theory and Techniques, Vol. MTT—26, pp. 774-778, October 1978.

39. Weiss, M.T., "A solid-state microwave amplifier and oscillator using ferrites," Physics Review, Vol. 107, p. 317, July 1957.

40. Reed, E.D., "The variable-capacitance parametric amplifier," IRE Trans. Election Device, Vol. ED-6, pp. 216-224, April 1959.

41. Kinoshita, Y., and Maeda, M., "An 18 GHz double-tuned parametric amplifier," IEEE Trans. Microwave Theory and Techniques, Vol. MTT-18, pp. 1114-1119, December 1970.

42. Branner, G.R., and Chan, S.P., "A new technique for synthesis of broad-band parametric amplifiers," IEEE Trans. Microwave Theory and Techniques, Vol. MTT-21, pp. 437-444, July 1974.

43. Niehenke, E.C., "An environmentalized low-noise parametric amplifier," IEEE Trans. Microwave Theory and Techniques, Vol. MTT-25, pp. 992-994, December 1977.

44. Wilson, W.W., Dickman, R.L., and Berry, G.G., "Cryogenic parametric amplifier noise performance at 4.2 K," IEEE Trans. Microwave Theory and Techniques, Vol. MTT-28, pp. 186-190, March 1980.

45. Gunn, J.B., "Microwave oscillations of current in III-V semiconductors," Solid-State Communications, Vol. 1, pp. 89-91, September 1963.

46. Narayan, S.Y., and Sterzer, F., "Transferred electron amplifiers and oscillators," IEEE Trans. Microwave Theory and Techniques, Vol. MTT-18, pp. 773-783, November 1970.

47. Perlman, B.S., Upadhyayula, C.L., and Marx, R.E., "Wide-band reflection-type transferred election amplifiers," IEEE Trans. Microwave Theory and Techniques, Vol. MTT-18, pp. 911-921, November 1970.

48. Copeland, J.A., "CW operation of LSA oscillator diodes—44 to 88 GHz," Bell System Tech. J, Vol. 46, pp. 284-287, January 1967.

49. Lee, C.A., Batdorf, R.L., Wiegmann, W., and Kaminsky, G., "The read diode—an avalanche, transit-time, negative resistance oscillator," Applied Physics Letters, Vol. 6, p. 89-91, 1965.

50. Read, W.T., "A proposed high-frequency negative diode," Bell System Tech. J, Vol. 37, pp. 401-446, March 1958.

51. Knerr, R.H., and Murray, J.H., "Microwave amplifiers using several IMPATT diodes in parallel," IEEE Trans. Microwave Theory and Techniques, Vol. MTT-22, pp. 569-572, May 1974.

52. Prager, H.J., Chang, K.K.N., and Weisbrod, S., "High-power, high efficiency silicon avalanche diode at ultra high frequencies," Proc. IEEE, Vol. 55, pp. 586-587, April 1967.

53. Coleman, D.J., and Sze, S.M., "A low-noise metal semiconductor metal (MSM) microwave oscillator," Bell System Tech. J, pp. 1675-1695, May/June 1971.

54. Sheingold, D., and Pouliot, F., "The hows and whys of log amps," Electronic Design 3, p. 52, 1 February 1974.

55. Helfrick, A., "Build high-gain, wide-range log amps," Electronic Design 6, p. 116, 15 March 1974.

56. Garver, R.V., and Rosado, J.A., "Broad-band TEM diode limiting," IEEE Trans. Microwave Theory and Techniques, Vol. MTT-10, pp. 302-310, September 1962.

57. Rodriguez, M.J., and Weissman, D., "A microwave power limiter," IEEE Trans. Microwave Theory and Techniques, Vol. MTT-10, pp. 219-220, May 1962.

58. Emergy, F.E., "Solid-state limiting amplifiers," Watkins-Johnson Company, Technical Notes, Vol. 5, No. 5, September/October 1978.

59. Suhl, H., "The nonlinear behavior of ferrites at high microwave signal levels," Proc. IRE, Vol. 44, pp. 1270-1284, October 1956.

60. Garver, R.V., and Tseng, D.Y., "X-band limiting," IRE Trans. Microwave Theory and Techniques, Vol. MTT-9 (correspondence), p. 202, March 1961.

61. Stitzer, S., and Goldie, H., "X-band frequency-selective YIG limiters," Air Force Avionics Laboratory, Technical Report AFAL-TR-77-204, November 1977.

62. Helszajn, J., Murray, R.W., Davidson, E.G.S., and Suttrie, R.A., "Waveguide subsidiary resonance ferrite limiters," IEEE Trans. Microwave Theory and Techniques, Vol. MTT-25, pp. 190-196, March 1977.

63. Emtage, P.R., and Stitzer, S.N., "Interaction of signals in ferromagnetic limiters," IEEE Trans. Microwave Theory and Technique, Vol. MTT-25, pp. 210-213, March 1977.

64. Uebele, G.S., "Characteristics of ferrite microwave limiters," IRE Trans. Microwave Theory and Techniques, Vol. MTT-7, pp. 18-23, January 1959.

65. Brown, J., "Ferromagnetic limiters," Microwave Journal, Vol. 4, p. 74, November 1961.

66. Stitzer, S.N., Goldie, H., and Carter, P.S., Jr., "X-band YIG limiters for FM/CW radar," Microwave Journal, Vol. 20, p. 35, December 1977.

67. Kotzebue, K.L., "Frequency-selective limiting," IRE Trans. Microwave Theory and Techniques, Vol. MTT-10, pp. 516-520, November 1962.

68. Chien, R.C.S., "Frequency-selective limiter and its application in a filter bank receiver," Technical Note 1973-1, Lincoln Laboratory, MIT, 13 April 1973.

69. Giarola, A.J., Jackson, D.R., Orth, R.W., and Robbins, W.P., "A frequency selective limiter using magnetoelastic instability," Proc. IEEE letter, Vol. 5, pp. 593-594, April 1967.

70. Jackson, D.R., and Orth, R.W., "A frequency selective limiter using nuclear magnetic resonance," Proc. IEEE, Vol. 55, pp. 36-45, January 1967.

71. Baghdady, E.J., "Theory of stronger signal capture in FM reception," Proc. IRE, Vol. 46, pp. 728-738, April 1958.

72. Jones, J.J., "Hard limiting of two signals in random noise," IEEE Trans. Information Theory, Vol. IT-9, pp. 34-42, January 1963.

73. Ho, I.T., and Siegman, A.E., "Passive phase distortionless parametric limiting with varactor diodes," IRE Trans. Microwave Theory and Techniques, Vol. MTT-9, pp. 459-472, November 1961.

74. Gianola, A.J., "Third order intermodulation products in a ferrite frequency FSL—two signal theory," Proc. IEEE Letters, Vol. MTT-54, pp. 2011-2013, December 1966.

75. Bogan, Z.I., and Tsui, J.B.Y., "Computer modeling of diode circuits and its applications," Air Force Technical Reports, AFAL-TR-77-266, March 1978.

CHAPTER 14
MICROWAVE MIXERS AND DETECTORS

14.1 INTRODUCTION

In general, a microwave mixer is a three-port device: the input (signal) port, the local oscillator (LO) port, and the output port. A mixer is primarily used to convert the input signal frequency [generally referred to as the radio frequency (RF)] to an intermediate frequency (IF). The IF is usually easier to amplify and process than the RF input signal. For example, the RF may be in the gigahertz range while IF can be in the megahertz range where narrow bandwidth filters and high performance amplifiers are available. A mixer is a nonlinear device; however, it is usually treated as a linear device in a receiver, because an ideal mixer will simply shift the input frequency to another frequency and retain all the information. An ideal mixer should not generate any other spurious frequencies, but in a real mixer, spurious frequencies will be generated.

In a receiver design, these spurious frequencies must be accounted for in order to avoid ambiguity frequency outputs. To minimize the spurious responses, the input, output, and LO frequencies must be chosen carefully. To suppress some of the spurious frequencies, special techniques are used in the mixer designs. Mixers can also be used for other applications such as phase detectors, microwave switches, and phase inverters.

A microwave detector is a two-port device which will convert microwave energy to video signals. It exists in almost all microwave receivers and also in logarithmic amplifiers. The signal processing circuitry usually cannot process the microwave energy directly but it can process the video signals. Therefore, even in the simplest microwave receiver, a detector is often required. A detector is generally a microwave diode which is a nonlinear device. It rectifies the RF signal and passes the signal through a low-pass filter. At the output of the detector is a video signal whose amplitude is related to the RF input signal strength. The frequency and phase information in the RF signal is lost after the detector.

14.2 OPERATIONAL PRINCIPLE OF MIXERS[1-4]

The analysis of a mixer is similar to that of an amplifier operating in a nonlinear region (Chapter 13). This is also the general approach to any nonlinear device. The mixing action can be produced from this nonlinear transfer function such as in a diode. The current in the diode can be expressed as

$$I = a_0 + a_1V + a_2V^2 + \ldots + a_nV^n \qquad (14.1)$$

where I is the current flowing through the diode; a_0, a_1, ... a_n are constants; and V is the applied voltage across the diode.

In a mixer, the applied voltage includes the LO ($V_0 \sin\omega_0 t$) and the signal ($V_s \sin\omega_s t$). Thus,

$$V = V_0 \sin\omega_0 t + V_s \sin(\omega_s t + \theta) \qquad (14.2)$$

where θ is an arbitrary phase angle.

Substituting Equation 14.2 into 14.1

$$I = a_0 + a_1 [V_0 \sin\omega_0 t + V_s \sin(\omega_s t + \theta)]$$
$$+ a_2 [V_0 \sin\omega_0 t + V_s \sin(\omega_s t + \theta)]^2$$
$$+ \ldots$$
$$+ a_n [V_0 \sin\omega_0 t + V_s \sin(\omega_s t + \theta)]^n \qquad (14.3)$$

The first two terms in Equation 14.3 represent a dc component and the original signal. The third term can be expressed as

$$a_2 [V_0 \sin\omega_0 t + V_s \sin(\omega_s t + \theta)]^2$$
$$= a_2 V_0^2 \sin^2\omega_0 t + a_2 V_s^2 \sin^2(\omega_s t + \theta)$$
$$+ a_2 V_0 V_s \{\cos[(\omega_0 - \omega_s)t - \theta] - \cos[(\omega_0 + \omega_s)t + \theta]\} \qquad (14.4)$$

The last term in Equation 14.4 generates the desired frequency shift $\omega_0 - \omega_s$ (which can be represented as $\omega_s - \omega_0$ if $\omega_s > \omega_0$) and $\omega_0 + \omega_s$. The procedure to generate $\omega_0 - \omega_s$ is often called down converting while the generation of $\omega_0 + \omega_s$ is often called up converting. From Equation 14.1 it is obvious that there are many other frequencies which will be present at the output of the mixer.

Another way of looking at the mixer action is that the diodes in the mixer can be considered as sampling switches. In general, the LO power in a mixer is higher than that of the signal power. As a first order approximation, the LO power should be 10 dB greater than the highest RF input signal level anticipated. For example, the LO power required could range from 7 dBm to 23 dBm. The LO will control the diodes as an on-off switch and sample the signal waveform accordingly. The sampling effect can be explained as shown in Figure 14.1. For simplicity the LO input to the diode can be considered as a square wave, and it will control the conductivity of the diode. The output of the diode is shown in Figure 14.1d. The Fourier analysis of the diode output will reveal the two important frequency components $\omega_0 - \omega_s$ and $\omega_0 + \omega_s$. Of course, other frequency components are also present in the mixer output.

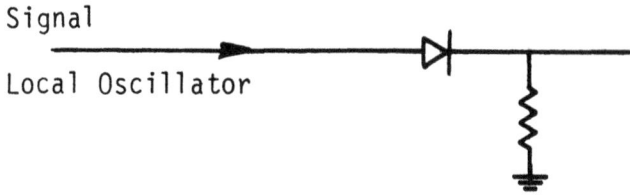

(a) Simple Diode Mixer

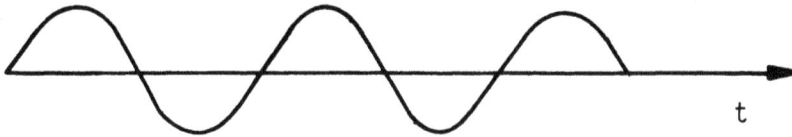

(b) Signal Input ω_s

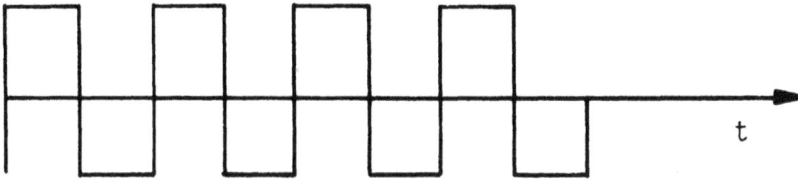

(c) Local Oscillator Input ω_o

(d) Diode Output

(e) $\omega_o - \omega_s$

(f) $\omega_0 + \omega_s$

Figure 14.1 Mixer as a Sampling Switch

14.3 PROPERTIES OF MIXERS [5-11]

The conversion loss of a mixer is defined as the ratio of the IF output power to signal input power. It is one of the most important parameters, since it measures the efficiency with which a mixer converts RF energy to IF energy. It depends on the losses in the RF circuitry surrounding the mixer, the quality of the nonlinear device, the match between the RF circuit and the diodes, and the bias level to the diodes.

As usual, the noise figure is defined as the ratio of the signal-to-noise (S/N) at the input of the mixer to the S/N ratio at the output of the mixer. In general, the noise figure of a mixer is equal to or slightly higher ($\sim$0.5 dB) than its conversion loss. In case there is an IF amplifier following the mixer as an integrated part of the mixer, the noise figure of the IF amplifier will also be included. The noise figure of the mixer is given approximately by

$$NF = L_c + NF_{IF} \tag{14.5}$$

where L_c is the conversion loss of the mixer in dB, and NF_{IF} is the noise figure of the IF amplifier in dB.

The conversion loss is a very important factor in receiver applications, especially to the receivers without the RF amplifier in front of the mixer, since the first component at the input of a receiver has the most influence on the noise figure of the receiver. For example, if a mixer has a 6 dB conversion loss, then the noise figure of the receiver must be higher than 6 dB. If there is an RF amplifier ahead of the mixer, then the noise figure of the receiver is primarily determined by the amplifier.

Isolation of a mixer is a measure of circuit balance within the mixer. It is equal to the ratio of the power input at one port to the power leak at another port. Since there are three ports (RF, IF, and LO) in a mixer, isolation between any two of the three ports are important. However, in many mixers, only the isolations between the RF-IF and LO-IF are given because the IF is the desired signal and usually further signal processing is followed. If the LO and RF signals leak in the IF circuit, it should be filtered out. The isolation between the RF and LO is also very important. Poor isolation means that the LO signal can reach the RF port with relatively small insertion loss, and the power may radiate through the receiving antenna.

Another problem may exist in a mixer assembly, as shown in Figure 14.2, if there is poor isolation between RF and LO ports. In Figure 14.2, one LO is shared between two mixers; this is a common practice in channelized receivers. Referring to Figure 14.2, the input frequency to mixer M_1 is f_1 while to M_2 is f_2. Let us assume that the desired output frequencies are $f_3 - f_1$ and $f_3 - f_2$, respectively. However, if there is poor isolation between the RF and LO ports, frequency f_1 from M_1 may reach the

LO port of M_2 and generate spurious frequencies at its output and vice versa. This effect can be greatly reduced by adding filters F_3 (dotted blocks) with center frequency at f_3 in the LO path.

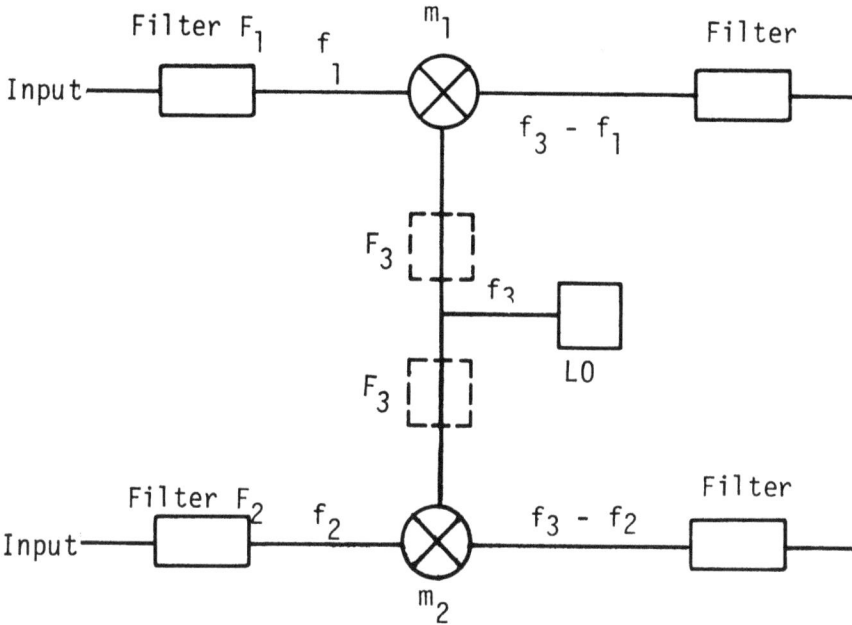

Figure 14.2. A Single LO is Shared Between Two Mixers

The dynamic range of a mixer can be loosely defined as the power range that a mixer can provide under normal operation. Its lower limit is the noise floor while the upper limit is the nonlinear transfer response. At the upper limit, there is the 1 dB compression point and third order intercept point. Therefore, as far as the dynamic range is concerned, the performance of a mixer is almost identical to that of an amplifier (see Chapter 13). That is why a mixer is often treated as a linear device in a receiver. The third order intermodulation frequencies generated in a mixer are different from that of an amplifier because the output frequency is shifted as shown in Figure 14.3. Here the mixer is used as a down converter. The IF frequencies are

$$f_{I1} = f_{LO} - f_{R1}, \quad f_{I2} = f_{LO} - f_{R2} \qquad (14.6)$$

where f_{LO} represents the LO frequency, and f_{R1} and f_{R2} represent input frequencies 1 and 2, respectively.

The third order intermod frequencies are

$$f_{M1} = f_{LO} - (2f_{R1} - f_{R2}) \qquad (14.7)$$

$$f_{M2} = f_{LO} - (2f_{R2} - f_{R1}) \qquad (14.8)$$

Their amplitudes determine the third order intercept point of a mixer.

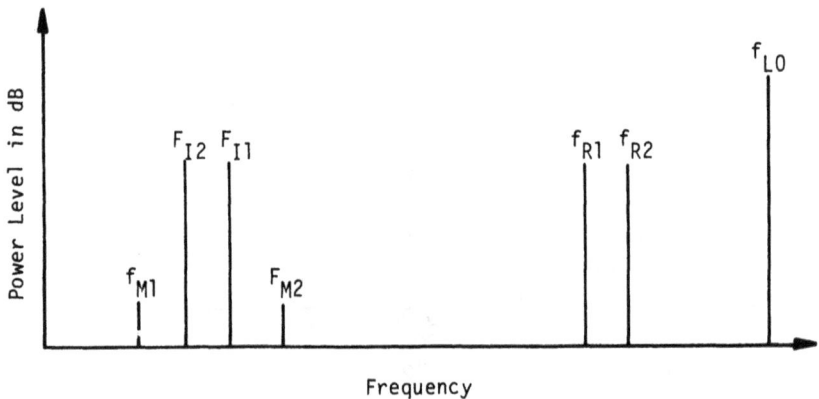

Figure 14.3. Third Order Intermod Generated in a Mixer

14.4 SPURIOUS RESPONSES[12-22]

Although the desired output signals from the mixer are $\omega_0 - \omega_s$ and $\omega_0 + \omega_s$, it is obvious from Equation 14.3 that there are many other frequencies at the output of the mixer also. All these undesired frequencies are called spurious. However, these frequencies are very much predictable. The frequencies at the output of a mixer can be written as

$$f_0 = Mf_1 + Nf_2 \qquad (14.9)$$

where f_1 and f_2 are the input frequencies, M and N are integers (either positive or negative), and the output frequency f_0 must be a positive value.

Although Equation 14.9 is extremely simple, it is tedious to calculate all the output frequencies. A simple computer program can be used to determine the spurs. However, a spur chart is often used to predicate the spurious frequencies. There are different ways (References 12-14) to produce a spur chart; each will be particularly suited for some special applications. Here, the spur chart generated by Brown (Reference 12) will be discussed because of its simplicity.

Assuming $f_2 > f_1$ and dividing Equation 14.9 by f_2, one can obtain

$$\frac{f_0}{f_2} = M \frac{f_1}{f_2} + N \tag{14.10}$$

Plotting f_0/f_2 versus f_1/f_2 with different M,N values, the spur chart can be obtained. It should be noted that the desired output frequencies are

$$f_0 = -f_1 + f_2 \text{ for } M = -1, N = 1 \tag{14.11}$$

and

$$f_0 = f_1 + f_2 \text{ for } M = 1, N = 1 \tag{14.12}$$

In order to keep f_0 positive, the spur chart for the down conversion case (Equation 14.11) is limited in the following region

$$0 \leqslant \frac{f_0}{f_2} \leqslant 1 \quad \text{and} \quad 0 \leqslant \frac{f_1}{f_2} \leqslant 1 \tag{14.13}$$

In order to demonstrate how the spur chart is generated, a limited number of M and N will be chosen. Let

$$M = -2, -1, 0, 1, 2$$

and $\hspace{10cm}$ (14.14)

$$N = -2, -1, 0, 1, 2$$

Any one value of M will combine with all five values of N. Thus, there is a total of 25 combinations. Substituting all these 25 combinations into Equation 14.10 and the line sections within the boundaries

$$0 \leqslant \frac{f_1}{f_2} \leqslant 1 \quad \text{and} \quad 0 \leqslant \frac{f_0}{f_2} \leqslant 2 \tag{14.15}$$

are plotted. The results, both the down and up convert cases, are shown in Figure 14.4. A spur chart for $f_0 = f_2 - f_1$ containing all the spurious frequencies generated up to the 6,6 order is shown in Figure 14.5. In order to simplify the notation let $f_1 = L$ and $f_2 = H$. The squares on the output lines $f_2 - f_1$ as shown in Figure 14.5 are a preferable chosen area for down conversion because there is no spur past these regions. However, if a wide bandwidth IF is desired, there is always some spurius response included. As a general rule, the higher the order of the spurs, the smaller their amplitude.

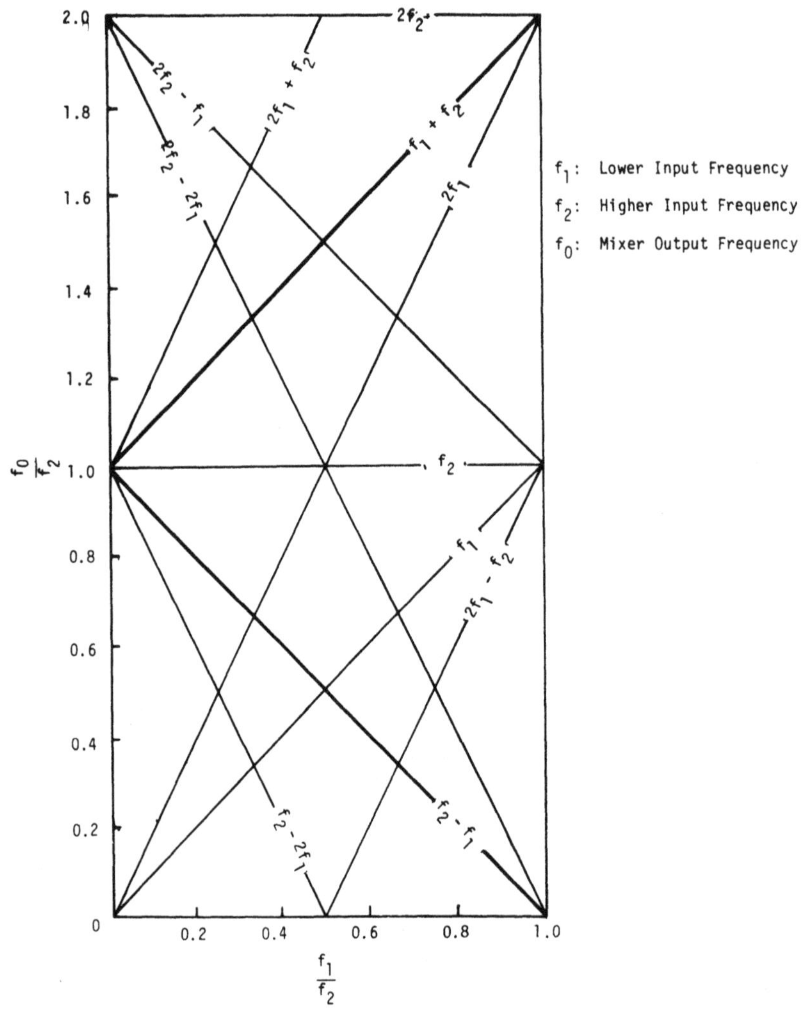

Figure 14.4. *Up Converter and Down Converter Spurious Chart*

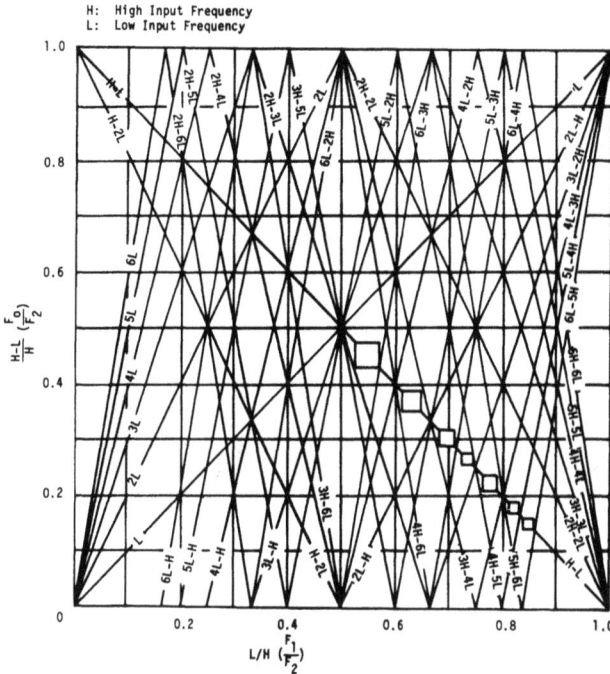

Figure 14.5. Down Converter Spurious Effects Chart

There are many papers (**References 15-22**) trying to theoretically predict the levels of the intermodulations and spurious responses of mixers. If the spurious responses and their amplitudes can be calculated, the performance of a frequency converter can be readily predicted. However, most of the papers deal with simple diode and neglect the inductances and capacitances in the circuitry; they can hardly predict the performance of an actual mixer which usually contains a multiple number of diodes. Even for a simple diode, the amplitudes of the spurious responses depend on the LO power level. The LO power in a real mixer is also critical to the generation of the spur levels. Many times the amplitudes of the spurs of a certain commercial mixer are listed in a table referred to as a spur table. An example is given in Table 14.1. This table includes three mixers. The LO power to the three mixers are +7, +17, and +27 dBm. The input powers are at 0 dBm (shaded area) and −10 dBm. The numbers in the table represent the power level in dB below the desired signal at certain spur products.

14.5 SINGLE DIODE MIXERS

The mixers can generally be divided into five groups according to their structure; single diode mixers, single balanced mixers, double balanced mixers, image rejection mixers, and image enhanced mixers.

TABLE 14-1. TYPICAL FREQUENCY INTERMODULATION PERFORMANCE
(Reference 49 Courtesy of Watkins-Johnson Co.)

M1/M1D/M1E

	0	1	2	3	4	5	6	7	8
7	>90	>90	>90	>90	>90	>90	>90	>90	>90
6	>90	>90	>90	>90	>90	>90	>90	>90	>90
5	>90	>90	>90	>90	>90	>90	>90	>90	>90
4	86 >90	80 >90	86 >90	88 >90	77 >90	75 >90	68 >90	65 >90	88 >90
3	67	64	69	50	71	74	69	72	77
2	73	71	76	67	76	80	77	85	90
1	24	0	74	75	86	66	82	78	84
0		0	84	70	71	64	69	64	74

M1/M1D/M1E: f_R at 0 dBm; f_L at +7/+17/+27 dBm respectively for the Models M1, M1D, and M1E.

M1/M1D/M1E: f_R at −10 dBm; f_L at +7/+17/+27 dBm respectively for the Models M1, M1D, and M1E.

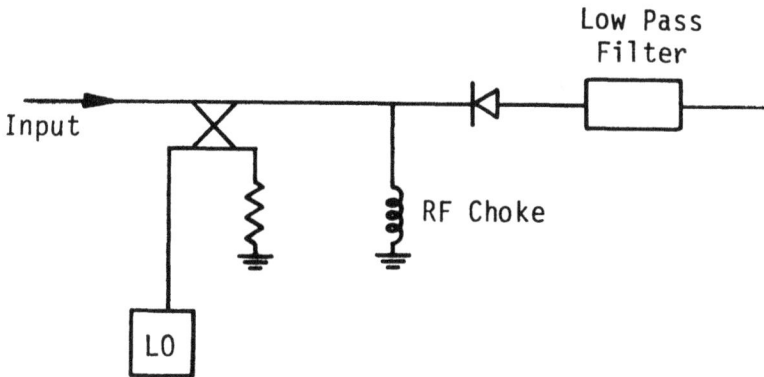

Figure 14.6. A Single Diode Mixer

The single diode mixer is the simplest diode mixer. As shown in Figure 14.6, a diode terminating a transmission line is a mixer. The LO power is applied through a directional coupler. The impedance match between the diode and the input transmission line determines the voltage standing wave ratio (VSWR) for the signal input, and the amount of LO power appearing at the input of the signal port.

A low-pass filter is usually used at the IF output to block the RF and LO frequencies. An RF choke is needed at the input side of the diode to prevent loss of IF energy in the input port. In this arrangement the output frequency must be lower than both the RF and LO frequencies.

The performance of a single diode mixer is relatively poor. The isolation between the RF and LO ports is generally low and the VSWR is relatively high. There is no special circuitry to reduce the amplitudes of spurious responses. More LO power is necessary as a result of the relatively poor coupling factor of the directional coupler.

14.6. SINGLE BALANCED (BALANCED) MIXERS

A single balanced mixer contains two single diode mixers as shown in Figure 14.7. A 3 dB hybrid is used to supply the RF and LO power to the two mixer diodes. The 3 dB hybrid can be either 90 or 180 degrees. The two diodes are connected in opposite directions. The single balanced mixer offers better performance over a single diode mixer, including reduced spurious responses, cancellation of the dc component at the IF output, and convenient separation of RF and LO inputs.

If the 3 dB hybrid is 90 degrees as discussed in Chapter 10, the use of the hybrid will improve the input VSWR of a device. In Figure 14.7, if a 90 degree hybrid is used and the two diodes are well balanced, then the RF

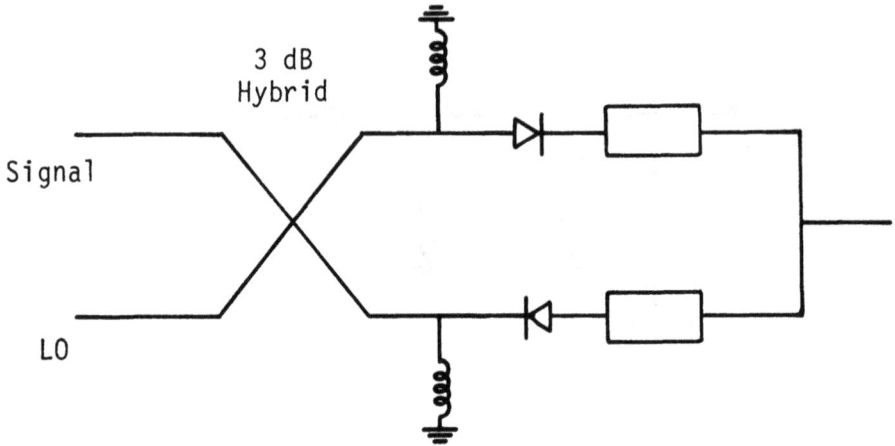

3 dB
Hybrid

Signal

LO

Figure 14.7. Single Balanced Mixer

input reflected from the diodes will emerge from the LO port, while the reflected LO power will come out the RF input port. Thus, the input VSWR of this kind of mixer is rather low (typically less than 1.5) at either of the RF and LO ports. However, the isolation between the RF and LO ports is rather poor (typically around 7 dB). The degree of harmonic and intermodulation suppression for this mixer is rather complicated and depends on the particular intermodulation product of interest. In general, there is a suppression of harmonics and intermodulations products of both RF and LO signals. Filters usually are required to minimize the spurious responses.

If a 180 degree hybrid is used in Figure 14.7, the mixer will suppress the spurious responses derived from the even harmonics of one of the input signals. It is usually designed to suppress the spurs derived by the even harmonics of the LO signal as shown in Figure 14.8. The LO power will be in phase at the diodes while the RF signal will be 180 degrees out-of-phase. At the output of the mixer, another 180 degree hybrid is used. The LO effect will be reduced because the two e_L inputs at the difference terminal (Δ) are 180 degrees out-of-phase, but the signal e_s inputs are in phase. Reversing the LO and RF input ports, the mixer will suppress the even harmonics of the input signal. The degree to which the harmonics and intermodulation products associated with those harmonics are suppressed depends on the degree of balance. This is influenced by the balance of the hybrids and the match of the mixer diodes.

The RF input reflected by the two diodes due to mismatch will emerge from the RF input port, while the reflections of the LO input will come out

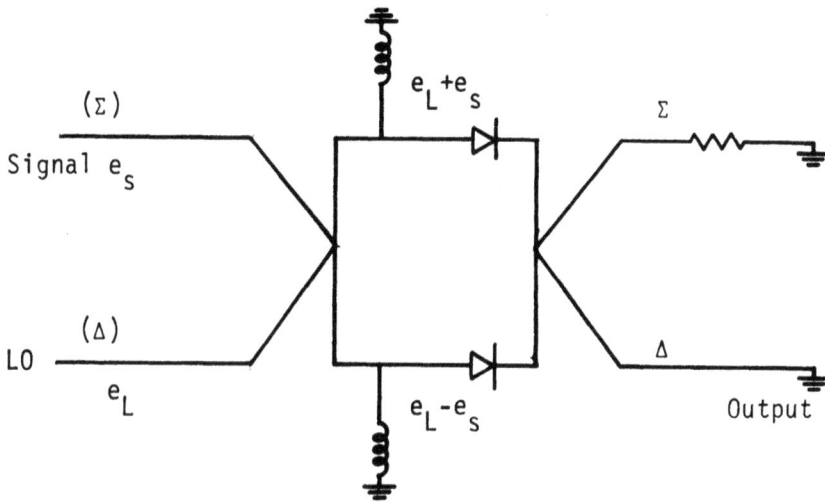

Figure 14.8. A Single Balanced Mixer with 180 Degree Hybrids

the LO port. As a result, the isolation between the RF and LO of this type of balanced mixer is very good (typically 20 dB or more). This same property, however, causes the VSWR for both the LO and signal ports to depend on the diode match to the transmission line. This VSWR is relatively poor (typically 2.0). Usually, filters are required to reduce the spurious responses at the output of the mixer.

14.7. DOUBLE BALANCED MIXERS (23-26)

Double balanced mixers using four mixer diodes to suppress spurious responses are the most commonly used (References 23 and 24) because they provide superior performance over single balanced mixers. As early as 1939, Caruthers (Reference 25) discussed various four diode modulator circuits. A double balanced mixer is basically two single balanced mixers connected in parallel with a 180 degree phase difference between them. A double balanced mixer circuit is shown in Figure 14.9. The symmetry provided by the four diode array formed by uniting two mixers ensures complete isolation between the RF and LO ports when all diodes are perfectly balanced. As a result, all three ports (RF, LO, and IF) are well isolated over wide frequency range. For mixers under 500 MHz, typical isolations are greater than 30 dB. The isolation will deteriorate at higher frequency because it is difficult to maintain the circuit symmetry required to provide the isolations. The isolations are typically greater than 12 dB for mixers up to 12 GHz.

Since the circuit in Figure 14.9 is symmetrical, the RF and LO ports are interchangeable. The double balanced mixer suppresses all the spurious responses derived from the even harmonics of both the RF and LO frequencies.

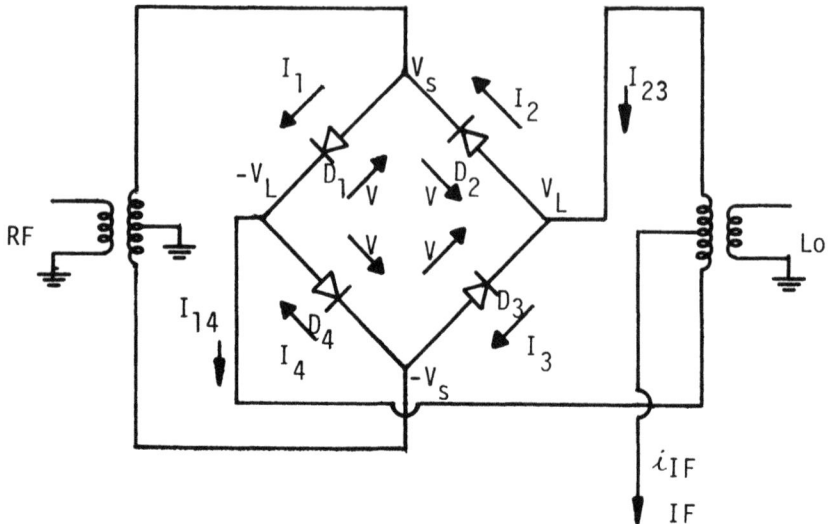

Figure 14.9. Double Balance Mixer

The spur table (Table 14.1) contains the results of double balanced mixers. The results show that the even harmonic spurs are generally less than the neighboring odd spurs of lower order.

The two-tone third order intermodulation product in a double balanced mixer is lower than that of a single balanced mixer, since less RF voltage appears across each diode for a given input power level. In other words double balanced mixers have higher dynamic range.

In Figure 14.9, when the voltage V is fed across the four diodes D_1, D_2, D_3, and D_4, the diode currents i_1, i_2, i_3, and i_4 can be written as (see Equation 14.68)

$$I_1 = I_s \left[\exp(\alpha V) - 1 \right] \tag{14.16}$$

$$I_2 = I_s \left[\exp(\alpha V) - 1 \right] \tag{14.17}$$

$$I_3 = I_s \left[\exp(\alpha V) - 1 \right] \tag{14.18}$$

$$I_4 = I_s \left[\exp(\alpha V) - 1 \right] \tag{14.19}$$

where α is the diode slope parameter and I_s is the saturation current.

The voltage V can be written as

$$V = V_L \cos\omega_L t \pm V_s \cos\omega_s t \tag{14.20}$$

where V_L and V_s are the amplitude of the LO and RF signals, respectively;

and ω_L and ω_s are the angular frequencies of the LO and RF signals, respectively.

The differential conductance for each diode may be written as

$$g_1 = \frac{dI_1}{dV} = \alpha I_s \exp(\alpha V) \tag{14.21}$$

$$g_2 = \frac{dI_2}{dV} = \alpha I_s \exp(\alpha V) \tag{14.22}$$

$$g_3 = \frac{dI_3}{dV} = \alpha I_s \exp(-\alpha V) \tag{14.23}$$

$$g_4 = \frac{dI_4}{dV} = \alpha I_s \exp(-\alpha V) \tag{14.24}$$

Since, in general, the LO power is higher than the RF power, it is assumed that the conductance of the diodes is modulated by the LO power alone. Thus, for the conductance calculations in Equations 14.21 through 14.24, the voltage V can be written as

$$V \simeq V_L \cos\omega_L t \tag{14.25}$$

then

$$g_1 = \alpha I_s \exp(\alpha V_L \cos\omega_L t) \tag{14.26}$$

$$g_2 = \alpha I_s \exp(\alpha V_L \cos\omega_L t) \tag{14.27}$$

$$g_3 = \alpha I_s \exp(-\alpha V_L \cos\omega_L t) \tag{14.28}$$

$$g_4 = \alpha I_s \exp(-\alpha V_L \cos\omega_L t) \tag{14.29}$$

The current in the diode can be approximated by

$$I_1 = g_1 V_1 \tag{14.30}$$

$$I_2 = g_2 V_2 \tag{14.31}$$

$$I_3 = g_3 V_3 \tag{14.32}$$

$$I_4 = g_4 V_4 \tag{14.33}$$

Here V_1, V_2, V_3, and V_4 will include the signal and be expressed as

$$V_1 = V_L \cos\omega_L t + V_s \cos\omega_s t \tag{14.34}$$

$$V_2 = V_L \cos\omega_L t - V_s \cos\omega_s t \tag{14.35}$$

$$V_3 = V_L \cos\omega_L t + V_s \cos\omega_s t \tag{14.36}$$

$$V_4 = V_L \cos\omega_L t - V_s \cos\omega_s t \tag{14.37}$$

Substituting Equations 14.26-14.29 and 14.34-14.37 into Equations 14.30-14.33 obtains

$$I_1 = \alpha I_s \exp(\alpha V_L \cos\omega_L t) [V_L \cos\omega_L t + V_s \cos\omega_s t] \tag{14.38}$$

$$I_2 = \alpha I_s \exp(\alpha V_L \cos\omega_L t) [V_L \cos\omega_L t - V_s \cos\omega_s t] \tag{14.39}$$

$$I_3 = \alpha I_s \exp(-\alpha V_L \cos\omega_L t) [V_L \cos\omega_L t + V_s \cos\omega_s t] \tag{14.40}$$

$$I_4 = \alpha I_s \exp(-\alpha V_L \cos\omega_L t) [V_L \cos\omega_L t - V_s \cos\omega_s t] \tag{14.41}$$

The current I_{23} and I_{14} in Figure 14.9 are

$$
\begin{aligned}
I_{23} &= I_2 + I_3 \\
&= \alpha V_L I_s [\exp(\alpha V_L \cos\omega_L t) + \exp(-\alpha V_L \cos\omega_L t)] \cos\omega_L t \\
&\quad - \alpha V_s I_s [\exp(\alpha V_L \cos\omega_L t) - \exp(-\alpha V_L \cos\omega_L t)] \cos\omega_s t
\end{aligned}
\tag{14.42}
$$

$$
\begin{aligned}
I_{14} &= I_1 + I_4 \\
&= \alpha V_L I_s [\exp(\alpha V_L \cos\omega_L t) + \exp(-\alpha V_L \cos\omega_L t)] \cos\omega_L t \\
&\quad - \alpha V_s I_s [\exp(\alpha V_L \cos\omega_L t) - \exp(-\alpha V_L \cos\omega_L t)] \cos\omega_s t
\end{aligned}
\tag{14.43}
$$

The current from the IF port is

$$
\begin{aligned}
I_{IF} &= I_{14} - I_{23} \\
&= 2\alpha V_s I_s [\exp(\alpha V_L \cos\omega_L t) - \exp(-\alpha V_L \cos\omega_L t)] \cos\omega_s t \\
&= 4\alpha V_s I_s [\sinh(\alpha V_L \cos\omega_L t)] \cos\omega_s t
\end{aligned}
\tag{14.44}
$$

Using the formula

$$\sinh(A\cos\theta) = 2 \sum_{k=0}^{\infty} N_{2k+1}(A) \cos(2k+1)\theta \tag{14.45}$$

where $N_k(A)$ are the modified Bessel functions of the first kind of order k. (Here $N_k(A)$ is used to represent the Bessel function rather than the conventional $I_k(A)$, in order not to be confused with the current I.) Equation 14.44 can be written as

$$I_{IF} = 4\alpha V_s I_s \, [2N_1(\alpha V_L)\cos\omega_L t\cos\omega_s t \, + \, 2N_3(\alpha V_L)\cos3\omega_L t\cos\omega_s t$$

$$+ 2N_5(\alpha V_L)\cos5\omega_L t\cos\omega_s t \, + \, \ldots]$$

$$= 4\alpha V_s I_s\{N_1(\alpha V_L)\,[\cos(\omega_L - \omega_s)t \, + \, \cos(\omega_L + \omega_s)t]$$

$$+ \, N_3(\alpha V_L)\,[\cos(3\omega_L - \omega_s)t \, + \, \cos(3\omega_L + \omega_s)t]$$

$$+ \, N_5(\alpha V_L)\,[\cos(5\omega_L - \omega_s)t \, + \, \cos(5\omega_L + \omega_s)t]$$

$$+ \, \ldots\} \hspace{3cm} (14.46)$$

From Equation 14.46 it can be seen that the total current at the IF port only contains frequency terms $mf_L \pm f_s$, where m is an odd integer. Thus, the harmonics derived from the even harmonics are suppressed. Similar analysis can be applied to single balanced mixer.

In order to improve the dynamic range of double balanced mixers, sometimes two diodes connected in series instead of one are used in each of the four branches of Figure 14.9. Although the intermodulation level of the mixer can be improved, more LO power is required to drive the mixer, since it must provide twice the voltage to control two diodes in series. This is one of the major disadvantages in high dynamic range mixers, especially in a system while LO power is scarce.

14.8 IMAGE REJECTION MIXERS[27]

For a given IF frequency ω_I, and LO frequency ω_L, there are two RF signals ($\omega_s = \omega_L \pm \omega_I$) that can produce the same IF output as shown in Figure 14.10. If one of these is considered to be the desired signal frequency, the other one is commonly referred to as an image, because they are

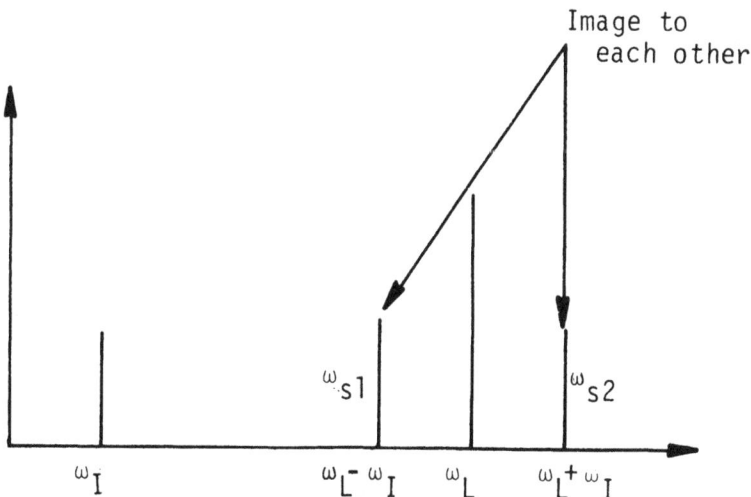

Figure 14.10. Image Frequency in a Mixer

mirror images with respect to the LO frequency. It should be noted that image frequency only exists in down converters and not in up converters. In a receiver, especially with a sweeping LO frequency, there is always the question of whether the detected signal is on the high or low side of the LO frequency. The obvious solution to the image problem is to insert a filter in front of the mixer to block one of the images. Another solution is to use an image rejection mixer.

An image rejection mixer will separate the RF inputs from the two sides of the LO frequency. It has two IF outputs; one contains the signal, the other one contains the image. Two mixers are often combined together to form an image rejection mixer as shown in Figure 14.11. The RF signal is fed to the mixer through a 90 degree hybrid, while the LO signal is applied through an in phase power divider. The IF outputs of the two mixers are combined through another 90 degree hybrid. With this arrangement the signal frequency appears at one of the IF outputs and the image at the other one. The concept of this approach is rather attractive; however, an image rejection mixer of decent performance requires good match between the two mixers.

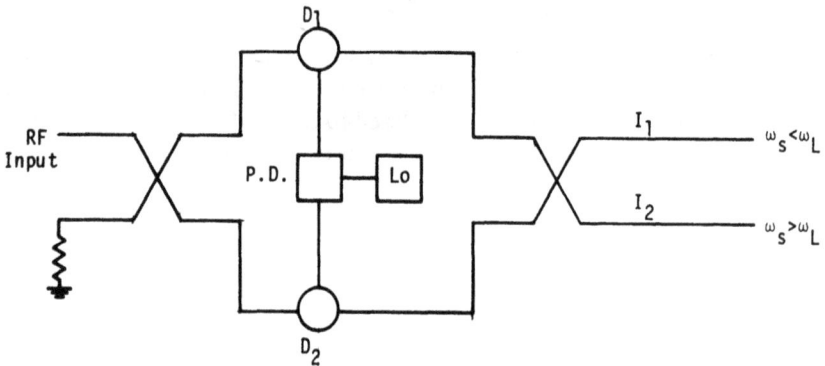

Figure 14.11. Image Rejection Mixer

The analysis of the image rejection mixer will be discussed here. Although the two mixers in Figure 14.11 are generally single balanced ones, single diode mixers are used in the analysis because of the simplicity. The conductances of mixers are (referring to Equation 14.26)

$$g_1 = g_2 = \alpha I_s \exp(\alpha V_L \cos\omega_L t) \qquad (14.47)$$

where α is the diode slope parameter, I_s is the diode saturation current, and V_L and ω_L are the amplitude and angular frequency of the LO signal, respectively.

The voltage across the diodes is

$$V_1 = V_L \cos\omega_L t + V_s \cos\omega_s t \qquad (14.48)$$

$$V_2 = V_L\cos\omega_L t + V_s(\cos\omega_s t - \frac{\pi}{2}) = V_L\cos\omega_L t + V_s\sin\omega_s t \quad (14.49)$$

where V_s and ω_s are the amplitude and angular frequency of the RF signal, respectively.

The current at the output of the diode mixers is

$$I_1' = g_1 V_1$$

$$= \alpha I_s (1 + \alpha V_L \cos\omega_L t + \frac{\alpha^2 V_L^2 \cos^2\omega_L t}{2} + \ldots)$$

$$\bullet \ (V_L \cos\omega_L t + V_s\sin\omega_s t)$$

$$= \alpha I_s (V_L \cos\omega_L t + V_s \cos\omega_s t)$$

$$+ \ \alpha^2 V_L I_s \cos\omega_L t (V_L \cos\omega_L t + V_s \cos\omega_s t)$$

$$+ \ \ldots$$

$$= \alpha V_L I_s \cos\omega_L t + \alpha V_s I_s \cos\omega_s t$$

$$+ \ \frac{\alpha^2 V_L^2 I_s}{2} \ (\cos 2\omega_L t +)$$

$$+ \ \frac{\alpha^2 V_L V_s I_s}{2} \ [\cos(\omega_L + \omega_s)t + \cos(\omega_L - \omega_s)t]$$

$$+ \ \ldots \qquad (14.50)$$

$$I_2' = \alpha V_L I_s \cos\omega_L t + \alpha V_s I_s \sin\omega_s t$$

$$+ \ \frac{\alpha^2 V_L^2 I_s}{2} \ (\sin 2\omega_L t)$$

$$+ \ \frac{\alpha^2 V_L V_s I_s}{2} \ [\sin(\omega_L + \omega_s)t - \sin(\omega_L - \omega_s)t]$$

$$+ \ \ldots \qquad (14.51)$$

In the previous derivation, the exponential term in Equation 14.47 is expressed in power series.

Since only the terms containing $\omega_L - \omega_s$ are of interest, the rest of the terms will be neglected and I_1' and I_2' can be simplified as

$$I_1' = A \cos(\omega_L - \omega_s)t \qquad (14.52)$$

$$I_2' = -A \cos(\omega_L - \omega_s)t \qquad (14.53)$$

where A is a constant equals $a2V_LV_sI_s/2$.

The currents from the second 90 degree hybrid can be written as

$$I_1 = I_1' + I_2' \; \underline{/{-\dfrac{\pi}{2}}} \qquad (14.54)$$

$$I_2 = I_1' \; \underline{/{-\dfrac{\pi}{2}}} + I_2' \qquad (14.55)$$

If ω_s is lower than ω_L ($\omega_s < \omega_L$)

$$I_1 = A \cos(\omega_L - \omega_s)t - A \sin[(\omega_L - \omega_s)t - \dfrac{\pi}{2}]$$

$$= 2A \cos(\omega_L - \omega_s)t \qquad (14.56)$$

$$I_2 = A \cos[(\omega_L - \omega_s)t - \dfrac{\pi}{2}] - A \sin(\omega_L - \omega_s)t = 0 \qquad (14.57)$$

If ω_s is higher than ω_L ($\omega_s > \omega_L$)

$$I_1 = A \cos(\omega_s - \omega_L)t + A \sin[(\omega_s - \omega_L)t - \dfrac{\pi}{2}] = 0 \qquad (14.58)$$

$$I_2 = A \cos[(\omega_s - \omega_L)t - \dfrac{\pi}{2}] + A \sin(\omega_s - \omega_L)t$$

$$= 2A \sin(\omega_s - \omega_L)t \qquad (14.59)$$

From Equations 14.56 to 14.59, it can be concluded that if $\omega_s = \omega_L - \omega_I$ the output appears at I_1 and if $\omega_s = \omega_L + \omega_I$ the output appears at I_2, which means the signal and its image are separated into the two IF outputs. This mixer can be used to improve the instantaneous RF bandwidth of a receiver.

14.9 IMAGE ENHANCED MIXERS[28, 29, 30]

An input signal to a mixer will produce a spurious response at its image frequency. Let $\omega_s = \omega_L \pm \omega_I$ then its image is at $\omega_L \mp \omega_I$. If ω_s mixes with the second harmonics of ω_L, the following results will be obtained

$$2\omega_L - \omega_s = 2\omega_L - (\omega_L \pm \omega_I) = \omega_L \mp \omega_I \qquad (14.60)$$

which indicates that any signal that mixes with the second harmonic of the LO frequency will generate its own image frequency. There is useful energy in the image frequency. This energy at the image frequency can be recovered to produce additional power, and thereby decrease conversion loss. This is the basic operating principle behind the image enhanced mixer. The insertion loss of a mixer can be reduced by 1-2 dB through the image enhanced approach.

For single balanced mixer with 180 degree hybrid, the generated image appears at the RF input port. A filter placed at the RF input port passing only the RF bandwidth of interest will reflect the image energy back into the mixer. By adjusting the electric length between the filter and the mixer, the image can be reflected back to produce IF energy in the proper phase to minimize the mixer conversion loss. This phasing is critical, since degradation in the conversion loss can also result from improperly tuning the distance between the RF filter and the diodes.

The above approach is suitable only for narrow band mixers. For wide bandwidth mixers, especially for mixers with overlapping signal and image bandwidths, filtering at the RF input port is not practical. A different approach with two single balanced mixers arranged as in Figure 14.12 can reflect the image frequency. The RF inputs are fed in phase to the two mixers, while the LO inputs and IF outputs are coupled through 90 degree hybrids. This configuration causes the image signals, generated by the two balanced mixers at the RF input ports, to be 180 degrees out-of-phase with each other. This results in a maximum image current at the input port and

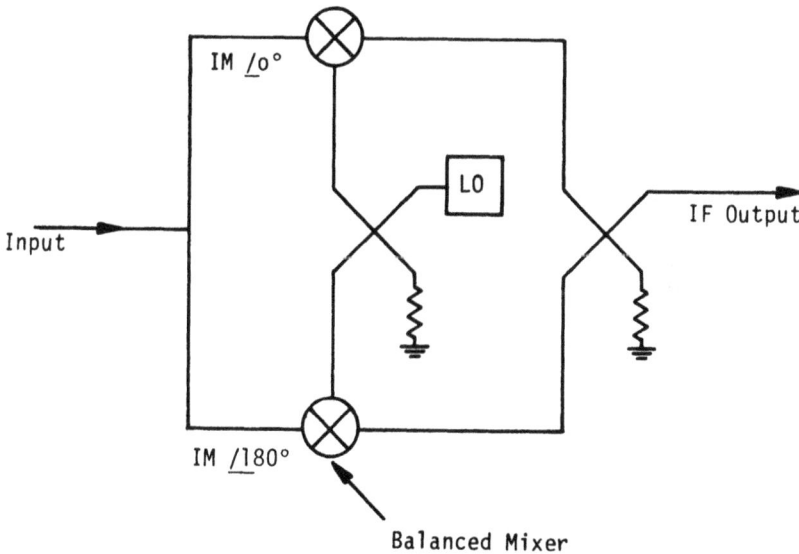

Figure 14.12. Image Enhanced Mixer with Two Balanced Mixers

is equivalent to a short circuit condition for the image current. This short circuit condition is not bandwidth limited because the electric length between the diodes of the two balanced mixers is very short, and there is essentially no phase differences between them.

The 180 degree phase difference of the image signal can be simply explained as follows. The two LO frequencies are $\cos\omega t$ and $\cos(\omega t - \pi/2)$ and their second harmonics, $\cos\omega t$ and $\cos(2\omega t - \pi)$, respectively, are 180 degrees out-of-phase. Since the image frequencies are generated by mixing the signal frequencies with the second harmonics of the LO frequency, they are 180 degrees out-of-phase because the second harmonics of the LO are 180 degrees out-of-phase.

14.10 HARMONICALLY PUMPED MIXERS[31-33]

Harmonically pumped mixing means that the IF output is generated by the signal frequency and the harmonics of the LO frequency (i.e., IF = $\pm \omega_s \mp n\omega_L$, where n is an integer and n< 1). These signals are always present in a conventional mixer. They are considered the spurious responses and are usually suppressed. In the harmonically pumped mixers, they are considered desired outputs and, therefore, are enhanced. Harmonic mixing has been used primarily at the millimeter (mm) wave frequencies where stable LO sources are either unavoidable or prohibitively expansive. Sometimes the harmonically pumped mixer is used in a superheterodyne receiver with wide input bandwidth and limited LO tuning range. In the latter application, fundamental mixing is used when the input frequency is low; and harmonic mixing is used when the input frequency is high. The sensitivity of this kind of receiver degrades as higher harmonic mixing is used.

If only even order harmonics are of interest, the antiparallel diode pair arrangement in Figure 14.13 can provide the following advantages:

1. Reduced conversion loss by suppressing the fundamental mixing products.
2. Lower noise figure through suppression of LO noise side-bands.
3. Suppression of direct video detection.
4. Inherent self-protection against large peak inverse voltage burnout.

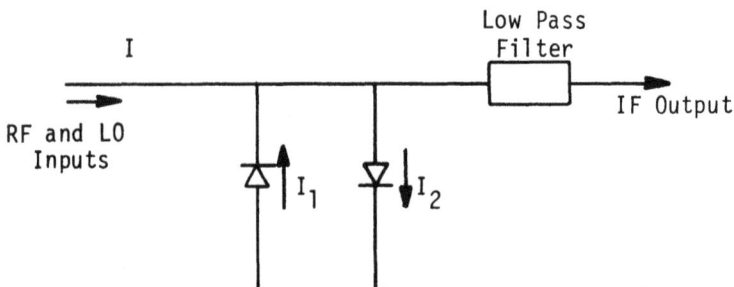

Figure 14.13. Antiparallel Diode Pair Mixer

The analysis of the antiparallel diode pair mixer will be discussed in a similar manner as a double balanced mixer. The differential conductance for each diode is (see Equation 14.21 through 14.24)

$$g_1 = \alpha I_s \exp(-\alpha V) \tag{14.61}$$

$$g_2 = \alpha I_s \exp(\alpha V) \tag{14.62}$$

The composite differential conduction is

$$\begin{aligned} g = g_1 + g_2 &= \alpha I_s \left[\exp(\alpha V) + \exp(-\alpha V)\right] \\ &= 2\alpha I_s \cosh \alpha V \end{aligned} \tag{14.63}$$

Assuming the conductance is controlled by the LO input only

$$\begin{aligned} g &= 2\alpha I_s \cosh(\alpha V_L \cos\omega_L t) \\ &= 2\alpha I_s \left[N_0(\alpha V_L) + 2N_2(\alpha V_L)\cos 2\omega_L t \right. \\ &\quad \left. + 2N_4(\alpha V_L)\cos 4\omega_L t + \ldots\right] \end{aligned} \tag{14.64}$$

where α is the diode slope parameter; and V_L and ω_L are the LO voltage and angular frequency, respectively. $N_n(\alpha V_L)$ are modified Bessel functions of first kind. The current I in the mixer can be written as

$$\begin{aligned} I = gV &= g(V_L \cos\omega_L t + V_s \cos\omega_s t) \\ &= A\cos\omega_L t + B\cos\omega_s t + C\cos 3\omega_L t + D\cos 5\omega_L t \\ &\quad + E\cos(2\omega_L + \omega_s)t + F\cos(2\omega_L - \omega_s)t + G\cos(4\omega_L + \omega_s)t \\ &\quad + H\cos(4\omega_L - \omega_s)t + \ldots \end{aligned} \tag{14.65}$$

It can be seen that the total current only contains frequency terms $mf_L \pm nf_s$ where $m + n$ is an odd integer (i.e., $m + n = 1, 3, 5\ldots$). The desired signals are the terms containing $2\omega_L + \omega_s$ and $2\omega_L - \omega_s$.

In the above approach, not only has the fundamental product been suppressed, but also all the mixing products containing the odd harmonics of the LO frequency. Thus, it is obvious that this antiparallel diode pair arrangement can only be used for even harmonic mixing.

14.11 OTHER APPLICATIONS OF MIXERS[34, 35]

Besides up or down converting one frequency to another, the mixers also have some other useful applications. Some of them are related to receiver work; others have no direct applications. Since they are important in general microwave laboratory work, some of them will be stated here.

A. Phase Detectors

The basic concept of phase detection is that the application of two identical frequency, constant amplitude signals to a mixer results in a dc output which is proportional to the phase difference between the two signals. Mathematically, the phase detection can be derived from the square term of of the mixer. For example, the input voltage

$$V_i = V_L \cos\omega t + V_s \cos(\omega t + \theta) \qquad (14.66)$$

where θ is the phase difference between the two input signals of the same frequency ω.

The square term at the output of the mixer is

$$V_i^2 = V_L^2 \cos^2 \omega t + V_s^2 \cos^2(\omega t + \theta) + 2V_L V_s \cos(\omega t)\cos(\omega t + \theta)$$

$$= \frac{V_L^2}{2} + \frac{V_L^2}{2}\cos(2\omega t) + \frac{V_s^2}{2} + \frac{V_s^2}{2}\cos(2\omega t + 2\theta)$$

$$+ V_L V_s \cos(2\omega t + \theta) + V_L V_s \cos\theta \qquad (14.67)$$

Equation 14.67 shows that the output of a mixer contains three dc terms: $V_L^2/2$, $V_s^2/2$, and $V_L V_s \cos\theta$. It can be shown from Equation 14.46 that in an ideal double balanced mixer the dc output contains only the $V_L V_s \cos\theta$ term which is generated from the very first term. If a low-pass filter is added to the output of the mixer, the IF output of the mixer is proportional only to $V_L V_s \cos\theta$. However, in a practical double balanced mixer, there might be some other dc components than $V_L V_s \cos\theta$ in the IF output which are generally referred to as dc offset. The dc offset should be kept at a minimum. As a general rule, the better the isolations between the ports of a double balanced mixer, the less the dc offset.

Phase detectors are often used in a phase locked loop to stabilize the frequency of an oscillator with respect to the stable reference signal which will be discussed in Chapter 15.

B. Double Balanced Mixers Used for Modulator and Current Controlled Attenuator

Using the LO port as input, the RF port as output, and the IF port as the control terminal (as shown in Figure 14.14), a double balanced mixer can be used in various ways as a signal modulator. The following modulations can be accomplished through the IF port:

1. If the IF input is a dc signal plus a modulating signal, the carrier frequency can be amplitude modulated.
2. If the IF input is an undirectional pulse, the carrier signal can be pulse modulated. The control pulse can be either positive or negative.

The maximum isolation can be provided equal to the isolation between the LO and RF ports.

3. For bi-phase modulation, reverse the polarity of the switching signal. Upon reversal, the output phase will shift by 180 degrees.

4. The amount of signal passing through the mixer from the LO port to the RF port is determined by the dc control current present at the IF port. Thus, a double balanced mixer can be used as an RF attenuator. Maximum attenuation is achieved with no dc current which corresponds to the isolation of the mixer. Attenuation will decrease with increasing the dc bias current to the IF port.

C. Simultaneous Signal Detectors

The mixer can be used as a simultaneous signal detector for an instantaneous frequency measurement (IFM) receiver. The discussion of this application can be found in Chapter 4.

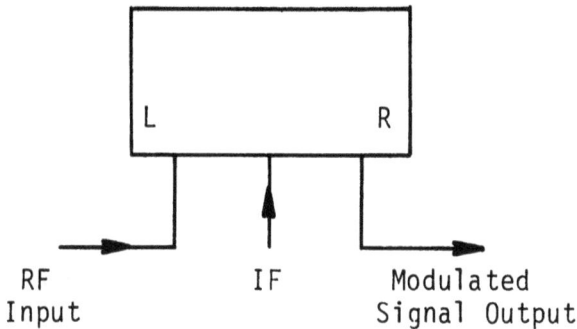

Figure 14.14. Double Balanced Mixer use as a Modulator

14.12 CHARACTERISTICS OF DETECTORS[36-39, 48]

The performance characteristics that are used to describe detector diodes are: current voltage relationship (I-V curve), tangential sensitivity (TSS), video resistance, voltage sensitivity, and figure of merit. Each of these terms will be discussed as follows:

A. I-V Curve

The current voltage relationship for diodes ideally obeys the well known diode equation.

$$I = I_s (\exp \frac{qV}{nkT} - 1) \qquad (14.68)$$

where I is the current through the diode, V is the voltage applied across the diode, I_s is the saturation current, n is the constant somewhat greater than unity but near unit, k is Boltzmann's constant $= 1.380 \times 10^{-16}$ erg/°K,

T is the absolute temperature, and q is the electronic charge = 1.601×10^{-19} Coul. For many calculations, n = 1 is used (i.e., in Equation 13.72). In the beginning of this chapter, the diode slope parameter α is used to replace q/nkT.

The I-V curve can be shown in Figure 14.15. Real diodes also exhibit parasitic resistance and capacitance. When the diode is used as a detector, it usually operates in the forward biased region.

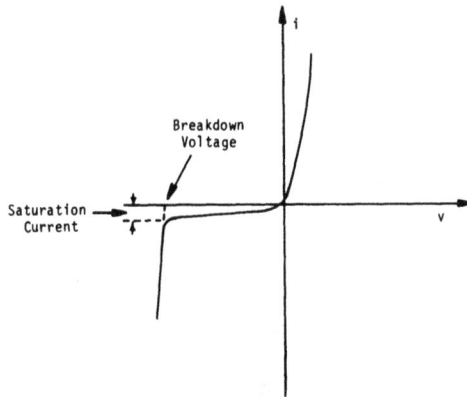

Figure 14.15. Typical Current Voltage Relationship of a Diode

B. Tangential Sensitivity (TSS)

Tangential sensitivity is the input signal level that will provide an 8 dB S/N at the output of the video amplifier. The TSS is discussed in detail in Chapter 2. The TSS does not depend on intrinsic diode parameters alone, but many other factors affect the measured TSS value of a given diode. The most important factors are:

1. RF frequency and bandwidth
2. Video bandwidth
3. Diode dc bias current
4. Test mount or circuit
5. Video amplifier noise figure following the detector

To obtain maximum sensitivity at any given frequency, the diode is usually biased in the forward direction. Bias, however, introduces noise in the diode and reduces the diode video resistance. These effects exert a competitive influence of TSS; therefore, the bias value must be stated for given TSS. A typical TSS versus dc bias current of some detector diodes is shown in Figure 14.16. The TSS increases with the increase of dc bias current. By further increasing the bias current, the TSS starts to decrease.

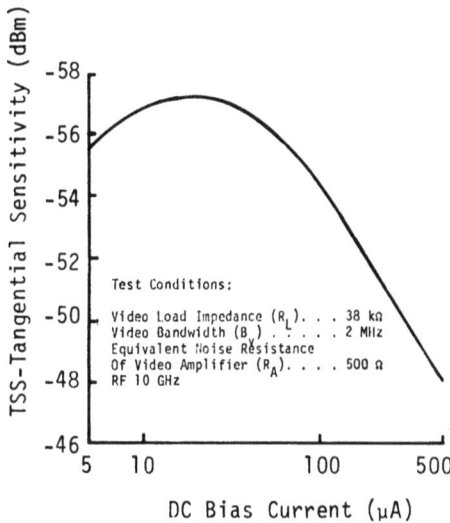

Figure 14.16. Effect of dc Bias on TSS (Based on Reference 36)

If the value of n = 1.08 is used in Equation 14.68, the TSS of a Schottky diode can be expressed as (References 36 and 48).

$$\text{TSS (dBm)} = -107 + 5 \log B_V + 10 \log I_d$$

$$+ 5 \log \left[R_A + \frac{28}{I_d}\left(1 + \frac{f_N}{B_V} \ln \frac{B_V}{f_L}\right)\right]$$

$$+ 10 \log \left[1 + \frac{R_s C_{j(i)}^2 f^2}{I_d} \right] \tag{14.69}$$

where B_V is the video bandwidth in Hz, I_d, is the diode dc bias current in μA, f_N is the diode flicker noise corner frequency in Hz, f_L is the video circuit low frequency 3 dB point in Hz, R_s is the diode series resistance in Ω, $C_{j(i)}$ is the diode junction capacitance in pf at the bias current I_d, f is the operating frequency in GHz, and R_A is the amplifier equivalent series noise resistance in $K\Omega$.

C. Equivalent Circuit

The equivalent circuit of a detector diode is shown in Figure 14.17. L_s and R_s are the series inductance and resistance, respectively. C_p is the package capacitance. C_j and R_j are the junction capacitance and resistance, respectively. C_p and L_s are circuit elements that depend on the

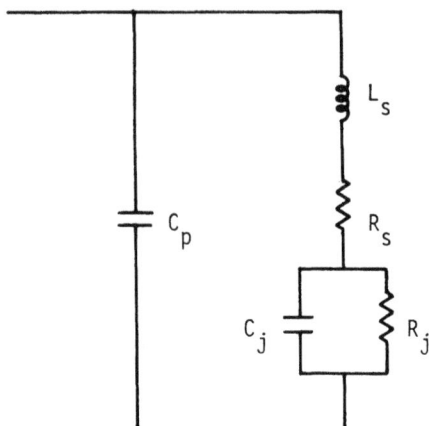

Figure 14.17. Equivalent Circuit of Detector Diode

diode geometry and the circuit configurations in which the diode is used. The resistors R_s, R_j, and capacitor C_j are dependent on diode material and the junction property.

D. Video Resistance (R_V)

This resistance consists of two parts: the diode series resistance (R_s) and the junction resistance (R_j)

$$R_v = R_s + R_j \qquad (14.70)$$

The junction resistance R_j is simply the small signal low frequency dynamic resistance of the diode, and it is a function of the dc bias current. R_j can be obtained by differentiating the diode I-V relationship and is given by

$$R_j = \frac{nkT}{q(I_d + I_s)} \qquad (14.71)$$

where I_d is the bias current.

E. Voltage Sensitivity (γ)

This parameter specifies the slope of the output video voltage versus the input power, i.e.,

$$V = \gamma P_i \qquad (14.72)$$

The value of γ depends on the load resistance, signal level, and RF frequency. It is particularly sensitive to the signal level which must be kept in the square law range of the diode. The higher the value of γ, the more sensitive the diode is.

F. Figure of Merit

The figure of merit is an old "measure" of the "supposed efficacy" of a video detector diode and is given by

$$M = \frac{\gamma}{\sqrt{R_v + R_A}} = \frac{\beta R_v}{\sqrt{R_A + R_v}} \qquad (14.73)$$

where R_A is the video amplifier series noise resistance, and β is the current sensitivity of the diode ($\beta = \gamma/R_V$).

However, in other papers, (Reference 37), the figure of merit is defined as

$$M = \frac{\gamma}{\sqrt{R_V}} = \beta \sqrt{R_V} \qquad (14.74)$$

which means $R_A = 0$ in Equation 14.73. This condition assumes that the noise contributed from the video amplifier is negligible compared to that of the diode which is usually true. If the condition under which the diode performance is evaluated is specific, there should not be any confusion caused by Equation 14.73 and 14.74.

14.13 VIDEO BANDWIDTH (B_V)[38, 39]

The video bandwidth of a detector must be wide enough to pass the video information. However, in order to maintain high receiver sensitivity, the video bandwidth should be no greater than necessary to recover the modulation information. The video bandwidth required for pulse recovery depends on the nature of the information to be gained from the pulse. For example, sometimes peak pulse detectability is more significant than pulse shape; other times leading edge information (rise time) and pulse width are more desirable. The formal approach in finding the required bandwidth is through the definition of effective bandwidth which is defined as

$$B_e^2 = \frac{(2\pi)^2 \int_{-\infty}^{\infty} f^2 |S(f)|^2 df}{\int_{-\infty}^{\infty} |S(f)|^2 df} = \frac{(2\pi)^2}{E} \int_{-\infty}^{\infty} f^2 |S(f)|^2 df \qquad (14.75)$$

where $S(f)$ is the Fourier transform of the input signal envelope and E is the signal energy.

The deviation of the pulse position (ΔT) is related to the effective bandwidth (B_e) by

$$\Delta T = \frac{1}{B_e \sqrt{2E/N_0}} \qquad (14.76)$$

where N_0 is noise power per unit bandwidth.

However, there are several empirical equations which are often used to obtain the required video bandwidth. If maximum pulse detectability, which means maximum S/N ratio is desired, the bandwidth is inversely proportional to the pulse width (PW)

$$B_V \cong \frac{1}{PW} \qquad (14.77)$$

for a rectangular pulse and rectangular low-pass filter. If a simple RC filter is used to provide the video bandwidth, the maximum S/N occurs when the filter's upper 3dB video frequency is approximately

$$f_{u(3\,dB)} \cong \frac{0.25}{PW} \qquad (14.78)$$

In other applications, it may be desirable to resolve the risetime (tr) of the pulse. The risetime of a pulse is usually defined from 0.1 to 0.9 of its peak value. The required upper 3 dB frequency is

$$f_{u(3\,dB)} \cong \frac{0.35}{tr} \qquad (14.79)$$

which is usually higher in frequency than that required for maximum S/N ratio.

If the video circuit is ac coupled, the low frequency cut-off of the video bandwidth will affect the amplitude of the pulse. The amplitude of the pulse will droop. The amplitude droop and the low frequency 3 dB point is related by

$$f_{L(3\,dB)} \cong \frac{droop\ \%}{600\ PW} \qquad (14.80)$$

If the video amplifier is dc coupled $f_{L(3\,dB)} = 0$, there is no amplitude droop problem.

14.14 TYPES OF DIODE DETECTORS[40-46]

Generally, the detector diodes can be classified into two types (Reference 40): the square law detectors (also referred to as the small signal type) and the linear detectors (also referred to as the large signal type or peak detector). It should be noted that although one type of the detector is called linear, it is indeed a nonlinear device which rectifies the RF signal. In a square law detector, the output voltage is proportional to the square of the input voltage or directly proportional to the input power. The relation is shown in Equation 14.72 and is rewritten as

$$E_0 = \gamma P_i \qquad (14.81)$$

where γ is a constant and P_i is the input power, while in a linear detector the output voltage

$$E_0 = kE_i \qquad (14.82)$$

where k is a constant and E_i is the input voltage.

A detector will contain both the square law region and the linear region. Figure 14.18 shows a typical detector output voltage versus the input power. At low input power level, the detector operates at the square law region, while at high input level the detector operates at the linear region. If the detector is used in a receiver with enough RF amplification and the signal is hard limited at the input of the detector, the detector will operate either in the linear or square law region depending on the input power to the detector. If the input signal is not hard limited, the detector will operate over both regions. In general, the difference between the linear and square law detector is small and many times they are chosen for mathematic reasons (i.e., the analysis of IFM receivers in Chapter 4).

There are many different solid-state devices that can be used as detectors (Reference 41). They are Schottky diodes, tunnel diodes, thermoelectric detectors, and space charge limited dielectric diodes. The most popular detectors for microwave applications are Schottky and tunnel diodes.

Schottky diodes use the Schottky barrier (References 42 and 43) which is a metal-to-semiconduct junction. The current in the Schottky barrier junction depends on majority carrier conduction, in contrast to the minority carrier operation of ordinary p-n junction diodes. Since there are no minority carrier storage effects involved, these devices are potentially capable of operation up to frequencies approaching the reciprocal of the dielectric relaxation time τ_d of the semiconductor crystal. For a practical device, this frequency will be in the order of 1000 GHz.

Schottky diodes can be made through two approaches; one is called the point contact diode; the other is called planar Schottky diode (or hot carrier diode). The point contact diode is fabricated by pressing a fine metal point into the surface of the semiconductor, sometimes followed by an electrical or mechanical forming operation. A hot carrier diode is fabricated by depositing a metal film on a prepared surface of the

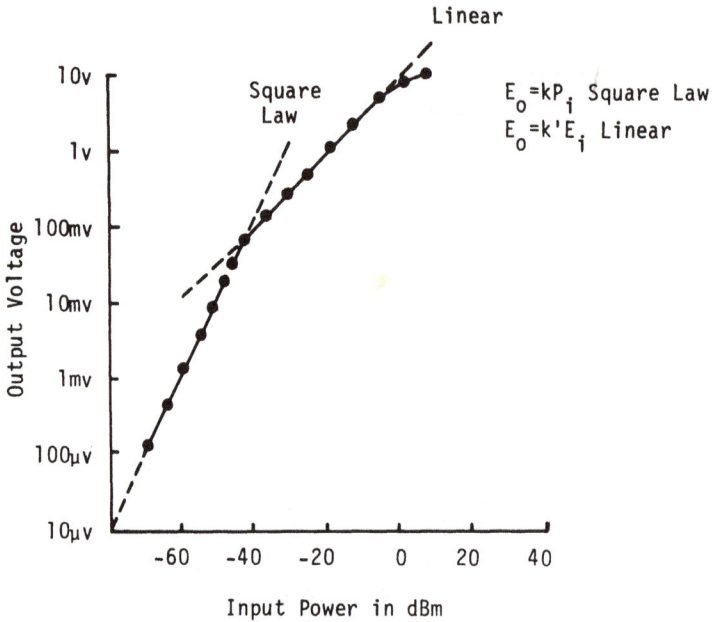

Figure 14.18. *Typical Detector Output Voltage versus Input Power (Based on Reference 44)*

semiconductor. A photometallurgically built Schottky diode is described in Szente et. al., paper (Reference 44) as shown in Figure 14.19. In Figure 14.19, A is a heavily doped, single crystal silicon waver; B is a lightly doped epitaxial silicon layer; C is a layer of protective glass (S_iO_2); D is the metal semiconductor junction; and E is a layer of suitable metal deposited on the surface of the semiconductor. The choice of the metal is the most important factor in controlling the Schottky barrier height and consequently the saturation current. F is the contact metal which is plated at the top and the bottom of the device. Some obvious advantages of the hot carrier diode over the point contact diode are the closer control of geometry and better resistance to mechanical shock. It is difficult, however, to make hot carrier diodes with capacitance as small as that of a point contact diode. This capacitance combining with the series resistance of the diode determines the upper operation frequency of the diode.

A tunnel diode is essentially a p-n junction whose density is made purposely high in order to produce a very narrow junction across which electrons can tunnel easily. The tunnel diode is also discussed in Chapter 13 as an amplifier. The I-V characteristics is shown again in Figure 14.20. When operated as detectors, tunnel diodes are ordinarily biased in the forward direction at a voltage somewhat lower than the voltage corresponding to the peak current I_p. Tunnel diodes which are designed for a lower peak current than the usual tunnel diode and operated on the reverse portion of the I-V curve

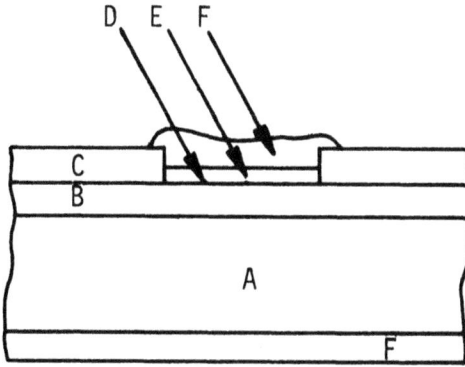

(a) Silicon Wafer   (d) Metal Semiconductor Junctions
(b) Epitaxially Grown Silicon Layer   (e) Metal
(c) Protective Glass   (f) Contact Metal

Figure 14.19. Cross Section View of a Schottky Diode Structure (Based
              on Reference 44)

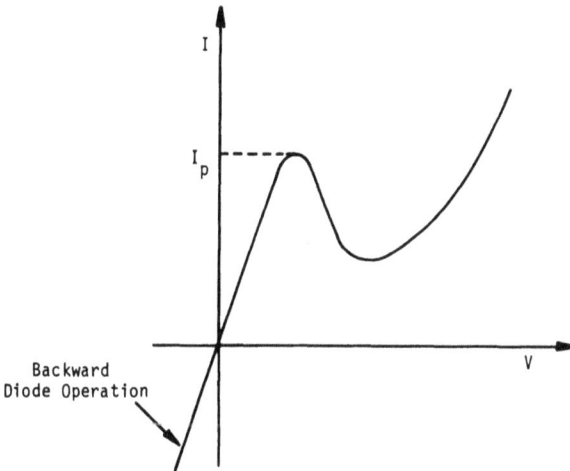

Figure 14.20. I-V Curve of a Tunnel Diode

(Reference 45) as shown in Figure 14.20 is referred to as backward diodes.
Sareen (Reference 46) compared the performance of Schottky diodes versus
tunnel diodes and drew the following conclusions:
  1. If properly biased, Schottky diodes have better sensitivity.
  2. Tunnel diodes have faster risetime, especially at higher power levels
     (0 dBm). The risetime of Schottky diodes can be improved by sacri-
     ficing its sensitivity through loading the diode with resistance.

3. It is easier to build dc coupled detector logarithmic amplifiers by using tunnel diodes, because no dc bias is required. In order to dc couple a Schottky detector, the dc bias must be eliminated from the detector. A common method for doing this is illustrated in Figure 14.21. The dc bias effect is subtracted by the differential amplifier. Even with this arrangement, the error introduced is quite large.
4. Tunnel diodes have better tracking accuracy versus temperature than that of Schottky diodes.
5. Tunnel diode input VSWR can be kept below 2.5 through proper circuit design. The input VSWR of Schottky can be improved by increasing the dc bias current which in turn decreases the detection sensitivity.

Sareen concluded that in system applications where the overall parameters need to be optimized, a tunnel diode is superior to the Schottky detector. However, if only one or two of the discussed parameters are important, a Schottky diode may be the logical choice.

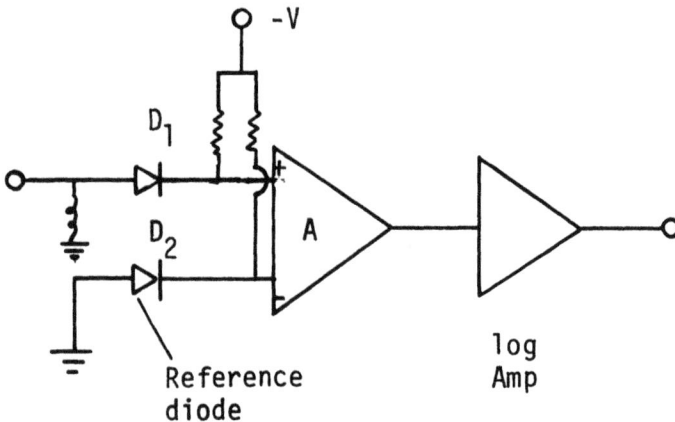

Figure 14.21. DC Coupled Logarithmic Amplifier with dc Biased Detectors

14.15 DIODE DETECTOR CIRCUITRY[44, 47, 48]

There are basically two kinds of circuits designed around the detector diodes: one kind does not provide the bias current for the diode (References 44 and 47); the other provides the diode with bias current. A simple circuit for the unbiased diode is shown in Figure 14.22. The resistor R is used to match the diode to the input impedance and C_b is an RF bypass capacitor. Using the resistor R to improve input impedance match will usually cause reduction in sensitivity due to power absorbed by the resistor. However, sensitivity loss due to reflections from an unmatched

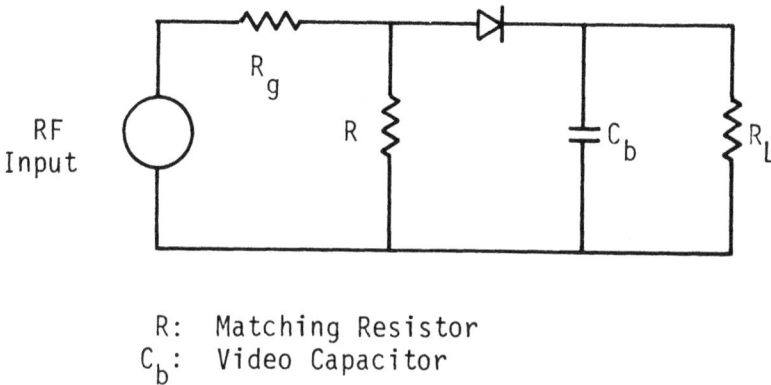

R: Matching Resistor
C_b: Video Capacitor

Figure 14.22. A Simple Detector Circuit

diode is sometimes worse than that due to the matching network. The capacitor C_b and the load resistor R_L will form a low-pass filter that determines the bandwidth of the video circuits.

Bias is often provided to improve the sensitivity and sometimes the input VSWR of a Schottky diode (References 36, 40, and 48). A typical video circuit with dc biasing is shown in Figure 14.23. At RF, the bypass capacitor C_b appears as a short circuit, and the RF chokes (RFC_1 and RFC_2) appear as open circuits. The dc bias is fed through RFC_1 and RFC_2. The input RF filter is optimized to match the signal source impedance to the diode's RF impedance over a specified bandwidth and, ideally,

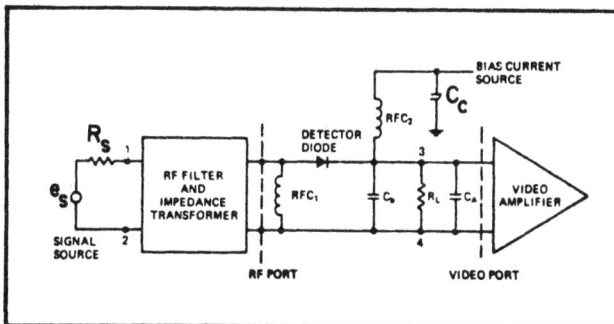

Figure 14.23. Typical Video Receiver (Reference 36, Courtesy of Hewlett Packard Co.)

all of the available signal power is delivered to the diode. The video signal is extracted from the modulated RF signal and appears across the video load resistance R_L. At video frequencies, RFC_1 acts as a short circuit, while C_b and the dc bias filters consisting of RFC_2 and C_c, both appear as high impedance. As shown in Figure 14.16, if the bias current is increased to 20 μA, the TSS of the detector is approximately at its maximum. However, increasing the bias current above 20 μA will sometimes widen the video bandwidth, improve the input impedance matching, and raise the compression point (Figure 14.18) which in turn extends the dynamic range. Therefore, the bias current for the Schottky diode is determined by the overall performance of the detector rather than the sensitivity alone.

REFERENCES

1. Reynolds, J.F., and Rosenzweig, M.R., "Learn the language of mixer specification," Microwaves, p. 72, May 1978.
2. Fukuchi, S.M., and Mouw, R.B., "Broad-band double balanced mixer/modulators," Parts I, II, Microwave Journal, March/May 1969.
3. Mills, H.D., "On the equation i = $i_0[exp\alpha(V - Ri) - 1]$," IBM Journal, pp. 553-554, September 1967.
4. Rutz-Philipp, E.M., "Power conversion in nonlinear resistive elements related to interference phenomena," IBM Journal, pp. 544-552, September 1967.
5. Kerr, A.R., "A technique for determining the local oscillator waveforms in a microwave mixer," IEEE Trans. Microwave Theory and Techniques, Vol. MTT-23, pp. 828-831, October 1975.
6. "17 most asked questions about mixers," Q&A No. 1 in a series Mini-Circuit, A Division of Scientific Components Corp., 2625 E. 14th St., Brooklyn, New York 11235.
7. Cheadle, D., "Selecting mixers for best intermod performance," Microwaves, p. 48, p. 58, November/December 1973.
8. Neuf, D., and Brown, D., "What to look for in mixer specs," Microwaves, p. 48, November 1974.
9. Cheadle, D.L., "Measure mixer noise with your power meter," Microwaves, March 1975.
10. Howson, D.P., "Minimum conversion loss and input match conditions in the broad-band mixer," The Radio and Electronic Engineer, Vol. 42, pp. 237-242, May 1972.
11. Cohen, J., "Mixer conversion loss versus local oscillator waveshape," Microwave Journal, p. 71, October 1978.
12. Brown, T.T., "Mixer harmonic chart," Electronic Buyers' Guide, pp. R-58-R-59, June 1953.
13. Skolnik, M., Radar Handbook, Chapter 5, McGraw-Hill, 1970.
14. Huang, M.Y., Buskirk, R.L., and Carlile, D.E., "Select mixer frequencies painlessly," Electronic Design, pp. 103-109, 12 April 1976.

15. Pollack, H.W., and Engelson, M., "An analysis of spurious response levels in microwave receivers," Microwave Journal, p. 72, December 1962.
16. Nitzberg, R., "Spurious frequency rejection," IEEE Trans. Electromagnetic Compatibility, Vol. EMC-6, pp. 33-36, January 1964.
17. Orloff, L.M., "Intermodulation analysis of crystal mixer," Proc. IEEE, Vol. 52, pp. 173-179, February 1964.
18. Lepoff, J.H., and Cowley, A.M., "Improved intermodulation rejection in mixers," IEEE Trans. Microwave Theory and Techniques, Vol. MTT-14, pp. 618-632, December 1966.
19. Herishen, J.T., "Diode mixer coefficients for spurious response prediction," IEEE Trans. Electromagnetic Compatibility, Vol. EMC-10, pp. 355-363, December 1968.
20. Yousif, A.M., and Gardiner, J.G., "Multifrequency analysis of switching diode modulators under high level signal conditions," Radio and Electronic Engineer, Vol. 41, V.D.C. 621.376.233, January 1971.
21. Beane, E.F., "Prediction of intermodulation levels as function of local oscillator power," IEEE Trans. Electromagnetic Compatibility, Vol. EMC-13, pp. 56-63, May 1971.
22. Bogan, Z.I., and Tsui, J.B.Y., "Computer modeling of diode circuits and its applications," Air Force Avionics Laboratory Technical Report, AFAL-TR-77-266, March 1978.
23. Pflieger, R., "A new MIC double balanced mixer with RF and IF band overlap," IEEE G-MTT International Symposium Digest Technical Papers, pp. 301-303, June 1973.
24. Ogawa, H., Aikawa, M., and Morita, K., "K-band integrated double balanced mixer," IEEE Trans. Microwave Theory and Techniques, Vol. MTT-28, pp. 180-184, March 1980.
25. Caruthers, R.S., "Copper oxide modulators in carrier telephone systems," Bell System Technical Journal, Vol. 18, pp. 315-337, April 1939.
26. Watson, H.A., *Microwave semiconductor devices and their circuit applications,* McGraw-Hill, New York, 1969.
27. Cochrane, J.B., "Thin film mixers team up to block out image noise," Microwaves, March 1977.
28. Salah, A.A.M., *Theory of resistive mixer,* M.I.T. Press, Boston, Massachusetts, 1971.
29. Dickens, L.E., and Maki, D.W., "A new phased type image enhanced mixer," IEEE G-MTT Int. Symp. Digest Tech., pp. 149-151, 1975.
30. Dickens, L.E., and Maki, D.W., "An integrated circuit balanced mixer, image, and sum enhanced," IEEE Trans. Microwave Theory and Techniques, Vol. MTT-23, pp. 276-281, March 1975.

31. Barber, M.R., "Noise figure and conversion loss of the Schottky barrier mixer diode," IEEE Trans. Microwave Theory and Techniques, Vol. MTT-15, pp. 629-635, November 1967.

32. Schneider, M.V., and Snell, W.W., Jr., "Harmonically pumped stripline down converter," IEEE Trans. Microwave Theory and Techniques, Vol. MTT-23, pp. 271-275, March 1975.

33. Cohn, M., Degenford, J.E., and Newman, B.A., "Harmonic mixing with an antiparallel diode pair," IEEE Trans. Microwave Theory and Techniques, Vol. MTT-23, pp. 667-673, August 1975.

34. Kurtz, S.R., "Mixers as phase detectors," Watkins-Johnson Technical Notes, Vol. 5, No. 1., January/February 1978.

35. *Mixer switches hybrid transformers,* Catalog No. 1, Watkins-Johnson Company.

36. "Hot carrier diode video detectors," Hewlett-Packard, Application Note 923.

37. Lucas, W.J., "Tangential sensitivity of a detector video system with r.f. preamplification," Proc. IEEE, Vol. 113, pp. 1321-1330, August 1966.

38. Burdic, W.S., *Radar signal analysis,* Chapter 5, Prentice Hall, 1968.

39. Skolnik, M.I., *Introduction to radar systems,* Chapter 10, McGraw-Hill Book Company, 1962.

40. "Dynamic range extension of Schottky detectors," Hewlett-Packard, Application Note 956-5.

41. Cowley, A.M., and Sorenson, H.O., "Quantitative comparison of solid-state microwave detectors," IEEE Trans. Microwave Theory and Techniques, Vol. MTT-14, December 1966.

42. Watson, H.A., *Microwave semiconductor devices and their circuit applications,* Chapter 11, McGraw-Hill Book Company, 1969.

43. Ziel, A. van der, *Solid-state physical electronics,* McGraw-Hill Book Company, 1968.

44. Szente, P.A., Adam, S., and Riley, R.B., "Low-barrier Schottky diode detectors," Microwave Journal, p. 42, February 1976.

45. Burrus, C.A., "Backward diodes for low level millimeter wave detection," IEEE Trans. Microwave Theory and Techniques, Vol. MTT-11, pp. 357-362, September 1963.

46. Sereen, S., "Schottky versus tunnel diode detectors in crystal video applications," Aertech Application Note.

47. "Impedance matching technique for mixers and detectors," Hewlett-Packard, Application Note 963.

48. "The 33800 series mixer/detector module," Hewlett-Packard, Application Note 921.

49. "Mixers, Switches, Hybrids, Transformers" Watkins-Johnson Company.

CHAPTER 15
OSCILLATORS AND FREQUENCY SYNTHESIZERS

15.1 INTRODUCTION[1-3]

A microwave oscillator is a one-port device which generates microwave energy. Its primary application in a receiver system is to be used as a local oscillator (LO) for a mixer. In many cases, the mixers are used simply to up or down convert the input signal to a different frequency for further processing; the LOs needed for these applications must generate very stable frequency and with low noise. In other applications the receiver is required to scan over a wide frequency range (i.e., a superheterodyne receiver). It is the LO that must sweep over this wide frequency range. If the receiver is required to search either continuously or in steps, the LO of the mixer usually must fulfill these requirements. In a compressive receiver as mentioned in Chapter 6, the LO must sweep a given bandwidth in a given time interval linearly. Sometimes the LO must cover a wide bandwidth in a very short time (i.e., over one gigahertz frequency range under 1 microsecond of time). Many different LO designs are available to fulfill these different operational requirements.

All of the devices mentioned in Chapter 13 that are suitable for microwave amplifiers can be made into oscillators. As in the microwave amplifiers, field effect transistors (FETs) have become the most popular device to be used in oscillators (References 2 and 3). This chapter will discuss the stability of oscillators and also different kinds of oscillators such as crystal, yttrium iron garnet (YIG) tuned, and voltage controlled oscillators (VCO).

There are two important categories of solid-state oscillators: the feedback oscillators and the negative resistance oscillators. In a feedback oscillator, an amplifier is required; and a fraction of the output is fed back via a frequency selective network to the input terminal. Transistor oscillators are usually feedback types. The negative resistance oscillator requires a device that exhibits a negative differential resistance (dV/dI). Tunnel diodes, avalanche diodes, and Gunn diodes are one-port negative resistance devices. The oscillators will start oscillation from white noise associated with the device and gradually build up to a steady-state.

To stabilize the frequency of an oscillator, phase locking loops (PLL) can be utilized which phase locks a microwave source to a very stable source. The frequency stability of the oscillator will be almost as good as the reference source. The PLL will be discussed here, since it is a common practice in microwave oscillators. Finally, synthesizers will be discussed.

15.2 STABILITY OF OSCILLATORS[4-11]

The stability of an oscillator is very important. It determines the capability of how accurate the receiver can measure the input signal. For example, if one desires to measure a signal at 10 GHz with 100 KHz accuracy and a mixer is used in the receiver to down convert the input frequency, the LO frequency stability must be better than 10^{-5} (100 KHz/10 GHz). Otherwise, the inaccuracy from the LO will be mixed into the input signal and the desired accuracy can no longer be obtained. The spurious response generated by the LO and the mixer must be below a certain level, or it will appear at the output of the mixer and be measured as a real signal. The result will be very confusing.

The stability of an oscillator can be divided briefly into two parts: long and short term. The long-term stability is usually expressed as parts per million of frequency change per hour, day, or even year. It represents a reasonably predictable phenomenon due to temperature change and the aging of device used in the frequency control element. This long-term stability is usually measured in the time domain as frequency versus time.

The short-term frequency fluctuations contain all elements causing frequency changes about the nominal frequency of less than a few seconds duration. Fluctuations of this nature are more conveniently viewed in the frequency domain than in the time domain. The short-term stability is more important in an oscillator and will be discussed here.

An ideal sine wave source can be described as (References 7 and 8)

$$V(t) = V_0 \sin 2\pi\nu_0 t \qquad (15.1)$$

where V_0 is the nominal amplitude, and ν_0 is the nominal frequency.

It should be noted that ν will be used to represent the frequency of the oscillation and f will be used to represent the fluctuations of the frequency ν or can be referred to as the Fourier frequency.

To account for the noise components at the output of a real sine wave generator, Equation 15.1 can be modified as

$$V(t) = [V_0 + E(t)] \sin[2\pi\nu_0 t + \phi(t)] \qquad (15.2)$$

where $E(t)$ is the amplitude fluctuation, and $\phi(t)$ is the phase fluctuations.

To determine the extent of the noise components $E(t)$ and $\phi(t)$, a measurement technique will be discussed. Since it is usually easier to represent these noises in the frequency domain than in the time domain, spectrum analysis will be used to characterize the noise. Figure 15.1a shows a sine wave which is perturbed for short instances by noise. These types of noise are loosely referred to as "glitches." This same signal can be represented in the frequency domain as in Figure 15.1b. This kind of plot is called the power spectrum. The power spectrum has a high value at ν_0 and lower value at ν_s where $\nu_0 = 1/T_0$ and $\nu_s = 1/T_s$ where T_0 and T_s are the time intervals shown in Figure 15.1a. From these figures, it should be noted that the noise glitches have a somewhat constant repetition rate.

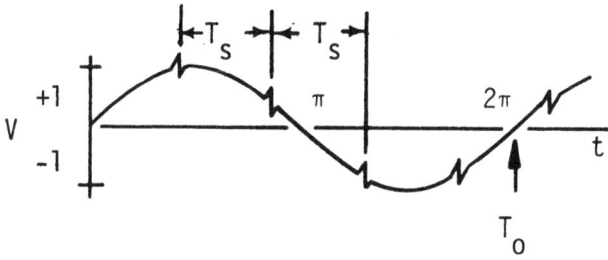

(a) Time Domain

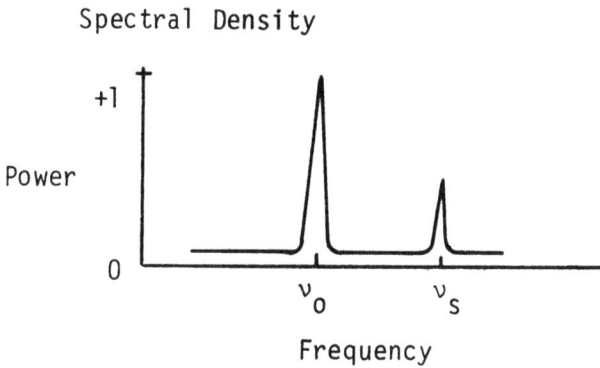

(b) Frequency Domain

Figure 15.1. Sine Wave Perturbed for Short Instances by Noise

Some noise will cause the instantaneous frequency to "jitter" around v_0, with the probability of being higher or lower than v_0. Usually a "pedestal" is associated with the v_0 as shown in Figure 15.2. A typical radio frequency (RF) signal iwth the sideband spectrum is shown in Figure 15.3. The sideband incudes the pedestal as well as some noise of constant repetition rate.

The power spectrum of V(t), often called the RF spectrum, is very useful in many applications. Unfortunately, in a given RF spectrum, it is impossible to determine whether the power at different Fourier frequencies is a result of amplitude fluctuations E(t) or phase fluctuations ϕ(t). The RF spectrum can be separated into two independent spectra: spectral density of E(t) and spectral density of ϕ(t). The phase spectral density is of primary interest.

The spectral density of phase fluctuations is denoted by $S_\phi(f)$ where f is the Fourier frequency. In general, the spectral density of E(t) is negligibly small and the total modulation of the phase fluctuation is also small (mean-square value is much less than one rad^2); thus, the RF spectrum has approximately the same shape as the phase spectral density. However, a

Spectral Density

+1

Power

0

ν_0

Frequency

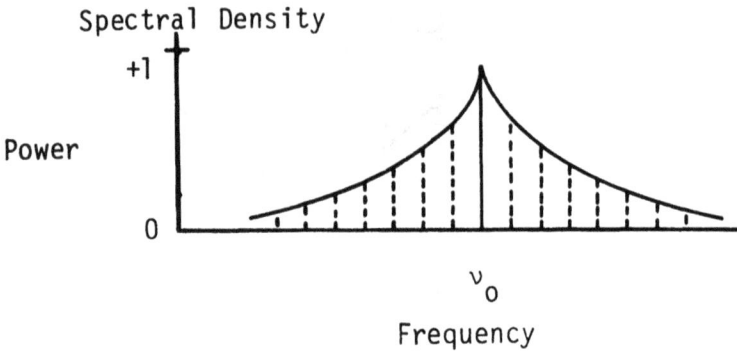

Figure 15.2. Frequency ν_0 with Jitter Around It

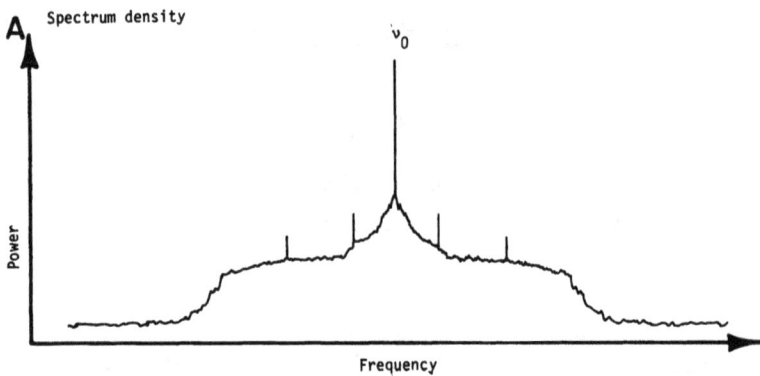

Spectrum density

A

ν_0

Power

Frequency

Figure 15.3. Typical RF Source with Side-Band Spectrum (Reference 8 Courtesy of Hewlett Packard Co.)

main difference in the representation of RF spectrum and phase spectral spectrum is that the RF spectrum includes the fundamental signal (Figure 15.3), and the phase spectral density does not (Figure 15.4). Another major difference is that the RF spectrum is a power spectral density and is measured in units of watts/Hz. The phase spectral density involves no power measurement of the signal and the unit is rad^2/Hz. It is tempting to think of $S_\phi(f)$ as a power spectral density because in practice it is measured by passing the signal V(t) through a phase detector and measuring the detector output power spectrum. The measurement technique makes use of the relation that for small deviation ($\delta\phi \ll 1$ rad) the phase spectral density is given by

$$S_\phi(f) = \left(\frac{V_{rms}(f)}{K}\right)^2 \frac{rad^2}{Hz} \qquad (15.3)$$

where $V_{rms}(f)$ is the root-mean-square noise voltage per $\sqrt{Hz}$ at a Fourier frequency f, and K is the sensitivity (volts per radian) at the phase quadrature output of a phase detector which is used to compare the two oscillators.

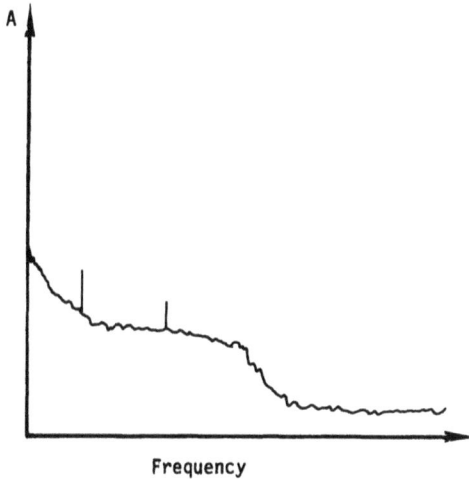

Figure 15.4. *Phase Spectral Density (Reference 8, Courtesy of Hewlett Packard Co.)*

The spectral density of frequency fluctuations is also an important quantity and can be derived from the phase spectral density. Since frequency is the time rate of change of phase in a sine wave it follows that $v(t)$ can be derived from Equation 15.2. By taking the derivative of the argument of sine term one can obtain

$$2\pi v(t) = \frac{d}{dt}[2\pi v_0 t + \phi(t)]$$

$$= 2\pi v_0 + \frac{d\phi(t)}{dt} \qquad (15.4)$$

Rearranging Equation 15.4

$$v(t) - v_0 \equiv \delta v(t) = \frac{1}{2\pi}\frac{d\phi(t)}{dt} \qquad (15.5)$$

where $\delta v(t)$ is a change in frequency at time t.

It is useful to denote $\delta v(t)/v_0$ with respect to the normal frequency v_0. The quantity $\delta v(t)$ is called the normalized frequency fluctuations and is denoted by y(t). Thus

$$y(t) = \frac{\delta v(t)}{v_0} = \frac{1}{2\pi v_0} \frac{d\phi(t)}{dt} \tag{15.6}$$

Taking the Laplace transformation Equation 15.5, the following result can be written

$$\delta v(s) = \frac{S}{2\pi} \phi(S) \tag{15.7}$$

For $S = j2\pi f$, the spectral density is

$$|\delta v(f)|^2 = \left|\frac{2\pi f}{2\pi}\right|^2 |\phi(f)|^2 \tag{15.8}$$

The spectral density of absolute frequency fluctuations can be represented by $S_v(f)$ which is related to the phase spectral density $S_\phi(f)$ as

$$S_v(f) = \frac{(2\pi f)^2}{(2\pi)^2} S_\phi(f) = f^2 S_\phi(f) \tag{15.9}$$

where $S_v(f)$ is the spectral density of absolute frequency fluctuation and S represents the Laplace transformation of the quantity in the subscripts.

Many times the absolute frequency fluctuation of the spectral density is not of the primary interest, as is the normalized frequency fluctuation. This can be accomplished by measuring the spectral density of y(t). From equations 15.6. and 15.7 the following relation can be obtained.

$$S_y(f) = \frac{1}{v_0^2} S_v(f) = \frac{f^2}{v_0^2} S_\phi(f) \quad \frac{1}{Hz} \tag{15.10}$$

In many cases, however, it is not the spectral density of the equivalent modulating source that is of interest but rather the actual side-band power of phase fluctuations with respect to the carrier level. As an expression, this is

$$\mathcal{L}(f) = \frac{\text{power density (one phase modulation side-band)}}{\text{carrier power}} \tag{15.11}$$

This spectral density, Script $\mathcal{L}(f)$, is defined as the ratio of the power per hertz of bandwidth at a frequency f from the carrier in one phase noise side-band to the carrier power.

From small angle modulation theory

$$\mathcal{L}(f) = \frac{S_\phi(f)}{2} \tag{15.12}$$

$\mathcal{L}(f)$ is often expressed in decibels relative to the carrier per hertz (dBc/Hz) which is calculated as

$$\mathcal{L}(f)_{dB} = 10 \log \frac{S_\phi(f)}{2} \tag{15.13}$$

The quantity of $\mathcal{L}(f)$ can be measured directly at the RF frequency if the side-band levels are within the dynamic range of the measuring spectrum analyzer.

Another measurement method is to utilize a double balanced mixer with the unknown source and a reference source set in phase quadrature (90 degree) at the input as shown in Figure 15.5. The two oscillators are at the same frequency in long term as guaranteed by the PLL. At quadrature, the difference frequency is zero and the average voltage output is zero volts.

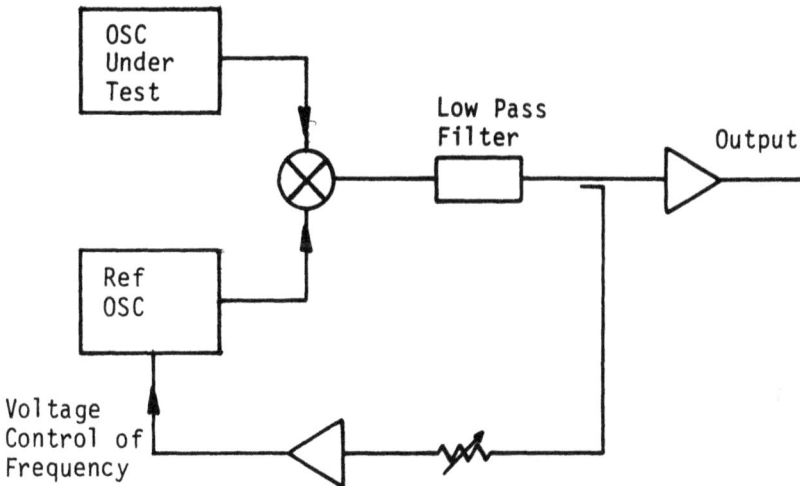

Figure 15.5. Two Oscillator Phase Noise Measurements

For phase fluctuations much less than 1 radian, the voltage fluctuations at the mixer output are related to the phase fluctuation by the equation

$$\phi = \frac{V}{K} \tag{15.14}$$

where ϕ is the phase fluctuation, V is the output voltage of the mixer, and K is the sensitivity of the mixer in volts/radian.

The slope at the zero crossing in volts/radian is K and sinusoidal beat signals are equal to the peak voltage of the signal. The beat signal as viewed on an analyzer is the rms value and is 3 dB less than the peak. In terms of the ratio of the side-band voltage to the beat signal voltage

$$S_\phi(f) = V_s - V_B - 3\,dB \tag{15.15}$$

$$\mathcal{L}(f) = V_s - V_B - 6\,dB \tag{15.16}$$

where V_s is the side-band voltage in dB corrected for bandwidth and analyzer characteristics, and V_B is the beat signal rms level in dB.

The underlying assumption is that the reference source has much lower phase noise than the unknown source. It is possible to compare two "identical" sources and assume that the phase noise of either one is 3 dB less than the measured values.

Another effect which causes a frequency shift in an oscillator is often referred to as frequency pulling (Reference 11). When the load of the oscillator changes, the frequency and the power level also changes. The pulling figure is defined by the frequency deviation from the load matched condition so that the voltage standing wave ratio (VSWR) becomes 1.5. If the load of an oscillator varies over a relatively wide range in a certain system, an isolator is often used to reduce the frequency shift caused by the pulling effect. When the supply voltage to the oscillator varies, the output frequency may also change. This phenomena is referred to as a pushing effect which can be expresed in Hz/V. In some oscillators there are internal voltage regulators to reduce the pushing effect.

15.3 CRYSTAL OSCILLATORS(12-14)

Since the stability of a frequency source is very important, it is essential to have stable oscillators. Although Cesium and Rubidium frequency standards have an accuracy of 10^{-11} resonating at fixed microwave frequencies, they are not commonly used as oscillators in microwave receivers because of the complicated design and cost. There is no known variable microwave frequency source that has very high accuracy. The common design is to phase lock a microwave source to a high accuracy frequency source at low frequency range. The low frequency source is generally referred to as the reference frequency source. Some of the reference sources will be discussed here.

Crystal oscillators have been the most popular stable frequency sources. The basic operating principle of a crystal oscillator is that a crystal is connected in the feedback circuit of an oscillator, and the mechanical vibration of the crystal stabilizes the oscillator frequency. Although there are many crystalline substances which have the basic requirements of a reference element, quartz has been the most widely accepted material due to its many desirable characteristics.

The crystal structure and different cuts of quartz are mentioned in the section on crystal filters (Section 12.13). A quartz is a piezoelectric material. In a piezoelectric material, a mechanical strain can produce an electric polarization which can be measured as voltage. The inverse effect also exists that a voltage can produce a mechanical movement. The equivalent circuit of a crystal is shown in Figure 15.6. The physical constants of the crystal determines the equivalent values of R_1, C_1, L_1, and C_0. R_1 is the result of bulk losses; C_1 is the motional capacitance; L_1 is determined by the mass; and C_0 is made up of the electrodes, the holder, and the leads. When operating at a far off resonance, the structure is simply a capacitance C_0; but at the precise resonant frequency, the circuit becomes a capacitor and resistor in parallel. The reactance of the crystal approaches zero at the point of series resonance and reaches a maximum at the antiresonance frequency f_A as shown in Figure 15.7.

An area typically chosen for operation of the oscillator is either near series resonance or at the more inductive area of parallel resonance. The series resonant circuit as shown in Figure 15.8 utilizes the characteristics of the crystal where the reactance is slightly inductive (above f_s in Figure 15.7). Series capacitance is then added to obtain a tuned circuit. The series capacitor is adjustable so the phase of the feedback can be changed slightly to fine tune the oscillator frequency. The parallel resonant mode adds capacitance in parallel as shown in Figure 15.9. This circuit typically operates at the highest point on the reactance wave (Figure 15.7), hence the crystal reactance is more inductive. A parallel capacitor is used for the fine adjustment.

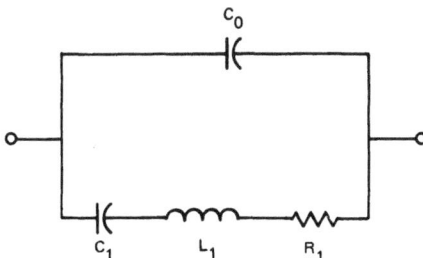

Figure 15.6. Electrical Equivalent Circuit for a Crystal Resonator (Reference 13, Courtesy of Hewlett Packard Co.)

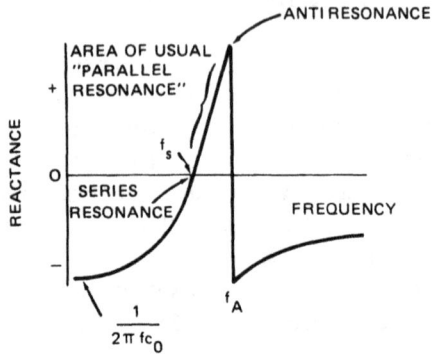

Figure 15.7. *Reactance of Crystal Varies with the Frequency of Operation Near Resonance (Reference 13, Courtesy of Hewlett Packard Co.)*

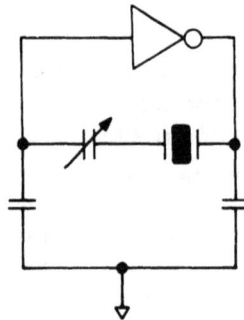

Figure 15.8. *Series Resonant Oscillator (Reference 13, Courtesy of Hewlett Packard Co.)*

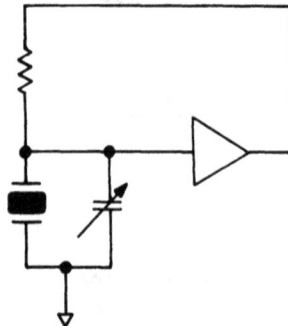

Figure 15.9. *Parallel Resonant Circuit (Reference 13, Courtesy of Hewlett Packard Co.)*

The long-term stability of a crystal oscillator is shown in Figure 15.10. The gradual change in frequency over days or months is known as aging. This occurs for various reasons (e.g., the physical properties of the crystal mounting may change or the crystal coefficient of elasticity may change when subjected to stress). To maintain an accurate frequency, periodic oscillator adjustments must be made to compensate these effects. Generally, the frequency of an oscillator can be varied a few cycles by slightly changing the phase of the feedback signal. This change usually can be accomplished by adjusting trimming capacitors in Figures 15.8 and 15.9.

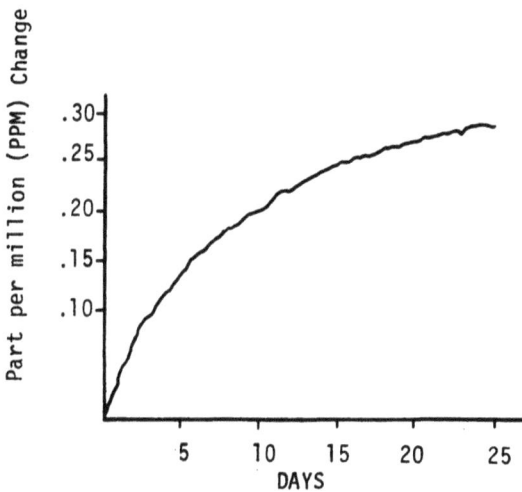

Figure 15.10. Time Doman Stability of the Fractional Frequency Change Over Time (Days) Starting from a Point of Calibration (Reference 13, Courtesy of Hewlett Packard Co.)

Short-term frequency stability of time domain stability is measured to represent the phase noise. Time domain stability is typically specified for a 1 second average. Shorter or longer averaging times may be required in an accuracy computation for some applications. Figure 15.11 shows the time domain measurement from 10^{-3} to 10^3 seconds. The spectral density of $\mathcal{L}(f)$ (Equation 15.13) is shown in Figure 15.12. At 10 Hz away from the carrier the side-band energy is below -120 dB in comparison with the carrier frequency.

A major influence on the crystal frequency is that of operating over variations in temperature. The temperature coefficient depends on the cuts of quartz and operating temperature. To improve the frequency stability versus temperature, temperature compensation circuits and oven temperature controlled crystals can be used.

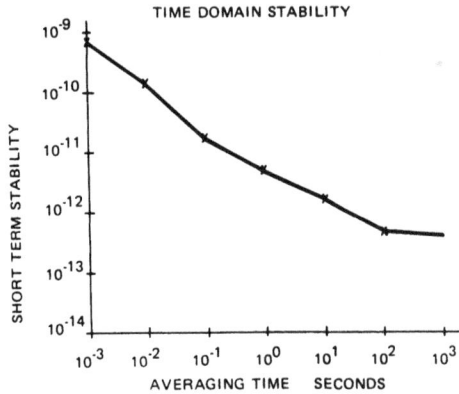

Figure 15.11. Time Domain Stability (Short Term) for Specific Averaging Times (Reference 13, Courtesy of Hewlett Packard Co.)

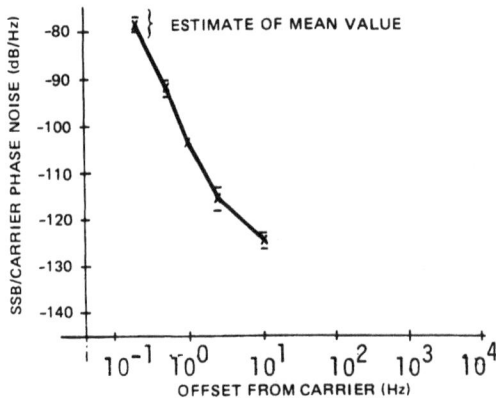

Figure 15.12. Phase Spectral Density at Specific Offsets from the Carrier (Reference 13, Courtesy of Hewlett Packard Co.)

15.4 SURFACE ACOUSTIC WAVE (SAW) OSCILLATORS[15-21]

Although the quartz crystal oscillators have many of the desirable performance features, their operation frequencies are generally under 100 MHz. If oscillators of higher frequencies are desirable, either frequency multipliers have to be used or an oscillator at microwave is phase locked to a stable oscillator at lower frequency. It is often desirable to have oscillators stabilized at microwave frequency directly. The approaches that have been investigated are SAW controlled oscillators, dielectric resonator stabilization, and microwave cavity controlled ones. SAW devices have the potential of functioning up to 2 GHz, while dielectric resonators and microwave cavities can work from 2 GHz and up.

The configuration of a SAW oscillator is shown in Figure 15.13. The amplifier uses a SAW delay line in its feedback circuit. The amplification must be high enough to compensate the total loss in the feedback loop. This configuration will generate multiple output frequencies as a comb generator. The frequency of oscillation must satisfy the condition (Reference 21)

$$2n\pi = 2\pi f \frac{l}{v} + \phi_e = 2\pi f \tau + \phi_e \qquad (15.17)$$

where n is an integer, l is the SAW line length, v is the velocity of SAW, ϕ_e is the remaining loop phase shift, and τ is the delay time of the SAW line.

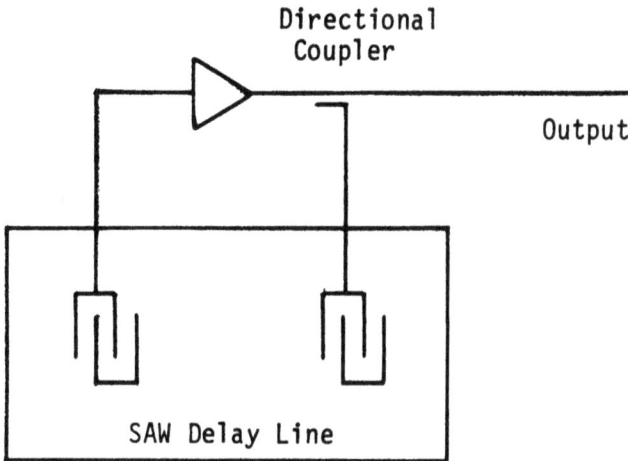

Directional
Coupler

Output

SAW Delay Line

Figure 15.13. Schematic Diagram of a SAW Oscillator

Assuming the delay line is stable, the frequency deviation is

$$\Delta f = -\Delta\phi_e \frac{v}{2\pi l} \qquad (15.18)$$

The oscillator stability increases with greater length l for a given value of n. The output frequencies are separated by $1/\tau$. A single frequency oscillator can be made through the transducer geometry. A typical transducer to produce a single mode is shown in Figure 15.14. The key feature in obtaining a single controllable frequency of oscillation is that the length of the narrow band transducer be approximately equal to the center-to-center separation $N\lambda$. From practical consideration the value of

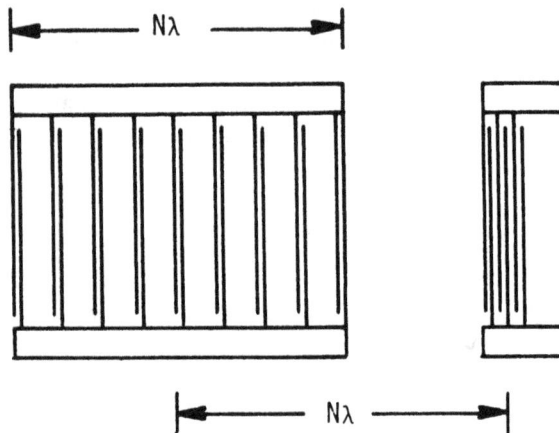

Figure 15.14. Transducer Configuration for Single-Mode SAW Controlled Oscillator (Based on Parker and Schulz, Reference 18)

N is limited to less than approximately 2000. The equivalent Q of the SAW delay line is proportional to the product of the delay time between the two transducers and the center frequency of the transducer.

The temperature stability is comparable to bulk wave devices. The short-term stability of SAW oscillators is already competitive for frequencies above 100 MHz; however, aging in SAW oscillators is worse than in bulk wave resonators.

15.5 DIELECTRIC AND CAVITY CONTROLLED OSCILLATORS[22-28]

Some dielectric materials have high temperature stability and high-Q values at microwave frequency. They can be incorporated into microwave integrated circuit (MIC) transistors and result in a highly stabilized, low-noise microwave power source. For example, $B_aO - TiO_2$ ceramic has excellent dielectric properties. Its temperature stability is a few ppm/°c, the relative dielectric constant, ε_r is 39 with an unloaded Q value of 7000 at 6 GHz. A relative dielectric constant of 39 results in approximately 80 percent of the energy stored within the resonator.

Since some of the energy is external to the resonator, it can be coupled by locating the resonator near a microstrip or strip line. Shielding must be provided to avoid radiation losses, and care must be exercised in mounting the resonator to avoid degradation of performance. Resonant frequency and unloaded Q value depend not only on resonator dimension, but also on substrate thickness and on air gap thickness. Thus, frequency tuning can be accomplished through conductor or dielectric mechanical adjustable screws.

A block diagram of a stabilized FET oscillator is shown in Figure 15.15. The oscillation frequency depends on the length of the open-ended microstrip line connected to the gate terminal of the active subnetwork. The dielectric resonator is placed in the vicinity of a 50Ω microstrip line, which is connected to the drain output terminal of the active subnetwork and constitutes a band rejection filter. The resonant frequency of the dielectric disk is designed to equal the unstabilized oscillation frequency. Thus, the frequency will be stabilized by the dielectric resonator. An oscillator has been constructed with $BaO-TiO_2$ as the stabilized dielectric material. The oscillator frequency is at 6 GHz of 100 mW output power at 17 percent efficiency and a frequency temperature coefficient as low as 2.3 ppm/°C. Frequency modulated (FM) noise level was improved more than 30 dB by the dielectric resonator stabilization.

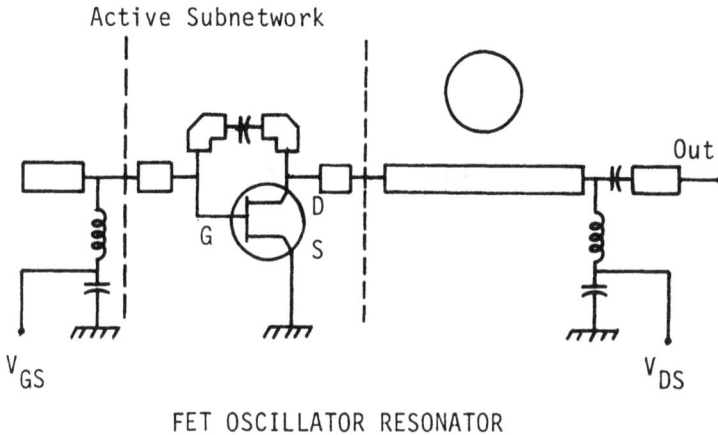

Figure 15.15. *Block Diagram of a Dielectric Resonator Stabilized Oscillator (Based on Abe et. al., Reference 23)*

Cavity of high-Q value can also be used to stabilize the output frequency of an oscillator especially for oscillators with high output power. The cavity can be located between the oscillating device and the load as a directly coupled frequency control circuit similar to the dielectric controlled resonator. There is another approach in which the cavity is coupled through automatic frequency control circuit. One simple automatic frequency control design uses a cylindrical cavity where two orthogonal TE_{111} modes are excited. One mode is tuned slightly lower than the other, and diode detectors with opposite polarity sample the power from the two modes such that the voltage characteristics of the two detectors combine into a discriminator. The output from the discriminator is used to control the frequency of the oscillator. The cavity controlled oscillator is often above 4 GHz, because the size of the cavity is reasonably small. At lower frequency the size of the cavity is usually large.

15.6 YIG TUNED OSCILLATORS[29-34]

A YIG sphere, when installed in the proper magnetic environment with suitable coupling, will behave like a tunable microwave cavity with Q on the order of 1000 to 8000. These YIG spheres can be made into filters as discussed in Section 12.10. In an oscillator, the spectral purity is directly related to circuit Q. Therefore, when a YIG sphere is used as the frequency determining element in a negative resistance oscillator, it provides excellent AM and FM noise characteristics. The active element in the YIG tuned oscillator can be either bulk-effect devices or high frequency transistors. Since the advance of FET technology, most of the YIG tuned oscillators up to 20 GHz use FET devices.

A basic circuit of a YIG tuned oscillator is shown in Figure 15.16. The YIG sphere used as the tuning element is connected to the source of the FET. In order to generate a negative conductance over the frequency range of interest, various feedback circuits have been considered. However, for YIG-tuned oscillators, it is desirable that an inductive admittance be presented to the oscillator in order to avoid spurious oscillations with the inductive coupling loop. This requirement almost always limits the FET in a common gate configuration with a common lead inductor (Lg) as the feedback element. One way to expand the negative resistance region of the device, which in turn widens the operation frequency range, is to add an output matching network at the drain.

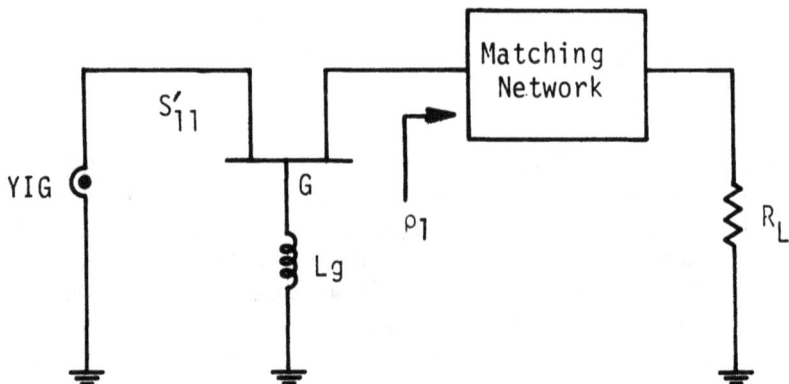

Figure 15.16. Basic Circuit for YIG-Tuned Oscillators

The equivalent circuit of the YIG resonator consists of a parallel resonator circuit in series with the coupling loop inductance as shown in Figure 15.17. The input impedance of the resonator is (Reference 30)

$$Z_r = j\omega L_c + \frac{\left(\frac{j\omega\omega_0}{Q_u}\right)\Big/ G_0}{\omega_0^2 - \omega^2 + \left(\frac{j\omega\omega_0}{Q_u}\right)} \tag{15.19}$$

where

$$G_0 = 1/\mu_0 V K^2 \omega_m Q_u \tag{15.20}$$

$$L_0 = 1/G_0\omega_0 Q_u \tag{15.21}$$

$$C_0 = 1/L_0\omega_0^2 \tag{15.22}$$

$$\omega_0 = 2\pi\gamma_e H_0 \tag{15.23}$$

and

$$Q_u = \frac{H_0 - \frac{1}{3}4\pi M_s}{\Delta H} \tag{15.24}$$

and where

$$\omega_m = \gamma_e 4\pi M_s$$

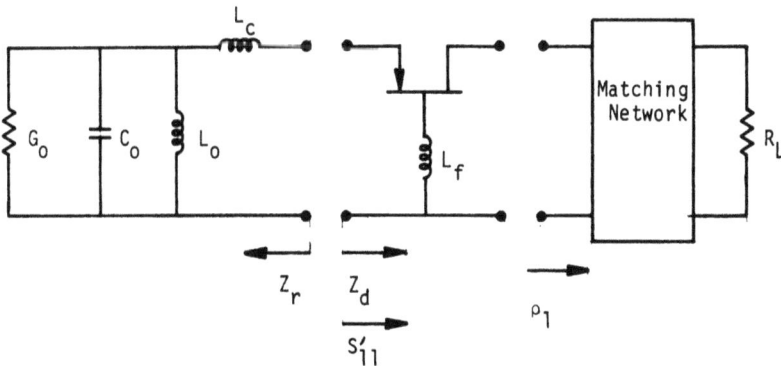

Figure 15.17. Equivalent Circuit of the YIG Resonator in an FET Oscillator (Based on Trew, Reference 30)

γ_e is the gyromagnetic ratio (2.8 MHz/Gauss), V is the sphere volume, K is 1/(loop diameter), $4\pi M_s$ is the saturation magnetization (1750 Gauss for pure YIG), Q_u is the unloaded Q of the resonator, H_0 is the magnetic biasing field, μ_0 is the free space permeability, and ΔH is the sphere line width. The resonant frequency is linearly related to the magnetic biasing field according to the expression

$$f_0 = \gamma_e H_0 \qquad (15.25)$$

Equations 15.21 to 15.23 show that both the equivalent inductance and capacitance of the resonator are functions of frequency and, therefore, functions of magnetic biasing field and tuning circuit. The two important factors of the YIG resonator are high-Q, multioctave, and linear tuning current.

If the matching network at the load is characterized by a reflection coefficient ϱ_l (see Figure 15.17), then

$$S_{11}' = S_{11} + \frac{S_{12}\,S_{21}\,\varrho_l}{1 - S_{22}\varrho_l} \qquad (15.26)$$

where S_{11}, S_{12}, S_{21}, and S_{22} are the S-parameters of the FET device. The objective is to choose a ϱ_l at all frequencies of interest so that $|S_{11}'| > 1$ for a particular feedback inductance. It is desirable to have a low magnitude reflection coefficient in midband with feedback provided by increasing reflection coefficient at the band edges. Although low-pass, high-pass, and band-pass filter designs all have been found to work, high-pass filters provide the best performance. In general, a constant k filter design as shown in Figure 15.18 is adequate to enhance the negative conductance over the designed range.

The gate feedback inductor is selected to produce a negative conductance at the high frequency region. At the low frequency end the negative conductance is provided by feedback from the drain filter. In this manner, the negative conductance bandwidth can be extended to greater than octave coverage.

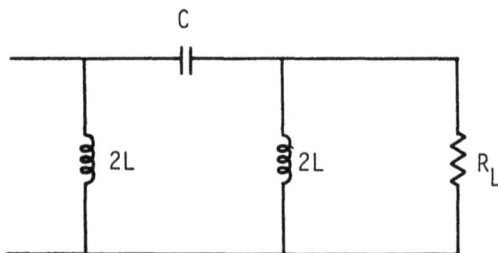

Figure 15.18. FET Oscillator Drain Filter Circuit (Based on Trew, Reference 30)

In general, the tuning curves (tuning current versus output frequency) of YIG tuned FET oscillators are extremely linear and will deviate from the ideal straight line ±0.05 percent, typically. The power output remains flat within ±1.5 to ±3.0 dB over the tuning range.

The YIG sphere is installed on a mechanically stable nonconducting mounting rod. The rod is adjusted to locate the sphere in its proper position within the magnetic field and coupling loop. Sometimes the temperature of the sphere is maintained by a heater to minimize frequency drift.

In many YIG tuned oscillators, a low inductance FM coil is added to the main coil. The coil is in close proximity to the YIG sphere and is used for fine tuning the oscillator for phase locking applications or providing frequency modulation. Since tuning the YIG oscillator is done by changing the biasing current which in turn changes the magnetic field around the YIG sphere, the tuning speed is relatively slow. In general, it takes milliseconds to tune a YIG oscillator from one end of the frequency to the other.

A YIG tuned oscillator can be phased locked to a reference source to obtain the required frequency stability.

15.7 VOLTAGE CONTROLLED OSCILLATORS (VCOs)[35-49]

In a VCO, a varactor diode usually serves as a voltage controlled capacitor in a tuning circuit to control the frequency of the oscillator. The active device in a VCO can be a Gunn or Impatt diodes or a transistor with appropriate biasing and feedback circuitry. The major feature of a varactor tuned oscillator is its extremely fast tuning speed which can be expressed in nanoseconds. The limiting factor of the tuning speed is the ability of the external voltage driver circuit to change the voltage across the varactor diode. This is primarily controlled by the driver impedance and the bypass capacitors in the tuning circuit.

Tuning curves for VCOs are basically nonlinear due to both the capacitance versus voltage characteristic of the varactor and the varying impedance of the negative resistance device. The curve, however, is quite smooth and monotonic (the output frequency is single valued at any voltage and continuously increasing for a continuously increasing tuning voltage). In order to tune the frequency of a VCO very rapidly over a wide range, the Q of the tuning circuit cannot be very high. In general, the Q of a VCO is below 50. The FM noise, varying approximately inversely proportional to the loaded Q of the timing circuit, will be relatively high. In comparison with an YIG tuned oscillator, the FM noise of a VCO is about 20 dB higher. There are some technical terms and definitions which are related to the performance of VCO that will be discussed here.

A. Tuning Curve and Linearizer

VCOs generally have a nonlinear voltage versus frequency tuning characteristics which will be discussed in detail later in this section. Often the tuning characteristics can be represented by a logarithmic function as

$$f_0 = \exp(aVi) + b \qquad (15.27)$$

where f_0 is the output frequency, a and b are constants, and V_i is the control voltage.

A typical curve is shown in Figure 15.19. A linearizer may be added to the VCO, if linear tuning is required. The linearizer has an E_i versus E_{out} (where $E_{out} = V_i b$; the control voltage on the VCO) transfer function which matches the VCO's nonlinear tuning curve through piecewise linear sections as shown in Figure 15.20. The number of the piecewise linear sections depends on the requirement of the application. Special amplifiers and circuits can be built to accomplish this purpose (see Section 6.7).

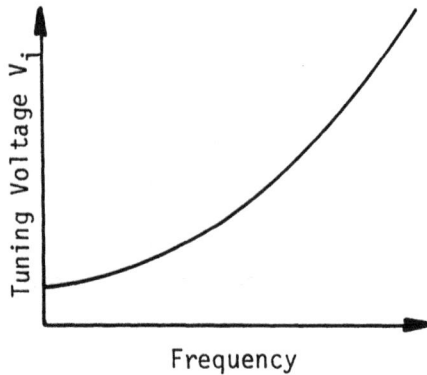

Figure 15.19. Typical VCO Tuning Curve

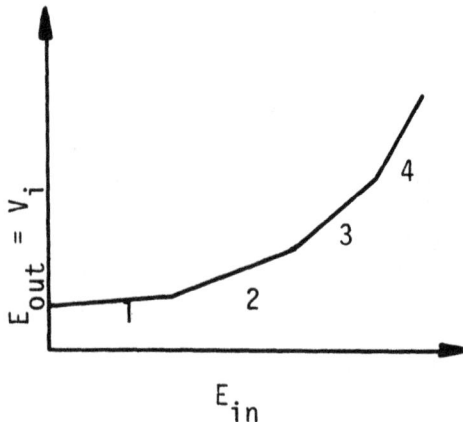

Figure 15.20. Linearizer Transfer Function

B. Settling Time and Post Tuning Drift

Although the VCO can be frequency tuned extremely fast, it also has transient effects when it is tuned from one frequency to another. During the transient, the frequency is difficult to measure and is even hard to define. Therefore, only the initial and final frequencies are of interest. In order to evaluate the frequency response, a step voltage is applied to the VCO and the frequency change after the step function is determined.

Settling time is the time required to tune the frequency within a specified tolerance of a desired final value, after the step change in tuning voltage. Post tuning drift is the frequency shift versus time between a certain specified time interval, after the step change in tuning voltage.

Figure 15.21a shows the VCO tuning voltage as a function of time while Figure 15.21b shows a typical VCO frequency response versus time. The time Δt_0 may be considered as a settling time. The post tuning drift is shown as Δf and is defined as the frequency change between arbitrarily defined times t_1 and t_2. Time t_1 has been defined from 10 μs to 1 second. Short-term post-tuning drift could be arbitrarily defined as frequency change over the period from 10 μs to 1 second and long-term post-tuning drift for the time period from 1 second to 1 hour.

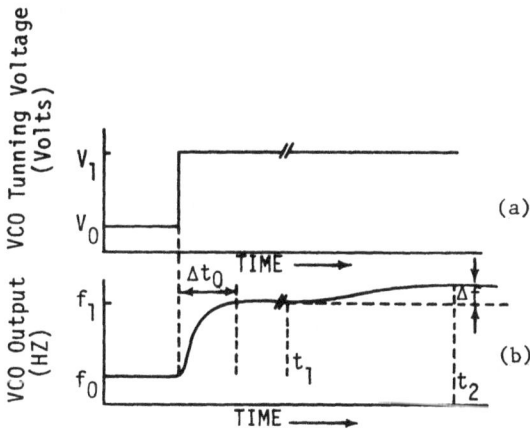

(a) Tuning Voltage   (b) Frequency Output

Figure 15.21. Typical VCO Frequency versus Time Response (Reference 49, Courtesy of Hewlett Packard Co.)

A varactor diode, the tuning element of the VCO, is a semiconductor p-n junction device in which its capacitance is a function of the applied voltage. Varactors are normally operating under reverse biased conditions, where the junction resistance is very high and can be neglected. The equivalent circuit is a capacitor and resistor in series. The resistor includes the intrinsic resistance of the p, n materials and ohmic contact resistance of the electrodes. The capacitance depends on the total diode area and the width of the depletion layer which is caused by moving the current carriers away from the junction by the applied voltage. The depletion layer width is a function of the applied voltage. Therefore, the capacitance of the diode is controlled by the applied voltage. The junction capacitance can be related to the applied voltage by (References 39 and 40)

$$C(V) = \frac{C(O)}{\left(1 + \frac{V}{\phi}\right)^{\gamma}} \qquad (15.28)$$

where $C(O)$ is the junction capacitance at zero volts, V is the applied voltage, ϕ is the contact potential, and γ is a constant and approximately equal to 0.5. The value of γ depends mainly on the doping of the device. If $\gamma = 2$, then the tuning curve of the VCO is approximately linear.

A hyperabrupt junction varactor can be used to linearize the tuning curve of a VCO through the control of the constant γ. The difference between an abrupt junction and hyperabrupt junction varactor is that the concentration of the "n" material is constant across the depletion region in an abrupt diode, and it is nonlinear in a hyperabrupt diode as shown in Figure 15.22a and b. As the reverse voltage is increased, the characteristics of the nonlinear "n" material cause a greater capacitance change in the hyperabrupt varactor than in the abrupt junction varactor. The junction capacitance of a hyperabrupt junction can be approximated by changing Equation 15.28 to

$$C(V) = \frac{C(O)}{\left(1 + \frac{V}{\phi}\right)^{2}} \qquad (15.29)$$

which means $\gamma = 2$ instead of 0.5.

In addition to the nearly linear tuning curve, the hyperabrupt varactor tuned oscillator offers the following advantages over the abrupt varactor:

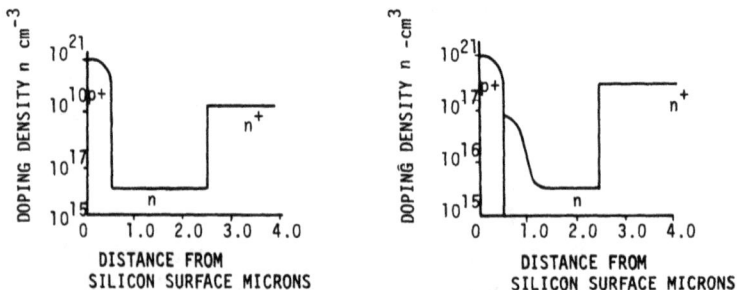

(a) Doping Density for an Abrupt   (b) Doping Density for a
        Junction Varactor   Hyperabrupt Varactor

Figure 15.22. Doping Densities of Two Kinds of Varactors (Beach, Reference 39, Courtesy of Watkins-Johnson Co.)

lower post-tuning drift, faster settling time, less fine-grain modulation sensitivity variation, smaller size, and lower cost.

Theoretically, the varactor diode can be connected either in parallel or in series to form the frequency tuning circuit. However, only the series tuning circuit is practical since microwave transistors typically have a large output capacitance and require a varactor of high capacitance to obtain a wide-band tuning. A high capacitance varactor with high-Q and large tuning range is not available for microwave frequency.

A series tuned VCO is shown in Figure 15.23. The transistor is used in the common collector configuration with the varactor tuning network and load connected to the base. The common collector configuration provides the best performance against other modes of operation. The oscillation frequency of the circuit in Figure 15.23 is

$$F = \frac{1}{2\pi \sqrt{LC_T}} \qquad (15.30)$$

where

$$C_T = \frac{C_t C(V)}{C_t + C(V)} \qquad (15.31)$$

where $C(V)$ is the varactor capacitance, C_t is the transistor input capacitance, and L is the effective series inductance from the tuning circuit and the load.

Substituting the varactor capacitance formula

$$C(V) = \frac{C(O)}{\left(1 + \frac{V}{\phi}\right)^\gamma} \qquad (15.32)$$

and Equation 15.31 into Equation 15.30, and utilizing the conditions that $V \gg \phi$ and $(V + \phi)^\gamma \gg C(O)\phi^\gamma/C$, then the oscillation frequency can be reduced to

$$f = \frac{V^{0.5\gamma}}{2\pi \sqrt{LC(O)\phi^\gamma}} \qquad (15.33)$$

For the hyperabrupt junction (γ = 2), Equation 15.33 provides a linear relationship between f and V. In general, the frequency versus applied voltage V is still represented by Equation 15.27.

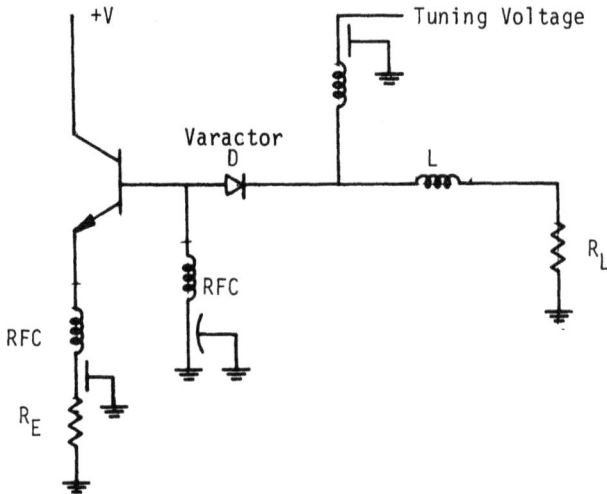

Figure 15.23. Simplified Series Tuned VCO

15.8 OSCILLATORS STABILIZED WITH PHASE LOCKED LOOPS (PLLs) (INDIRECT SYNTHESIZERS) (50-55, 74)

A frequency synthesizer translates the stable frequency of a precision frequency standard to one of thousands or millions of frequencies over a spectrum range. Two methods have been used in frequency synthesis: the direct and indirect method. Direct synthesis is accomplished by multiplexing, dividing, and mixing, while the indirect method derives its frequency from one or more phase locked VCOs. The stability and accuracy in either case is derived from the references sources. The PLL will be discussed first; and in the next section, the direct synthesizers will be discussed.

A PLL is a feedback system whose function is to force the output frequency of a VCO to be coherent with a certain reference frequency. "Coherent" means highly correlated in both frequency and phase. In a PLL the basic technique used is to compare the frequency and phase of the reference signal to the output of a VCO. If the two signals differ in frequency and/or phase, an error voltage is generated and applied to the VCO, causing it to correct in the direction required for decreasing the difference. The correction procedure continues until lock is achieved; after that the VCO will continue to track the reference signal. The general requirements are that the PLL must very quickly lock-up to the reference signal and yet be capable of ignoring short-term jitter inadvertantly introduced to the reference.

There are several different kinds of PLLs and they can be used repeatedly in a circuit to generate a tunable output frequency in increments of the reference frequency f_r. One of the PLLs is shown in Figure 15.24. The output frequency f_0 from the VCO is divided by N through a programmable divider which means the value of N can be changed through external control. The phase detector compares the instantaneous phase difference between the reference frequency f_r and f_0/N. In order to compare the phase, f_r must equal f_0/N or $f_0 = Nf_r$. Therefore, when N changes, the output frequency f_0 changes. For example, if the reference $f_r = 100$ KHz and N = 99 to 198, the output frequency f_0 can change from 9.9 ∼ 19.8 MHz at 100 KHz steps. To generate high frequency, a frequency multiplier can be placed at the output of the VCO as shown in Figure 15.24.

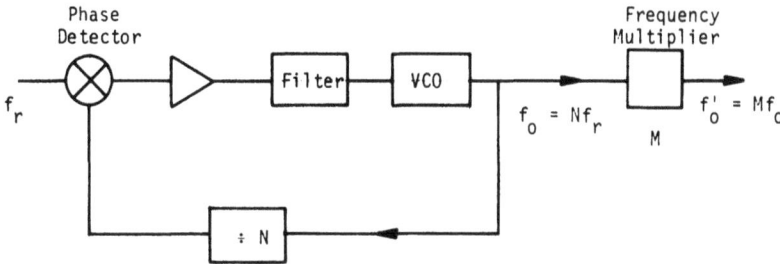

Figure 15.24. Basic Phase Locked Loop

In microwave applications there is no readily available divider to divide the microwave frequency. In this case, the most common PLL is the one shown in Figure 15.25. There are two input frequencies f_r and f_m. If f_0 is greater than f_m, then the mixer output frequency $f_{IN} = f_0 - f_m$. The difference frequency f_{IN} is low enough that the divider can operate at this frequency. For the phase detector to work, the following condition must be held:

$$\frac{f_0 - f_m}{N} = f_r$$

or

$$f_0 = f_m + Nf_r \tag{15.34}$$

In order to keep f_0 stable, both f_m and f_r should be stable. Sometimes f_m can be generated by multiplying the output frequency of a VCO with a PLL (i.e., the output frequency from Figure 15.24). In such arrangements, the frequency f_m can be one of many values generated from the reference f_r, and the output frequency f_0 is more versatile.

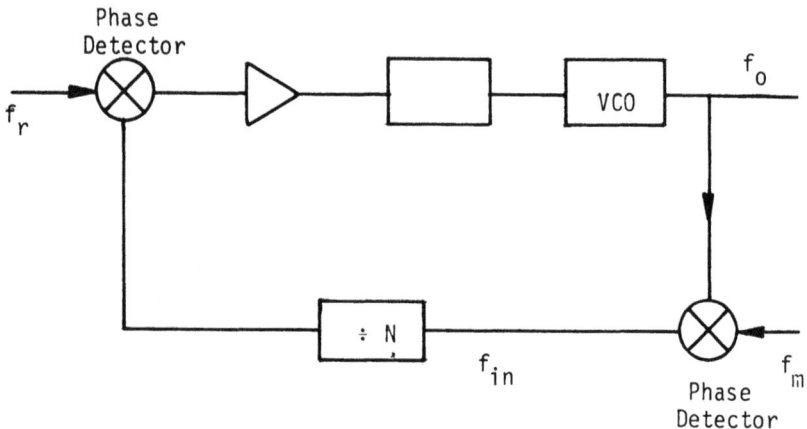

Figure 15.25. Phase Lock Loop by Frequency Mixing Down

The characteristics of the PLL can be evaluated through Laplace transformation. The PLL can be redrawn as shown in Figure 15.26. The VCO acts as an integrator in the circuit, since frequency is the rate change of phase. It can be represented by K_V/s where $1/s$ is the Laplace notation for integration. The output phase $\Theta_0(s)$ can be expressed as

$$\Theta_0(s) = K_p \frac{K_V}{s} F(s) \left[\Theta_i(s) - \frac{1}{N} \Theta_0(s) \right] \tag{15.35}$$

or

$$\frac{\Theta_0(s)}{\Theta_i(s)} = K_p K_V F(s) \left[\frac{1}{S + \dfrac{K_p K_V F(s)}{N}} \right] \tag{15.36}$$

where $\Theta_i(s)$ is the input phase, K_p is the sensitivity of the phase detector (volts/radian), K_V is gain constant of the VCO, and $F(s)$ is the transfer function of the filter.

For simplicity, let us assume that there is not filter required, then $F(s) = 1$ and Equation 15.36 can be written as

$$\frac{\Theta_0(s)}{\Theta_i(s)} = K_p K_V \left[\frac{1}{S + \dfrac{K_p K_V}{N}} \right]$$

(15.37)

This transfer function has a simple RC response with $RC = K_p K_V/N$. It has a 6 dB/octave roll-off with a corner frequency of $\omega_a = K_p K_V/N$, which means the VCO's output phase will start to track the reference output phase to about ω_a. Increasing the bandwidth ω_a can improve the lock range of the PLL. However, the values of K_p and K_V are limited in actual devices. For this simple case, the PLL is always stable.

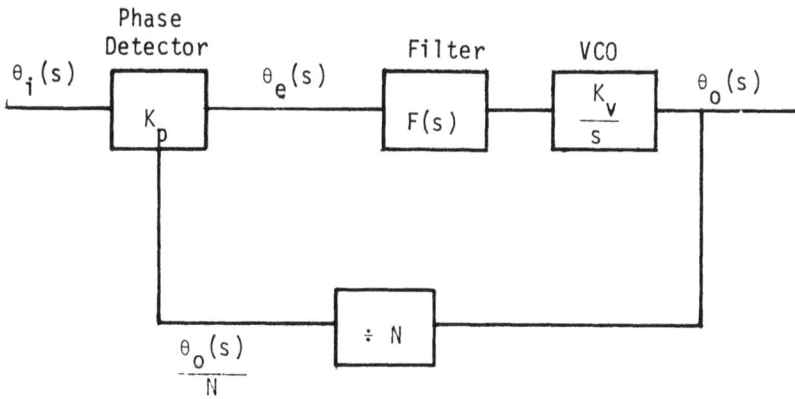

Figure 15.26. Basic Phase Locked Loop

In general, there is a low-pass filter in the PLL. The response of the PLL can be solved through conventional control theory. For second order functions of S (which is the commonly discussed case), there are damping factors and natural frequencies which determine the overshoot and stability of the PLL. The characteristics of the PLL and the speed of the programmable counter determines the phase lock speed. Usually, the response time of the indirect synthesizer is in the microseconds or milliseconds range.

Detailed discussions of PLL can be found in Reference 55 and a design example is in Reference 54.

From the discussion above, it is obvious that the VCO must sweep through a certain frequency range in order to accomplish the phase lock. When the VCO frequency is close to the reference frequency, it will phase lock to the reference signal. Therefore, in a frequency synthesizer, every time a new frequency is dialed, the VCO will sweep until it reaches the phase lock again. In some synthesizers, there is a coarse frequency tuning circuit. Whenever a new frequency is programmed, the coarse frequency tuning circuit will tune the VCO to a frequency close to the desired value; and the PLL will tune the VCO to the desired value.

15.9 DIRECT FREQUENCY SYNTHESIZERS[56-58]

The indirect synthesizer discussed in the last section employs PLLs. The frequency switching speed is limited by the loop bandwidth which seldom exceeds a few megahertz; thus the switching speed is relatively slow. If fast frequency switching is desirable, a direct synthesizer may be the solution since it is not limited by the bandwidth of the PLL.

The direct synthesizer generates frequency by successive frequency mixing, filtering, and dividing/multiplexing operations. Many fixed frequencies are available all the time in normal operations, and the output frequency is obtained by selecting the proper frequencies in the mixer. As discussed in Chapter 14, there are many spurious frequencies coming from the output of a mixer; it is very important to keep the undesired frequencies below a certain level with respect to the desired output. In order to reduce these spurious frequencies, filters following the mixers are used. The frequency selection speed depends primarily on the RF switching speed (the transient noise settling time should be included) and the RF propagation time. The main RF propagation delay is through the filters. In general, the operation speed of direct synthesizer can be from ten to several hundred nanoseconds.

The operating principle of a direct synthesizer can be best explained through an example. The example given in References 57 and 58 will be discussed here and is shown in Figure 15.27. A stable reference oscillator is fed through a harmonic generator. At the output of the harmonic generator, multiple frequencies which are harmonic related are generated. A bank of filters can be used to separate the frequencies and they are usually related by $f + \Delta f$, $f + 2\Delta f, \ldots, f + N\Delta f$. These frequencies are fed into a switching matrix. The switch matrix output frequencies f_x and f_y can be separately selected by external logic control. The f_x output is fed directly to a mixer. The f_y frequency output is first divided by N and then fed into the other input of the mixer. The output of the mixer can be selected as $f_x + f_y/N$, which can be any one of the N^2 frequencies.

There are many different arrangements that can be made by a direct frequency synthesizer. Reference 57 discusses a binary system and a octal system. In the binary system, there are two input frequencies and one output at each switch. Successive mixing and dividing/multiplexing are following the switches. The analysis has shown that the binary system is superior to octal system in terms of volume, weight, and reliability.

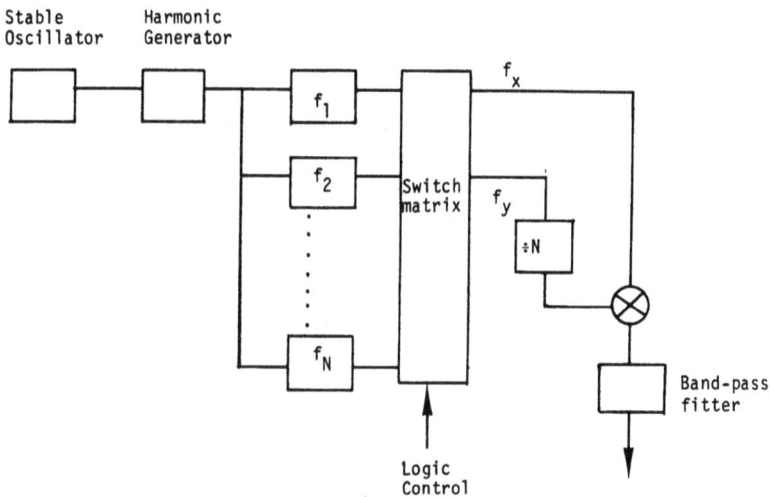

Figure 15.27. Basic Circuit Frequency Synthesizer (Based on Foster and Simon Reference 58)

The above discussion clearly reveals that the direct frequency synthesizer requires many frequency sources. These sources must be on all the time although they will not be used all the time. As a matter of fact, they will be used only a small fraction of the time. Therefore, the direct frequency synthesizer design is very complex, including a large quantity of amplifiers, mixers, multipliers, dividers, and filters and has relatively poor efficiency. Also, many of the reference frequencies are generated through VCOs with PLLs. Taking this point into consideration, one can see that the direct synthesizer is even more complicated. However, it can provide extremely fast switching speed. The direct frequency synthesizer has been made feasible through the recent advances in microwave technology (i.e., SAW filters and MICs).

15.10 HARMONIC GENERATION/STEP RECOVERY DIODES
(SNAP-BACK DIODES)[59-65]

In the phase locked loop and direct frequency synthesizer, there are many frequency dividers and multipliers. Most of the frequency dividers work at frequencies below approximately 1 GHz and are discussed in many articles (i.e., Reference 50). To operate in a microwave range, frequency multiplexers are usually used and they will be discussed here.

A step recovery diode is often used as a frequency multiplier. The step recovery diode is a p-n junction charge-storage diode which means that the holes from the p-type region are injected into the n-type region and electrons from the n-type region are injected into the p-type region during the forward bias condition. During the reverse-bias half circle, the injected charge will be recovered. In order for this phenomena to occur, two conditions must be met: (1) the carriers must not be combined before they are withdrawn, and (2) the carrier must not diffuse so far from the junction that they cannot be retrieved. Relatively long minority carrier lifetime is required to fulfill the first condition. In silicon the lifetime ranges $10^{-8} \sim 10^{-6}$ seconds which is considered relatively long, and that is why most step recovery diodes are made of silicon. The second condition can be fulfilled through large built-in electric fields on both sides of the junction. These build-in fields are produced by steep gradients in the impurity distribution.

In the forward bias condition, the step recovery diode conducts and minority carrier injection occurs. When the direction of the applied voltage reverses, the stored minority carriers are depleted, and a very abrupt step (transition period) in current occurs. Assuming that a sinusoidal waveform is applied to a step recovery diode, the current versus time in comparison with an ideal diode are shown in Figure 15.28. If the step recovery diode is properly design, the transition period can be made very small (in the few to several hundred picoseconds range a picosecond = 10^{-12} sec.) This transition period will generate harmonics of the basic input frequency. Therefore, the step recovery diode is generally used in frequency multipliers and in comb generators. A comb generator produces all the harmonics of the fundamental frequency.

The maximum efficiency (Reference 65) that can be achieved for a nonlinear resistance (i.e., an ideal rectifier) harmonic generator is $1/n^2$ where n is the harmonic number. The efficiency is limited by the loss inherent in the resistance. However, a step recovery diode can be considered as a nonlinear capacitance device. The maximum efficiency of a harmonic generator with nonlinear capacitance can theoretically reach 100 percent, since there is no loss in a capacitor. The step recovery diode in a harmonic generator can be either connected in series or in parallel as shown in Figure 15.29. Matching networks are usually added at the input of the diode, and a filter is added at the output to eliminate the unwanted frequencies. In

(a) A Simple Diode Circuit

(b) Input Voltage

(c) Current In An Ideal Diode

(d) Current In a Step Recovery Diode

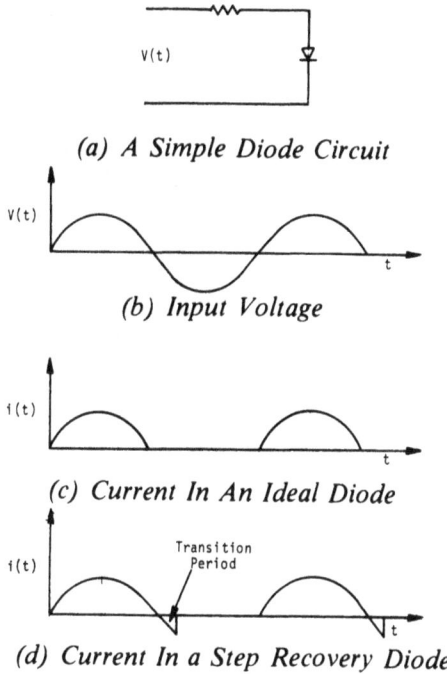

Figure 15.28. Current Flows in Ideal Diode and Step Recovery Diode

case multiple frequencies are required simultaneously, such as in the direct frequency synthesizer, a bank of filters can be placed at the output of the diode.

In many designs multiple step recovery diodes are used (References 60-63) to enhance certain harmonic outputs or improve performance. The important parameters concerning a harmonic generator are the power output capability, conversion efficiency, output spectral purity, and stability. These parameters are discussed in Reference 60.

15.11 INJECTION LOCKED OSCILLATORS[66-73]

A free running oscillator can be stabilized by injecting a stable signal externally. If two oscillators with frequencies are very close or harmonically closely related, one frequency can be used to inject lock the other oscillator. The injection locked oscillator is often used for oscillators whose frequency cannot be tuned easily through external control, since oscillators with externally electronic tuning can be used in conventional phase locked loop as discussed earlier in this chapter. For example, Gunn and IMPATT oscillators can be frequency stabilized through injection lock. The FM noise of the oscillator can be improved through injection lock; however, the AM noise cannot be improved by this scheme. In general, the injection power is much less than the oscillator output power especially for fundamentally frequency injection lock. The fundamental

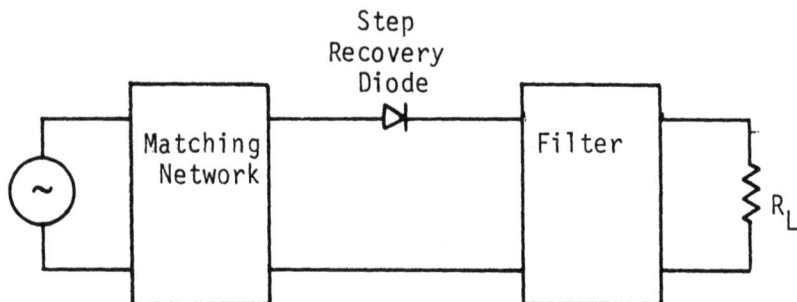

(a) Series Connection

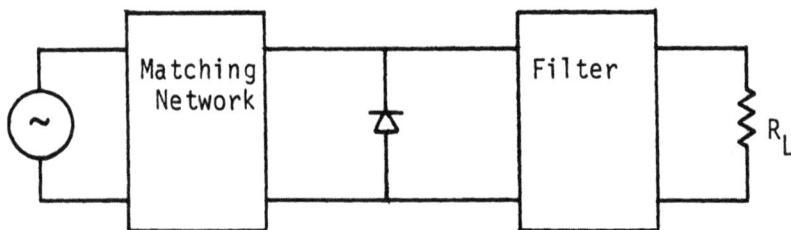

(b) Parallel Connect

Figure 15.29. Basic Harmonic Generator with Step Recovery Diode

frequency injection locking oscillator can be considered as a very narrow bandwidth amplifier. However, it differs from a conventional amplifier, because when there is no input signal, there is still output power from the oscillator.

Figure 15.30 shows a fundamental frequency injection locked oscillator with a circulator to isolate the input and output frequency. Experimentally, it has been shown that the locking range was proportional to the square root of the injection power, provided the injection power is kept sufficiently low. The injection lock can be connected in cascade as shown in Figure 15.31 to improve locking gain which is defined as the ratio of output power to the injected power. The nonreciprocal property of the circulators will reduce the undesired interactions between active devices. The locking gain of a three-stage cascade oscillator could reach 25 dB.

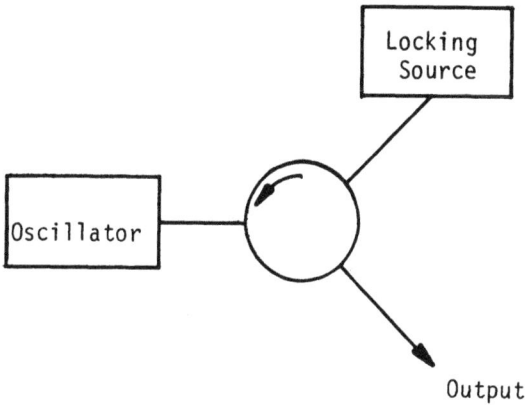

Figure 15.30. Fundamental Frequency Injection Locked Oscillator

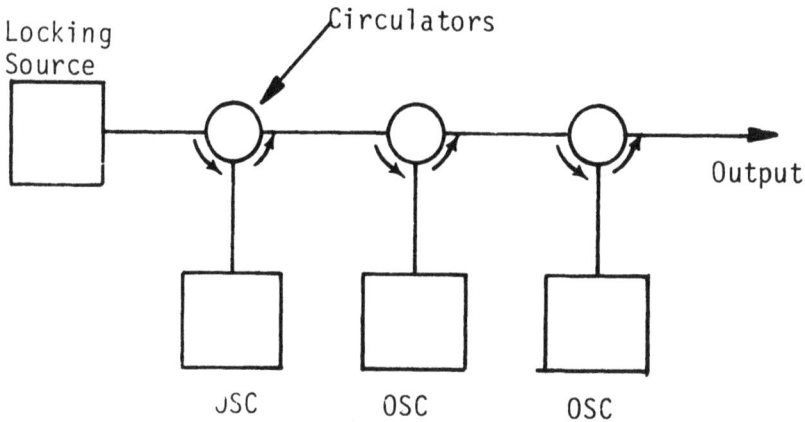

Figure 15.31. Cascade Injection Locked Oscillator

In a subharmonically injection locked oscillator, the output frequency of the oscillator can be injected locked with a frequency which is approximately equal to a subharmonic of the output frequency. For continuously operated IMPATT diodes, subharmonically phase lock with ratios as high as 9:1 were demonstrated (Reference 69). A basic subharmonically injection locked oscillator is hown in Figure 15.32. Since the locking frequency is usually lower than the oscillator frequency, low-pass filters or band-pass filters should be used to separate the locking frequency from the oscillator frequency. Directional couplers in the circuit are used only to measure the injection power. The locking range in megahertz versus the locking gain for different frequency ratio are shown in Figure 15.33. It is clearly shown that the higher the frequency ratio, the less the locking gain and the narrower the locking range.

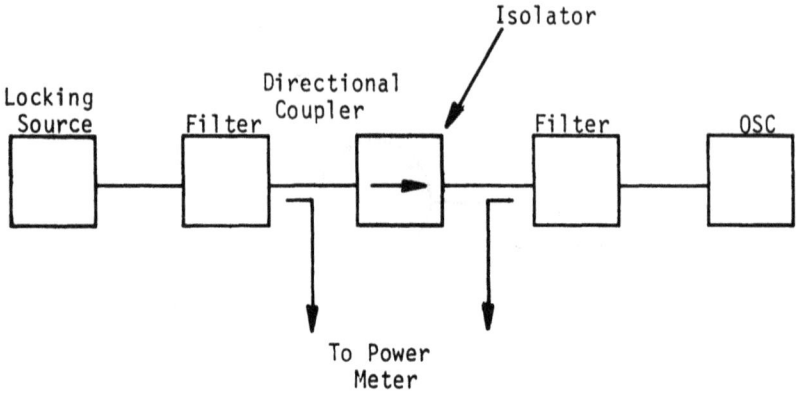

Figure 15.32. Basic Subharmonically Injection Phase-Locked Oscillator

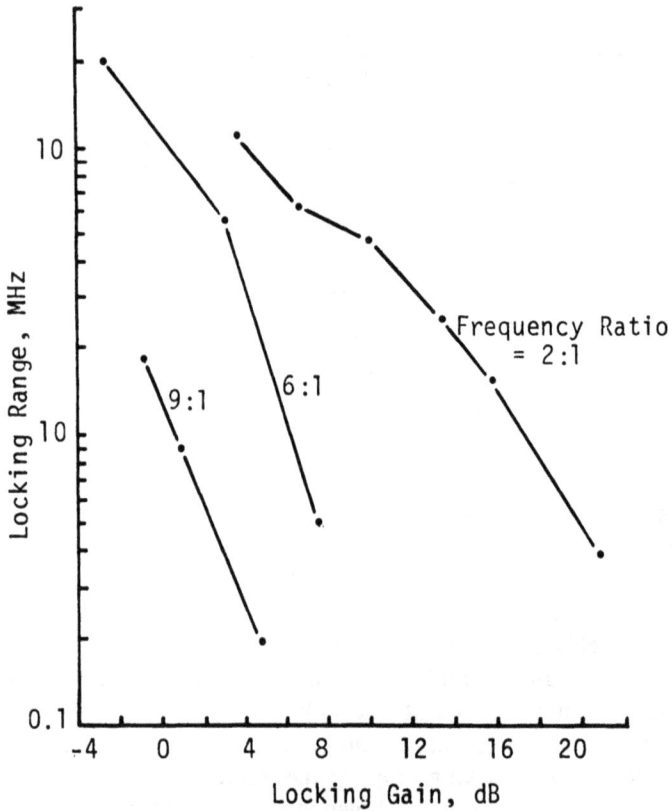

Figure 15.33. Locking Range/Locking Gain Plot (Based on Chien and Dalman, Reference 69)

REFERENCES

1. Hamilton, S., "Microwave oscillator circuits," Microwave Journal, p. 63, April 1978.
2. Meada, M., Kimura, K., and Kodera, H., "Design and performance of x-band oscillators with GaAs Schottky-gate field effect transistor," IEEE Trans. Microwave Theory and Techniques, Vol. MTT-23, pp. 661-667, August 1975.
3. Mitsui, Y., Nakatani, M., and Mitsui, M., "Design of GaAs MESFET oscillator using large-signal s-parameters," IEEE Trans. Microwave Theory and Techniques, Vol. MTT-25, pp. 981-984, December 1977.
4. Cutler, L.S., and Searle, C.L., "Some aspects of the theory and measurement of frequency fluctuations in frequency standards," Proc. IEEE, Vol. 54, pp. 136-154, February 1966.
5. Barnes. J.A., Chi, A.R., Cutler, L.S., Healy, D.J., Leeson, D.B., McGuinigal, T.E., Mullen, J.A., Smith, W.L., Sydnor, R., Vessot R.F.C., and Winkler, G.M.R., "Characterization of frequency stability," NBS Technical Note 394, U.S. Department of Commerce/National Bureau of Standards, October 1970.
6. Shoaf, J.H., Halford, D., and Risley, A.S., "Frequency stability specifications and measurement: high frequency and microwave signals," NBS Technical Note 632, U.S. Department of Commerce/National Bureau of Standards, January 1973.
7. Howe, D.A., "Frequency domain stability measurement: a typical introduction," NBS Technical Note 679, U.S. Department of Commerce/National Bureau of Standards, March 1976.
8. "Understanding and measuring phase noise in the frequency domain," Application Note 207, Hewlett-Packard Company, October 1976.
9. Lance, A.L., Seal, W.D., Mendoza, F.G., and Hudson, N.W., "Automated phase noise measurements," Microwave Journal, p. 87, June 1977.
10. Abrams, S., "Sideband noise problems in receiver testing," Microwave Journal, p. 51, January 1980.
11. Malcolm, B.G., "Frequency pulling: what happens to oscillator stability with load variations," Microwave Systems News, p. 75, June/July 1975.
12. Ballato, A., "Crystal oscillators with increased immunity to acceleration fields," IEEE Trans. Sonics and Ultrasonics, Vol. SU-27, pp. 195-201, July 1980.
13. "Fundamentals of quartz oscillators," Application Note 200-2, Hewlett-Packard Company.
14. Ziegler, R.R., "Know your crystal oscillators," Microwave Journal, p. 44, June 1976.

15. Lewis, M., "The surface acoustic wave oscillators—a natural and timely development of quartz crystal oscillator," Proc. of the 28th Annual Frequency Control Symposium, pp. 304-314, May 1974.

16. Van Der Windt, J., and Grassi, R., "To specify a time-base oscillator you'd better know your crystals," Electronic Design, p. 126, 22 November 1978.

17. Karrer, H.E., and Dias, J.F., "Surface acoustic wave oscillators," Proc. of the 28th Annual Frequency Control Symposium, pp. 266-269, May 1974.

18. Parker, T.E., and Schulz, M.B., "Stability of SAW controlled oscillators," Proc. Ultrasonic Symposium 1975, IEEE Cat. No. 75, CHO 994-4SU.

19. Ivanek, F., Kvarna, Y., Plourde, J.K., and Henaff, J., "Frequency stabilization of microwave oscillators for radio-relay systems," Microwave Journal, p. 44, September 1978.

20. Abe, H., Parker, T.E., and Ivanek, F., "Frequency stabilization of microwave oscillators for radio-relay systems," Microwave Journal, p. 65, October 1978.

21. Salmon, S.K., "Practical aspects of surface-acoustic-wave oscillators," IEEE Trans. Microwave Theory and Techniques, Vol. MTT-27, pp. 1012-1018, December 1978.

22. Plourde, J.K., Linn, D.F., Tatsuguchi, I., and Swan, C.B., "A dielectric resonator oscillator with 5 PPM long-term frequency stability at 4 GHz," IEEE MTT-S International Microwave Symposium Digest, pp. 273-276, June 1977.

23. Abe, H., Takayama, Y., Higashisaka, A., and Takamizawa, H., "A highly stabilized low-noise GaAs FET integrated oscillator with a dielectric resonator in the C band," IEEE Trans. Microwave Theory and Techniques, Vol. MTT-26, pp. 156-162, March 1978.

24. Makino, T., and Hashima, A., "A highly stabilized MIG gunn oscillator using a dielectric resonator," IEEE Trans. Microwave Theory and Techniques, Vol. MTT-27, pp. 633-638, July 1979.

25. Bushnell, T.R., and Buswell, R.N., "Cavity-controlled oscillators," Watkins-Johnson Technical Notes, Vol. 4, No. 6, November/December 1977.

26. Schunemann, K., and Knochel, R., "On the matching of transmission cavity stabilized microwave oscillators," IEEE Trans. Microwave Theory and Techniques, Vol. MTT-26, pp. 147-155, March 1978.

27. Nagano, S., and Ohnaka, S., "A low-noise 80 GHz silicon IMPATT oscillator highly stabilized with a transmission cavity," IEEE Trans. Microwave Theory and Techniques, Vol. MTT-22, pp. 1152-1159, December 1974.

28. Yokouchi, H., "Cavity stabilization of NW oscillators," Microwave Journal, p. 68, December 1978.

29. Golio, J.M., and Krowne, C.M., "New approach for FET oscillator design," Microwave Journal, p. 59, October 1978.

30. Trew, R.J., "Design theory for broad-band YIG-tuned FET oscillators," IEEE Trans. Microwave Theory and Techniques, Vol. MTT-27, pp. 8-14, January 1979.

31. Papp, J.C., "YIG-tuned FET oscillator design 8-18 GHz," Watkins-Johnson Company, Technical Notes, Vol. 6, No. 5, September/October 1979.

32. Hegboer, T.L., and Emergy, F.E., "YIG tuned GaAs FET oscillators," IEEE MTT-S, International Symposium Digest, pp. 48-50, 1976.

33. Ollivier, P.M., "Microwave YIG-tuned transistor oscillator amplifier design: application to C-band," IEEE Journal Solid State Circuits, Vol. SC-7, pp. 54-60, February 1972.

34. Ruttan, T., "X-band GaAs FET YIG-tuned oscillator," IEEE International Microwave Symposium MTT-S, pp. 264-265, 1977.

35. "Voltage controlled oscillators," Frequency Sources, Inc., 1974.

36. Buswell, R.N., "Voltage controlled oscillators in modern ECM systems," Watkins-Johnson Technical Notes, Vol. 1, No. 6, November/December 1974.

37. Fuller, J.B., "Frequency-agile local oscillator," Microwave Journal, p. 31, May 1974.

38. Buswell, R.N., "Linear VCOs," Watkins-Johnson Company, Technical Notes, Vol. 3, No. 2, March/April 1976.

39. Beach, R.M., "Hyperabrupt varactor-tuned oscillators," Watkins-Johnson Company, Technical Notes, Vol. 5, No. 4, July/August 1978.

40. Beach, R., "High speed linear oscillators," Microwave Journal, p. 59, December 1978.

41. Chalmers, E.G., and Carman, R.E., "Wide-band VCO uses Gunn varactor tuning," Microwave System News, p. 75, January 1979.

42. Dydyk, M., "A step-by-step approach to high-power VCO design," Microwaves, p. 54, February 1979.

43. Wagner, W., "Oscillator design by device line measurement," Microwave Journal, p. 43, February 1979.

44. Pergal, F., "Detail a Colpitts VCO as a tuned one-port," Microwaves, p. 110, April 1979.

45. Traynor, R.J., and Schuerch, W., "New VCO test equipment aids manufacturing," Microwaves, p. 57, January 1981.

46. Bissegger, C.A., Haddad, G.I., and Peterson, D.R., et al., "Voltage controlled oscillator (VCO) technology," Avionics Laboratory Technical Report, AFWAL-TR-80-1044, Vols. I and II, Wright-Patterson Air Force Base, Ohio, May 1980.

47. Corbey, C.D., Davies, R., and Gough, R.A., "Wide-band varactor-tuned coaxial oscillators," IEEE Trans. Microwave Theory and Techniques, Vol. MTT-24, pp. 31-39, January 1976.
48. Niehenke, E.C., and Hess, R.D., "A microstrip low-noise x-band voltage-controlled oscillator," IEEE Trans. Microwave Theory and Techniques, Vol. MTT—27, pp. 1075-1079, December 1979.
49. "Measuring the tuning step transient response of VCOs to 18 GHz," Hewlett-Packard Application Note 174-13, November 1974.
50. "Phase locked loop systems," Motorola Semiconductor Products, Inc., second edition, August 1973.
51. Gardner, F.M., *Phase Lock Techniques,* John Wiley and Sons, 1979.
52. Shelton, G.B., "Phase locking increases range of wide-range phase controller," Electronic Design 16, p. 96, 2 August 1973.
53. Tipon, P.G., "New microwave synthesizers that exhibit broader bandwidths and increased spectral purity," IEEE Trans. Microwave Theory and Techniques, Vol. MTT-22, pp. 1246-1254, December 1974.
54. Kurtz, S.R., "Mixers as phase detectors," Watkins-Johnson Technical Notes, Vol. 5, No. 1, January/February 1978.
55. Wetenkamp, S.F., and Wong, K.J., "Transportation lag in phase-locked loops," Watkins-Johnson Technical Notes, Vol. 5, No. 3, May/June 1978.
56. Alley, G.D., and Wang, H.C., "An ultra-low noise microwave synthesizer," IEEE Trans. Microwave Theory and Techniques, Vol. MTT-27, pp. 969-974, December 1979.
57. Caffee, L., Clausen, W., Hill, H., Kettle, J., Thurmond, R., and Stirling, R., "High speed frequency synthesizer," Air Force Avionics Laboratory Technical Report, AFAL-TR-79-1044, Wright-Patterson Air Force Base, June 1977.
58. Foster, C.E., and Simon, T.G., "Microwave frequency synthesizers," Watkins-Johnson Company, Technical Notes, Vol. 2, No. 3, May/June 1975.
59. Moll, J.L., Krakauer, S., and Shen, R., "P-N junction charge-storage diodes," Proc. of IRE, pp. 43-53, January 1962.
60. Krakauer, S.M., "Harmonic generation, rectification, and lifetime evaluation with the step recovery diodes," Proc. IRE, pp. 1665-1676, July 1962.
61. Hamilton, S., Hall, R., "Shunt mode harmonic generation using the step recovery diode," Microwave Journal, p. 69, April 1967.
62. "Ku-Step recovery mulitpliers," Application Note 928, Hewlett-Packard.
63. Watson, H.A., *Microwave Semiconductor Devices and Their Circuit Applications,* Chapters 7 and 8, McGraw-Hill, Inc., 1969.

64. Racy, J., "How to select varactors for harmonic generation," Micro-waves, p. 54, November 1973.
65. Page, C.H., "Harmonic generation with ideal rectifiers," Proc. IRE, Vol. 46, pp. 1738-1740, October 1958.
66. Adler, R., "A study of locking phenomena in oscillators," Proc. IRE, pp. 351-357, June 1946.
67. Huntoon, R.D., and Weiss, A., "Synchronization of oscillators," Proc. IRE, pp. 1415-1423, December 1947.
68. Oltman, H.G., and Nonnemaker, C.H., "Subharmonically injection phase-locked Gunn oscillator experiments," IEEE Trans. Microwave Theory and Techniques, Vol. MTT-17, pp. 728-729, September 1969.
69. Chien, C.H., and Dalman, G.C., "Subharmonically injected phase-locked impatt-oscillator experiments," Electronics Letters, Vol. 6, No. 8, pp. 240-241, April 1970.
70. Kurokawa, K., "Injection locking of microwave solid-state oscilla-tors," Proc. IEEE, Vol. 61, pp. 1386-1410, October 1973.
71. Young, J.C.T., and Stephenson, I.M., "Measurement of the large-signal devices using an injection locking techniques," IEEE Trans. Microwave Theory and Techniques, Vol. MTT-22, pp. 1320-1323, December 1974.
72. Berceli, T., "FM distortion in single and cascade injection locked diode oscillators," IEEE MTT-S, pp. 92-95, International Microwave Symposium Digest, 1977.
73. Pavlids, D., Hartnagel, H.L., and Tomizawa, K., "Dynamic consid-erations of injection locked pulsed oscillators with very fast switching characteristics," IEEE Trans. Microwave Theory and Techniques, Vol. MTT-26, pp. 162-169, March 1978.
74. Manassewitsch, V. "Frequency Synthesizer Theory and Design", John Wiley & Sons, 1980.

CHAPTER 16
EVALUATION OF MICROWAVE RECEIVERS

16.1 INTRODUCTION [1]

The evaluation of microwave receivers was not a serious problem before the development of wide-band receivers with relatively fine frequency resolution. For instance, the most important performance characteristic of a crystal video receiver is the sensitivity, which can be measured by observing the output of the receiver while changing the frequency or the pulse width (PW) of the input signal. Although the amount of data may be very large in order to cover the entire frequency range and PW range, it is still a manageable task. If an instantaneous frequency measurement (IFM) receiver covers 2000 MHz instantaneous bandwidth with 1 MHz frequency resolution, a dynamic range from -60 dBm to 0 dBm, and a PW ranging from 100 ns to continuous wave (CW), the amount of input required to evaluate the complete performance of the receiver can be extremely high. Even for a fixed PW, the 2000 resolution cell and 60 dB dynamic range with 1 dB amplitude resolution requires 1.2×10^5 (2000 $\times$ 60) points of input data. It is not only taking considerable time and effort to collect the output data, but presenting them in a meaningful manner is not a trivial task. If the simultaneous signal problem of this IFM recover (for example, with only two simultaneous signals) then 1.44×10^{10} (the square of 1.2×10^5) data points can be generated which is out of hand. If only some sampling points are measured, then statistical evaluation of the receiver will be involved (Reference 1). The presentation of the results will become even more complicated. For channelized, compressive and Bragg cell receivers, the problem is more eminent, since these receivers are designed to handle multiple signals.

Since many of the wide-band microwave receivers with fine frequency resolution and the capability to handle simultaneous signals are still in the development stage, the evaluation of these receivers has not been fully investigated. This chapter does not provide the answers to the receiver measurement problem. It only makes some suggestions and discusses some possible measurement schemes. Many of these measurements suggested have already been used to evaluate wide receivers in the laboratory. There is room for improvement, and different testing schemes should be initiated. The calibration of the measurement equipment (i.e., power level versus frequency) can also be tedious. From the possible number of data points available in measuring receivers and the required calibration procedures, one can readily realize that automatic testing equipment is almost an essential requirement, even if only a fraction of the total data points will be taken. The receiver measurement procedures including frequency accuracy, sensitivity, dynamic range, and other encoding (measurement) capabilities will be discussed here. The receivers under test are assumed to have digital outputs, since it is impractical to process the outputs from a wide-band receiver through only analog computations. A digital processor following the receiver to analyze the signal is the only known approach, although they are still in the development stage. Therefore, it is reasonable to assume that all wide-band receivers have binary digital outputs. The outputs include frequency, pulse amplitude (PA), PW, angle of arrival (AOA), and time of arrival (TOA). Receiver testing discussed here only requires two signal generators. For more advanced testing, a sophisticated electronic environment simulator must be used which is not within the scope of this book.

16.2 BASIC MEASUREMENT SETUP

The basic setup discussed here is intended to be used to measure all the parameters. Sometimes only a portion of the setup will be used, other times all the equipment will be used. The basic setup suggested is shown in Figure 16.1. The system contains two radio frequency (RF) signal generators, two pulse modulation units, and a computer. The computer is the central control unit. It controls the frequencies of the RF signal generators and the two pulse modulation units. The pulse modulation units will in turn control the PA, PW, and TOA of the input signals to the receiver. This setup can generate pulsed and CW signals and simulate AOA information over one quadrant for a receiver having an amplitude comparison scheme to obtain AOA. The equipment can be calibrated by the computer, and calibration data is stored in the memory. This setup

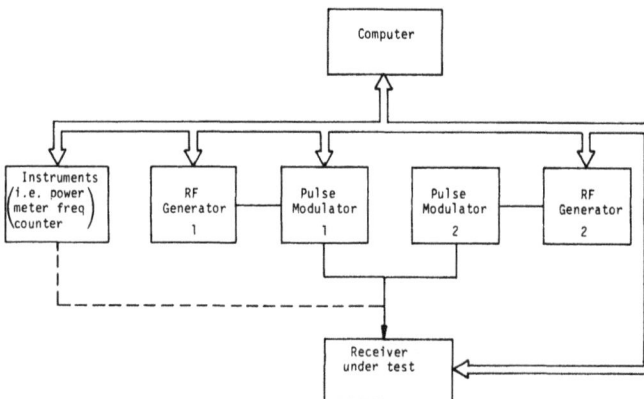

Figure 16.1. Basic Receiver Measurement Setup

lacks the capability to simulate AOA input information for a receiver with a phase comparison scheme. The two RF signals generated from two generators are combined together to the input of the receiver except for the AOA measurement. For the AOA measurements, each of the generated signals is separated into two paths and applied to two different input ports of the receiver to simulate the inputs from two adjacent quadrant antennas. The outputs of the receiver are also connected to the computer.

The computer controls the input of the receiver and at the same time measures the output of the receiver. By comparing the input and the output of the receiver, the performance of a receiver can be determined.

16.3 FREQUENCY ACCURACY MEASURMENT

It is assumed that the receiver under test is a wide-band receiver with fine frequency resolution. The frequency measurement capability of the receiver can be determined by using only one of the two signal generators. The PW and PA of the input signal will be chosen in some proper range first, then the frequency of the signal generator can be changed from one end of the receiver to the other end in steps approximately corresponding to the resolution of the receiver. The output frequency word of the receiver is monitored at each step.

The next important task is how to present the data. The simplest way is to print every input and output frequency. However, this approach is diffi-cult to provide a perspective view of the performance of the receiver that one can easily comprehend. A different approach is shown in Figure 16.2. The error frequency is plotted against the input frequency. The error frequency is defined as the difference between the output and the input frequencies. If the receiver performance is perfect, then the error frequency is always 0. From this plot one can readily read the performance of any receiver at any given PW and PA. In Figure 16.2, when the error frequency is too large to

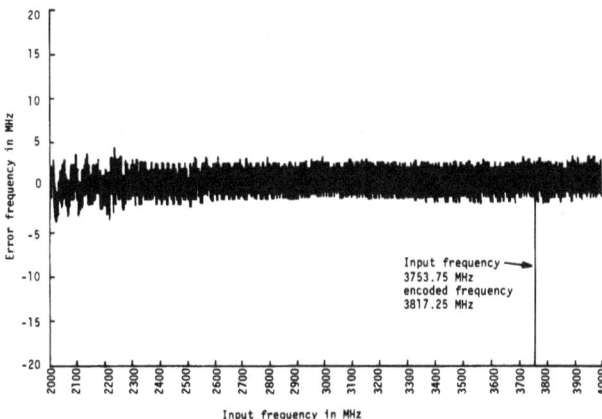

Figure 16.2. Frequency Measurement Accuracy of a Receiver (Reference 1)

be plotted, a vertical line is drawn and the input and output frequencies are printed. This large error frequency can be investigated further, if desired.

In addition to the above frequency measurement, it is often desirable to know the receiver capability to separate two simultaneous signals in frequency. In order to minimize the PW effect which will spread the spectrum, two relatively long pulsed signals should be used as the input signals. The two input signals should be equivalent in amplitude. First, the two signals are set far apart in frequency and the receiver reads both frequencies correctly. Then move one signal close to the other one and monitor the receiver frequency output. If the receiver reports anything but the two correct frequencies, the results should be considered erroneous. The minimum frequency separation where the receiver can read both signals correctly is the frequency resolution on two simultaneous signals. This test should be carried out at different reference frequencies, and an average frequency separation obtained from all the tests should be reported.

It should be noted that one can start the two signal frequency resolution measurement by initially setting the two signals at the same frequency, and then move one away from the other until the receiver measures the two frequencies correctly. However, in this approach, there is the possibility that the receiver will read the two frequencies correctly, but further separating the two signals will report erroneous data again. By separating the two signals still further away from each other, the receiver will report correct frequencies again. This phenomena exists in some wide-band receivers depending on the frequency encoding scheme used which is discussed in Section 5.4. Closing two frequencies together is an easier way to evaluate the frequency measurement capability of a receiver than separating two frequencies apart.

In the above measurements, it is possible that for the same input signal, the output data from the receiver may be different. This is especially true when the input signal (or signals) is at the boundary of two adjacent frequency readings. Under this condition, the average value and the standard deviation should be reported. If N measurements are taken from the same input conditions, and f_i are the measured data, the average value is

$$f_{avg} = \frac{1}{N} \sum_{i=1}^{N} f_i \qquad (16.1)$$

and the standard deviation is

$$\Delta f = \sqrt{\frac{1}{N} \sum_{i=1}^{N} (f_i - f_{avg})^2} \qquad (16.2)$$

16.4 SENSITIVITY AND MINIMUM PULSE WIDTH CAPABILITY MEASUREMENTS

The sensitivity of the receiver can often be determined through two approaches. The first one is to measure the tangential sensitivity (TSS). This is an analog approach. The output from the video detector or video amplifier following the detector is displayed on the oscilloscope. The TSS can be determined by direct observation as discussed in Chapter 2. However, in almost all the receivers the outputs from the video detector/ amplifier cannot be easily monitored. In addition, the TSS cannot be directly related to the probability of detection.

The other approach is to measure the probability of detection directly. Before the measurement the false alarm rate must be determined first. The false alarm rate is measured by terminating the input of the receiver with a matched load an observing the data ready/threshold break at the output of the receiver. The data ready or threshold break is usually a single bit (or flag) from the output of the receiver. Whenever there is output from the data ready, it means the receiver has received an input signal. The number of outputs measured per unit time from the data ready with the receiver input terminated is the false alarm rate. In practical measurement, there might be some difficulties in determining the false alarm rate. The most annoying difficulty is that there is no available control in the receiver to change the false alarm rate. The only way to change it is by adjusting the threshold inside the receiver which, in general, is not readily available. The difficulty caused by this shortcoming is that the false alarm rate is extremely difficult to determine. For the receiver to perform satisfactorily, the false alarm rate must be low. For example, if the data ready bits has zero output in one hour, the actual value of the false alarm rate cannot be determined. In order to generate some meaningful average data, extremely long observation time is required which might be impractical. Turning on and off some microwave sources or power switches near the receiver may disturb the readings. One possible remedy to this problem might be using a calibrated noise source at the input of the receiver, and generating a false alarm rate that is easy to observe (such as a few to several hundred data ready outputs per second or minute). The false alarm rate with a termination can be extrapolated from the data measured. This problem needs to be studied further.

The probability of detection can be determined by measuring the same data ready/threshold break bit, with an input signal of known pulse repetition frequency (PRF). By increasing the input signal level, the output rate from the data ready will also increase. For example, if the input PRF is 1 KHz, the data ready output is approximately 0.9 KHz, and the input power level is at -70 dBm. The probability of detection is approximately 90 percent at this input level.

Since the frequency accuracy is one of the most important parameters in microwave receivers, the probability of detection is sometimes measured through the frequency measurement. The automatic test setup controlled by the computer is required to make such a measurement. Using the example in Section 16.3, the receiver bandwidth is from 2 to 4 GHz with 1 MHz frequency resolution. If the frequency measurement results in a ± 5 MHz accuracy for signal level from − 60 dBm to 0 dBm, one can increase the input signal strength from a lower to a higher power level (below −60 dBm) and monitor the frequency output. Let use assume that the signal level starts from −75 dBm. The frequency is stepping through the 2 to 4 GHz range with 1 MHz step and monitoring the percentage of the frequency accuracy within ± 5 MHz. At each frequency, increase the signal strength and read the frequency report for a number of pulses (say 100 pulses). One can arbitrarily choose a certain percentage of frequency reports that are within the normal measurement accuracy and increase the signal strength until this condition is satisfied. Then change the input frequency and repeat the above measurement procedure again at each frequency. The final result will be sensitivity versus frequency with the specified measurement condition (i.e., above 90 percent of the input signal) at each frequency within ± 5 MHz error (the normal operation condition). Figure 16.3 shows a typical sensitivity versus frequency plot. In many receivers the threshold is chosen high enough that whenever the input signal crosses the threshold, the correct frequency will be reported. With such a receiver, the sensitivity measured against frequency resolution is not necessary.

In all the above measurements, the PW is kept at a convenient range so that the frequency measurement is not affected by the PW selected.

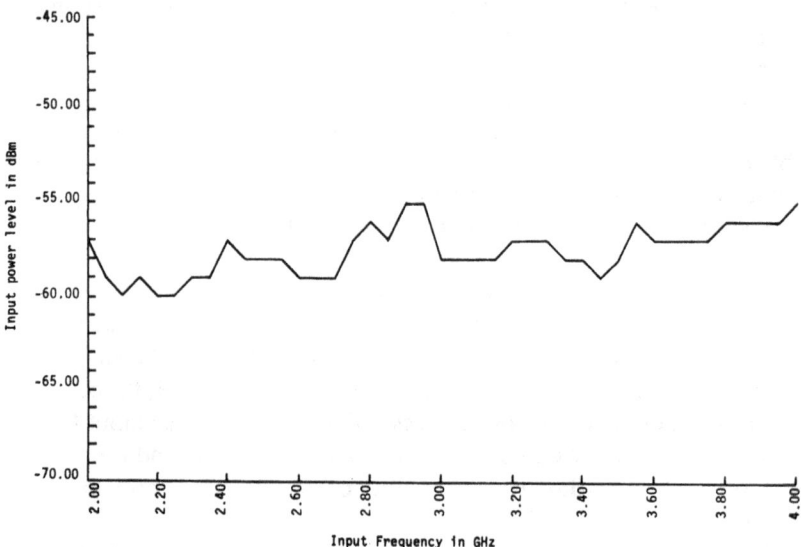

Figure 16.3. Sensitivity versus Frequency

The minimum PW the receiver can handle is usually determined by the frequency measurement, since the short PW will cause a spectrum spreading effect in the frequency domain. The measurement is quite similar to the sensitivity measurement mentioned before. Start from a relatively short pulse and vary the frequency over the desired range while the frequency accuracy is monitored. Increase the PW and repeat the same measurement. At a certain PW, the frequency accuracy measured will be in a predetermined percentage of the normal tolerance. For example, a 100 ns PW can be measured within ± 5 MHz of the input frequency over 90 percent of the time for the receiver mentioned above. In this measurement the input power level is set at some convenient level not to complicate the measurement. It should be noted that a short pulse may cause multiple frequency reports in receivers that can process simultaneous pulses. Of course, this condition is also considered unacceptable. When the PW increases, the multiple signal phenomenon will disappear.

16.5 DYNAMIC RANGE MEASUREMENTS

The lower limit of the dynamic range is the sensitivity of the receiver which has been discussed in Section 16.4. If the upper limit of the receiver is measured, the dynamic range can be determined. As mentioned in Chapter 2, there are a few different definitions of dynamic range. Their corresponding measurements will be discussed here.

If the receiver has the capability to measure PA, the PA encoding circuit can be used to determine the upper limit of the dynamic range. The input PA versus output amplitude generated by the receiver is measured. When the input is increased by 1 dB, the output should also increase by 1 dB if the receiver works in the linear region. However, when the input power is relatively high, the PA measurement circuit will be saturated; and the output will deviate from the linear region as shown in Figure 16.4. Around the knee of the curve is the upper limit. The lower limit of the curve is close to the noise floor of the receivers. If the receiver uses an amplitude comparison scheme to measure the AOA of the input signals, this measurement is very important because it determines the dynamic range of the AOA measurement capabilities.

Another method to determine the upper limit of the dynamic range is through frequency measurement. If one signal is applied to the receiver, the receiver should report only one signal with the correct frequency. If more than one signal is reported, spurs are generated by the receiver. If the frequency reported is out of the normal error range, then the signal is too strong and some measurement circuits in the receiver are saturated. Both of these cases limit the upper range of the dynamic range. This dynamic range is sometimes referred to as the single signal spur free dynamic range.

If the receiver can receive simultaneous signals, the following tests are usually necessary. In these tests both signal generators in Figure 16.1 will be used.

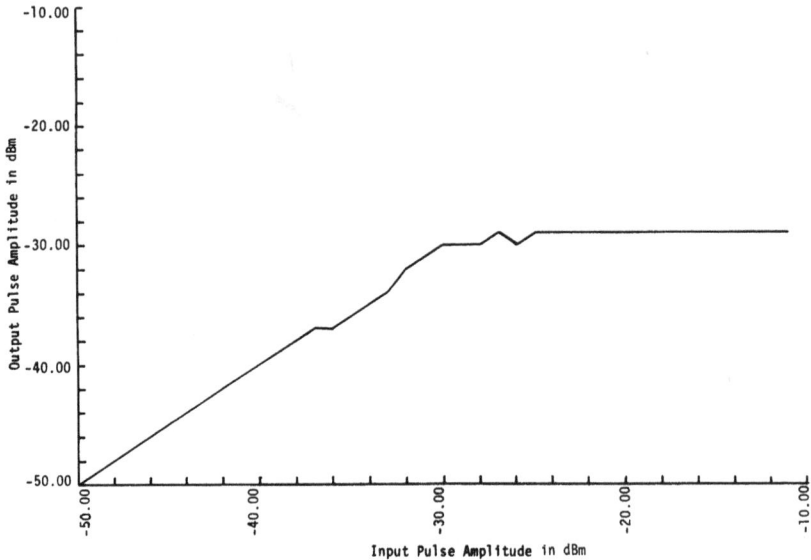

Figure 16.4. Input Pulse Amplitude versus Output Pulse Amplitude

In this test, one of the input signals is set at a relatively strong level while the second one is weak and separated from the first one in frequency more than the two signal frequency resolution of the receiver (Section 16.3). In order to avoid the spectrum spreading problem caused by pulsed signals which may interfere with the measurement, CW signals or long pulsed signals are often used. The receiver should receive both signals correctly. The signal strength of the strong one is gradually increased, while the weak one is decreased until the receiver can receive only the strong one. The maximum amplitude difference where the receiver can read both signals is the dynamic range. This dynamic range depends on the frequency separation of the two signals. Sometimes this is referred to as the instantaneous dynamic range. An attenuator controlled by the input signal strength in front of the receiver cannot improve the instantaneous dynamic range.

The last dynamic range test measures the third order intermod product of the receiver. Two signals of equal amplitudes separated by more than the two frequency resolutions of the receiver, are applied to the input of the receiver. If the amplitudes of the two signals are increased, a two-tone intermod will be generated; the receiver will report more than two signals separated by the same frequency. From the sensitivity level to the maximum signal amplitudes where the receiver still reads two signals correctly is the dynamic range. This dynamic range is often referred to as the two signal spur free dynamic range. This is different from the single signal spur free dynamic range which could be wider.

In all four measurements, the data should be taken at different frequencies. For the two signal tests, not only should the absolute frequency be changed, but also the frequency separation between the two signals. Therefore, the amount of data generated can be very high. It is reasonable to make only a few measurements and report the worst case, the average and the best results obtained, if the receiver is well designed and the performance does not vary substantially over the frequency range. When the dynamic range of the receiver is given, the kind of dynamic range should also be given, since dynamic ranges have different values.

16.6. THROUGHPUT RATE, PULSE AMPLITUDE (PA), PULSE WIDTH (PW), AND ANGLE OF ARRIVAL (AOA) MEASUREMENTS

The throughput rate of a receiver is measured by using an input signal with PW very close to the minimum that the receiver can handle, because the throughput rate often depends on it. The longer the pulse, the slower the throughput rate; but when the PW is under a certain value, the throughput rate is independent of it. This is the throughput rate of interest. The measurement is usually relatively simple, and it is often easier to measure manually than through the automatic measurement system. Most of the time one signal generator is enough; however, in some specially designed receivers, multiple signal generators may be required. Increasing the PRF of the input pulse and monitoring the data ready/threshold break of the receiver, the data ready should equal the PRF. When the input PRF reaches a certain value, the data ready report will drop to approximately half of the PRF value. This means that the receiver will process one signal but miss the following one. The maximum PRF that can be received without missing pulses is the throughput rate. This test is often independent of the signal frequency.

The PA and PW measurements are straightforward. The PA measurement has already been mentioned in determining the dynamic range of the receiver. The input signal strength (or PA) is increased in uniform steps in dB and the output is monitored and plotted. This procedure should be repeated in several frequencies.

Similarly, the PW can be measured by increasing the input signal PW and monitoring the receiver output at the same time. The input PW versus output PW can be plotted. The PW is usually plotted on a log versus log scale. The measurement should also be carried out at a few different frequencies and power levels. Usually, the PW is not measured in uniform steps.

The AOA information can only be measured in a complete system with an antenna installed. The evaluations of the AOA should be carried out in an anechoic chamber. Using signal generators directly connected to the receivers cannot effectively evaluate the AOA measurement capability of the system. It can only provide the information on how the AOA measured is related to the signal strengths of two adjacent receiver channels in an amplitude comparison system. In this measurement, the setup in Figure 16.1 should be modified slightly. One signal generator is used, and the output is fed into two separate pulse modulating units which are synchronized in time. In other words, the RF, PW, and PRF are the same at the outputs of the two pulse modulation units, but the amplitudes are different. They are applied at the two separate inputs of the receiver to simulate pulses coming from different antennas. The tests results should be plotted as amplitude difference versus output angle rather than input angle versus output angle. If multiple signals will be used in the test, another signal generator with the same arrangement should be added at the input of the receiver.

For a phase comparison AOA measurement system, a laboratory evaluation is even more involved. The input signal is split into a number of paths. The phase difference between the different paths must be tightly controlled to simulate an incoming signal. The angle of the signal is changed by varying the phase. Such an experimental setup to cover a wide frequency range is difficult to fabricate and tedious to calibrate. To measure an AOA phase comparison system, a radiating source in an anechoic chamber is probably more practical.

REFERENCES

1. Shaw, R.L., and Tsui, J.B.Y., "IFM receiver test and evaluation," Technical Report Air Force Avionics Laboratory, AFAL-TR-79-1049, Wright-Patterson Air Force Base, Ohio, April 1979.

APPENDIX

Physical Constants

Constant	Symbol	Value	Unit
Permittivity of free space	ε_o	8.854×10^{-12}	Farad/meter
Permeability of free space	μ_o	1.257×10^{-6}	Henry/meter
Speed of light in vacuum	c	3×10^8	Meters/second
Free-space wave impedance	$\sqrt{\mu_o/\varepsilon_o}$	376.7	Ohms
Boltzmann constant	k	1.38054×10^{23}	Joule/$^\circ$K
Electron charge	e	1.6021×10^{-19}	Coulomb

INDEX

521